Molecular Structure and Biological Activity of Steroids

Edited by

Martin Bohl
Manager
TRIPOS Molecular Science Group
Evans & Sutherland Computer GmbH
Munich, Germany

William L. Duax
Research Director
Medical Foundation of Buffalo, Inc.
Buffalo, New York

CRC Press
Taylor & Francis Group
Boca Raton London New York

CRC Press is an imprint of the
Taylor & Francis Group, an informa business

First published 1992 by CRC Press
Taylor & Francis Group
6000 Broken Sound Parkway NW, Suite 300
Boca Raton, FL 33487-2742

Reissued 2018 by CRC Press

Library of Congress Cataloging-in-Publication Data

Molecular structure and biological activity of steroids / editors,
Martin Bohl, William L. Duax.
p. cm.
Includes bibliographical references and index.
ISBN 0-8493-6955-X
1. Steroids--Structure-activity relationships. I. Bohl, Martin.
II. Duax, William L., 1939–
[DNLM: 1. Molecular Structure 2. Steroids--physiology.
3. Structure-Activity Relationship. QU 85 M718]
QP752.S7M65 1992
612.4'05--dc20
DNLM/DLC
for Library of Congress 92-40

A Library of Congress record exists under LC control number: 92004024

Publisher's Note
The publisher has gone to great lengths to ensure the quality of this reprint but points out that some imperfections in t original copies may be apparent.

Disclaimer
The publisher has made every effort to trace copyright holders and welcomes correspondence from those they have be unable to contact.

ISBN 13: 978-1-315-89570-3 (hbk)
ISBN 13: 978-1-351-07480-3 (ebk)

Visit the Taylor & Francis Web site at http://www.taylorandfrancis.com and the
CRC Press Web site at http://www.crcpress.com

EDITORS

Martin Bohl, Ph.D., is manager of the TRIPOS Molecular Science Group at Evans & Sutherland Computer GmbH, Munich, Germany.

Dr. Bohl obtained his training at the Friedrich Schiller University of Jena, receiving the Diplom-Chemiker degree in 1974 and the Ph.D. degree in 1979. After three years research work at the Jenaer Glaswerk, he joined the Academy of Sciences of the G.D.R., Central Institute of Microbiology and Experimental Therapy, Jena. During an eight-year period, his major interests included computations of steroid structure and structure-activity relationships of steroids. Since 1989 he has been working in the field of molecular modeling within the TRIPOS group.

William L. Duax, Ph.D., is the Research Director of the Medical Foundation of Buffalo, Buffalo, New York, Associate Research Professor, Department of Biochemistry, Roswell Park Division of the State University of New York at Buffalo, and Adjunct Associate Professor, Medicinal Chemistry Department, School of Pharmacy, State University of New York at Buffalo.

Dr. Duax graduated from St. Ambrose College, Davenport, Iowa, with a B.A. degree in 1961 and obtained his Ph.D. in Physical Chemistry in 1967 from the University of Iowa, Iowa City. He spent one year as a postdoctoral fellow at Ohio University and came to the Medical Foundation as a Research Associate in 1968. Dr. Duax was appointed Chairman of the Crystallography Department in 1969, Head of the Molecular Biophysics Department in 1970, Associate Research Director in 1983, and Research Director in 1988.

Dr. Duax is a member of the American Crystallographic Association, The Biophysical Society, American Chemical Society, American Association for Cancer Research, American Association for the Advancement of Science, The Protein Society, The New York Academy of Sciences, and the Molecular Graphics Society.

He served as President of the American Crystallographic Association in 1986 and is currently its Executive Officer. He was a member of the U.S. National Committee for Crystallography from 1980 to 1991. He is on the Board of Members of the American Institute of Physics, is Director of distribution of the Cambridge Structural Database to academic users in the United States, was a Charter Member of the Commission on Small Molecules of the International Union of Crystallography, and is its current Chairperson. Dr. Duax was commissioned to the Honorable Order of Kentucky Colonels in 1982 and received a Distinguished Alumni Award from St. Ambrose College in 1983. He has been the recipient of many research grants from the National Institutes of Health and currently holds a MERIT (Methods of Extending Research in Time) award from the N.I.H., Institute of Arthritis and Metabolic Disease.

Dr. Duax is co-author of 220 manuscripts and 39 chapters dealing with the structure and function of steroid hormones, ion transport antibiotics, and CNS active agents. He is co-editor of a two-volume *Atlas of Steroid Structure* and a text on *Molecular Structure and Biological Activity*. He has presented 130 invited lectures at national and international conferences and leading research laboratories throughout the world, and is co-author of 230 abstracts of scientific presentations at national and international meetings. Dr. Duax has been on the organizing committees of dozens of symposia concerned with the three-dimensional structure and biological activity of steroids, drugs, and antibiotics.

CONTRIBUTORS

Martin Bohl, Ph.D.
Manager
TRIPOS Molecular Science Group
Evans & Sutherland Computer GmbH
Munich, Germany

Philip A. Cole, M.D., Ph.D.
House Officer
Brigham and Women's Hospital and Harvard Medical School
Boston, Massachusetts

Jean-Christophe Doré, Ph.D.
Research Scientist
Laboratory of Computer Modelling
Centre National de Recherche Scientifique and Museum National d'Histoire Naturelle
Paris, France

William L. Duax, Ph.D.
Research Director
Medical Foundation of Buffalo
Buffalo, New York

Oldrich Hanč, Ph.D.
(Deceased)
Research Institute of Pharmacy and Biochemistry
Prague, Czechoslovakia

Nobuo Ikekawa, Ph.D.
Professor
Department of Science and Engineering
Iwaki Meisei University
Iwaki, Japan

Seiichi Ishizuka, Ph.D.
Senior Researcher
Teijin Institute for Bio-Medical Research
Hino, Japan

Kenneth S. Korach, Ph.D.
Senior Research Endocrinologist and Chief, Receptor Biology Section
Laboratory of Reproductive and Development Toxicology
National Institute of Environmental Health Sciences
Research Triangle Park, North Carolina

Sándor Mahó
Research Fellow
Research Laboratory of Organic Chemistry
Chemical Works of G. Richter Ltd.
Budapest, Hungary

Kenji Mori, Ph.D.
Professor
Department of Agricultural Chemistry
The University of Tokyo
Tokyo, Japan

Jean-Paul Mornon, Ph.D.
Director of Research
Department of Macromolecular Biology
Centre National de Recherche Scientifique URA 09
Paris, France

Tiiu Ojasoo, M.A.
Personal Assistant
Department of Innovation
Roussel-Uclaf
Paris, France

Zdeněk Peřina, Ph.D.
Research Scientist
Research Institute of
Pharmacology and Biochemistry
Prague, Czechoslovakia

Jean-Pierre Raynaud, Ph.D.
President
Aribio
Paris, France

Cecil H. Robinson, Ph.D.
Professor
Department of Pharmacology and
Molecular Sciences
Johns Hopkins University School
of Medicine
Baltimore, Maryland

**George J. Schroepfer, Jr.,
M.D., Ph.D.**
Professor
Departments of Biochemistry and
Cell Biology and of Chemistry
Rice University
Houston, Texas

S. Stoney Simons, Jr., Ph.D.
Chief
Steroid Hormones Section
National Institutes of Health
Bethesda, Maryland

Richard E. Thomas, Ph.D.
Associate Professor
Department of Pharmacy
The University of Sydney
Sydney, Australia

Zoltán Tuba, Ph.D.
Department Head
Research Laboratory of Organic
Chemistry
Chemical Works of
G. Richter Ltd.
Budapest, Hungary

Irena Urbánková, Ph.D.
Information Scientist
Science Information Department
Research Institute of Pharmacy
and Biochemistry
Prague, Czechoslovakia

William K. Wilson, Ph.D.
Assistant Director
Laboratory of Basic Medical
Science
Department of Biochemistry
and Cell Biology
Rice University
Houston, Texas

Takao Yokota, Ph.D.
Professor
Department of Biosciences
Teikyo University
Utsunomiya, Japan

TABLE OF CONTENTS

Chapter 1

STEROID STRUCTURE AND FUNCTION FROM X-RAY CRYSTALLOGRAPHIC STUDIES

William L. Duax

TABLE OF CONTENTS

I. INTRODUCTION

The characteristic actions of each steroid hormone are known to be contingent on the high-affinity binding of the steroid to specific receptor proteins in the target tissues.[1-3] Furthermore, hormonal function is modulated by interaction of the steroids with synthesizing and metabolizing enzymes as well as binding proteins in the blood and some specific tissues. Steroid function can be altered by synthetic hormones or antihormones that selectively interfere with protein interactions. While receptor binding is essential if certain hormonal responses are to take place, it is not sufficient. Many antihormones bind to steroid receptors but fail to elicit the characteristic responses. Structural similarities between the many natural and synthetic steroids result in a complex pattern of competition for the binding sites on receptor proteins, binding proteins, and metabolizing enzymes. Consequently, a thorough understanding of the molecular basis of these steroid-protein interactions is essential for the rational design of steroid drugs.

Crystallographic data on over 1000 steroids[4-5] provide information concerning preferred conformations, relative stabilities, and substituent influence on the interactive potential of steroid hormones. Systematic conformational analysis of subsets of this data having common structural features suggests that steroids crystallize in global minimum energy conformations or local minimum energy conformations that are less than a few kilocalories per mole above the global minimum.[6-8]

Comparison of patterns in conformational preference and intermolecular interactions observed in single crystal studies with data on the binding affinity of the same steroids for protein targets can contribute to a better understanding of the molecular basis of hormone action and lead to the design of more potent and selective agonists and antagonists of those actions. Ultimately, X-ray analysis of the receptors, binding and transport proteins, and enzymes involved in the synthesis and metabolism of steroids and the complexes of these proteins with steroids will be required to test proposed models and mechanisms and permit truly rational drug design. This review includes examples of comparative conformational analyses of crystallographic data on the progesterone side chain and the 4-ene-3-one A ring, a proposed empirical model for steroid receptor binding based on structure-activity comparisons, and a description of efforts to model the steroid binding pocket in receptors, binding proteins, and the proposed active sites of enzymes that have steroids as their primary substrate.

II. PROGESTERONE SIDE-CHAIN CONFORMATIONAL ANALYSIS

A number of steroid hormones including progestins and corticoids have a two-carbon keto side chain. It is highly probable that the 17β-side-chain

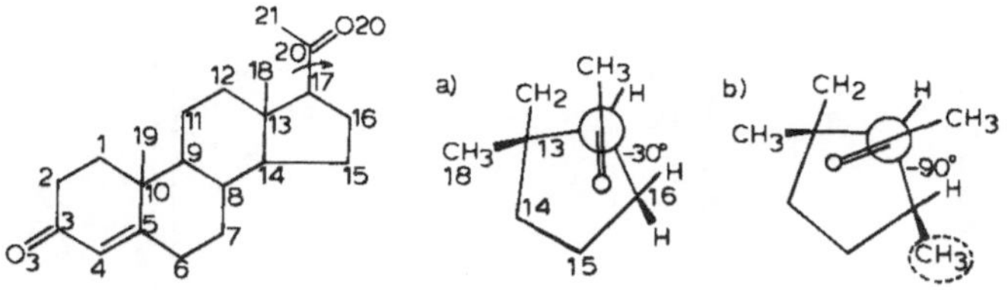

FIGURE 1. (a) Chemical formula for progesterone; Newman projections C(20) → C(17) illustrating two conformations of the progesterone side chain in which the C(16)–C(17)–C(20)–0(20) torsion angle is (b) −30° and (c) −90°.

conformation influences the interaction of these hormones with metabolizing, binding, and receptor proteins. For this reason, the preferred conformation of the side chain, the change in energy during rotation about the C(17)–C(20) bond (Figure 1), and the influence of substitution upon side-chain conformation have been the subject of numerous investigations.

Wellman and Djerassi[9] concluded that for steroids having the D ring and side chain of progesterone, the two most stable side-chain conformations are those shown in Figure 1. Modeling studies and CD spectra indicated that the conformer with a C(16)–C(17)–C(20)–0(20) torsion angle ϕ of approximately −30° (Figure 1a) was 1.1 kcal/mol lower in energy than the conformer for which ϕ is −90° (Figure 1b). In the case of the 16β-methyl-substituted structures, double maxima in the CD spectra suggested an equilibrium between the two conformers.

Crystallographic data on 285 estranes, androstanes, and pregnanes have been assembled and systematically presented in the first two volumes of the *Atlas of Steroid Structure*.[4,5] The total number of estrane, androstane, and pregnane structures entered into the Medical Foundation of Buffalo Steroid Database now number 611. This database contains 175 crystallographically independent observations of the conformation of steroids having the 17β side-chain substituted with the carbonyl oxygen found in progesterone. A histogram of the C(16)–C(17)–C(20)–0(20) torsion angle ϕ reveals a marked preference for a single conformation (Figure 2), in which the value ϕ for 163 structures lies between 0 and −45° and has an average value of −22.2°. The variation in observed conformation from the average position is due to a combination of inter- and intramolecular factors. Five crystallographically independent observations of progesterone itself[10-14] have ϕ values of −6°, −7°, −15°, −16° and −24° (average = 13 ± 7°). While this distribution demonstrates that there is a shallow energy minimum extending from −5° to −25°, the fact that observations are not uniformly distributed in that range may indicate that the distribution is sensitive to subtle intramolecular or intermolecular factors. In Figure 3 the distribution of selected subpopulations of the 175 observations are compared. The distribution of structures having no other substitution in the D ring covers the range observed for progesterone and has

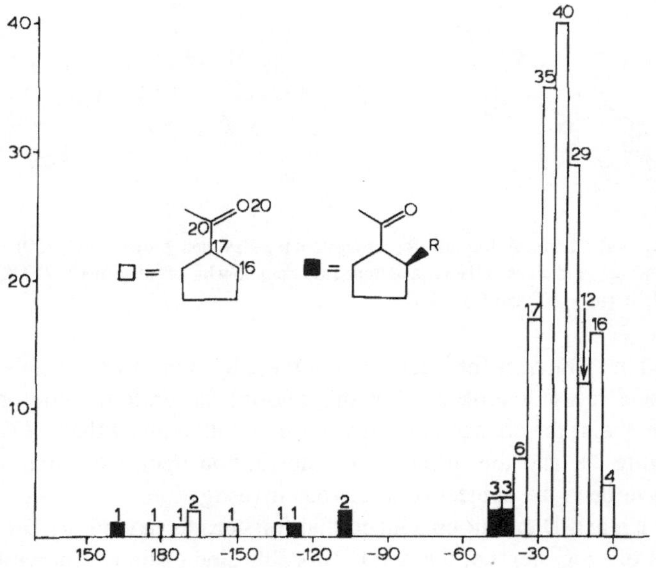

FIGURE 2. Crystallographically observed distribution of the 17-side-chain conformation in 175 pregn-20-one structures presented as a histogram of φ, the torsion angle C(16)–C(17)–C(20)–0(20). The dark squares represent structures having non-hydrogen substituents in the 16β position.

an average value for C(16)–C(17)–C(20)–0(20) of $-16 \pm 7°$. 17-Acetoxy substituents (Figure 3b) appear to enhance the stability of conformers on one end of this range with an average C(16)–C(17)–C(20)–0(20) torsion angle of $-21 \pm 3°$. The combination of 17α-hydroxy-21 ester substituents (Figure 3a) shifts the conformation even further from the range observed in the unsubstituted steroids (Figure 3d) resulting in an average value of φ of $-27 \pm 4°$.

Examination of the ten structures having φ values outside the normal range reveals that in all cases the variation is due to intramolecular factors and not crystal packing forces. The distribution of observed conformations of the eight 16β-substituted analogs is somewhat unusual.[7] Four of these structures have 17β side-group orientations at the low extreme in the range of the unsubstituted analogs (mean $\phi = -43 \pm 3°$, conformer ϕ_1, see Figure 2), but in three other structures the mean value of φ is $-115 \pm 11°$ and in one case $\phi = \pm 162°$. Although the number of observations is limited, the addition of the 16β substituent seems to shift the location of the lowest minimum energy orientation and reduce the difference between minima so that additional orientations have significant populations. This is consistent with the earlier proposal by Wellman and Djerassi[9] from CD spectra that two conformations of the 17β side-chain were present in the solutions of 16β-substituted progesterone compounds. Although the actual conformations pre-

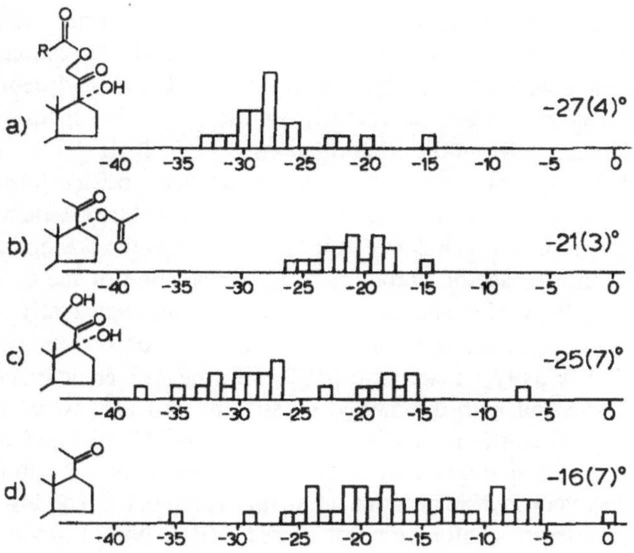

FIGURE 3. Crystallographically observed distribution of 17 side chain conformation in subpopulations of pregn-20-one steroids as defined by ϕ, the C(16)–C(17)–C(20)–0(20) torsion angle. 21-Acetoxy (a), 17-acetoxy (b), and 17,21-dihydroxy (c) substituents are observed to shift the population of conformers relative to that observed in structures having no D-ring or side-chain substitution (d). Average values of ϕ for the subpopulations are given on the right of the figure.

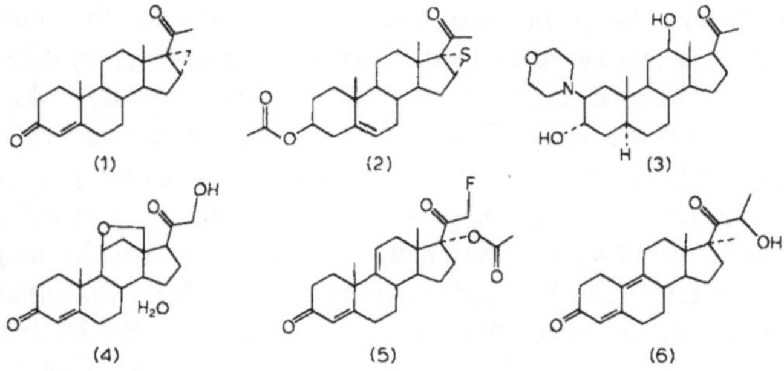

FIGURE 4. Structures in which intramolecular factors stabilize an unusual conformation of the progesterone side chain.

dicted may not have been precisely correct, conformers ϕ_1 and ϕ_2 are probably the forms responsible for spectral observations.

The other six structures having side-chain conformations that lie outside the range $-50° < \phi < 0°$ are illustrated in Figure 4. Each of these structures contains an unusual feature that appears to stabilize a conformation that would not otherwise be preferred. The introduction of a three-membered ring at the

C(16)–C(17) positions in steroids 1 and 2[15] stabilizes a conformation in which the carbonyl is nearly trans to the C(16)–C(17) bond. An intramolecular hydrogen bond donated by the 12β-hydroxyl to the 20-carbonyl stabilizes the conformation $\phi = -150°$, observed in steroid 3.[16] This suggests that the energy difference between the conformations $\phi = -20$ and $\phi = -150°$ is less than the energy of a single hydrogen bond. The epoxide formation in steroid 4 draws the C(18) away from the D ring.[17] This reduces steric hindrance so that the position in which $\phi \simeq -130°$ is as stable as the position in which $\phi \simeq -30°$. In the case of steroid 5,[18] the combination of the C(9)–C(11) double bond in the C ring and the 21-fluoro substituent apparently alters the balance of intramolecular interactions in favor of a conformation in which $\phi = -173°$.[18] Finally, in steroid 6 (RU27988, the 21R enantiomer), steric hindrance associated with the methyl substituent on C(21) is calculated to cause the normal conformation ($\phi \simeq -30°$) to be 12 kcal/mol higher in energy than the crystallographically observed conformation.[19] Compound 6 (Figure 4) has very low affinity for all steroid receptors.[19] The low affinity and unusual observed conformation of steroid 6 (RU27988) support the contention that the minimum-energy conformation of the progesterone side chain ($\phi \simeq -30°$) is the biologically active conformation.

Molecular mechanics calculations provide information concerning variation in potential energy as a function of variation in molecular structure. The accuracy with which a given program successfully identifies the global minimum-energy conformation, the relative energies of local minimum conformations, and the height and shape of barriers to conformational transition are difficult for the average user of these programs to estimate.

The molecular mechanics program MM2p[20] was used to minimize the energy of 17 progesterone-related structures, starting with the crystallographic observations.[21] With the exception of conformers in which the value of ϕ is near $-120°$ and 180°, the C(17) side chains in the minimized structures are rotated an average of 15° toward the point at which the carbonyl eclipses the C(16)–C(17) bond ($\phi = 0°$). The published calculations[22] on a smaller subset of steroid data reveal similar systematic differences of smaller magnitude. The systematic difference between observed and calculated conformations of the O=C–C–C torsion angle must be due to a pattern in the intermolecular interactions in the crystal or to inaccuracy in the parameterization of the MM2p program. A careful examination of the crystal packing in 80 of the pregna-20-one analogs revealed no obvious pattern in intermolecular interactions.[7] There is no apparent correlation between the presence or directionality of hydrogen bonding involving side-chain substitutions and the observed conformation.[7]

Data from a variety of sources are taken into consideration in the development of parameters that influence the results of molecular mechanics calculations. Early *ab initio* calculations[23] on 2-butanone and microwave data on propionaldehyde[24] indicated a preference for eclipsing of the C–C single bond and the C=O bond. These data were taken into consideration in the development of parameters for the MM2p programs that predict the most stable progesterone side-chain orientation is one in which the C(16)–C(17)–C(20)–O(20) angle is very near 0°. More recent *ab initio* calculations on a larger molecular fragment, isopropyl methyl ketone, using a 6-31G basis set, indicate that steric/dipole interactions not present in 2-butanone are responsible for stabilizing the conformer in which the C–C–C=O angle is 17° away from the eclipsed position.[25] This value for the conformation of an isolated molecule of isopropyl methyl ketone is in nearly exact agreement with the average value observed for progesterone molecules and analogs without additional substituents on the D ring.

It is noteworthy that, as more *ab initio* calculations are used on larger molecular fragments, the conformation of isolated molecules predicted to be most stable approaches the average value for ensembles of the same fragment observed in X-ray crystal structure determinations. This is strong support for our previously drawn conclusion that crystal packing forces have a negligible effect on determining steroid conformation and that the crystallographically independent multiple copies of a given fragment can provide the most reliable information on conformational stability and flexibility.

The most recent version of Allinger's excellent series of molecular mechanics program, MM3, is described as providing information on aldehyde and ketone conformations that is of experimental accuracy.[26] The MM3 program predicts that the minimum energy conformation of methyl isopropyl ketone is one in which the carbonyl eclipses a methyl group and has a torsion angle of −4°,[26] whereas the *ab initio* calculation predicts a torsion angle of −17°[25] for this molecule; the average value for the progesterone side-chain conformation in crystal structure determinations of 39 steroids that have no other substitution in the D ring or side chain is 16.1°. It would appear that the MM3 calculations will not be significantly better than the MM2 series at predicting the conformation of the progesterone side chain moiety and that a bias towards an eclipsed conformation across a $C(sp^2)$–$C(sp^3)$ bond remains.

III. 4-EN-3-ONE A-RING CONFORMATION

The 4-en-3-one A ring, characteristic of many endogenous steroid hormones, has conformational flexibility that appears to influence the selectivity of receptor interaction. Generally, the 4-en-3-one A ring is observed in a conformation in the range from an ideal 1α,2β-half-chair (Figure 5a) to an ideal 1α-sofa (Figure 5b). In 191 of 237 steroid structures having the 4-en-3-one composition,[4,5] the A rings have this "normal" conformation in which C(1) is below and C(2) is above or in the plane of the conjugated double bonds (Figure 5a and b).

In 46 of the 4-en-3-one steroids studied crystallographically, most of which are synthetic steroids not known to be naturally occurring, the A ring

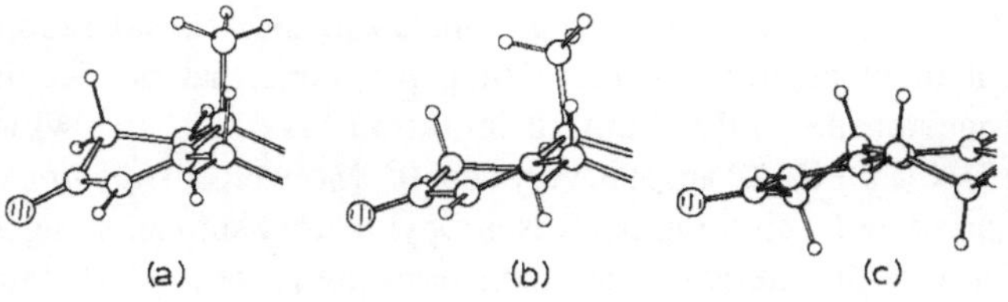

FIGURE 5. Of 237 steroids having 4-en-3-one composition, the A rings of 191 range between two conformations: (a) the 1α,2β-half-chair and (b) the 1α-sofa. Only 46 have the unusual inverted 1β,2α-half-chair conformation (c).

is observed in an unusual inverted conformation. In this conformation C(1) is *above* and C(2) *below* the plane of the en-one group (Figure 5c). Certain features appear to favor the inverted conformation: a nonstandard 9α,10β-configuration, a bulky 2β substituent, an additional C(9)–C(10) double bond, 19-nor composition, and a 17α-ester substituent. While the significant influence on the A-ring conformation of bulky substituents at the 2β position[28] and chiral changes at C(9) and C(10)[27] are easily rationalized, the influence of additional unsaturation in the B and C rings and long-range effects associated with 17α-ester formation are not so readily explained. The distance of the C(2) atom from the plane of atoms C(3), C(4), C(5), and C(10) is a sensitive measure of A-ring conformation, C(2) is above this plane when the A ring has the inverted 1β,2α conformation. In Figure 6 the observed values for the deviation of atom C(2) from this plane are plotted vs. a measure of the conjugation of the en-one system, the torsion angle τ [O(3)–C(3)–C(4)–C(5)]. The distinct conformational patterns of the 4-en-3-one (□), 4,9-dien-3-one (△), and 4,9,11-trien-3-one (○) structures are illustrated. The structures used in Figure 6 include all 4,9-dien-3-one and 4,9,11-trien-3-one structures in the MFB database (28 and 14 structures, respectively) and all 4-en-3-one structures that have all-trans steroid ring configurations and no other substituents or unsaturation in the A or B ring. The majority of 4-en-3-one structures (75 of 77) that have no other substitution or unsaturation in the B and C ring have normal 1α,2β-A-ring conformations that vary in a range centered about the point of perfect conjugation of the 4-en-3-one (180°). Of the 28 structures having 4,9-diene-3-one composition, 90% are observed in the inverted 1β,2α conformation. The normal 1α,2β conformation is observed in only three of these structures. A careful examination of the structures having normal and inverted conformations suggests that stabilization of the inverted conformer results in part from the presence of steric interaction between the hydrogen substituents on C(1) and C(11). The introduction of an additional double bond at C(11)–C(12) produces a further shift in the relative energy between the inverted and normal conformation reflected in the observed ratio of normal to inverted conformers in the 4,9,11-trien-3-one structures (Figure 6b).

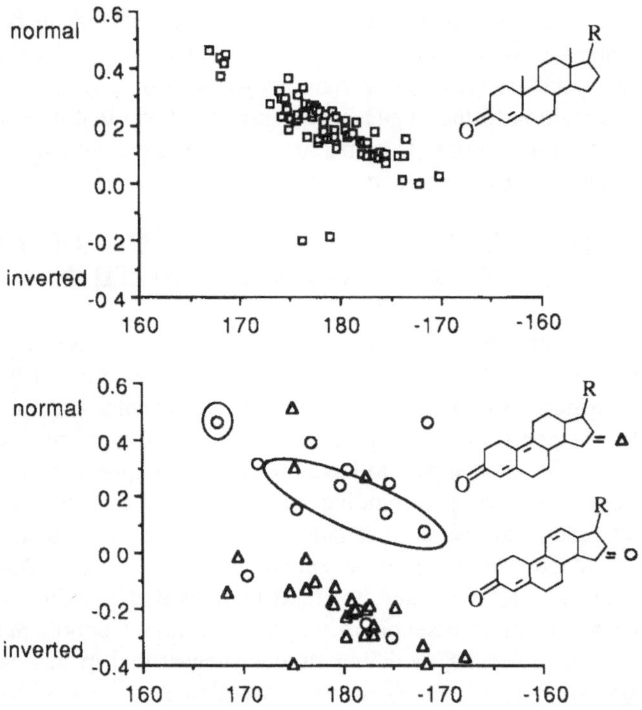

FIGURE 6. A-ring conformation as a function of the distance in Ångstroms of the C(2) atom from the plane of the atoms C(3), C(4), C(5), and C(10) vs. the conjugation of the 4-en-3-one system as measured by the torsion angle τ, 0(3)–C(3)–C(4)–C(5). While 97% of the 4-en-3-one A rings (□) have normal conformations, top, 90% of the 4,9-diene-3-one (△) are inverted and a 10:4 distribution is observed for the 4,9,11-trien-3-one (○), bottom.

The presence of a stable inverted A-ring conformation in a 4-en-3-one steroid containing a 19β-methyl substituent was first noted in the X-ray crystal structure determination of medroxyprogesterone acetate, MPA,[27] a compound having high affinity for the human progesterone receptor. We have proposed that in the case of MPA, the A-ring inversion involves long-range conformational transmission effects[29] associated with the 17α-substituent combined with the proximal influence of the 6α-methyl group.[30] The 6α-methyl group[31] alone does not appear to stabilize the inverted A ring, but the presence of both the 17α-acetate and the 6α-methyl[27] appears to have that potential.

In the propionate ester of 17α-hydroxy-4-pregnen-3,20-dione (MMP) the A-ring was found to be a disordered mixture of the normal and the inverted forms in the solid state. To determine whether the disorder in the propionate derivative is static or dynamic, full-intensity data on a single crystal of MMP were collected at 85, 264, and 323 K. The percentage occupancy of the inverted form was found to be 72% at 85 K, 52% at 264 K, and 48% at 323 K. Analysis of the change in population as a function of temperature and a careful examination of close contacts in the crystal suggest that the conformer having the normal A-ring conformation is the lower-energy form and that unfavorable crystal packing interactions involving this form induce a conformational transition to the higher-energy inverted form that does not have such unfavorable interactions. This conclusion is consistent with interpretation of nuclear magnetic resonance data on MPA.[32]

IV. STEROID MOLECULAR STRUCTURE, RECEPTOR BINDING, AND HORMONE ACTION

Examination of the structures of compounds having high affinity for estrogen, progestin, and corticoid receptors suggests that receptor binding is primarily the result of a tight association between the receptor and the steroidal A ring.[33-35] Antagonists that compete for estrogen, progesterone, and glucocorticoid receptor sites appear to have the A-ring composition and conformation necessary for receptor binding but lack the D-ring conformational features and functional groups that induce or stabilize subsequent receptor functions.[36] Antagonists may also be compounds with A-ring conformations appropriate for binding but other structural features that interfere with subsequent receptor functions essential to activity. In contrast, androgen receptor binding data and molecular modeling studies suggest that in this case a D-ring binding/A-ring acting model is more appropriate.[37] The possible means by which the D ring of estrogen, progestins, and corticoids might control activity include (1) inducing or stabilizing an essential conformational state in the receptor (allostery), (2) influencing the aggregation state of the receptor, or (3) participating in a direct interaction with DNA or chromatin.[38]

A. ESTROGEN RECEPTOR BINDING AND ACTIVITY

There is extensive overlap in binding site competition between steroids that bind strongly to the progesterone receptor (Pr), the androgen receptor (Ar), the glucocorticoid receptor (Gr), and the mineralocorticoid receptor (Mr). In contrast to this, compounds with high affinity for these receptors show little or no affinity for the estrogen receptor (Er).[39] The principal structural difference between steroids that bind well to the estrogen receptor and those that bind well to all the other receptors is the presence in the former of the phenolic A ring. Other compounds that bind to the estrogen receptor include nonsteroidal synthetic estrogens,[40] clinically useful anticancer agents,[41] micotoxins,[42] and single one- or two-ring compounds.[43]

Most investigators have concluded that, when binding to the receptor site, a phenolic ring mimics the steroid A ring.[40,34] The potent nonsteroidal estrogen diethylstilbestrol (DES) has two phenolic rings capable of mimicking the A ring of estradiol. Metabolites of DES are of interest due to uncertainty concerning the form responsible for the carcinogenic properties of DES and as additional probes of the structure-activity relationships of estrogens. It is

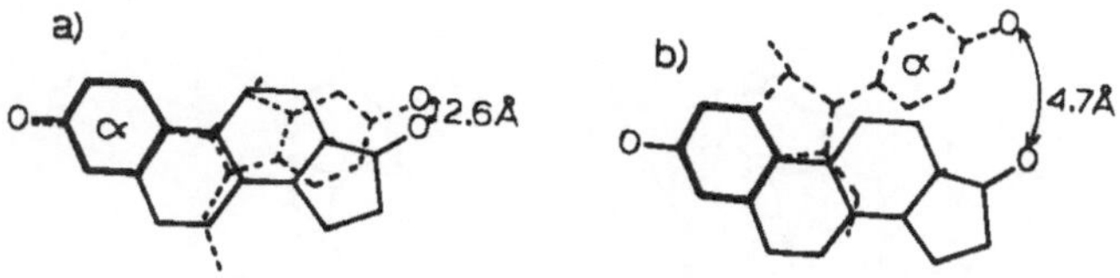

FIGURE 7. Comparison of the conformations of IA with estradiol reveal that the best fit is achieved when the α-ring of IA is superimposed on the A ring of estradiol with the double bond in the β-ring overlapping the C(8)–C(9) bond in estradiol.

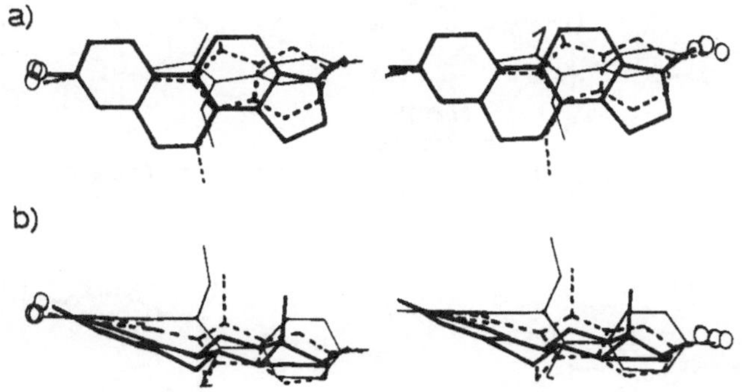

FIGURE 8. Stereo views illustrating the best fit of the hydroxyl groups and the hydrophobic bulk of the structures of estradiol (dark solid lines), DES (light solid lines), and IA (dashed lines).

evident that the precise location of steric bulk and functional groups relative to the phenolic ring will govern binding and activity.

The subtlety of the details of molecular control is exemplified by the case of the racemate of indenestrol A (IA). Only one enantiomer of (IA) has high affinity for the estrogen receptor.[44] Model studies using the X-ray crystallographically observed structure of IA suggest that it is the α-ring that mimics the estrogen A ring[45] (Figure 7). The structural basis for the enantiomeric effect in IA must be related to the topology of the molecule in the region of the double bond. On the basis of a comparison of the crystallographically observed structures of estradiol, DES, and IA (Figure 8), the C3-S enantiomer of IA, with structural features most closely resembling DES and estradiol, was proposed to be the active enantiomer. These predictions were subsequently verified by experiment. Receptor binding studies of the monohydroxy analog of IA support the proposal that the α-ring initiates receptor binding.[46] The X-ray crystal structure determination of the absolute configuration of the active enantiomer of indenestrol A revealed it to be the C(3)-S enantiomer.[47] That the A-ring binding/D-ring acting model has predictive capabilities is a persuasive argument for its essential correctness.

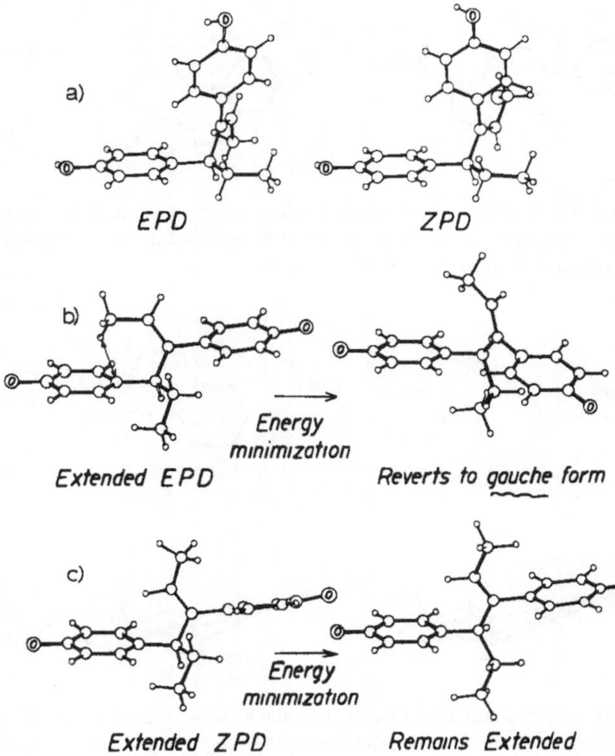

FIGURE 9. (a) Crystallographically observed conformations of ZPD and EPD. (b) Results of energy minimization of extended models of the inactive (EPD). (c) Active (ZPD) isomers.

Crystallographic analysis of another DES analogue, pseudo-DES, together with empirical energy calculations, has provided additional information about the importance of overall conformation to estrogenic activity.[48] Pseudo-DES exists as the Z and E isomers, ZPD and EPD. While both forms bind to the estrogen receptor, only the Z form has appreciable estrogenic activity. X-ray structure determinations revealed that both EPD and ZPD have bent conformations, completely unlike the conformations of estradiol (Figure 9a and b). When the crystallographically observed conformations of EPD and ZPD were subjected to energy minimization the structures retained their overall conformations with only minor changes in individual torsion angles that averaged less than 4°.

The consistency between the crystallographic results and the energy calculations suggests that the bent form is the minimum energy conformation of ZPD and EPD. However, previous studies indicate that this conformation is unlikely to be compatible with any significant degree of estrogenic activity. For this reason it became important to determine whether an extended DES-like conformation might constitute another stable conformer and account for

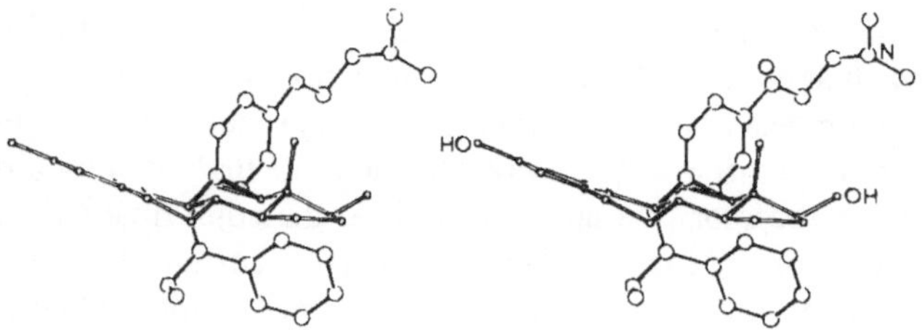

FIGURE 10. Stereo diagram comparing the structures of estradiol, small circles, and tamoxifen, large circles.

the estrogenic activity of ZPD. The crystallographically observed structures of ZPD and EPD were transformed to the DES-like extended conformation and subjected to minimization using a modified version of the programs of Allinger.[20,21] As a result of extremely close contacts between the hydrogens on C(9) and the α-ring in EPD, the extended conformation is energetically unfavorable and the molecule refined to a bent conformation resembling in overall shape the crystallographically observed structure. In contrast to this, the extended conformation of the more active Z isomer does not incorporate intolerable non-bonding interactions, and the refined extended structure had an energy about 2 kcal higher than the observed bent structure. This conformer could (Figure 9c) satisfactorily account for the observed activity.

Estrogen antagonists such as *trans*-tamoxifen compete for binding to estrogen receptors and elicit little or none of the characteristic hormonal response. The 4-hydroxy metabolite of *trans*-tamoxifen has been shown to be a potent competitor for the estrogen receptor and thus responsible for the antagonist properties useful in breast cancer therapy.[49] This observation has led most investigators[35,50] to conclude that, at least in the case of the 4-hydroxy derivative of tamoxifen, it is the hydroxylated ring that mimics the estradiol A ring in receptor interaction. The molecular fit achieved by overlapping the α-ring of tamoxifen and the A ring of estradiol is illustrated in Figure 10. The antiestrogen has the phenolic ring required for high-affinity binding to the receptor, but lacks a hydrogen bonding group corresponding to the O(17) hydroxyl of estradiol that appears to be required for activity. The antiestrogen also possesses a bulky substituent extending nearly perpendicular to the steroid skeleton that may interfere with a conformational change in the receptor that is postulated to be essential to subsequent macromolecular interaction and activity.[37]

B. PROGESTERONE RECEPTOR BINDING AND ACTIVITY

Examination of the chemical structures of steroids whose affinity for the progesterone receptor in rabbit, mouse, sheep, or human uterus is equal to or higher than that of progesterone itself[39,51-53] indicates that extensive structural variation is compatible with high-affinity binding.[27] The only structural feature common to all compounds with high affinity for the progesterone receptor is the steroid ring system and 4-en-3-one composition. However, many steroids that have the 4-en-3-one composition have little or no affinity for the uterine progesterone receptor (e.g., testosterone). Since a 4-en-3-one A ring appears to be necessary but not sufficient for high-affinity binding to the progesterone receptor, the author examined the conformations of the 4-en-3-one A rings of the highest-affinity binders in search of some unusual electronic, geometric, or stereochemical feature that might explain their enhanced binding. The conformational flexibility of the 4-en-3-one A ring has been described in detail in Section III. The A rings of all of the naturally occurring 4-en-3-one steroids that have been studied crystallographically have the normal conformation in which C-1 is below and C-2 is above the plane of the en-one system (Figures 5 and 6). The inverted A-ring conformation in which C-1 is above and C-2 is below the plane of the en-one group has only been observed in semisynthetic steroids that have unnatural chirality, unusual substitution, or certain types of bond unsaturation. Many of these compounds of unnatural composition and unusual conformation exhibit enhanced affinity for the progesterone receptor.

Introduction of a double bond at the 9-10 position induces A-ring inversion in 16 of the 17 structures for which X-ray analysis has been undertaken. Many of the 4,9-dien-3-one structures exhibit a very high affinity for the progesterone receptor.[39] The correspondence between the consistency of inverted conformation and of high affinity for the receptor suggests that the key to receptor binding is the presence of a 4-en-3-one ring *in the inverted conformation.*

Addition of a third double bond at C(11)–C(12) destabilizes the inverted A ring and produces a molecule able to exist in a number of distinctly different conformations.[54] The flexibility is probably the result of an effort to balance three forces: the constraints of planarity to increase conjugation throughout the system from O-3 to C-12, unfavorable interactions in two regions (1-3 hydrogen interactions at C-4 and C-6), and 1-4 hydrogen interactions at C-1 and C-11. It is worth noting that while these steroids generally have lower affinity than the 4,9-diene steroids for the progesterone receptor, they exhibit more effective competition for other steroid receptors than do the 4,9-dienes.[39] This suggests that different conformers of the more flexible 4,9,11-trien-3-one structures satisfy the requirements of different receptors.

Other structures for which the inverted A-ring conformation appears to be relatively stable include the potent progestin medroxy progesterone acetate[27] and the weaker progestin 19-nortestosterone.[55] The high progesterone receptor affinities of steroids having the inverted A-ring conformation suggest that this conformation is optimal for binding. The most potent progestins appear to be constrained to the conformation that permits strongest association with the receptor; consequently, they are observed in this conformation in the solid state. Compounds with lower affinity for the receptor, including progesterone itself, are not constrained to the optimal conformation for receptor binding.

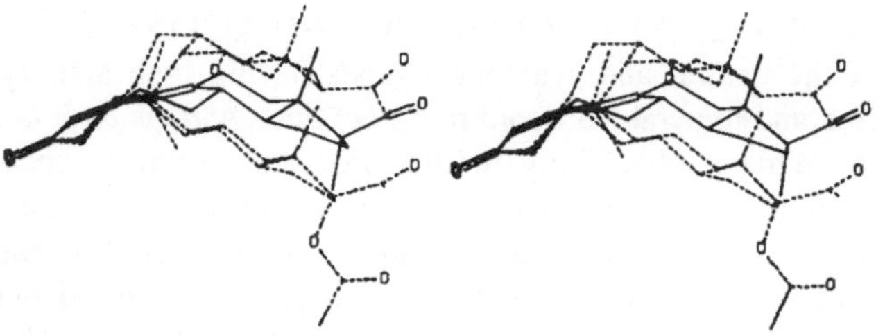

FIGURE 11. A stereo diagram illustrating superposition of the nearly identical A rings of high-affinity binders R5020(17,21-dimethyl-19-nor-4,9-pregnadiene-3,20-dione), solid lines, retroprogesterone, top dotted lines, and medroxyprogesterone acetate, bottom dotted lines, suggests a progestin receptor site that provides intimate, specific contact with the A ring but far less specific contact with the remainder of the steroid.

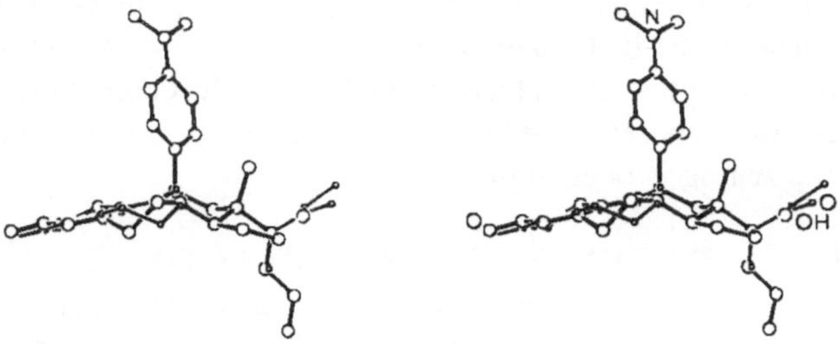

FIGURE 12. Stereo superposition of the crystallographically observed conformations of the potent progestin R5020 (small circles) and a derivative of the potent antiprogestin RU38486 (large circles) reveals that the A ring of each has the unusual inverted 1β,2α-half-chair conformation that is characteristic of many steroids having high affinity for the progesterone receptor.

In summary, examination of the compositions, configurations, and conformations of steroids with high affinity for the progesterone receptor suggests that there is a highly stereospecific association between the receptor and the A-ring region that is best satisfied by a 4-en-3-one ring in the inverted conformation. Association between the receptor and the D-ring end of the steroid either does not occur or is far less stereospecific (Figure 11).

A series of 11β-substituted 19-norsteroids have proven to be potent antiprogestins. The crystallographically observed conformations of these potent antiprogestins reveal the same unusual inverted conformation of the A ring[56] (Figure 12). This further supports the A-ring-binding/D-ring-acting model of steroid action.[33]

The similarity in steric and electronic properties of the A rings of potent progestins and antiprogestins may account for their competition for a common binding site on the receptor. Although very bulky 11β-phenyl substituents can be tolerated by the receptor and may, in fact, contribute to enhanced binding, some appear to interfere with the conformational change in the receptor that is essential for enhanced activity.

C. CORTICOID RECEPTOR BINDING AND ACTIVITY

Most of the natural and synthetic ligands having high affinity for the glucocorticoid and mineralocorticoid receptors have A rings with 4-en-3-one composition. While high affinity for the progestin receptor appears to be favored for steroids with an inverted 1β,2α-conformation of the A ring, optimum binding to the glucocorticoid and mineralocorticoid receptors may depend upon more subtle differences in the shape of the steroid A ring and the additional participation of the oxygen substituent at C(11) in the B ring. 9α-Fluorocortisol and certain prednisolone derivatives have greater affinity for the glucocorticoid receptor than does the endogenous hormone cortisol.[39] Crystallographic analysis of these steroids revealed a change in A-ring orientation that places the O(3) far below the plane of the B,C, and D rings.[57] This may contribute to their enhanced receptor binding. In contrast to this, the additional rings generated by epoxide formation in the endogenous mineralocorticoid, aldosterone, flatten the A ring and bring the O(3) very near the general plane of the B, C, and D rings.[58] The fact that 9α-fluorocortisol has higher affinity for the mineralocorticoid than the glucocorticoid receptor[39] suggests that this proposed dependency of receptor binding on relative orientation of the A ring is oversimplified.

D. ANDROGEN RECEPTOR BINDING AND ACTIVITY

The binding affinity of hundreds of steroids for the androgen receptor in rat prostate[59,60] suggests that a 17β-hydroxy and a 3-oxo group are essential for high-affinity binding. Most evidence suggests that if any portion of the androgen is critical to receptor binding it is the 17β-hydroxy-substituted D ring. Steroidal antiandrogens that are of sufficient potency to be used clinically also possess a 17β-hydroxyl-substituted D ring.[61,62] The 17β-hydroxyl group is usually flanked by an 18-methyl group and a hydrophobic substituent in the 17α- or 16β-position. These groups may be directly involved in receptor interaction or have some influence on the orientation and strength of the interaction of the 17β-hydroxyl group with the receptor.

One of the most active antiandrogens is the nonsteroidal compound flutamide.[63] The X-ray crystal structure determination of hydroxy-flutamide (HYF) (unpublished results) revealed that it is an extended planar molecule, in which 16 of the 20 nonhydrogen atoms are essentially coplanar. A reasonably good match between the steric and electronic features of the potent androgen agonists 17β-hydroxy-17α-methyl-4,9,11-estratrien-3-one (R1881) and HYF is achieved when the phenyl ring of HYF is aligned with the B ring of R1881 and the C(9)-atom of HYF is fitted to C(17) (Figure 13). This aligns the flat HYF molecule with the planar 3,9,11-triene portion of R1881. The positioning of the hydroxyl substituent on C(9) of HYF, flanked by two methyl groups, bears a close resemblance to the position of the 17β-hydroxyl between the 18- and 17α-methyl groups in R1881.

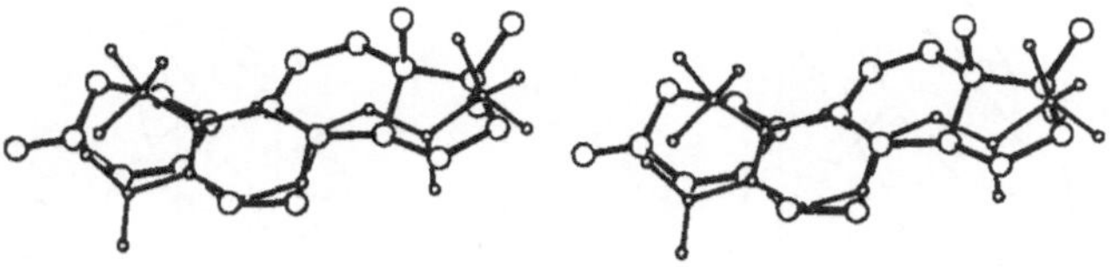

FIGURE 13. Stereo illustration comparing the crystallographically observed structures of the potent androgen 17β-hydroxy-17α-methyl-4,9,11-estratrien-3-one (R1881, large circles) and the antiandrogen hydroxyflutamide (HYF, small circles).

Since the outline of the flutamide molecule appears to fall within that of the androgen agonists, it would appear to function in a passive manner, failing to have some feature essential to induction or stabilization of receptor transformation or requisite macromolecular interaction that follows hormone binding. In contrast the antagonist action of tamoxifen and RU38486 {11β-[4-(dimethylamino) phenyl]- 17β-hydroxy-17α- (2-propenyl) -estra-4,9-dien-3-one} could be active, a result of the fact that the very bulky hydrophobic substituent, projected perpendicular to the general plane of the steroid, blocks a necessary conformational change in the receptor.

V. STEROID-PROTEIN INTERACTIONS AND BINDING

Information concerning the molecular details of steroid-protein interactions can be obtained from X-ray structure determination of the protein complex with appropriate steroid agonists and antagonists, and from X-ray study of model complexes of steroids with amino acids and nucleic acids.

The crystal structure of three enzymes that have steroids as their primary substrate have been reported.[64-66] The steroid-binding site has been tentatively identified in one on the basis of imaging a bound inhibitor and inferred in the others on biochemical grounds. The structure of uteroglobin, a progesterone-binding protein,[67] provides a useful model for some steroid-binding proteins despite the fact that there appears to be no steroid present in its presumed binding pocket. Although quantities of pure, stable, intact steroid receptor proteins sufficient to allow single crystal X-ray analysis are not yet available, the crystal structures of complexes between the DNA-binding domain of the glucocorticoid receptor and specific recognition elements from DNA have been reported.[68]

A. ENZYME STRUCTURE AND ACTIVE SITE ARCHITECTURE

The crystal structures of Δ^5-3-keto steroid isomerase,[64] cholesterol oxidase[65] and 3α,20β-hydroxysteroid dehydrogenase[66] have been reported. Δ^5-3-Keto isomerase (EC 5.3.3.1) catalyzes the isomerization of Δ^5-3-ketosteroid to Δ^4-3-ketosteroid by stereospecific direct transfer of a hydrogen atom from the C(4) position to the C(6) position. There is considerable evidence that this transfer is of the 4β-proton to the 6β-proton through an enolic inter-

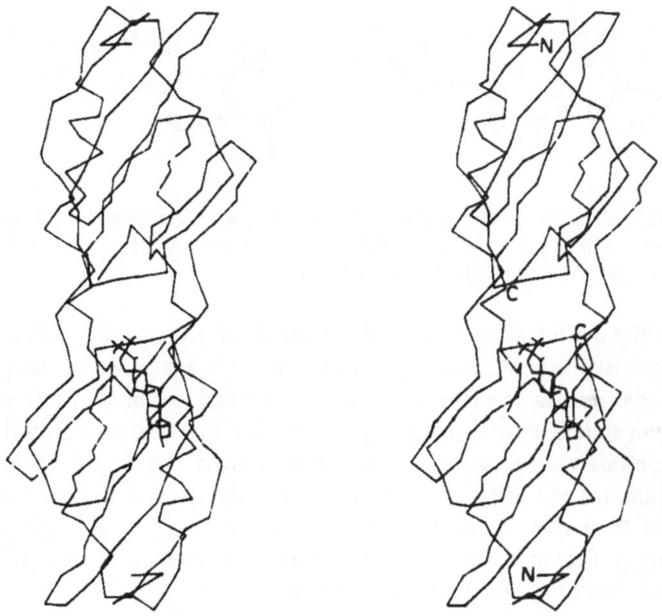

FIGURE 14. Stereo drawing showing the position of doxyl-DHT in one of the two subunits of Δ^5-3-ketoisomerase, viewed along the twofold axis of symmetry. The amino (N) and carboxyl (C) termini are identified. The doxyl moiety of the steroid protrudes into the cleft which communicates with the bulk solvent. A 10-residue α-helix is to the right of and below the steroid. (From Kuliopulos, A., Westbrook, E. M., Talalay, P., and Mildvan, A. S., *Biochemistry*, 26, 3927, 1987. With permission.)

mediate. A low-resolution (6 Å) crystal structure determination has been published and the probable steroid-binding site identified via a bound inhibitor 4-acetoxymercuric estradiol. The results of a higher-resolution study (2.5 Å) have been combined with the results of NMR studies and the preparation and analysis of activity of mutant forms of the enzyme to further define the probable active site of the enzyme.[69]

The isomerase enzyme is an elongated dimer in which the proposed steroid-binding site is surrounded by a distorted barrel of eight beta strands and a single alpha helix. Six beta strands and the alpha helix are from one monomer and two strands are contributed from the second monomer. The dimer interface is made up of the meeting of the barrel heads. While there are two similarly shaped barrels in the dimer, only one appears to contain the substrate with one end of the steroid at the surface of the barrel top (Figure 14).

NMR spectroscopy of doxyl dihydrotestosterone bound to the enzyme was used to determine the proximity of residues implicated in enzyme activity. Initial NMR and modeling studies[69] suggested that Tyr 55 was the proton donor that protonates the 3-keto group to form the en-ol intermediate and Asp

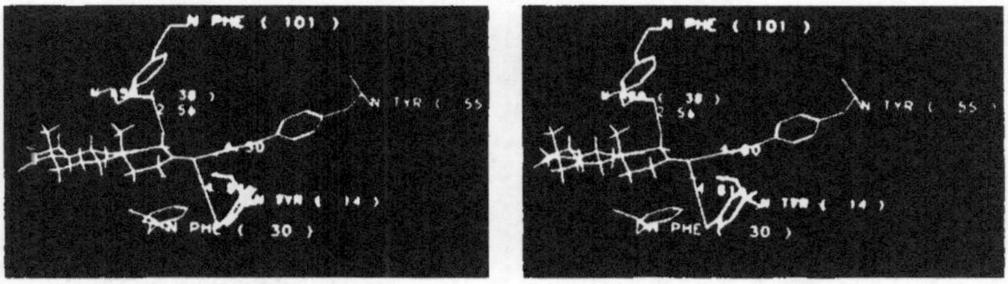

FIGURE 15. Details of the steroid binding site of Δ^5-3-ketoisomerase showing initial distances between critical active-site residues and the modeled substrate 5-androstene-3,17-dione. A carboxylate oxygen of Asp-38 (green) is in molecular contact (2.6 Å) with the 4β-proton of the substrate. (From Kuliopuluos, A., Mildvan, A. S., Shortle, D., and Talalay, P., *Biochemistry*, 28, 149, 1989.

38 is the protein acceptor that facilitates the hydrogen transfer from C(4) to C(6). Subsequent studies of mutant forms of the enzyme[70] confirmed the proposed role of Asp 38 in the hydrogen transfer but indicate that Tyr 14 rather than Tyr 55 is responsible for the protonation of the 3-keto group (Figure 15).

Cholesterol oxidase (EC 1.1.3.6) is a flavin adenine dinucleotide-dependant enzyme that catalyzes the oxidation and isomerization of steroids having a *trans* A-B ring junction. The structure of the enzyme from *Brevibacterium sterolicium* co-crystallized with the FAD co-factor has been determined at 1.8-Å resolution.[65] The 492 amino acids of the enzyme make up a two-domain structure. The FAD domain is composed of three noncontinuous segments of sequence that has a six-stranded β-sheet sandwiched between a four-stranded β-sheet and three α-helices, a topology similar to that seen in other FAD-binding proteins.[71] The steroid-binding domain is composed of two non contiguous segments of protein in which a six-stranded antiparallel β-sheet forms a "roof" over the postulated active site. The active site is at the interface of the two domains where the N-5 of the flavin ring system is in proximity to a cavity in the substrate domain that is occupied by relatively well-ordered water molecules. The amino acid residues binding this pocket are primarily hydrophobic near the pyrimidine ring, although the steroid-binding cavity is buried in the protein. It is proposed that three hydrophobic loops at the surface of the molecule that exhibit relatively high thermal motion are flexible enough to allow access of the substrate to the active site.[65] The only charged residue in the presumed active site, a glutamic acid, is suitably positioned to act as a proton acceptor in the proposed mechanism for isomerization. The N-5 atom of the FAD and a tyrosine residue lie on the opposite side of the substrate cavity from the glutamic acid site (Figure 16). Consequently, arrangement of proton donors and acceptors on the surface of the presumed binding pockets of cholesterol esterase and Δ^5-3-keto isomerase are very similar, supporting the contention[72] that the isomerization actions of the two enzymes proceed by similar mechanisms.

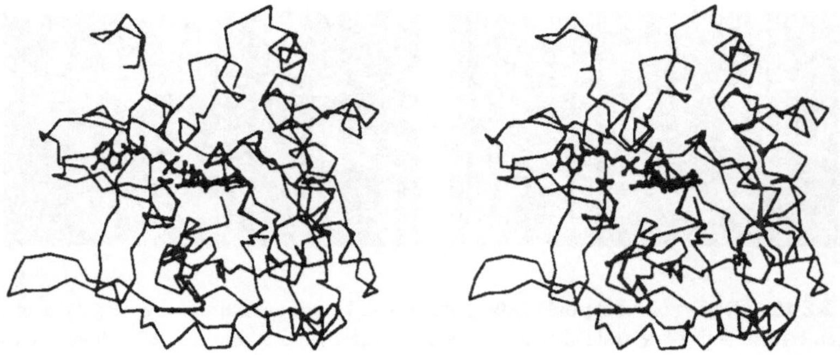

FIGURE 16. A stereo view of the C^{α} backbone of the steroid-binding domain of cholesterol oxidase showing the proposed entrance to the active site. The FAD molecule is shown in thick lines. The mobile loops believed to be situated near the proposed entrance are also shown in thick lines. (From Vrielink, A., Lloyd, L. F., and Blow, D. M., *J. Mol. Biol.*, 219, 533, 1991. With permission.)

3α,20β-Hydroxysteroid dehydrogenase (3α,20β-HSD, EC 1.1.1.53) from *Streptomyces hydrogenans* is a nicotinamide adenine dinucleotide [NAD(H)]-linked enzyme involved in the reversible oxidation of the 3α- and 20β-hydroxyl groups of androstane and pregnane derivatives.[73] Protein sequence analysis has revealed that 3α,20β-HSD is a member of the nonmetallo-short-chain alcohol dehydrogenase family that includes bacterial, insect, and mammalian dehydrogenases for steroids, prostaglandins, sugars, and other compounds.[74] The X-ray structure determination of the complex of the tetrameric enzymes and co-factors has been determined to 2.6-Å resolution.[66] Each subunit of the tetramer contains a co-factor binding site and a putative steroid-binding site. The 245 amino acid monomer has essentially a single domain.

The first 145 residues have the characteristic fold observed in many other NAD-dependent enzymes.[75] This fold is composed of a six-stranded parallel β-sheet with three α-helices on either side. The NAD(H) co-factor lies in a cleft generated by the N-terminal ends of α-helices B, C, and D, the carboxyl terminal end of the β-strand A, B, and C, and the mid section of βD and αE. The co-factor is not buried as deeply in the cleft of the 3α,20β-short-chain dehydrogenase as it is in the long-chain dehydrogenase structures. The NAD(H) is on one side of the β-sheet and a solvent-filled cavity large enough to accommodate a steroid molecule is on the other side. The nicotinamide end of the co-factor approaches the binding cavity and a hydrogen bond is formed between the C4 end of the nicotinamide ring and an arginine residue (Arg 16) that lies between the co-factor binding site and the presumed substrate binding site. Without disturbing the protein side chain from the refined position it is possible to place a cortisone molecule into the proposed binding site so that the 20 carbonyl is also within hydrogen-bonding distance of the arginine residue that is hydrogen bonded to C(4) of the nicotinamide (Figure 17). Although there are no documented cases of Arg participating in a hydride

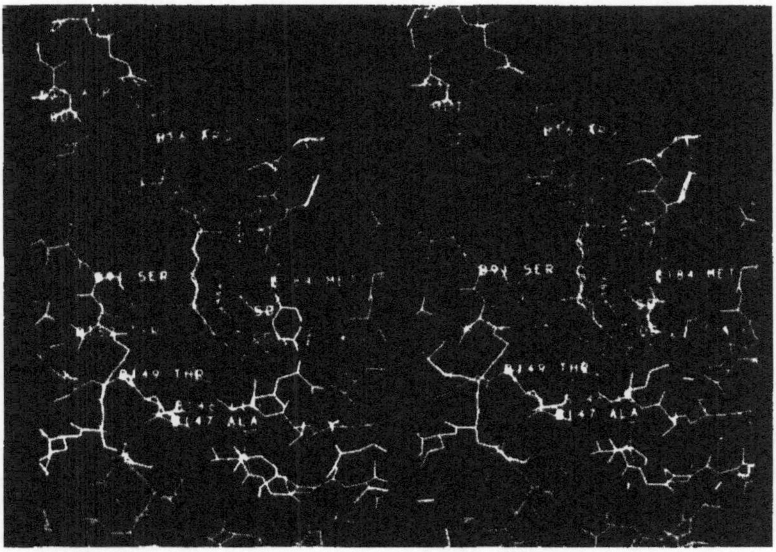

FIGURE 17. A view of the catalytic site in 3α,20β-hydroxysteroid dehydrogenase. A cortisone molecule is modeled in the pocket Dotted Van der Waals surfaces of cortisone atoms and proximal protein atoms demonstrate an excellent fit of the molecule in the proposed binding pocket. Some of the residues having potential interaction with the steroid are labeled. The nicotinamide end of the cofactor is visible at the upper left-hand corner of the figure.

transfer, that possibility clearly exists in 3α,20β-HSD. It is also possible that the Arg 16 side chain could swing away from the pocket, thereby permitting a closer approach of the steroid and the dinucleotide to each other and a direct hydride transfer through a much shorter distance. A structure of an enzyme-substrate/inhibitor complex could cast further light upon details of the mechanism.

It is noteworthy that there is a tyrosine residue in the steroid-binding site that is conserved in the short-chain dehydrogenase family. The presence of a tyrosine in the steroid-binding pocket is a common feature of all three of the enzymes for which steroids are substrate and X-ray structures have been determined.

B. STEROID-BINDING PROTEINS

The steroid-binding proteins of blood serum maintain a high concentration of circulating hormones that are protected from chemical and enzymic attack.[76] Unfortunately, the presence of a high percent of heterogeneous carbohydrate in the composition of the binding globulins limits the quality of single crystals that have been prepared.[77,78]

Uteroglobin, a progesterone-induced progesterone-binding globulin secreted by the rabbit uterus, contains no carbohydrate. It is a dimer of two

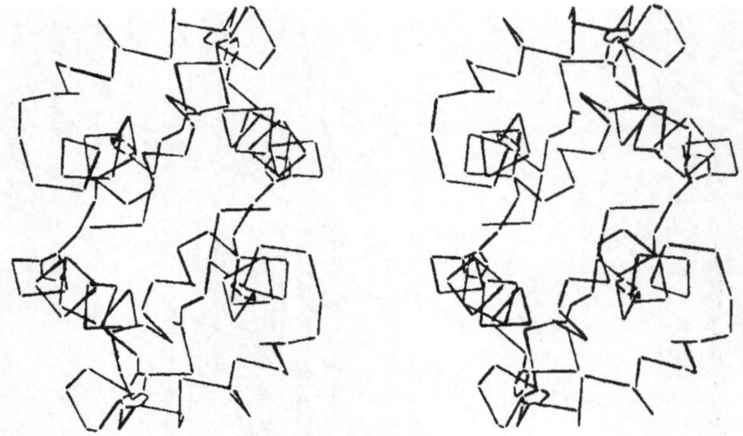

FIGURE 18. Stereo diagram of backbone of the uteroglobin structure with rotation axis passing through probable steroid-binding pocket. Taken from Mornon, et al. (From Morize, I., Surcouf, E., Vaney, M. C., Epelboin, Y., Buehner, M., Fridlansky, F., Milgrom, E., and Mornon, J. P., *J. Mol. Biol.*, 194, 725, 1987 With permission.)

identical amino acid chains of 70 residues each, linked by two disulfide bridges. A high-resolution X-ray crystal structure determination revealed the presence of a hydrophobic pocket in the center of the dimer that has dimensions corresponding to that of progesterone (Figure 18).[67] Tyrosine residues at either end of the oblong cavity could form hydrogen bonds to a progesterone molecule when bound. Because there was no progesterone in the uteroglobin samples used to prepare crystals and because it has not been possible to demonstrate binding or exchange of labeled progesterone molecules to uteroglobin unless the disulfide bonds are reduced, it has not been possible to demonstrate unequivocally that the central cavity is the progesterone binding site.

In an effort to determine the characteristics of the steroid-binding site in PBG, Westphal[79] measured the association constants for PBG of more than 100 steroids. Almost any structural change on any part of the progesterone molecule resulted in a reduction in binding affinity. Consequently, Westphal proposed a very tight fit of the progesterone molecule in a binding site with primarily hydrophobic interactions over most of the molecular surface.

Ultraviolet difference spectra suggest a specific interaction between the 4-en-3-one chromophore and tryptophan residues when progesterone is bound to PBG.[80] The spectral evidence of tryptophan interaction stimulated efforts to prepare crystalline complexes of progesterone and amino acids as models for steroid-receptor binding. Progesterone co-crystallizes readily with a number of hydrogen-bond donors, including indole.[13] Crystals of the 2:1 complex of indole and progesterone were reproducibly grown from a variety of solvents. In the crystal structure, indole molecules are hydrogen bonded to the

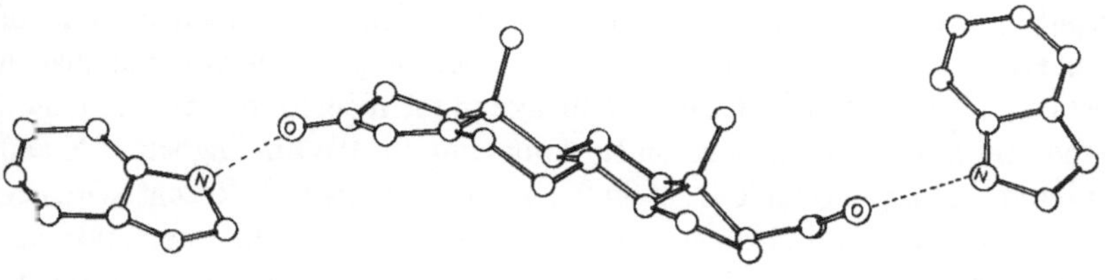

FIGURE 19. Hydrogen bonding observed between indole and progesterone in a 2:1 complex.

carbonyl oxygens O(3) and O(20) as illustrated in Figure 19. Although crystals of the complex have a pale pink color indicative of a chromophoric interaction, there is no evidence in the crystals of a π-π stacking interaction. The indole molecules do not stack over each other or over the 4-en-3-one chromophore. No significant structural differences have been detected when the 4-en-3-one chromophore of progesterone in the complex is compared with that in either of two polymorphic crystal forms of unsolvated progesterone. In the complex the indole molecule is nearly coplanar with the 4-en-3-one chromophore. In the model of the progestin uteroglobin complex structure, the tyrosine residue hydrogen bonded to the C(3)-carbonyl is also nearly coplanar with the A ring rather than stacked over it. It would appear that hydrogen bonding and coplanarity of π systems similar to those observed in the progesterone indole complex, and proposed in the uteroglobin model, may account for the observed spectral shifts and the high-affinity binding of progesterone to PBG.

C. RECEPTOR STRUCTURE AND STEROID FUNCTION

The steroid receptors are members of a super family of soluble proteins having extensive sequence homologies.[81,82] The family includes thyroid hormone receptors, retinoic acid receptors, and analogs of unknown specificity and function. These receptors range in size from 60 to 107 kDa. Analysis of the sequence of the receptors and preparation of mutants have led to the characterization of several domains in the steroid receptor: the DNA-binding domain, the steroid-binding domain, and an immunodominant domain that appears to have a role in transcriptional activation. The steroid-binding region is at the C-terminal end of the receptor. The degree of sequence homology observed in the steroid-binding region of the receptor parallels the competition for binding to different receptors by the steroid hormones.

It was originally proposed that the steroid's role in hormone activity was to effect a conformational change in the receptor essential to migration of the receptor from the cytosol to the cell nucleus.[2] Subsequent investigations have revealed that most steroid receptors are found in the nucleus in their unliganded form.[83] Consequently, the critical function of the steroid molecule is generally thought to be the induction of a change in the receptor required for specific DNA interaction. Because modified receptor molecules from which the steroid-binding domain has been deleted are found to be fully functional,[81] the steroid appears to act as a derepressor and the change it induces, one of uncovering the DNA interactive residues. There is growing evidence that in most cases the same conformational change in the receptor can be stimulated by the action of hormone antagonists competing for the binding site,[84,85] and that the receptor can bind to DNA in the absence of ligand.[86] Recent evidence that the steroid ligand changes the kinetic parameters of the interaction between the receptor and the DNA[87] indicates that *in vivo* the steroid's ultimate role may be to ensure the specificity of receptor DNA interaction.

Biochemical, spectroscopic, and gene mutation studies continue to be used in an effort to identify the specific amino acids that are involved in steroid-receptor binding. For example, a specific methionine residue (622) in the rat liver glucocorticoid receptor has been identified by several investigators[88,89] as being proximal to the site of attachment to the A-ring end of the bound steroid. A tryptophan residue in the same region (Trp 618) may well participate in steroid binding via an interaction, comparable to that seen in the progesterone/indole complex or observed in the binding pockets of the enzymes described in Section V.A above. A homologously located tryptophan residue is present in glucocorticoid, mineralocorticoid, and progesterone receptors.[90]

Two crystal structures of complexes of the DNA-binding domain (DBD) of the glucocorticoid receptor and a fragment of DNA have recently been reported.[68] The DBD is made up of approximately 80 amino acids and contains two sets of four cysteine residues. Each set of cysteine is found to be tetrahedrally coordinated to zinc ions, forming zinc fingers that make up a single globular domain. The DBD binds to the DNA as a dimer with the subunits lying in adjacent major grooves. The two complexes differ in the length of nucleic acid spacers in the glucocorticoid response element (GRE) of the DNA fragment, exhibit specific and nonspecific interactions, and support a model proposed to explain the role of DNA-induced dimerization in determining specific affinity of steroid receptors for different DNA targets.

Although steroids are implicated in determining the specific recognition between the receptor DBD and the genomic recognition element, there is no evidence of a direct interaction between the steroid and DNA. However, the specific interactions between the corticoid side chain and the nucleic acid component adenine observed in the stable crystalline complex between deoxycorticosterone and adenine,[91] where the carbonyl and hydroxyl substituents on the corticoid side chain form hydrogen bonds to the two nitrogens of adenine normally involved in Watson-Crick base pairing, might be critically involved, either in DNA sequence recognition or in the activation of transcription by the steroid/receptor complex.[91] Such a role for a small molecule in macromolecular interaction is not without precedent. The crystal structure determination of a complex between the trp repressor and its DNA operator region revealed the direct participation of the tryptophan molecule in DNA interaction, forming a 1.9-Å hydrogen bond to a phosphate oxygen, Figure

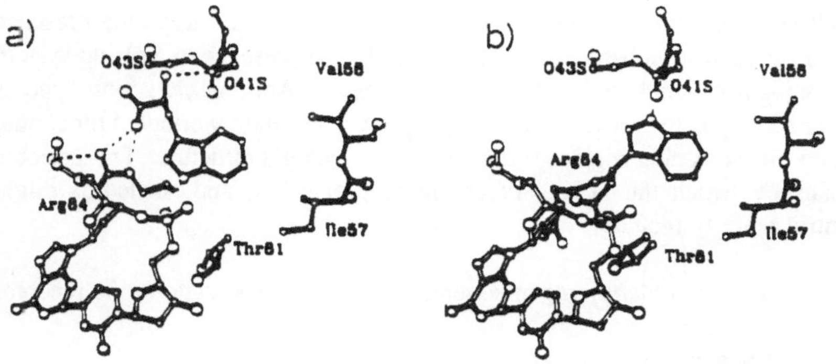

FIGURE 20. (a) Tryptophan environment in ternary complex of tryptophan, the tryptophan repressor, and a fragment of DNA reveals hydrogen bonds between tryptophan and both macromolecules. (b) Attempts to model interaction between the observed indole propionate tryptophan complex and DNA suggests that the indole propionate antagonist interferes with DNA interaction.

20a.[92-94] X-ray analysis of a complex of the trp repressor with indole propionate (IPA) revealed that the IPA binds in the same position as the indole but is "flipped over" (Figure 20b). As a result, the carboxyl group of IPA is oriented toward the DNA binding surface of the protein and is in a position where it sterically and electrostatically repels the phosphate backbone of both operator and nonoperator DNA.[93]

VI. SUMMARY

Crystallographic data on over 1000 steroids provide information concerning preferred conformation, relative stabilities, and substituent influence on the interactive potential of steroid hormones. Conformational analysis of these data and comparison with *ab initio* and empirical energy calculations indicate that the conformations observed in the solid state are at or near the global minimum energy conformation and provide a useful basis upon which to prepare empirical models for structure-function relationships.

Examination of the structures of compounds having high affinity for estrogen, progestin, and corticoid receptors strongly suggests that receptor binding is primarily the result of a tight association between the receptor and the steroidal A ring. Antagonists that compete for estrogen, progesterone, and glucocorticoid receptor sites may be expected to have the A-ring composition and conformation necessary for receptor binding but lack the D-ring conformational features and functional groups that induce or stabilize subsequent receptor functions. Antagonists might also be compounds with A-ring conformations appropriate for binding but have other structural features that interfere with subsequent receptor functions essential to activity. In contrast, androgen receptor-binding data and molecular-modeling studies suggest that in this case a D-ring binding/A-ring acting model is more appropriate.

Androgen receptor binding appears to be most dependent upon the presence of a 17β-hydroxy substituent and enhanced by the presence of a planar system of conjugated bonds in the B- and C-ring region. Androgenic action appears dependent upon the presence of an appropriate and suitably oriented functional group on the A-ring region of the steroid or analog's structure. The possible means by which the D ring of estrogen, progestins, and corticoids might control activity include:

1. Inducing or stabilizing an essential conformational state in the receptor (allostery)
2. Influencing the aggregation state of the receptor
3. Participating in a direct interaction with DNA in chromatin

Recent X-ray studies of enzymes for which steroids are primary substrates, steroid-binding proteins, and fragments of steroid receptors with amino acids and nucleic acids are providing additional insight into the nature of steroid receptor binding and macromolecular interactions, and the mechanism by which steroids exert their hormonal functions.

REFERENCES

1. **Jensen, E. V.,** Basic guides to the mechanism of estrogen action, in *Recent Progress in Hormone Research,* Greep, R. O., Ed., Academic Press, New York, 1962, 387.
2. **O'Malley, B. W. and Birnbaumer, L.,** *Receptors and Hormone Action,* Vol. 1, Academic Press, New York, 1978.
3. **Milgrom, E.,** Activation of steroid-receptor complexes, in *Biochemical Actions of Hormones,* Vol. 8, Litwack, G., Ed., Academic Press, New York, 1981, chap. 7.
4. **Duax, W. L. and Norton, D. A.,** *Atlas of Steroid Structure,* Vol. 1, Plenum Press, New York, 1975.
5. **Griffin, J. F., Duax, W. L., and Weeks, C. M.,** *Atlas of Steroid Structure,* Vol. 2, Plenum Press, New York, 1984.
6. **Duax, W. L., Weeks, C. M., and Rohrer, D. C.,** Crystal structures of steroids, in *Topics in Stereochemistry,* Vol. 9, Eliel, E. L. and Allinger, N. L., Eds., Wiley-Interscience, New York, 1976, 271.
7. **Duax, W. L., Griffin, J. F., and Rohrer, D. C.,** Conformation of progesterone side-chain: conflict between X-ray data and force field calculations, *J. Am. Chem. Soc.,* 103, 6705, 1981.
8. **Duax, W. L., Griffin, J. F., and Cheer, C.,** Steroid conformational analyses based on X-ray crystal structure determination, in *Analysis of Sterols and Other Biologically Significant Steroids,* Nes, W D. and Parish, E. J., Eds., Academic Press, San Diego, 1989, chap. 10.
9. **Wellman, K. M. and Djerassi, C.,** Optical rotatory dispersion studies. XCVIII. The conformation of the 17-acetyl group in steroids and its correlation with variable-temperature circular dichroism, *J. Am. Chem. Soc.,* 87, 60, 1965.
10. **Dideberg, O., Dupont, L., and Campsteyn, H.,** Structure cristalline et moleculaire du complexe 1:1 progesterone-resorcinal, *Acta Crystallogr Sect. B,* 31, 637, 1975.

11. **Compsteyn, H., Dupont, L., and Dideberg, O.,** Structure cristalline et moleculaire de la progesterone, *Acta Crystallogr. Sect. B*, 28, 3032, 1972.
12. **Serantoni, E. F., Krajewski, A., Mongiorgi, R., Riva di Sanseverino, L., and Cameroni, R.,** 4-Pregnen-3,20-dione (progesterone form II), *Cryst. Struct. Commun.*, 4, 189, 1975.
13. **Duax, W. L., Strong, P., Rohrer, D. C.,** unpublished data, 1980.
14. **Galdecki, Z., Grochulski, P., Wawrzak, Z., and Duax, W. L.,** Structure of progesterone hydroquinone monohydrate (1:1:1), $C_{21}H_{30}O_2 \cdot C_6H_6O_2 \cdot H_2O$, *J. Cryst. Spec. Res.*, 19, 983, 1989.
15. **Kamernitzky, A. V., Turuta, A. M., Fundlieler, I. N., Pavlov, V. A., Antipin, A. Yu., Struckhov, Yu. T., and Polishchuk, A. P.,** α-Epithioketosteroids: conformation and reactivity, *Tetrahedron*, 38, 165, 1982.
16. **Galdecki, A., Grochulski, P., Wawrzak, Z., Duax, W. L., and Strong, P. D.,** Structure of 3α,12β-dihydroxy-2β-morpholino-5α-pregnan-20-one, $C_{25}H_{41}O_4N$, *J. Cryst. Spec. Res.*, 19, 561, 1989.
17. **Duax, W. L., Griffin, J. F., Strong, P. D., Funder, J. W., and Ulick, S.,** Molecular structure of 18-deoxyaldosterone and its relationships to receptor binding and activity, *J. Am. Chem. Soc.*, 104, 7291, 1982.
18. **Galdecki, Z., Grochulski, P., Wawrzak, Z., Duax, W. L., and Strong, P. D.,** Structure of 21-fluoro-4,9(11)-pregnadiene-3,20-dion-17α-yl acetate, $C_{23}H_{29}O_4F$, *J. Cryst. Spec. Res.*, 20, 441, 1990.
19. **Raynaud, J. P., Brown, N. L., Coussedieri, C., Pottier, J., Delettre, J., and Mornon, J. P.,** Role of receptor binding and metabolism in progestin-induced responses in the uterus, in *Steroid Induced Uterine Proteins*, Beato, M., Eds., Elsevier/North-Holland Biomedical Press, Amsterdam, 1980, 217.
20. **Allinger, N. A. and Yuh, V. H.,** Descriptions of computational chemistry computer programs, *Quantum Chem. Program Exch.*, 13, 395, 1981.
21. **Duax, W. L., Fronckowiak, M. D., Griffin, J. F., and Rohrer, D. C.,** A comparison between crystallographic data and molecular mechanics calculations on the side chain and backbone conformations of steroids, in *Intramolecular Dynamics*, Jortner, J. and Pullman, B., Eds., D. Reidel, Dordrecht, The Netherlands, 1982, 505.
22. **Profeta, S., Jr., Kollman, P. A., and Wolff, M. E.,** Conformation of the progesterone side chain: resolution of the apparent conflict between X-ray data and force field calculations using MM2, *J. Am. Chem. Soc.*, 104, 3745, 1982.
23. **Wiberg, K. B. and Martin, E.,** Barriers to rotation adjacent to double bonds, *J. Am. Chem. Soc.*, 107, 5035, 1985.
24. **Butcher, S. S. and Wilson, E. B., Jr.,** Microwave spectrum of propionaldehyde, *J. Chem. Phys.*, 40, 1671, 1964.
25. **Wiberg, K. B.,** Rotation adjacent to double bonds. 2n-Propyl vs. isopropyl groups, *J. Am. Chem. Soc.*, 108, 5817, 1986.
26. **Allinger, N. L., Chen, K., Rahmon, M., and Pathiaseril, A.,** Molecular mechanics (MM3) calculation on aldehydes and ketones, *J. Am. Chem. Soc.*, 113, 4505, 1991.
27. **Duax, W. L., Cody, V., Griffin, J., Hazel, J., and Weeks, C. M.,** Steroid structure and function. II. Conformational transmission and receptor binding of medroxyprogesterone acetate, *J. Steroid Biochem.*, 9, 901, 1978.
28. **Duax, W. L., Eger, C., Pokrywiecki, S. A., and Osawa, Y.,** Crystalline molecular structures of two derivatives of 2β-hydroxytestosterone having an unusual A-ring conformation, *J. Med. Chem.*, 14, 295, 1971.
29. **Barton, D. H. R., Head, A. J., and May, P. J.,** Long-range effects in alicyclic systems. II. The rates of condensation of some triterpenoid ketones with benzaldehyde, *J. Chem. Soc.*, 935, 1957.
30. **Duax, W. L., Cody, V., and Hazel, J.,** Steroid structure and function. I. Conformational transmission in 17α-acetoxy progesterone, *Steroids*, 30, 471, 1977.

31 **Duax, W. L. and Strong, P. D.,** A-ring conformation in 17-hydroxy-6c-methylpregenolone, *Steroids,* 34, 501, 1979.
32. **Barrett, M. W., Farrant, R. D., Kirk, D. N., Mersh, J. D., Sanders, J. K. M., and Duax, W. L.,** The solution conformation of 17α-acetoxy-6α-methylprogesterone ('medroxyprogesterone acetate'): use of circular dichroism, nuclear overhauser effect difference and two-dimensional J. spectroscopy, *J. Chem. Soc., Perkin Trans.,* 2, 105, 1982.
33. **Duax, W. L., Cody, V., Griffin, J. F., Rohrer, D. C., and Weeks, C. M.,** Molecular conformation and protein binding affinity of progestins, *J. Toxicol. Envir. Health,* 4, 205, 1978.
34. **Duax, W. L. and Weeks, C. M.,** Molecular basis of estrogenicity; X-ray crystallographic studies, in *Estrogens in the Environment,* McLachlan, J. A., Ed., Elsevier, New York, 1980, 11.
35. **Duax, W. L., Griffin, J. F., Rohrer, D. C., Swenson, D. C., and Weeks, C. M.,** Molecular details of receptor binding and hormonal action of steroids derived from X-ray crystallographic investigations, *J. Steroid Biochem.,* 15, 41, 1981.
36. **Duax, W. L., Griffin, J. F., Weeks, C. M., and Wawrzak, Z.,** The mechanism of action of steroid antagonists: insights from crystallographic studies, *J. Steroid Biochem.,* 31, 481, 1988.
37. **Duax, W. L. and Griffin, J. F.,** Structural features which distinguish estrogen agonists and antagonists, *J. Steroid Biochem.,* 27, 271, 1987.
38. **Duax, W. L. and Griffin, J. F.,** The structure and receptor binding of steroid hormones, in *Advances in Drug Research,* Vol. 18, Testa, B., Ed., Academic Press, New York, 1989, chap. 2.
39. **Raynaud, J. P., Ojasoo, T., Bouton, M. M., and Philibert, D.,** Receptor binding as a tool in the development of new bioactive steroids, in *Drug Design,* Vol. 8, Ariens, E. J., Eds., Academic Press, New York, 1979, 160.
40 **Busetta, C., Courseille, C., Leroy, F., and Hospital, M.,** Structure cristalline et moleculaire de la testosterone hydrates étude de l'interaction steraide eau pour certains derives 17β-ol, *Acta Crystallogr. Sect. B,* 28, 3293, 1972.
41. **Binart, N., Catelli, M. G., Geynet, C., Puri, V., Hanel, R., Mester, J., and Baulieu, E. E.,** Monohydroxy-tamoxifen: an antiestrogen with high affinity for the chick oviduct oestrogen receptor, *Biochem. Biophys. Res. Commun.,* 91, 812, 1979.
42. **Mirocha, C. J., Christensen, C. M., and Nelson, G. H.,** Physiologic activity of some fungal estrogens produced by fusarium, *Cancer Res.,* 28, 2319, 1965.
43. **Mueller, G. and Kim, U.-H.,** Displacement of estradiol from estrogen receptors by simple alkyl phenols, *Endocrinology,* 102, 1429, 1978.
44. **Korach, K. S., Levy, L. A., and Sarver, P. J.,** Stereochemical analysis of stilbene estrogens: receptor binding and hormone responsiveness, in *Estrogens in the Environment II: Influences on Development,* McLachlan, J. A., Ed., Elsevier, New York, 1985, 43.
45. **Duax, W. L., Swenson, D. C., Strong, P. D., Korach, K. S., McLachlan, J., and Metzler, M.,** Molecular structure of metabolites and analogs of diethylstilbestrol and their relationship to receptor binding and biological activity, *Mol. Pharm.,* 26, 520, 1984.
46. **Anstead, G. M., Wilson, S. R., and Katzenellenbogan, J. A.,** 2-Arylindenes and 2-arylindenones: molecular structures and considerations in the binding orientation of the unsymmetrical nonsteroidal ligands to the estrogen receptor, *J. Med. Chem.,* 32, 2163, 1989.
47. **Korach, K. S., Chae, K., Levy, L. A., Duax, W. L., and Sarver, P. J.,** Diethylstilbestrol metabolites and analogs: stereochemical probes for the estrogen receptor binding site, *J. Biol. Chem.,* 264, 5642, 1989.
48. **Duax, W. L., Griffin, J. F., Weeks, C. M., and Korach, K. S.,** Molecular conformation, receptor binding and hormone action of natural and synthetic estrogens and antiestrogens, in *Environmental Health Perspectives,* Vol. 61, McKinney, J., Ed., Elsevier, New York, 1985, 111.

49. **Borgna, J. L., Coezy, E., and Rochefort, H.,** Mode of action of LN1643 (a triphenylbromoethylene antiestrogen) probable mediation by the estrogen receptor and high affinity metabolite, *Biochem. Pharmacol.*, 31, 3187, 1982.

50. **Durani, S., Agarwal, A. K., Saxena, R., Setty, B. S., Gupta, R. C., Kole, P. L., Ray, S., and Anand, N. J.,** Seco-oestradiols and some non-steroidal oestrogens: structural correlates of oestrogenic action, *J Steroid Biochem.*, 11, 67, 1979.

51. **Kontula, K., Jänne, O., Vihko, R., de Jager, E., de Visser, J., and Zeelen, F.,** Progesterone-binding proteins: *in vitro* binding and biological activity of different steroidal ligands, *Acta Endocrinol. (Copenhagen)*, 78, 574, 1975.

52. **Smith, H. E., Smith, R. G., Toft, D. O., Neergaard, J. R., Burrows, E. P., and O'Malley, B. W.,** Binding of steroids to progesterone receptor proteins in chick oviduct and human uterus, *J Biol. Chem.*, 249, 5924, 1974.

53. **Terenius, L.,** Affinities of progestogen and estrogen receptors in rabbit uterus for synthetic progestins, *Steroids*, 23, 909, 1974.

54. **Mornon, J.-P., Delettre, J., Lepicard, G., Bally, R., Surcouf, E., and Bondot, P.,** Interactions of hormonal steroids: progestogens, *J. Steroid Biochem.*, 8, 51, 1977.

55. **Precigoux, G., Busetta, B., Courseille, C., and Hospital, M.,** Structure cristalline et moleculaire de la nor-19 testosterone, *Acta Crystallogr. Sect. B*, 31, 1527, 1975.

56. **Surcouf, E.,** Contribution to the Study of Interaction of Protein Receptors and Substrates, Doctoral thesis, University Pierre and Marie Cure, Paris, 1982, 94.

57. **Weeks, C. M., Duax, W. L., and Wolff, M. D.,** A comparison of the molecular structure of six corticoids, *J. Am. Chem. Soc.*, 95, 2865, 1973.

58. **Duax, W. L. and Hauptman, H.,** The crystal structure and molecular conformation of aldosterone, *J. Am. Chem. Soc.*, 94, 5467, 1972.

59. **Cunningham, G. R., Tindall, D. J., and Means, A. R.,** Differences in steroid specificity for rat androgen binding protein and the cytoplasmic receptor, *Steroids*, 33, 261, 1979.

60. **Kirchhoff, J., Soffie, M., and Rousseau, G. G.,** Differences in the steroid-binding site specificities of rat prostate androgen receptor and epididymal androgen-binding protein, *J. Steroid Biochem.*, 10, 487, 1979.

61. **Hamada, H., Neumann, F., and Junkmann, K.,** Intrauterine antimaskuline beeinflussung von rattenfeten durch ein stark gestagen wirksames steroid, *Acta Endocrinol. Copenhagen*, 44, 380, 1963.

62. **Mangan, F. R. and Mainwaring, W. I. P.,** An explanation of the antiandrogenic properties of 6α-bromo-17β-hydroxy-17α-methyl-4-oxa-5α-androstane-3-one, *Steroids*, 20, 331, 1972.

63. **Wakeling, A. E., Furr, B. J. A., Glen, A. T., and Hughes, L. R.,** Receptor binding and biological activity of steroidal and nonsteroidal antiandrogens, *J Steroid Biochem.*, 15, 355, 1981.

64. **Westbrook, E. M., Piro. O. E., and Sigler, P. B.,** The 6Å crystal structure of Δ^5-3-ketosteroid isomerase: architecture and location of the active center, *J. Biol. Chem.*, 259, 9096, 1984.

65. **Vrielink, A., Lloyd, L. F., and Blow, D. M.,** Crystal structure of cholesterol oxidase from *Brevibacterium sterolicum* refined at 1.8Å resolution, *J. Mol. Biol.*, 219, 533, 1991.

66. **Ghosh, D., Weeks, C. M., Grochulski, P., Duax, W. L., Erman, M., Rimsay, R. L., and Orr, J. C.,** Three-dimensional structure of holo 3α,20β-hydroxysteroid dehydrogenase: a member of the short-chain dehydrogenase family, *Proc. Natl. Acad. Sci. U.S.A.*, in press.

67. **Morize, I., Surcouf, E., Vaney, M. C., Epelboin, Y., Buehner, M., Fridlansky, F., Milgrom, E., and Mornon, J. P.,** Refinement of the $C222_1$ crystal form of oxidized uteroglobin at 1.34Å resolution, *J. Mol. Biol.*, 194, 725, 1987.

68. **Luisi, B. F., Xu, W. X., Otwinowski, Z., Freedman, L. P., Yamamoto, K. R., and Sigler, P. B.**, Crystallographic analysis of the interaction of the glucocorticoid receptor with DNA, *Nature,* 352, 497, 1991
69. **Kuliopulos, A., Westbrook, E. M., Talalay, P., and Mildvan, A. S.**, Positioning of a spin labeled substrate analogue into the structure of Δ^5-3-ketosteroid isomerase by combined kinetic, magnetic resonance and X-ray diffraction methods, *Biochemistry,* 26, 3927, 1987.
70. **Kuliopuluos, A., Mildvan, A. S., Shortle, D., and Talalay, P.**, Kinetic and ultraviolet spectroscopic studies of active site mutants of Δ^5-3-ketosteroid isomerase, *Biochemistry,* 28, 149, 1989
71. **Wierenga, R. K., Drenth, J., and Schulz, G. E.**, Comparison of the three-dimensional protein and nucleotide structure of the FAD-binding domain of *p*-hydroxybenzoate hydroxylase with the FAD as well as NADPH-binding domains of glutathione reductase, *J Mol. Biol.*, 167, 725, 1983.
72. **Smith, A. G. and Brooks, C. J. W.**, The substrate specificity and stereochemistry, reversibility and inhibition of the 3-oxo steroid Δ^4-Δ^5 isomerase component of cholesterol oxidase, *J. Biochem.*, 167, 121, 1977.
73 **Pocklington, T. and Jeffery, J.**, 3α-Hydroxysteroid: NAD oxidoreductase activity in crystalline preparations of 20β-hydroxysteroid. NAD oxidoreductase, *Eur. J. Biochem.*, 7, 63, 1968.
74. **Marekov, L., Krook, M., and Jornvall, H.**, Prokaryotic 20β-hydroxysteroid dehydrogenase is an enzyme of the 'short-chain, non-metalloenayme' alcohol dehydrogenase type, *Fed. Eur. Biochem. Soc.*, 266, 51, 1990.
75. **Birktoft, J. J. and Banaszak, L. J.**, Structure-function relationships among nicotinamde adenine dinucleotide dependent oxidoreductases, *Peptide Protein Res.*, 4, 1, 1984.
76. **Baulieu, E. E.**, Steroid hormone binding plasma proteins and their intra and extra-cellular congeners, in *Binding Proteins of Steroid Hormones,* Forest, M. G. and Pugeat, M., Eds., John Libby Eurotext, London, 1986, chap. 1.
77. **McPherson, A., Mickelson, K. E., and Westphal, U.**, Crystallization of corticosteroid binding globulin (CBG) and α_1-acid glycoprotein (AAG), *J. Steroid Biochem* , 13, 991, 1980.
78. **Koszelak, S., McPherson, A., Bouillon, R., and Van Ballen, H.**, Crystallization and preliminary X-ray analysis of the vitamin D-binding protein from human serum, *J Steroid Biochem.*, 23, 1077, 1985.
79. **Blanford, A. T., Wittman, W., Stroupe, S. D., and Westphal, U.**, Steroid-protein interactions. XXXVIII. Influence of steroid structure on affinity to the progesterone-binding globulin, *J. Steroid Biochem.*, 9, 187, 1978.
80. **Stroupe, S. D. and Westphal, U.**, Steroid-protein interactions. XXXII. Conformational changes in the progesterone binding globulin-progesterone complex, *Biochemistry,* 14, 3296, 1975.
81. **Evans, R. M.**, The steroid and thyroid hormone receptor superfamily, *Science,* 240, 889, 1988.
82. **Wahli, W. and Martinex, E.**, Superfamily of steroid nuclear receptors: positive and negative regulators of gene expression, *FASEB J.*, 5, 2243, 1991.
83. **Greene, G. L., Gilna, P., Waterfiled, M., Baker, A., Hort, Y., and Shine, J.**, Sequence and expression of human estrogen receptor complementary DNA, *Science,* 231, 1150, 1986.
84. **Formstecher, P., Lefebvre, P., and Dautrevaux, M.**, RU 486 stabilizes the glucocorticoid receptor in a non-transformed high molecular weight form in intact thymus cells under physiological conditions, *J. Steroid Biochem.*, 31, 607, 1988.
85. **Guiochon-Mantel, A., Loosfelt, H., Ragot, T., Bailly, A., Atger, M., Misrahi, M., Perricaudet, M., and Milgrom, E.**, Receptors bound to antiprogestin form abortive complexes with hormone responsive elements, *Nature,* 336, 695, 1988.

86. **Willmann, T. and Beato, M.,** Steroid-free glucocorticoid receptor binds specifically to mouse mammary tumour virus DNA, *Nature*, 324, 688, 1986
87. **Schauer, M., Chalepakis, G., Willmann, T., and Beato, M.,** Binding of hormone accelerates the kinetics of glucocorticoid and progesterone receptor binding to DNA, *Proc. Natl. Acad. Sci. U S.A.*, 86, 1123, 1989.
88. **Miller, N. R. and Simons, S. S., Jr.,** Steroid binding to hepatoma tissue culture cell-glucocorticoid receptors involves at least two sulfhydryl groups, *J. Biol. Chem.*, 263, 15217, 1988.
89. **Carlstedt-Duke, J., Strömstedt, P.-E., Persson, B., Cederlund, E., Gustafsson, J.-A., and Jornvall, H.,** Identification of hormone-interacting amino acid residues within the steroid-binding domain of the glucocorticoid receptor in relation to other steroid hormone receptors, *J. Biol. Chem.*, 263, 6842, 1988.
90. **Evans, R. M. and Hollenberg, S. M.,** Zinc fingers: gilt by association, *Cell*, 52, 1, 1988.
91. **Weeks, C. M., Rohrer, D. C., and Duax, W. L.,** Deoxycorticosterone adenine interactions in a crystalline complex, *Science*, 190, 1096, 1975.
92. **Lawson, C. L., Zhang, R.-G., Schevitz, R. W., Otwinowski, A., Joachimiak, A., and Sigler, P. B.,** Flexibility of the DNA-binding domains of *trp* repressor, *Proteins: Struct. Function Genet.*, 3, 18, 1988
93. **Lawson, C. L. and Sigler, P. B.,** The structure of *trp* pseudo repressor at 1.65a shows shy indole propionate acts as a *trp* inducer, *Nature*, 333, 869, 1988.
94. **Sigler, P. B., Otwinowski, Z., Schevitz, R. W., Zhang, R.-G., Lawson, C. L., Marmorsteinm, R. Q., Joachimiak, A., and Luisi, B.,** Crystallography of the *trp* Operator/Repressor System: Molecular Flexibility in Genetic Regulation, American Crystallography Association Annual Meeting, Abstr. 08, p. 50.

Chapter 2

^{1}H AND ^{13}C NMR SPECTROSCOPY OF STEROLS

William K. Wilson and George J. Schroepfer, Jr.

TABLE OF CONTENTS

I. INTRODUCTION

Cholesterol and related sterols represent important components of the cells of eukaryotic organisms. The isolation of sterols, structure determination, delineation of their modes of formation and metabolism, regulatory processes involved therein, and probing of the functions of sterols represent topics of continuing interest. These matters are of increasing importance in a number of fields including chemistry, biochemistry, pharmacology, cellular and molecular biology, nutrition, and medicine. Cholesterol and other sterols are, under normal circumstances, located primarily in membranous structures of cells and appear to be essential for normal cell function and cell replication in eukaryotes.[1] In higher animals, cholesterol serves as a precursor of steroid hormones and bile acids. 7-Dehydrocholesterol, an intermediate in the biosynthesis of cholesterol in animal cells, and ergosterol, a major sterol in yeasts, serve as precursors of the D vitamins. Cholesterol and its fatty acid esters are significant components of the lesions involved in the pathogenesis of atherosclerosis, a major cause of mortality and morbidity in this country. Major efforts have been directed toward the development of agents to inhibit the formation of cholesterol, to inhibit its esterification and absorption, and to modulate its deposition and removal from walls of arteries.

Oxygenated derivatives of cholesterol, of sterol precursors of cholesterol, and of other sterols have been shown to be potent regulators of 3-hydroxy-3-methylglutaryl coenzyme A (HMG-CoA) reductase and of sterol formation in cultured mammalian cells.[2,3] An oxysterol-binding protein has been isolated which appears to be specific for oxygenated sterols that regulate the levels of HMG-CoA reductase activity.[4-6] The potencies of individual oxysterols in lowering HMG-CoA reductase activity in cells have been correlated with their affinity to the binding protein.[6] One oxysterol, 3β-hydroxy-5α-cholest-8(14)-en-15-one (**1**), which is highly active in lowering reductase activity[6-10] and shows very high affinity for the oxysterol-binding protein,[6] also inhibits acyl coenzyme A:cholesterol acyltransferase (ACAT),[11] an important enzyme involved in the formation of cholesterol esters and in the intestinal absorption of cholesterol. Oral administration of **1** to animals lowers ACAT activity in jejunal microsomes,[12] inhibits the absorption of cholesterol,[13] and lowers the levels of cholesterol in blood.[14-16]

Oxygenated sterols also represent potential regulators of cell replication by virtue of their inhibitory effects on the formation of cholesterol critical to new membrane formation and, at higher concentrations, possibly on the formation of other important mevalonate-derived products such as dolichols, ubiquinones, prenyl moieties on critical regulatory proteins, and isopentenyl modifications of certain adenine residues in tRNA. A number of oxysterols have been shown to be present in blood and tissues,[17-19] (and references cited therein) and their biosynthesis and metabolism and definition of their biological actions are currently major areas of investigation.

In lower forms of animals and other eukaryotes, other oxygenated derivatives of sterols have been shown to have notable biological activity. These sterols include ecdysteroids in insects,[20] brassinolides (e.g., **2**) in plants,[21] oogoniols in the water mold *Achlya*,[22] the shark-repelling pavoninins in some fish (*Pardachirus*),[23] and the pesticidal petuniasterones in *Petunia* species.[24]

The biosynthesis and metabolism of cholesterol and other sterols are areas of considerable importance. Investigations of the chemical nature of sterol intermediates in the biosynthesis of cholesterol have been complicated by their low steady-state concentrations in most tissues, by the localization of the concerned enzymes in membranous structures of the cells (thereby complicating their purification), and by difficulties in the chemical syntheses and chromatographic separation of potential sterol intermediates. Despite these obstacles, considerable progress has been made, a portion of which has been reviewed previously.[25] Extensions of these investigations led to the demonstration that 14α-hydroxymethyl sterol precursors of cholesterol are potent inhibitors of HMG-CoA reductase and sterol biosynthesis.[26-29] Since the formation of 14α-hydroxymethyl sterols can be considered as an obligatory step in the biosynthesis of cholesterol, these oxysterols represent potential regulators of HMG-CoA reductase in any cells capable of the *de novo* synthesis of cholesterol.

Advances in all of these areas of investigation are critically dependent upon definitive determinations of structures for the concerned sterols. In this work we illustrate advances in the application of NMR in the structure elucidation of sterols.

Nuclear magnetic resonance (NMR) spectroscopy is a powerful technique for determining the molecular structure of sterols. NMR is a non-destructive method that requires only a modest quantity of sample. Data can be collected on samples as small as 5 μg,[30] although milligram quantities may be needed for a full structure determination. Useful NMR results can often be obtained even on mixtures[31] and rather highly contaminated samples.[32] Sample preparation is very simple, and spectra can be obtained in times ranging from minutes to hours on spectrometers that are widely available. Unlike other spectroscopic techniques, NMR shows a discrete signal for virtually every carbon (^{13}C) and hydrogen (^{1}H) atom of a sterol molecule. The position (chemical shift or shielding) of each signal is a function of the location of nearby atoms (and their electrons). Fine structure in the signal (spin couplings) can generally be resolved and used to establish the configuration of functional groups as well as the conformation of rings. The ^{1}H and ^{13}C spectra together provide a fingerprint of a sterol molecule much more distinctive than that obtained by any other spectroscopic or physical method except X-ray crystallography. This combination of simplicity, versatility, and power makes NMR an extraordinarily valuable method for the characterization and structure determination of sterols.

The impetus for determining the structure of a sterol usually stems from an interest in its biological function or activity. NMR spectra of sterols and

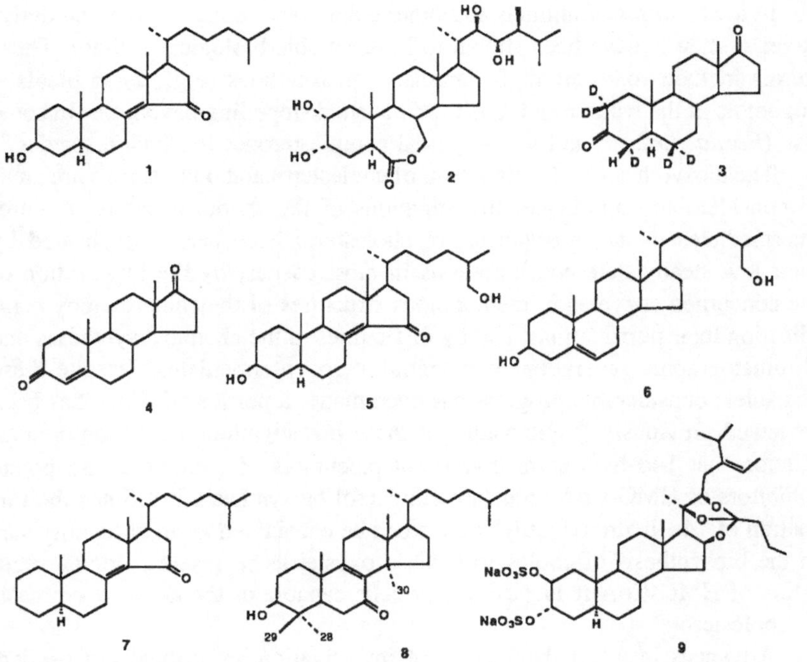

FIGURE 1. Chemical structures of sterols and steroids discussed in Sections I to III.

related compounds in membranes and other biological environments are very informative,[33,34] but these experiments are better suited to the determination of intermolecular relationships and dynamics than to elucidation of structure or conformation. X-ray crystallography gives a precisely defined conformation, but the crystal structure may deviate subtly from the sterol conformation in the fluid environment of a biological system. A compromise between these extremes is the moderately precise NMR solution conformation, usually obtained on a purified sterol in an organic or aqueous solvent.

Covered here are structure determination and conformational analysis of sterols, which are loosely understood to be lipophilic steroids with a side chain of ~8 to 10 carbon atoms. The side chain makes the NMR spectra of sterols somewhat more complex than those of steroid hormones. However, NMR methods for structure elucidation are similar for all steroids, and examples will also be given for other classes of steroids. Chemical structures of sterols and steroids discussed in Sections I to III are shown in Figure 1.

II. SPECTRAL TECHNIQUES

Unlike X-ray crystallographic results, which can be unambiguously condensed to a set of atomic coordinates, NMR analysis of a sterol typically employs a variety of spectral techniques, the results of which are combined to establish the chemical structure. Although incorrect structures are occasionally reported by X-ray crystallographers, the possibilities for errors and poor judgment in NMR are considerably greater. Therefore, some familiarity with experimental details of NMR techniques is a prerequisite for judging the rigor of signal assignments and structure arguments, which may in turn form the underpinnings of a proposed biosynthetic pathway or a hypothesis relating chemical structure to biological activity.

A decade ago, collecting NMR data on a sterol seldom meant more than recording the ^{1}H spectrum and perhaps the ^{13}C and SFORD spectra. Since then, hundreds of new NMR pulse sequences have been introduced, and some forethought is required in choosing a set of experiments that give definitive answers in the least amount of time. We will illustrate how combinations[35] of one-dimensional (1D) and two-dimensional (2D) experiments can be used efficiently to elucidate the structure and conformation of sterols.

A survey of multipulse NMR experiments[36-50] applicable to sterols is given in Table 1. Basic knowledge of 1D and 2D multipulse NMR techniques is assumed in the following discussions, as these subjects have been amply reviewed.[36-39] Briefly, it will be recalled that a typical 2D NMR experiment consists of systematically increasing (incrementing) some time period t_1 in a pulse sequence and collecting the free-induction decay (FID) for each increment of t_1. For example, in a ^{1}H–^{13}C correlation (HETCOR) experiment, a series of, say, 50 FIDs are collected and Fourier transformed. The result can be viewed as an array of 50 horizontally displayed 1D ^{13}C-like spectra, whose resolution and sensitivity depend chiefly on the acquisition time and number of scans. If instead of looking across in the F_2 or horizontal dimension, we look down along the F_1 dimension, we will see for each ^{13}C frequency a set of 50 intensity values that can be regarded as an FID. Transformation of the "fabricated" FID corresponding to each ^{13}C frequency gives the 2D spectrum, which is usually displayed as a contour map. A cross section in F_1 corresponds to one of these transformed FIDs and shows the ^{1}H spectrum of the hydrogen(s) directly bonded to the carbon having the given F_2 frequency. Other pulse sequences give different spectra in the F_1 and F_2 dimensions (Table 1).

Resolution and sensitivity are critical to the success of 2D experiments. Whereas resolution in F_2 is generally adequate, resolution in the vertical or F_1 dimension is often coarse due to limitations of computer memory, signal decay from relaxation, and availability of spectrometer time for collecting FIDs for many increments of t_1. The sensitivity of many 2D experiments (HETCOR, COSY, 2D-J) can be roughly estimated from the sensitivity of the 1D spectrum of the F_2 nucleus (^{1}H or ^{13}C) with N scans. The total experiment time (excluding disk writing operations and dummy scans) is the product of N, the number of increments, and the recycle time.

The practicability of spectral techniques depends on such spectrometer hardware features as magnet field strength, probe design, flexibility of the

TABLE 1
Multipulse NMR Experiments for Sterols

Acronym	Full name	Axis labels[a] F₂	Axis labels[a] F₁	Sample required[b] (mg)	Ref.[c]
	Spectral Editing (1D)				
DEPT	Distortionless enhancement by polarization transfer	^{13}C		2	37
INEPT	Insensitive nucleus enhancement by polarization transfer	^{13}C		2	37
POMMIE	Phase oscillations to maximize editing	^{13}C		2	37
APT	Attached proton test	^{13}C		3	37
GASPE	Gated spin echo	^{13}C		3	37
SEMUT	Subspectral editing by a multiple quantum trap	^{13}C		3	37
SFORD	Single-frequency off resonance decoupling (obsolete)	^{13}C		8	40
	^{1}H–^{1}H Correlation (2D)				
COSY	^{1}H–^{1}H correlated spectroscopy	^{1}H	^{1}H	1	41
COSYDEC	COSY with decoupling in F_1	^{1}H	^{1}H	1	42
SECSY	Spin-echo correlated spectroscopy	^{1}H	$\Delta\delta_H$	1	43
RELAY	Homonuclear relayed coherence transfer	^{1}H	^{1}H	1	41
NOESY	Nuclear Overhauser and exchange spectroscopy	^{1}H	^{1}H	1[d]	44,45
	^{1}H–^{13}C Correlation (2D)				
HETCOR	^{1}H–^{13}C heteronuclear (shift) correlated spectroscopy	^{13}C	^{1}H	8	41,46
HMQC	Heteronuclear multiple quantum coherence spectroscopy	^{1}H	^{13}C	2	47
LR HETCOR	Long-range HETCOR	^{13}C	^{1}H	20	48
HMBC	Heteronuclear multiple bond correlation spectroscopy	^{1}H	^{13}C	5	49
	^{13}C–^{13}C Correlation (2D)				
INADEQUATE	Incredible natural abundance double quantum transfer experiment	^{13}C	DQ	100[e]	41,50
	Other Experiments				
2D J	Two-dimensional J-resolved spectroscopy	^{1}H	J_{HH}	1	43
NOEDS	Nuclear Overhauser difference spectroscopy	^{1}H		30[f]	44

TABLE 1 (continued)
Multipulse NMR Experiments for Sterols

[a] Axis labels: ^1H and ^{13}C, chemical shift (δ); $\Delta\delta_H$, difference in ^1H chemical shifts of the correlated protons (see Reference 43); DQ, double quantum frequency; J_{HH}, ^1H-^1H spin couplings (Hz).
[b] Rough estimates for a 1-h experiment on a mid-field spectrometer. Amounts for ^1H–^{13}C correlation experiments assume low digital resolution in F_1. Amount of material required will vary considerably according to the exact hardware and software used and according to the purpose of the experiment.
[c] References (mainly to secondary sources).
[d] Overnight acquisition usually needed regardless of sample size.
[e] Requires ~48-h acquisition.
[f] Sample size required can be highly variable (see References 51 and 52).

pulse programming and phase cycling, and accessibility to a computer for rapid transformation of large 2D data sets. Instrumental choices can be divided into low-field (80 to 100 MHz for ^1H) spectrometers with electromagnets and spectrometers having superconducting magnets at either mid-field (200 to 400 MHz for ^1H) or high-field (≥500 MHz for ^1H). Table 2 gives rough estimates of the amount of material needed for NMR experiments depending on the type of spectrometer. The feasibility of any experiment will vary according to the features and design of the spectrometer.

For sterols available in 20-mg quantities, the authors routinely acquire a basic series of four experiments that require about 1 h of mid-field spectrometer time (including transformation and plotting): ^1H, ^{13}C, DEPT (135°), and HETCOR. Sometimes a quaternary carbon spectrum, 90° DEPT, COSY, or long-range HETCOR experiment is also needed. These experiments are illustrated below for 3β-hydroxy-5α-cholest-8(14)-en-15-one (**1**), a potent regulator of cholesterol metabolism. Excluding the long-range HETCOR, the full set of experiments can be carried out on a 5-mg sample in ~12 h (or 6 h if coarse resolution in F_1 can be tolerated in the HETCOR spectrum).

A. 1D EXPERIMENTS

The familiar ^1H spectrum is perhaps the simplest of all routine NMR spectra to acquire, although its complexity makes it the most laborious to plot and fully interpret. The mid-field (300 MHz) ^1H spectrum of the 15-ketosterol **1** (Figure 2) shows a few well-resolved downfield signals, several sharp upfield signals corresponding to methyl resonances, and a methylene envelope that is much improved from those of low-field spectra but still largely uninterpretable, even under resolution enhancement. Only the signals at the downfield edge of the methylene envelope can be interpreted with the aid of resolution enhancement. The spectrum of **1**, which is representative of ^1H spectra of sterols with few functional groups, is particularly congested in the δ 1.0 to 1.6 region, in which 18 multiplets are located. Each of these multiplets is ~20 to 40 Hz wide (0.04 to 0.08 ppm at 500 MHz) and consists of numerous

TABLE 2
One-Hour NMR Experiments as a Function of Amount of Sample and Type of Spectrometer[a]

	Material Available			
Spectrometer available	**100 μg**	**1 mg**	**5 mg**	**20 mg**
Low field (1.9–2.4 tesla) 80–100 MHz for ^1H, FT	^1H[b]		^{13}C[b] INEPT[b] APT[b]	SFORD[b]
Mid-field (4.7-9.4 tesla) 200–400 MHz for ^1H	^1H	COSY	DEPT ^{13}C	HETCOR[c] LR HETCOR[c]
High field (≥11.7 tesla) ≥500 MHz for ^1H	^1H	COSY	DEPT ^{13}C HMQC[d]	HETCOR[c] LR HETCOR[c] HMBC[d]

[a] Amount of sample required to produce a usable albeit noisy spectrum with a 5-mm ^1H-^{13}C dual probe optimized for ^{13}C. Assumes solubility in 0.3 ml of a suitable NMR solvent.

[b] Assumes the availability of a microsample probe (1.7 mm ID tube); otherwise, ~3 times as much sample is needed for a 1-h experiment in a 5-mm tube.

[c] Assumes low resolution in F_1 (~20 increments).

[d] Assumes a probe optimized for ^1H acquisition.

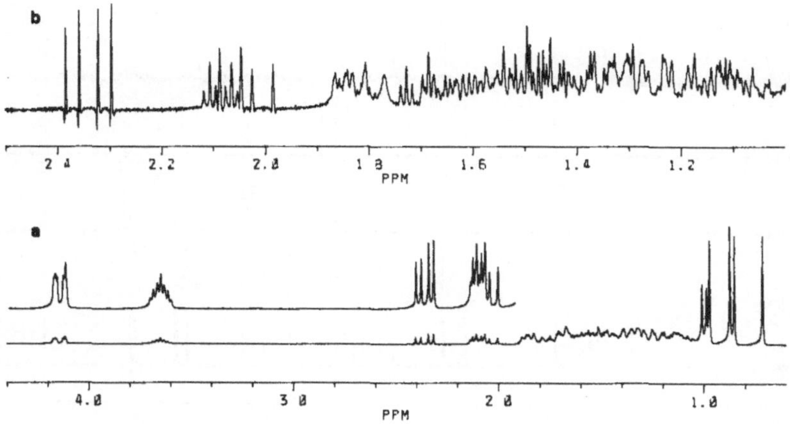

FIGURE 2. (a) ^{1}H NMR spectrum of 3β-hydroxy-5α-cholest-8(14)-en-15-one (**1**). Acquired in $CDCl_3$ solution (20 mg/0.45 ml) at 300 MHz: 64 scans, acquisition time 3 s, recycle time 3.1 s, pulse width 4.0 μs (42° flip angle), no line broadening, digital resolution 0.001 ppm (0.3 Hz). (b) Resolution enhancement of a portion of the spectrum (LB −2, GB 0.4, Bruker Aspect 3000 software).

peaks. Eight of these signals are from the saturated side chain. Even at high field, this envelope of overlapping peaks is often indecipherable, especially in comparison with the reasonably well-resolved ^{1}H spectra obtained for more highly funtionalized C_{19} steroids.[53] Relative to androstanes, pregnanes, and Δ^{22} sterols, the density of ^{1}H resonances in the δ 1.0 to 1.6 region is nearly doubled. Fortunately, these seemingly inaccessible ^{1}H chemical shifts and coupling constants can be measured indirectly from routine 2D experiments.

The 75-MHz proton-decoupled ^{13}C spectrum of 15-ketosterol **1** (Figure 3a) shows resolution of all 27 signals. Although low-field (20–25 MHz) spectrometers can resolve nearly all the ^{13}C peaks of sterols, higher fields are beneficial, especially for the resolution and quantitation of signals in isomeric mixtures.[31,54,55] The absence of spin couplings makes the ^{13}C spectrum very clean but at the expense of valuable coupling information.

The chemical shift information by itself seldom permits complete and unequivocal signal assignments or a definitive structure determination. Coupled and partially coupled ^{13}C spectra (e.g., SFORD) give additional information, but a far more efficient approach is provided by spectral editing experiments (see Table 1) that divide ^{13}C resonances into groups according to their number of directly bonded hydrogen atoms.

Depending on the magnitude of a key delay time or flip angle in the pulse sequence, spectral editing experiments show various combinations of carbon multiplicities as either normal, nulled, or inverted signals. In the case of DEPT, a 45° pulse gives CH, CH_2, and CH_3 signals in the same orientation. A 90° pulse gives only CH signals (CH_2 and CH_3 nulled), and a 135° pulse

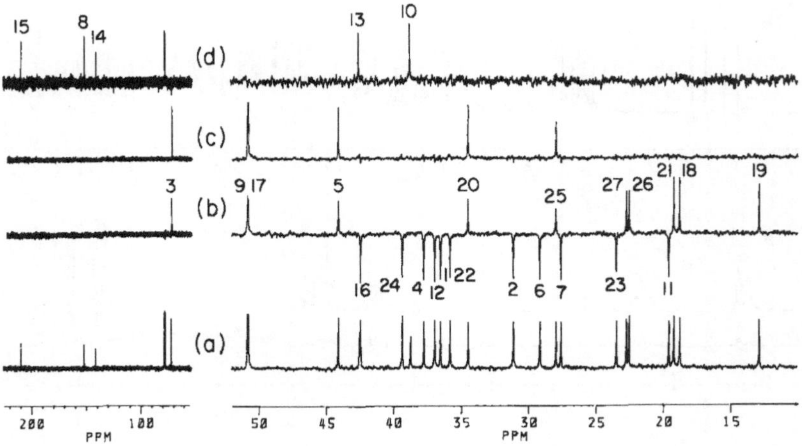

FIGURE 3. NMR spectra of 3β-hydroxy-5α-cholest-8(14)-en-15-one (**1**): (a) WALTZ-decoupled ^{13}C, (b) 135° DEPT, (c) 90° DEPT, and (d), QUATD. Spectra were acquired in $CDCl_3$ solution (20 mg/0.32 ml) at 75 MHz: 100 scans [400 scans for (a)], acquisition time 0.82 s, line broadening 2 Hz, digital resolution <0.01 ppm Recycle times were 0.9 s for (a) (32° flip angle), 1.8 s for (b) and (c), and 3.8 s for (d).

shows CH_2 signals inverted from those of CH and CH_3; quaternary carbons are not observed in any DEPT experiment. A 135° DEPT spectrum of 15-ketosterol **1** is shown in Figure 3b. Because CH and CH_3 carbons are normally distinguished by their chemical shifts, this single experiment usually establishes the multiplicity of each signal. A combination of the 90° DEPT (Figure 3c) and 135° DEPT spectra may be needed occasionally to differentiate CH and CH_3 signals in 5β-sterols, 14β-sterols, methyl esters, mesylates, and other derivatives. The spectrum of quaternary carbons (Figure 3d), a SEMUT experiment known as QUATD in Bruker terminology, is also useful to confirm the location of unprotonated carbon signals in congested regions of the spectrum. These and other spectral editing experiments related to DEPT have been compared and discussed in detail.[37] Pulse sequences based on polarization transfer (DEPT, INEPT, POMMIE) give higher sensitivity, but the other experiments (APT, GASPE, SEMUT) show quaternary carbon signals and can be implemented on a wider range of spectrometers. Nearly all of the numerous variations of editing techniques give excellent results for sterols.

The ^1H, ^{13}C, and 135° DEPT spectra in conjunction with the HETCOR experiment described below are frequently sufficient for signal assignments and structure elucidation. Other 1D experiments are nevertheless helpful, especially when sample size is small or when a high-field spectrometer is not available. For example, decoupling difference spectroscopy,[56] although largely superseded by COSY, is occasionally superior to COSY in isolating ^1H multiplets from overlapping signals.[57,58]

Solvent shifts are also informative. Relative to ^1H sterol spectra in $CDCl_3$ or CD_3OD, spectra in pyridine-d_5 and benzene-d_6 often show improved dispersion,[31] and the aromatic solvent-induced shifts (ASIS)[59] can be used to assign the configuration of hydroxyl groups.[60,61] Although ^1H chemical shifts are similar in CD_3OD and $CDCl_3$,[32] ^{13}C methanol-induced shifts in $CDCl_3$ spectra can be substantial. This effect has been exploited for sterol signal assignments when more rigorous 2D methods were unavailable.[62]

An especially powerful experiment makes use of lanthanide-induced shifts (LIS).[63] When an organic-soluble complex of ytterbium or other lanthanide is added to a sterol solution, the lanthanide reversibly complexes with lone pairs of electrons on hydroxyls and other polar groups, and the ^1H and ^{13}C NMR signals are shifted in a systematic manner that permits straightforward calculation[64,65] of *relative* LIS. The magnitude of the lanthanide-induced shift (LIS) of each nucleus N can be described approximately by the simplified McConnell-Robertson equation $LIS_{calc} = k(3\cos^2\theta - 1)/r^3$ where k is a proportionality constant, r is the distance between the lanthanide atom and the substrate atom, and θ is essentially the angle defined by nucleus N, the lanthanide, and the complexing oxygen atom.

If all functional groups are blocked except for a single ketone or hydroxyl, the observed ^1H and ^{13}C LIS can be compared with those calculated for a particular configuration or conformation. Structures giving poor agreement between observed and calculated LIS can be ruled out. Given accurately measured LIS for ~40 ^1H and ^{13}C signals, this can be a powerful method of structure determination for conformationally rigid sterols. However, the complexity of calculations combined with various complications[65] has limited its use. Precision in LIS calculations is favored by a single (or at most 2 to 3) well-defined conformer(s) of a sterol-shift reagent complex. In particular, accurate LIS calculations are precluded for complexation with most hydroxyls that are primary or in the side chain, and precision may be lowered in the conformationally flexible D ring. Qualitative (semi-quantitative) interpretation of LIS based on neglect of the angular term of the McConnell-Robertson equation is risky, although this approximation is reasonable for 3β-hydroxysterols.

Measurement of spin-lattice relaxation times (T_1) to differentiate methylene and methine carbons has been superseded by DEPT, but relaxation times may be used to distinguish carbons in the side chain from those in the steroid nucleus. A short set of inversion-recovery experiments for ^1H and ^{13}C gives useful estimates of T_1 values for optimizing recycle times in 1D and 2D experiments. A ^1H inversion-recovery experiment in which methine signals are nulled may clarify the coupling pattern of an overlapped methylene resonance.[53] A shorter delay (τ) gives a partially relaxed spectrum in which methine signals appear inverted and methylene signals are largely nulled. Systematic measurements of relaxation times may also provide insight into the molecular motion of steroids.[66]

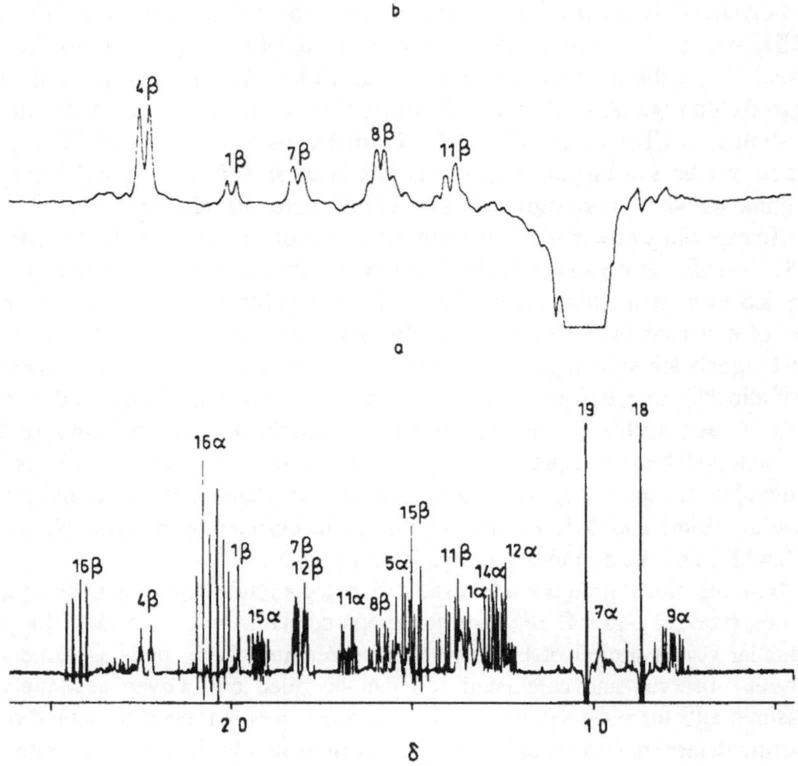

FIGURE 4. (a) 500 MHz ^1H NMR spectrum of partially deuterated dihydrotestosterone (**3**) in $CDCl_3$. (b) NOE difference spectrum obtained by irradiation of H-19 (δ 1.02) (From Holland, H. L , Xu, W., and Hughes, D. W., *J. Chem. Soc. Chem. Commun.*, 1760, 1989. With permission.)

Irradiation of a proton resonance may affect the relaxation of other protons within ~3 Å and thereby lead to small changes in their signal intensity. This change in intensity is the nuclear Overhauser effect (NOE), which is conveniently observed by difference spectroscopy (NOEDS).[44] The difference NOE peaks are usually positive for steroids, but negative NOEs have been reported.[67-70] Although NOEDS is a lengthy and technically demanding experiment, it has been used extensively to assign the stereochemistry of methylene protons.[31,51,53,71] Figure 4 depicts the results of an NOEDS experiment in which irradiation of H-19 of partially deuterated 5α-dihydrotestosterone (**3**) led to enhancement of signals corresponding to H–1β, H–4β, H–7β, H–8β, and H–11β.[72] Heteronuclear NOE experiments may also give valuable information.[37]

B. 2D EXPERIMENTS

A HETCOR spectrum[41] correlates the signal of each carbon atom with the chemical shifts of its directly bonded protons. This information clarifies

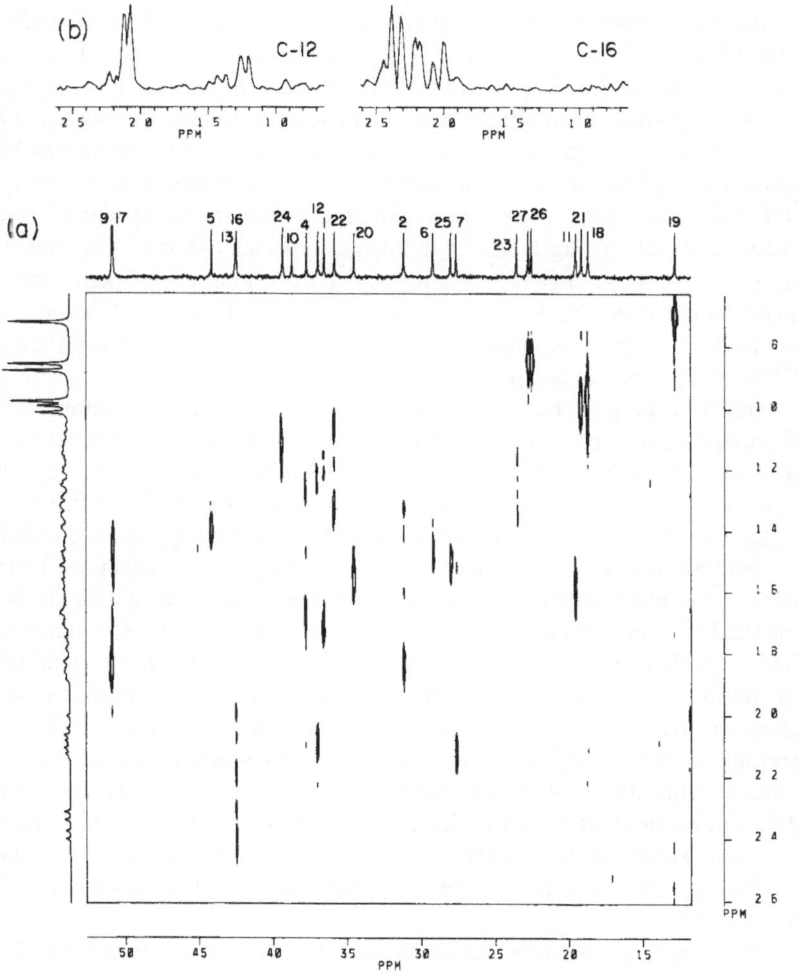

FIGURE 5. (a) Proton-decoupled HETCOR spectrum of 3β-hydroxy-5α-cholest-8(14)-en-15-one (**1**) with ^{1}H and ^{13}C spectra plotted along the F_1 and F_2 axes. (b) F_1 cross sections corresponding to C-12 and C-16. HETCOR spectrum acquired at 75 MHz in $CDCl_3$ solution (20 mg/0.32 ml): XHDEPTD, 50 increments of 24 scans each, acquisition time 0.2 s, recycle time 1.2 s, optimized for J_{CH} 130 Hz, data displayed as the power spectrum. F_2 window, 5 to 77 ppm (digital resolution 0.07 ppm); F_1 window, 0.6 to 2.6 ppm (digital resolution 0.016 ppm).

most ambiguities remaining in assigning the ^{13}C spectrum and also establishes the ^{1}H chemical shifts and assignments (except for stereochemistry) in the δ 1 to 2 region. A HETCOR spectrum of 15-ketosterol **1** with decoupling in F_1 is shown in Figure 5a. The ^{13}C and ^{1}H spectra of **1** are plotted along the F_2 and F_1 axes, respectively. The spectral window was set at δ 5 to 77 in the ^{13}C dimension and δ 0.6 to 2.6 in the ^{1}H dimension. The narrow F_1 window leads to good signal definition in the ^{1}H dimension after only 50 increments.

No significant information is lost from folding (aliasing) of the downfield peaks (δ 4.12, H-7β and δ 3.64, H-3α), which are easily identified from either the ^{1}H spectrum or a short HETCOR experiment with a larger F_1 window. The small window size described above is suitable for most sterols; a filter in the ^{13}C dimension strongly attenuates any protonated olefinic carbon signals folding into the spectrum, and the occasional peaks in the vicinity of δ 80 or δ -5 fold predictably into uncluttered regions of the spectrum. Such deliberate signal folding in 2D experiments, especially in the F_1 dimension, can greatly improve digital resolution and reduce total experiment time. It should be noted that peaks outside the F_1 window fold from the far edge of the spectrum without attenuation by any filter, whereas the orientation of folding in F_2 depends on spectrometer design.[73]

The ^{1}H chemical shifts are obtained by examining cross sections in F_1 corresponding to each carbon (methylenes), or by a computer printout of all significant 2D peaks (methyls and methines). Whereas precise chemical shifts are readily obtained for methyl ($\pm$0.003 ppm) and methine ($\pm$0.005 ppm) singlets, methylene correlations are weaker because they appear as doublets. (Decoupling in F_1 does not affect geminal couplings.) Cross sections of the methylene doublets often show distortion in shape and intensity (Figure 5b), a difficulty compounded by an artifact halfway between the two doublets. These complications limit the chemical shift accuracy for many methylene protons to $\pm$0.02 ppm at best. The reliability of methylene chemical shift values can be improved somewhat by using a narrower window in F_1, increasing the sensitivity, and displaying the power spectrum rather than the absolute value mode. A better solution is provided by a fully decoupled HETCOR experiment developed by Reynolds et al.[46] This modified COLOC experiment significantly improves both resolution and sensitivity of methylene correlations. This sequence has been applied successfully to several natural products.[46,74]

An even better solution for measuring ^{1}H chemical shifts is HMQC, a ^{1}H–^{13}C correlation experiment with proton detection.[47] The advantages of HMQC over HETCOR are improved sensitivity and better resolution[75] in the proton dimension (now F_2). The disadvantages are much lower resolution in the ^{13}C dimension and hardware requirements that preclude use of HMBC on most older spectrometers. Despite the congestion of ^{13}C signals in the δ 10–40 region, HMQC spectra of several steroids[52,69,75] and sterols[58] have been reported. Use of a narrow window in F_1 to judiciously fold selected ^{13}C signals improves ^{13}C resolution.[76] An HMQC spectrum of a 1-mg sample of androst-4-ene-3,17-dione (**4**) is shown in Figure 6a.[52] HMQC spectra have generally been acquired in polar solvents, often containing D_2O.

HETCOR and HMQC are of limited use in establishing carbon connectivities of sterols when the COSY or COSYDEC spectra are largely uninterpretable (see below). The ^{13}C-^{13}C correlation experiment (INADEQUATE) provides carbon connectivities but is highly inefficient, requiring ≥100 mg

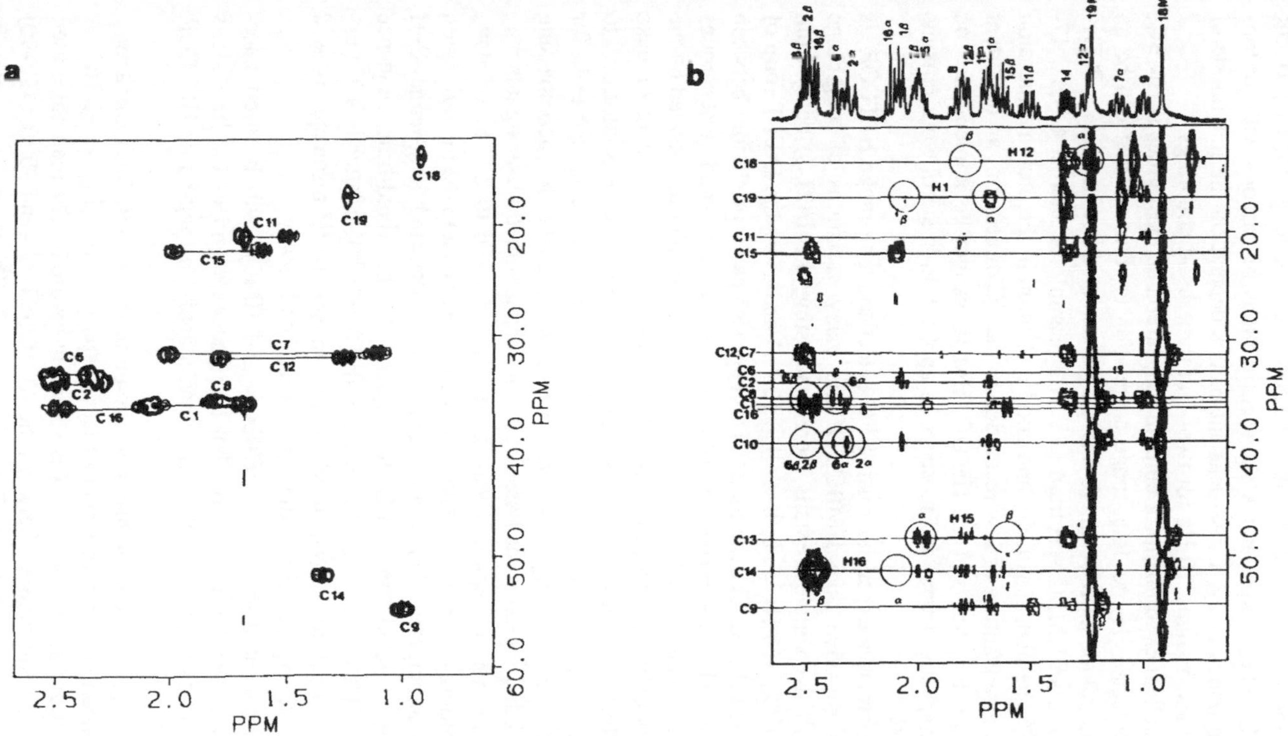

FIGURE 6. (a) HMQC spectrum of androst-4-ene-3,17-dione (**4**), 1 mg in 0.4 ml of $CD_3OD:D_2O$ (4.1). Spectral conditions: GE GN-500 spectrometer (500 MHz for ^1H), 128 increments of 32 scans each, recycle time ~4 s. (b) HMBC spectrum of androst-4-ene-3,17-dione (**4**), 1 mg in 0.4 ml of $CD_3OD:D_2O$ (4:1). Spectral conditions GE GN-500 spectrometer (500 MHz for ^1H), 128 increments of 128 scans each, recycle time ~3 s. (From Zeng, B., Pollack, R. M., and Summers, M. F., *J. Org. Chem.*, 55, 2534, 1990. With permission)

of sample and a weekend of spectrometer time. Although INADEQUATE spectra of many sterols and steroids have been reported, carbon connectivities can be established by more efficient methods, such as long-range HETCOR and HMBC. These experiments show correlations of carbons with protons up to three bonds away. Cross-peak intensities depend on the magnitude of the J_{CH} values, roughly 10 Hz for two- and three-bond couplings.

An excellent experiment for determining carbon connectivities is HMBC, a proton-detected ^{1}H–^{13}C long-range correlation experiment that can be an order of magnitude more sensitive than its ^{13}C-detected counterparts.[77,78] Figure 6b shows an HMBC spectrum of a 1-mg sample of androst-4-ene-3,17-dione (**4**).[52] The high sensitivity and good resolution in the proton dimension offsets the disadvantage of low resolution in the ^{13}C dimension, an important consideration for sterols. Like HMQC, HMBC cannot be implemented on most older spectrometers, and its scope of application in sterol work remains to be explored.

The alternative experiment using ^{13}C detection, long-range HETCOR, is much less sensitive than HMBC. Of the many variations of long-range HETCOR,[48] the simplest entails merely adjusting the $1/(2J_{CH})$ delay from ~0.004 s ($^{1}J_{CH}$ 125 Hz) to 0.05 s ($^{3}J_{CH}$ 10 Hz) and using the INEPT form of the decoupled HETCOR experiment. This approach gave correlations between the eight methyl proton resonances of lanosterol derivatives and most carbons within three bonds (Figure 7).[79] The very coarse F_1 resolution dictated by the low sensitivity of the long-range HETCOR experiment was adequate to make the distinctions needed for signal assignment. No significance is attributed to the absence of a cross peak, as negligible intensities are expected even for some correlations with J_{CH} ~10 Hz.[48] A narrow window in F_1 encompassing only the methyl region is inadvisable, as methine and methylene signals may fold into the methyl region. Even for the larger δ 0.6 to 2.6 F_1 window, possible folding of olefinic and other downfield signals into the methyl region should be anticipated. Other approaches to ^{13}C-detected long-range C–H correlations[48] include the FLOCK sequence,[80] the fixed-time sequence COLOC,[81] FUCOUP,[48,82] and selective INEPT.[37,74,83] The latter is a 1D experiment in which low-power irradiation of an isolated ^{1}H resonance is used to observe ^{13}C signals of carbons up to three bonds away.

In subsequent discussions, HETCOR and HMQC will be used interchangeably under the assumption that the reader recognizes the limitations and merits of each method. Similar remarks apply to long-range HETCOR and HMBC.

COSY, the most commonly used proton-proton correlation experiment, shows connections between protons that are spin-coupled (J_{HH} > ~2 Hz). A COSY spectrum of 15-ketosterol **1** is shown in Figure 8. The read pulse used here is 60°, a compromise[50] between the 45° (COSY-45) and 90° (COSY-90) pulses most commonly used. This 300 MHz COSY spectrum is scarcely interpretable except for correlations of H-16α/β (δ 2.36 and 2.06), H-12β (δ

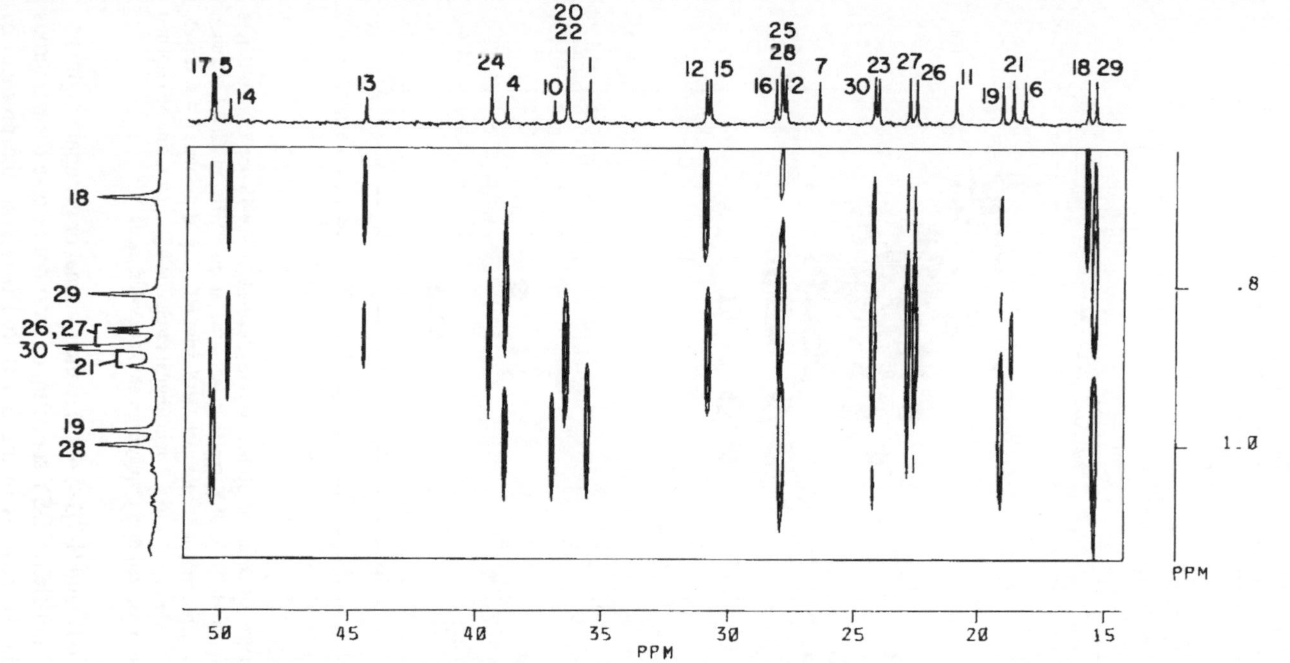

FIGURE 7. Long-range HETCOR spectrum of dihydrolanosterol (numbering as shown for **8** in Figure 1). Acquired at 75 MHz in $CDCl_3$ solution (~40 mg/0.32 ml): XHCORRD, 18 increments of 256 scans each, optimized for J_{CH} 10 Hz; other conditions as in Figure 5.

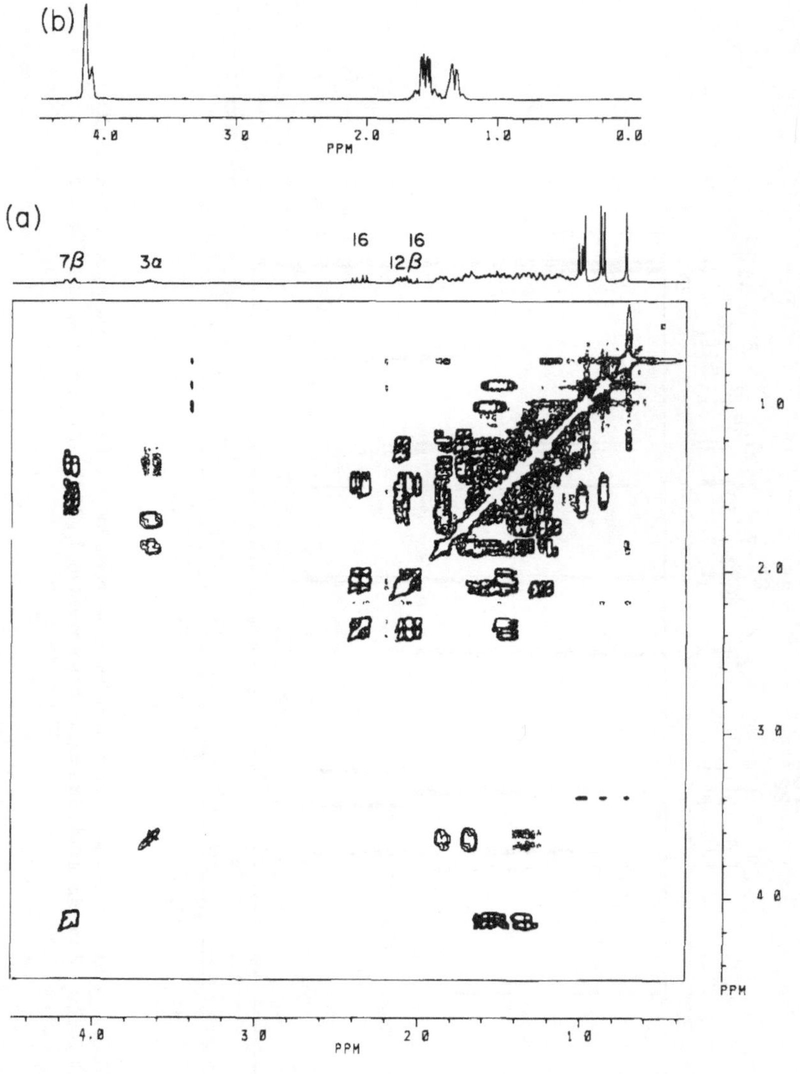

FIGURE 8. (a) COSY spectrum of 3β-hydroxy-5α-cholest-8(14)-en-15-one (**1**). (b) Cross section corresponding to H-7β. COSY spectrum acquired at 300 MHz in $CDCl_3$ solution (20 mg/0.45 ml): 256 increments of 8 scans each, read pulse 60°, acquisition time 0.4 s, recycle time 0.9 s, F_1 and F_2 windows −0.1 to 4.9 ppm (digital resolution 0.01 ppm), sine bell squared apodization (π/16), symmetrized data displayed in absolute value mode.

2.11), H-3α (δ 3.64), and H-7β (δ 4.13) to their vicinal or geminal neighbors. Although even a high-field COSY spectrum of **1** cannot resolve the numerous cross peaks of the 18 protons in the δ 1.0 to 1.6 region, the problem of overlap is largely overcome by the COSYDEC experiment, which gives a spectrum decoupled in F_1.[42] Except for a few tightly coupled spin systems in the δ 1.0 to 1.6 region, most cross peaks are resolved, even at mid-field. Furthermore, coupling patterns are often revealed, thus permitting estimates of J values and assignments of ^1H stereochemistry. The value of the COSY-DEC experiment in steroid work has been pointed out only recently.[42]

Other modifications of COSY are also useful in sterol work. The phase-sensitive COSY improves resolution considerably at no expense of additional pulses for complete phase cycling. Long-range couplings (J_{HH} < 2Hz) that sometimes appear when digital resolution in F_1 is high, are readily visualized by adding fixed delays to the COSY pulse sequence (COSY-LR).[41] Homonuclear relay coherence transfer (RELAY, RCT, or RCOSY) spectra, which show long-range correlations resulting from transfer (relay) of magnetization from nucleus to nucleus, have been used in the ^1H assignments of estrogens.[84] Numerous other ^1H–^1H correlation experiments and related multiple quantum experiments have been reviewed,[41] including the SECSY experiment, which has occasionally been applied to steroids.[85] The utility of such spin-locked experiments as HOHAHA and ROESY in sterol NMR seems to have been barely[86] explored.

NOESY may be viewed as a modification of the COSY pulse sequence in which a mixing time for development of NOE is incorporated.[44,45] This 2D version of NOEDS gives a spectrum similar to COSY, but cross-peak intensities are a function of proximities in space rather than spin coupling. Several examples of NOESY spectra of steroids have been reported[35,52,87,88] despite some opinions that NOESY is unsuitable for small molecules.[45] Discrepancies between NOESY and NOEDS results for testosterone have led to conflicting assignments of H-16α and H-16β.[53,87]

Another class of 2D experiments known as J spectroscopy places chemical shifts along F_2 and spin couplings along F_1. The ^1H homonuclear (2D J) experiment,[43,50] shows the ^1H resonances centered at their F_2 chemical shift with their spin couplings spread out along F_1. The F_2 projection is a proton-decoupled ^1H spectrum, but it is too distorted in the case of sterols to be of much value. Initially it appeared that J spectroscopy would disentangle the congested ^1H spectra of sterols and steroids,[53] but artifacts of strong coupling[43,50] and other difficulties have severely limited its utility.[32,51] Despite its demonstrated[67,85] value for measuring ^1H coupling constants, 2D J spectroscopy and its heteronuclear counterpart are now rarely[89] used in steroid NMR.

C. PROTON COUPLING CONSTANTS

Special attention is given to the measurement of proton coupling constants because of their value in conformational analysis and structure elucidation. As shown in Table 3, vicinal coupling constants show remarkably little variability in rings A, B, and C of typical 5α-steroids.[51] The signals of ring protons, which are generally coupled to three or four vicinal and geminal neighbors, usually consist of fewer than the theoretical 8 or 16 lines because two or more of the J values are often nearly identical.

TABLE 3
Vicinal ^1H–^1H Coupling Constants in Cyclohexanoid Rings of Steroids[a,b]

Type	J_{exptl}, Hz	J_{calcd} for ±5°[c]	Example
Axial-equatorial	3.5–7	±1.0	1α–2α
Equatorial-equatorial	2–5	0.8	1β–2α
Axial-axial (secondary)	13–14[d]	0.2	1α–2β
Axial-axial (≥1 tertiary)	11–12[d]	0.1	8β–9α

[a] Adapted from data in Table VI of Reference 51, with permission.
[b] Coupling constants from cross sections of HETCOR spectra of 16 steroids; digital resolution ±1.5 Hz.
[c] Calculated change in J upon variation of the dihedral angle by ±5°.
[d] These ranges may overlap in other cases.

Some of these resonances are sufficiently resolved that coupling constants can be measured directly from peak spacing in the ^1H spectrum, often to an accuracy of ±0.1 Hz. However, if two J values are similar, the accuracy may be as low as ±1 Hz. Analysis of complex spectral regions is facilitated by resolution enhancement (see Figure 2b) and by profiles of characteristic splitting patterns (see Section III.C). However, multiple t patterns are sometimes deceptive. For example, if strong coupling is present in an ABX-type system, i.e., $\Delta\nu_{AB} < 3J_{AB}$, then J_{HH} values measured from peak spacings of X are likely to be incorrect.[57,90] Thus, couplings to closely spaced geminal protons (e.g., the C-4 and C-11 protons in 3β-hydroxy-Δ^5-sterols) or to one of a pair of accidentally isochronous vicinal protons should be measured at high field and/or with the aid of iterative spin simulation.[51,57]

Most ^1H resonances of sterols are buried in the methylene envelope. If only a few couplings are needed, 1D techniques, such as difference spectroscopy[56,57] and inversion-recovery editing,[53] may suffice. A change in solvent to benzene[67] or pyridine[91] may shift certain signals adjacent to functional groups downfield into uncluttered regions of the spectrum. However, a reasonably complete set of coupling constants requires such 2D methods as COSYDEC,[42] coupled HETCOR,[51] HMQC,[75] or J spectroscopy.[53,67,85] Because of limited digital resolution and artifacts, the precision of J values from 2D spectra is substantially lower than those from 1D spectra. Table 4 summarizes the methods available for obtaining ^1H–^1H coupling constants.

D. DATA ACQUISITION

Chemical shift comparisons often must be made with high precision, such as in determining the configuration of a C-24 alkyl group or in differentiating closely spaced carbons for signal assignments. To assure accurate and reproducible chemical shifts, some care is required in the preparation of the NMR sample and in the recording and processing of the data. Other precautions can dramatically improve sensitivity and resolution, thereby reducing the

TABLE 4
Spectral Techniques for Obtaining ^1H–^1H Coupling Constants[a]

Spectrum or technique	Requirements
^1H spectrum[b]	No overlapping signals
COSYDEC spectrum	Coupling pattern resolved and well-defined
Phase-sensitive COSY spectrum	No overlapping signals
Cross sections of HMQC (HETCOR) spectrum	
1D difference spectroscopy[c]	Suitable coupled signal for irradiation
Cross sections of 2D J spectrum	No interfacing artifacts from strong coupling

[a] Iterative spin simulation of multiplets is needed in cases of strong coupling. All 2D experiments require fine digital resolution.

[b] Isolation of resonances in overlapping multiplets may be achieved by a change in NMR solvent or by inversion recovery editing (Reference 53). Resolution enhancement and peak matching with templates of profiles (Reference 32) is also helpful.

[c] Includes decoupling difference spectroscopy, selective population inversion (SPI), and selective population transfer (SPT), see Reference 56.

spectrometer time needed and strengthening the arguments in a structure determination. The following discussion assumes an NMR spectrometer with a superconducting magnet and a 5-mm switchable ^1H–^{13}C probe operating at ambient temperature.

For reproducible chemical shifts, the sample should be free of solvent impurities, which can lead to solvent-induced shifts, and the concentration should be ≤0.2 molar. Fluctuating temperatures also lead to variability in chemical shifts. Acquiring data at slightly elevated temperature[32] (e.g., 30°C) gives better temperature control and also facilitates comparison with data collected on spectrometers with electromagnets. However, recording spectra at elevated temperature is impractical on busy multiuser spectrometers because of the excessive time required for the probe and the sample to reach thermal equilibrium.

Temperature and solvent impurities can affect the reference signal, which is normally taken as internal tetramethylsilane or the deuterated solvent (or its protio impurity). In making chemical shift comparisons, two sets of ^{13}C chemical shifts that should be identical (±0.03 ppm) may differ by a fixed increment (e.g., 0.20 ± 0.03 ppm). This increment is usually attributable to recording of the two spectra under slightly different conditions of solvent impurities, temperature, or concentration (or perhaps simply careless referencing).

If a sterol sample is much less than 1 mg, large signals from water, the deuterated solvent, and various trace impurities may obscure nearby spectral features. Such problems can be minimized by use of a deuterated solvent that is very pure, dry, and of high isotopic composition and by pretreatment of the NMR tube with D_2O.[30] Traces of undissolved solids and paramagnetic impurities, which seldom affect ^{13}C resolution noticeably, should be removed for high resolution ^1H spectra. Dilute (<0.02 molar) solution and a generous sample volume also slightly improve resolution. Alternatively, sensitivity can be improved considerably at a small sacrifice in resolution by using a minimal sample volume (~0.3 ml).

Spectrometer settings also affect sensitivity, resolution, and peak definition. Zero-filling to produce a digital resolution of 0.001 ppm for ^1H and 0.01 ppm for ^{13}C improves the precision of chemical shift and coupling constants. Digital resolution can often be dramatically increased in HETCOR and other 2D experiments by using a small window and allowing for unimportant signals to be folded. Adequate acquisition times (3 s for ^1H and 1 s for ^{13}C) facilitate resolution enhancement. For concentrated samples, a small flip angle (~30°) combined with a ~1-s acquisition time gives a ^{13}C spectrum in which most signals are nearly the same height and quaternary carbons are only slightly less intense (Figure 3a). Such a spectrum permits easy recognition of doubled peaks and sterol impurities, which can be roughly quantitated. For dilute samples, a larger flip angle (50 to 60°) may be used at the expense of diminished quaternary carbon signals that may be lost in the noise or confused with minor impurities. No recycle delays are needed for routine ^1H and ^{13}C spectra is the above acquisition times are used. Polarization transfer experiments with ^{13}C detection (DEPT, HETCOR) need a total recycle time of ~1.5 s; the recycle delay is set as the difference between 1.5 s and the acquisition time. The QUATD experiment requires a longer recycle time (~4 s).

Resolution enhancement is a powerful method of separating overlapping signals in ^1H spectra and for transforming nondescript multiplets into (apparent) first-order patterns. This is routinely accomplished by Gaussian multiplication of the FID to change the usual exponential decay into an FID with a bulge in the middle. Artifacts and intensity distortions resulting from moderate resolution enhancement are easily identified and ignored, and reduction in sensitivity is modest. Figure 2b shows part of the ^1H resolution-enhanced spectrum of 15-ketosterol **1**. Resolution enhancement can also confirm that a ^1H or ^{13}C signal of double intensity represents two resonances and to analyze mixtures of epimers.[54,55] A DEPT spectrum will occasionally show no signal at the chemical shift of two unresolved ^{13}C peaks. Resolution enhancement of the DEPT spectrum usually reveals a negative and a positive peak, which cancelled to zero at lower resolution. There is, however, rarely any merit to resolving and identifying ^{13}C signals closer than 0.02 ppm, since minor solvent impurities or small variations in temperature can transpose such peaks.[79] General 2D experimental techniques, including resolution enhancement, have been reviewed.[72]

III. SIGNAL ASSIGNMENTS

Assigning the NMR signals to individual carbons and hydrogens of a sterol molecule provides an exacting test of the correctness of a proposed sterol structure as well as insight into the conformation. Thus, making accurate signal assignments is an important part of the structure determination and conformational analysis of a sterol.

A. CHEMICAL SHIFT COMPARISONS

The method of chemical shift comparisons[40] is illustrated for (25*R*)-3β,26-dihydroxy-5α-cholest-8(14)-en-15-one (**5**), whose ^{13}C signals can be almost completely assigned by comparison with those of structurally similar sterols for which signal assignments have been established. First, the signals of **5** were grouped in descending order according to multiplicities determined from the DEPT spectrum. Then similar sterols (15-ketosterol **1** and 26-hydroxycholesterol, **6**) with established ^{13}C assignments were taken for comparisons. As functional groups have little effect on ^{13}C shieldings beyond the δ carbon, **1** was considered to be an accurate model for carbons 1 to 22, and **6** a good model for carbons 23 to 27. These ^{13}C signals for **1** and **6** were also grouped according to multiplicity and listed adjacent to the chemical shifts for **5**. As shown in Table 5, all signals could be assigned by inspection except for C-9 and C-17, which were too close to be distinguished. Apart from signals of polarizable carbons (sp^2 carbons and carbons bonded to a heteroatom), the corresponding chemical shifts of **1** and **5** matched very closely, especially after addition of a 0.04 ppm adjustment to each shielding of **1**.

In this example, the ^{13}C spectra of all three sterols were collected under similar conditions of temperature and concentration so that signals as close as 0.3 ppm could be confidently differentiated by comparisons. When spectral data recorded under different or unspecified conditions are taken for comparison, adding a fixed reference adjustment (see above) may improve the precision of comparisons. However, for spectra acquired in different solvents or for assignments at carbons close to a point of structural difference, discrimination between signals even 1 ppm apart may be difficult. The possibility of misassignments in the reference spectrum should also not be overlooked. Failure to recognize the limitations on chemical shift comparisons has led to numerous assignment errors.

Data compilations[32,34,92-101] presented in steroid NMR reviews (Table 6) constitute a rich source of chemical shifts for comparisons. Other sterol chemical shift data are fragmented among hundreds of reports, a few of which are listed in Table 7.[62,79,102-109] Some of these reports and reviews contain ^{1}H chemical shifts, for which comparisons are also fruitful. Computer data banks represent another source of NMR data.[92] Specialized collections of steroid chemical shifts are readily assembled and searched with microcomputer spreadsheets or databases.

B. SUBSTITUENT INCREMENTS AND PREDICTIVE METHODS

Addition of an alkyl or hydroxyl substituent to a molecule leads to a downfield shift of the α and β carbons and an upfield shift in the γ carbon.[110] This upfield shift is much larger (~3 to 5 ppm) when the γ substituent is

TABLE 5
Assignment of ^{13}C Signals of 5 by Comparison with Assigned Spectra of Sterols Having a Similar Side Chain (6) or a Similar Steroid Nucleus (1)[a]

Model (6) for carbons in side chain	Signals to be assigned for 5	Model (1) for carbons in steroid nucleus
	Quaternary carbon signals	
	208.07*	208.16 (C-15)
	150.74*	150.68 (C-8)
	140.23*	140.23 (C-14)
	42.56*	42.51 (C-13)
	38.73*	38.69 (C-10)
	Methine signals	
	70.91*	70.79 (C-3)
	50.84	50.82 (C-9)
	50.80	50.75 (C-17)
	44.10*	44.07 (C-5)
(C-25) 35.78	35.74‡	
	34.45*	34.43 (C-20)
	Methyl signals	
	19.17*	19.17 (C-21)
	18.78*	18.73 (C-18)
(C-27) 16.49	16.47‡	
	12.92*	12.87 (C-19)
	Methylene signals	
(C-26) 68.52	68.37‡	
	42.46*	42.41 (C-16)
	37.75*	37.69 (C-4)
	36.95*	36.91 (C-12)
	36.51*	36.49 (C-1)
	35.81*	35.76 (C-22)
(C-24) 33.52	33.43‡	
	31.14*	31.07 (C-2)
	29.13*	29.10 (C-6)
	27.58*	27.55 (C-7)
(C-23) 23.39	23.18‡	
	19.56*	19.52 (C-11)

[a] Signals marked with * or ‡ were assigned by comparison with signals of **6** or **1**, respectively.

TABLE 6
Major Compilations of ^1H and ^{13}C NMR Chemical Shifts of Steroids[a]

Subject	Nucleus	Number of Spectra	Year	Ref.
Steroids (mainly C_{18} to C_{21})	^1H	344	1970	93
Steroids-General	^{13}C	413	1977	94
Steroids-General	^{13}C	92	1979	95
Natural products	^{13}C	95	1979	96
Sapogenins and saponins	^{13}C	130	1985	97
Steroids-General	^{13}C	105	1987	98
Cardenolides and bufadienolides	^{13}C	247	1987	99
Sterols and triterpene alcohols	^{13}C	112	1987	100
Sterols	^{13}C	48	1989	34
Bile acids	^1H,^{13}C	60	1988	101
Steroids (mainly C_{19} and C_{21})	^1H	166	1990	32

[a] Although many errors and inconsistencies from the original literature have been corrected in these data collections, some incorrect assignments have inevitably been propagated, especially in data taken from early papers (see Reference 79).

TABLE 7
Additional Sources of NMR Chemical Shifts of Sterols

Subject	Assigned Spectra ^{13}C	Assigned Spectra ^1H	Ref.
Cholestanes			
3β-Hydroxy-$\Delta^{8(14)}$-15-ketones, 9α-substituted	11	—	102
$\Delta^{3(14)}$ Allylic hydroxysterols	13	—	62[a]
3β-Hydroxy, monoenes and dienes	15	—	103[b]
3β-Hydroxy dienes in 5α and 5β series	12	8[c]	104
3β,26-Diols with various groups in rings B, C, and D	14	14	105
6-Methyl with various groups in ring A	13	—	106
Lanostanes and Plant Sterols			
Lanostanes, wide variety of functionality	30	30	79
Lanostane olefins and epoxides	21	—	107
$\Delta^{7,9,11,24}$-26-Acid; hydroxyl at various positions	24	—	108
Revised assignments for common plant sterols	12	32	109

[a] In Table IV of Reference 62, the column heading under TAI shifts should read 3β, 15β [2c].

[b] The chemical shift of C-6 in compound 4b of Reference 103 should be δ 29.5.

[c] Full ^1H assignments except for stereochemistry.

TABLE 8
Variability of Generalized ^{13}C Substituent Increments for Steroids[a-c]

<table>
<tr><th>Substituent</th><th></th><th>β</th><th>γ-Gauche</th><th>γ-Anti</th></tr>
<tr><td>Alcohol[d]</td><td>C(OH)-β-γ</td><td>7 ± 3</td><td>−5 ± 2</td><td>−1 ± 1</td></tr>
<tr><td>Ketone</td><td>C(=O)-β-γ</td><td>12 ± 4</td><td colspan="2">0 ± 4</td></tr>
<tr><td>C=C[d]</td><td>C=C-β-γ</td><td>0 ± 4</td><td colspan="2">−2 ± 3</td></tr>
</table>

[a] Generalizations mainly from data in Reference 94.
[b] δ(functionalized steroid) − δ(hydrocarbon steroid).
[c] This table illustrates the low precision of generalized β and γ effects for sterols and is not intended as a guide for signal assignments. Some substituent increments fall outside the ranges given here, e.g., if the affected carbon is quaternary.
[d] A discussion of the variation in substituent increments is found in Reference 138 (alcohols) and Reference 115 (olefins).

gauche (syn), a generalization called γ-gauche effect. Table 8 shows that the magnitude of the β and γ shifts in steroids is too variable to be useful for ^{13}C assignments. In order to improve the accuracy of estimated shifts, sets of specific substituent increments that denote the location of the substituent on the steroid nucleus have been developed. For example, substituent increments for a 3β-hydroxyl group are obtained by taking the chemical shift differences between 5α-androstan-3β-ol and 5α-androstane (Figure 9).[94] The chemical shifts for the nonpolarizable carbons of 5α-cholest-8(14)-en-15-one (7) can be estimated to $\sim \pm 0.2$ ppm by subtracting the substituent increments for a 3β-hydroxyl group from the chemical shifts of 3β-hydroxy-5α-cholest-8(14)-en-15-one (**1**). Comparison of the observed and estimated chemical shifts in conjunction with a spectral editing experiment allows all signals to be assigned (Table 9). This simple but valuable method generally provides more reliable estimates of chemical shifts than other predictive methods (see below).

Substituent increments, also called substituent effects, substituent chemical shifts, or substituent-induced shifts, are defined as δ(steroid with specified functional group) − δ(steroid without specified functional group). Chemical shifts can be estimated for multifunctional sterols by successive applications of substituent increments, provided that the functional groups are not too close. Estimates for multifunctional sterols are more accurate when assigned chemical shifts can be found for a model differing by only one or two functional groups so that successive applications of substituent increments are minimized. Composite substituent increments have been tabulated for certain combinations of functional groups, such as α,β-unsaturated ketones.[94] Chemical shift estimates may be quite accurate, as in the example in Table 9 of a hydroxyl group remote from other functionality, or rather inaccurate if the incremented functional group causes a conformational change or interacts

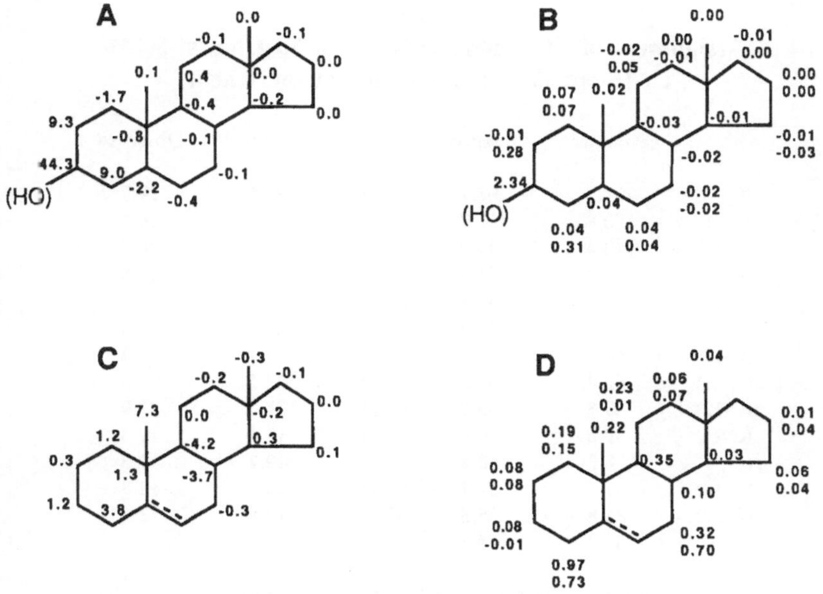

FIGURE 9. Substituent increments for a 3β-hydroxy group (A, ^{13}C increments; B, ^{1}H increments) and for a Δ^5 double bond (C, ^{13}C increments; D, ^{1}H increments). In B and D, upper number represent increments for β protons. (Data for A and C derived from Blunt, J. W., and Stothers, J. B., *Org. Magn. Reson.*, 9, 439, 1977 Data for B and D from Schneider, H.-J., Buchheit, U., Becker, N., Schmidt, G., and Siehl, U., *J. Am. Chem. Soc*, 107, 7027, 1985.)

with another functional group. Conformational changes are more likely for added double bonds than for added hydroxyl or methyl groups.

In addition to an extensive set of ^{13}C subsitutent increments in the Blunt and Stothers review,[94] tables of ^{13}C substituent increments have been compiled for specialized classes of steroids, including methyl 5β-cholan-24-oates,[34] cardenolides and bufadienolides,[99] sterols brominated in ring A or B,[111] 6-methylsteroids,[106] $\Delta^{8(14)}$ allylic alcohols,[62] epoxylanostanes,[107] and 9α-substituted sterols.[102] Because substituent increments are affected by the stereochemistry of ring junctions, separate tables of values must be consulted for 5β-sterols and 14β-sterols. ^{1}H and ^{13}C substituent increments for Δ^5 and 3β-hydroxyl functions are shown in Figure 9.

Proton substituent increments for H-18 and H-19 are known as Zürcher's rules and have long been used in the identification of sterols and steroids from low-field ^{1}H NMR spectra. A table of 344 of these values has been compiled.[93] Tables of substituent increments for all proton positions have recently become available.[32,51] Regularities[112] in a large collection of ^{1}H chemical shifts indicate that ^{1}H substituent increments, like their ^{13}C counterparts, are additive in the absence of interacting functional groups or conformational differences.[32] The general effects of functional groups on ^{1}H chemical shifts

TABLE 9
Assignment of ^{13}C Chemical Shifts of 5α-Cholest-8(14)-en-15-one (7) from Substituent Increments

	Estimated[a]	Observed		Estimated[a]	Observed
(C15)	208.2	208.2	(C-6)	29.5	29.7
(C-8)	150.8	151.7	(C-4)	28.7	28.6
(C-14)	140.4	139 9	(C-25)	27.9	28.0
(C-9)	51.2	51.3	(C-7)	27.6	27.8
(C-17)	50.8	50.8	(C-3)	26.5	26.4
(C-5)	46.3	46.2	(C-23)	23 5	23.5
(C-13)	42.5	42.6	(C-27)	22.7	22.7
(C-16)	42.4	42 5	(C-26)	22.5	22.5
(C-10)	39.5	39.7	(C-2)	21.8	21.9
(C-24)	39.3	39 4	(C-11)	19.1	19.3
(C-1)	38.2	38.4	(C-21)	19.2	19.2
(C-12)	37.0	37 1	(C-18)	18.7	18.8
(C-22)	35.8	35.8	(C-19)	12 8	12.8
(C-20)	34.4	34 5			

[a] Chemical shifts estimated by subtracting the substituent increment for the 3β-hydroxyl group in Reference 94 from the corresponding chemical shift of 15-ketosterol **1**. The substituent increments for C-13 and C-14 in Reference 94 are evidently transposed and have been corrected here.

TABLE 10
General Trends ^{1}H Substituent Increments for Steroids[a]

Substituent		β *(cis)*	β *(trans)*	γ *(cis)*	γ *(trans)*
Alcohol (eq)[b]	C(OH)–CHβ–CHγ	~0	0.3	<0.1	<0.1
Alcohol (ax)[b]		0.1	0.3	−0.2	0.5
Ketone	C(=O)–CHβ–CHγ	0.8		0.3	
C=C	C=C–CHβ=CHγ	0.7 (varies)		variable	

[a] Adapted from data in References 51 and 32. The precision and scope of these increments has not been fully investigated.
[b] eq, equatorial; ax, axial.

are shown in Table 10. As yet, few ^{1}H substituent increments are available for 3β-hydroxycholestanes and for olefin or ketone groups.

Acetylation shifts are a special case of substituent increments defined as δ(acetate)–δ(free sterol). Acetylation shifts for several 3-hydroxysterols are shown in Figure 10. Typically, the ^{13}C increment for acetylation at C-3 is ~1 ppm at the α carbon and ~ −4 ppm at the β carbon. Proton 3-acetylation shifts are quite small except at the α carbon. As acetylation seldom induces conformational changes, ^{13}C chemical shifts are normally identical for the acetate and free sterol except in the vicinity of the hydroxyl group. The

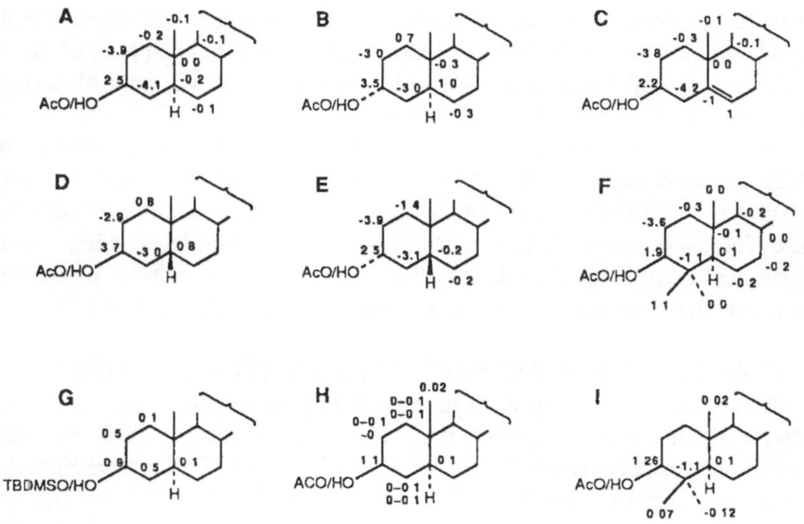

FIGURE 10. Selected sterol acetylation shifts (δ(acetate) − δ(free sterol)) and silylation shifts (δ(*tert*-butyldimethylsilyl ether) − δ(free sterol)). Upper numbers in H represent acetylation shifts for β protons. Data from References 32, 79, 100, 105, and unpublished observations.

resulting precision enables many closely spaced pairs of signals in either spectrum to be distinguished.[79] Benzoylation shifts are comparable in magnitude to acetylation shifts, but silylation shifts are much smaller and may differ in sign (Figure 10).

Predictive methods based only on steroid structures might appear to provide a simpler alternative to substituent increments and chemical shift comparisons. Such predictive schemes are based on the application of factor analysis to correlate a large collection of ^{13}C data with features of molecular structure. An early set of additivity parameters based on simple topological features gave good chemical shift estimates for steroid hydrocarbons.[113] Beierbeck et al.[114] reinterpreted this set of parameters according to the hypothesis that the upfield γ-gauche effect is due to loss of a hydrogen from the β carbon and extended the calculations to alcohols, amines, ketones, and olefins. The resulting ^{13}C chemical shift estimates generally agree to ±1 ppm with observed values, but occasionally such predicted values are rather unreliable.[115] Also, these predictions are seldom valid for carbons near a double bond or in a five-membered ring. Key assumptions of the predictive method of Beierbeck et al. have been criticized.[116] More recent predictive schemes[117,118] based on geometric and electronic parameters (obtained from molecular mechanics calculations) offer somewhat improved accuracy and generality at the expense of computational complexity.

Fundamental explanations of ^{13}C chemical shifts by classical mechanisms[116] indicate that empirical correlations of chemical shifts with topological and even geometric factors should be used cautiously. Exceptions[115] have been noted to proposed empirical correlations of ^{13}C chemical shifts with topological features of olefinic sterols.[119] Exceptions have also been pointed out[116] to the suggestion[120] that downfield shifts are observed for pairs of carbon atoms bearing hydrogens in an *anti* relationship.

Predictive methods have been more successful for acyclic systems, in which stereochemical effects on ^{13}C chemical shifts are attenuated by averaging over the distribution of rotamers. A tabulation of calculated chemical shifts for all possible C_7 to C_{11} sterol side chains facilitates either signal assignment or structure elucidation,[121] but such calculations are of no use for assigning stereochemistry in the side chain except possibly at C-22.[122]

C. CORRELATION SPECTROSCOPY AND OTHER METHODS

Rarely can all ^{13}C resonances be assigned by comparisons and substituent increments. The remaining ambiguities are usually settled by 2D correlation experiments, especially HETCOR (or HMQC) and COSY (or COSYDEC). For example, the two signals of $\Delta^{8(14)}$-15-ketosterol **5** at δ ~50.8 corresponding to C-9 and C-17 (Table 5) are correlated to protons at δ_H 1.86 (δ_C 50.84) and δ_H 1.46 (δ_C 50.80). The COSY column at δ 2.36 (H-16) shows the presence of a cross peak at δ ~1.5 and the absence of any correlation near δ 1.86, thus establishing the assignment of H-17 to δ_H 1.46 and C-17 to δ_C 50.80. Other helpful experiments include HMBC, COSYDEC,[42] and COSY-LR as well as such classical techniques as chemical derivatization, deuterium labeling,[40] T_1 measurements, and solvent- and lanthanide-induced shifts. The close agreement between observed and calculated LIS for dihydrolanosterol (Figure 11) illustrates the power of the LIS technique for establishing signal assignments.

For steroid samples too small for ^{13}C experiments, most of the ^1H signals can be assigned from high-field COSY, COSY-DEC,[42] and ^1H spectra.[32] Despite the many overlapping muiltiplet in the δ 1–2 region, resonances can often be recognized by the shape of their splitting pattern, which is remarkably constant, apart from some variation in the conformationally mobile D ring. Standardized profiles of splitting patterns have been presented for each proton position of Δ^4-3-ketosteroids,[32] thus extending and updating an earlier[93] set of diagrams for hydroxyl CH signals. Copied onto transparent plastic sheets, these profiles can be overlaid onto spectra to facilitate the recognition of a splitting pattern partially obscured by many other peaks. Profiles of splitting patterns for ring C protons are shown in Figure 12. Deviations from the standard profiles are generally minor in the absence of geminal or vicinal substituents (including unsaturation) or major conformational differences.[32] Whereas ^1H and ^{13}C chemical shift comparisons are strongly affected by γ substituents, these coupling constant comparisons are barely affected beyond the β position. The principal effects of α and β substituents are merely simplification of the coupling pattern.[32] Interpretation of the more complex spectra of sterols mainly by ^1H-based techniques appears to be feasible at high field in some cases.[58]

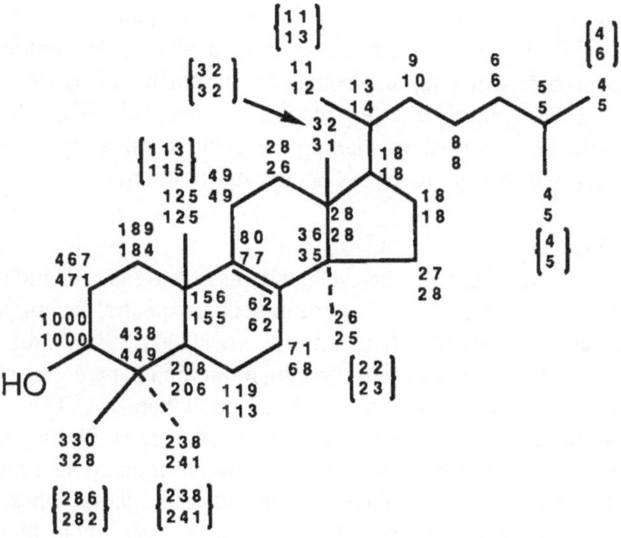

FIGURE 11. Observed (upper numbers) and calculated (lower numbers) LIS for dihydrolanosterol from Emmons, G. T., Wilson, W. K., and Schroepfer, G. J., Jr., *Magn. Reson. Chem.*, 27, 1012, 1989. All LIS are for ^{13}C except for the ^{1}H methyl LIS values shown in braces.

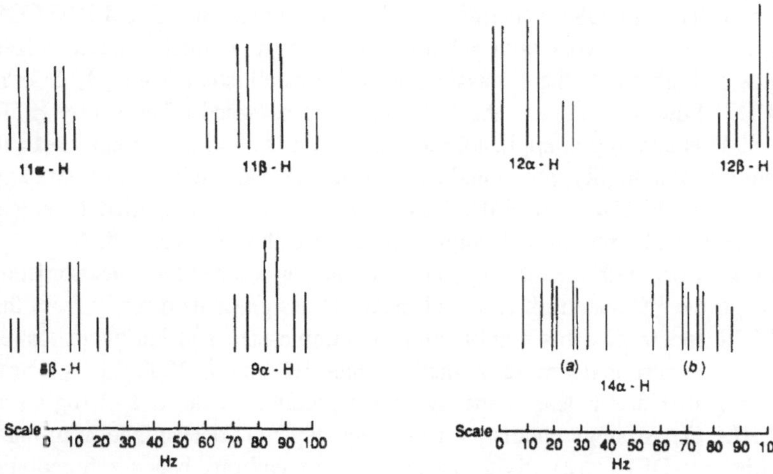

FIGURE 12. Templates for recognition of ^{1}H signals from their splitting patterns. Vertical scales not necessarily identical. Two templates for H-14α illustrate common variations resulting from slight differences in coupling constants. (Adapted from Kirk, D. N., Toms, H. C., Douglas, C., White, K. A., Smith, K. E., Latif, S., and Hubbard, R W. P., *J. Chem. Soc. Perkin Trans. 2*, 1567, 1990. With permission.)

Assigning the stereochemistry of ^{1}H geminal pairs can be done by NOEDS,[31,51,53,71] chemical shift comparisons, or HMBC.[52] Another method involves comparisons of coupling constants or profiles of splitting patterns (see above) in ^{1}H spectra of cross sections of coupled HETCOR spectra. Caution should be observed in making stereochemical assignments in the flexible D ring and in the vicinity of a *cis* ring junction.

D. STRATEGIES

Strategies for assigning signals depend on many factors, including the density of functional groups in the sterol molecule, spectral techniques available, and the accessibility of substituent increments and models for chemical shift comparison. Comparisons and substituent increments are generally more important for sterols with few functional groups, whereas COSY and HETCOR spectra play a larger role in the assignment of more highly functionalized sterols. Efficient use of comparisons and substituent increments demands the patience to organize and manipulate large arrays of numbers, whereas determining a structure from 2D correlation spectra is comparable to solving a puzzle. The LIS and solvent-induced shift techniques used previously to establish reliable assignments of many sterols[62,102,103] have largely been replaced by 2D methods, although the LIS method has been refined[64] and is still a valuable technique for sterols with few functional groups.[65,79]

Two major strategies[35] for utilizing 2D methods have emerged, one focusing on ^{1}H and COSY spectra[32,39,53,87] and the other on ^{13}C and HETCOR spectra.[51,71,79,123] In work with ~1-mg samples of moderately functionalized steroids at high field, the ^{1}H assignments are established from ^{1}H, COSY, and COSYDEC spectra, and the ^{13}C signals are assigned (if at all) by HETCOR.[32] This strategy is applicable to many steroids and presumably also to Δ^{22} sterols and highly functionalized sterols. A variation of this strategy incorporating HMQC and HMBC experiments has been applied to signal assignments and structure determination of the orthoester sterol **9**.[58]

In our work with 5 to 20 mg sterol samples on a 300 MHz spectrometer, we assign the ^{13}C spectrum first and make ^{1}H assignments directly from the HETCOR spectrum. This combination of sample size and hardware makes ^{13}C-based experiments more attractive than ^{1}H and COSYDEC spectra, which together are at least partially uninterpretable in the δ 1–2 region at 300 MHz. Our strategy entails the following steps: (1) determination of multiplicities by DEPT; (2) trivial assignments of carbons bearing functional groups; (3) assignments from substituent increments and chemical shift comparisons; (4) differentiation of the remaining carbons by HETCOR, COSY (or recently, COSYDEC), and other methods. This strategy, is illustrated for 7-ketodihydrolanosterol (**8**).

Application of steps 1 to 3 and consideration of acetylation shifts gave the assignments shown in Figure 13 but left three sets of signals unassigned. The signals in each set were differentiated largely by their correlated ^{1}H

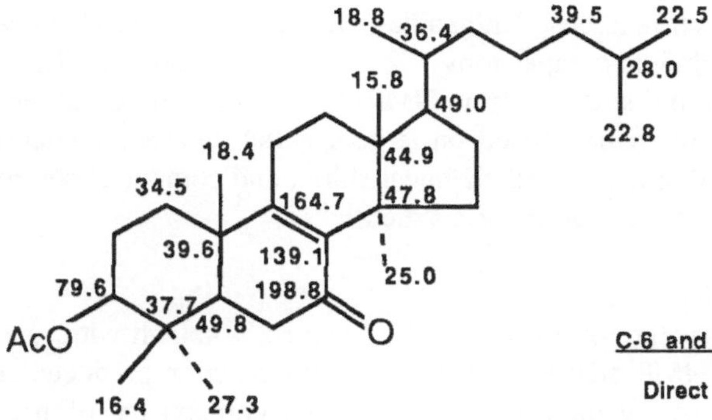

C-6 and C-22

	Direct	Long-range
δ_C 36.37	δ_H 1.0, 1.4	
δ_C 36.42	δ_H 2.4	

C-12, C-15, and C-16

	Direct	Long-range
δ_C 30.06	δ_H 1.8	δ_H 0.65 (H-18)
δ_C 31.98	δ_H 1.7, 2.1	
δ_C 28.75	δ_H 1.3, 1.9	

C-2, C-11, and C-23

	Direct	Long-range
δ_C 23.82	δ_H 1.8	
δ_C 23.69	δ_H 2.3	
δ_C 24.06	δ_H 1.1, 1.4	

FIGURE 13. Basis for assigning the final eight ^{13}C signals of 7-ketodihydrolanosterol (**8**) from HETCOR and long-range HETCOR information. (Derived from Emmons, G. T., Wilson, W. K., and Schroepfer, G. J., Jr., *Magn. Reson. Chem* , 27, 1012, 1989.)

chemical shifts from a low-resolution HETCOR spectrum of low sensitivity. Thus, the downfield C-6 allylic protons were easily distinguished from the C-22 protons, which are expected to be at δ_H 1.0 and 1.4 by analogy with other sterols having a C_8H_{17} side chain.[104] Of the ^{13}C methylene signals at δ_C 28-32, the one showing a long-range HETCOR connection to H-18 was identified as C-12. C-16 was assigned from its correlation to protons at δ 1.3 and 1.9.[104] The downfield position of the C-15 protons reflects their proximity and orientation to the 7-keto group, an effect analogous to the downfield shift of H-7β by a 15-keto group.[105,124] Of the third set of methylene carbons, C-23 was identified by its characteristic high-field ^1H signals (δ_H 1.1 and 1.4),[104] and C-2 was distinguished from C-11 by the correlation of the C-2 protons with H-3α in the COSY spectrum. In view of numerous incorrect ^{13}C assignments made by comparisons and now being discovered by 2D methods, ^1H chemical shift comparisons were used conservatively.

IV. STRUCTURE DETERMINATION

In practice, signal assignments and structure determinations are done concomitantly. Most methods described above for signal assignments can be applied to structure determinations as well. Such methods include substituent increments, 2D correlations, lanthanide-induced shifts, and chemical shift comparisons, including comparisons of 2D correlation spectra.[58] The process of elucidating steroid structure from NMR data is still carried out largely by hand, although automation based on large spectral databases, relationships between chemical structure and chemical shifts, and computer interpretation of 2D correlation spectra is being pursued.[125]

A. GROSS STRUCTURE

The gross structure of a new sterol is frequently known with reasonable certainty before NMR analysis begins. Synthetic reaction products, samples studied extensively by mass spectrometry, and components of previously characterized biological extracts usually fall in this category. In such cases, NMR is needed only for structure confirmation, which is readily done by comparisons and signal assignments from routine 1D and 2D experiments. More difficult structural problems are posed by the often unusual sterols isolated from plants and marine invertebrates.

If the structure of a sterol sample is completely unknown, a tentative molecular formula can frequently be established by NMR and low-resolution mass spectrometry. Functional groups, such as alcohols, ketones, esters, and double bonds are recognized from a combination of the ^{1}H and ^{13}C spectra. Careful examination of the ^{13}C and DEPT spectra establishes the number of carbon and hydrogen atoms (excluding hydrogens attached to heteroatoms). The number and type of heteroatoms can usually be determined from a combination of the NMR data and the mass spectrum, sometimes even in the absence of a molecular ion. If there are no uncertainties in the number of protons on heteroatoms, the number of rings can be calculated from the unsaturation number and the number of olefinic carbons in the ^{13}C spectrum.

Two strategies can be formulated for determining gross structure, one focusing on comparisons and substituent effects and the other on assembling substructures from 2D correlation spectra. The choice of strategy and methods used may be guided by the number of functional groups in the sterol (Table 11).

If few functional groups are present and the side chain is saturated, congestion in the δ 1.0 to 2.0 region of the ^{1}H and COSYDEC spectra will impede the substructure strategy, especially if high-field spectra are unavailable. However, in this case, large parts of the sterol are likely to have ^{13}C-chemical shifts nearly identical with ones reported in the literature. A matching structural fragment is normally found either by searching a database of steroid chemical shifts or by combining chemical intuition with analysis of the multiplicities and positions of distinctive signals.

The latter process can be systematized by listing the ^{13}C-chemical shifts grouped according to multiplicity in ascending or descending order and marking out data as they are assigned to a fragment. If plausible guesses of fragments for chemical shift comparisons are not forthcoming, one might

TABLE 11
Selection of NMR Structure Determination Procedures According to the Density of Functional Groups in a Sterol Molecule

	Many functional groups	Few functional groups
Primary strategy	Build and assemble substructures from ^1H–^1H coupling networks	Use comparisons and substituent effects to test proposed structures
Basic 2D experiment	COSYDEC	HMQC (or HETCOR)
C–C connectivities	From COSYDEC and HMQC	From HMQC and HMBC
^1H–^1H coupling constants	Many directly accessible	Most are relatively inaccessible[a]
Useful solvents	Pyridine-d_6, CD_3OD, D_2O	$CDCl_3$, pyridine-d_6,C_6D_6
LIS experiments	Seldom practical	May provide alternative strategy

[a] See Table 4 for feasible methods.

focus first on establishing the side chain and then on structural implications of the methyl and quaternary carbon signals. For the side chain, the number of carbon atoms and the presence of any functional groups can usually be determined from routine NMR and mass spectral results. Methyl group shieldings can then be matched against tables of side-chain chemical shifts[101,109,121] to identify the side chain. If few quaternary carbons are present, the chemical shifts of C-10 and C-13 can usually be established and used as a probe of functionality in the steroid nucleus. Similar probes are the signals of C-18 and C-19, which may be identifiable by their correlations to ^1H singlets and their COSYDEC correlations (e.g., to H-1α, H-9α, or H-12α).

The exact types and locations of functional groups can be established by using substituent increments, HMBC, or a combination of HETCOR and COSYDEC to rule out alternative possibilities. Such an approach has been used to locate the 12β-hydroxyl group of sapogenin **10** from substituent increment calculations and γ-gauche effects.[126] ^1H chemical shift comparisons for H-18 and H-19 (Zürcher values) also provide a simple but powerful probe of functionality and stereochemistry in the steroid nucleus.[93,127] The structure of **10** and other steroids from Section IV are shown in Figure 14.

If many functional groups are present, the gross sterol structure can often be derived from 2D correlation spectra. Strategies emphasizing COSY spectra have been discussed for small (~1 mg) samples of steroids.[32,39] Slightly larger sample quantities permit the acquisition of HMQC and HMBC spectra.[58,69] Together with COSY results, these spectra reveal numerous carbon connectivities, even across heteroatoms.[69] If much larger amounts of sample (≥100 mg) and spectrometer time are available, INADEQUATE will give directly and unequivocally nearly all carbon connectivities. The rather elaborate structure of paxisterol (**11**), a sterol with analgesic activity isolated from penicillin broth, was determined largely by INADEQUATE and other NMR experiments.[81]

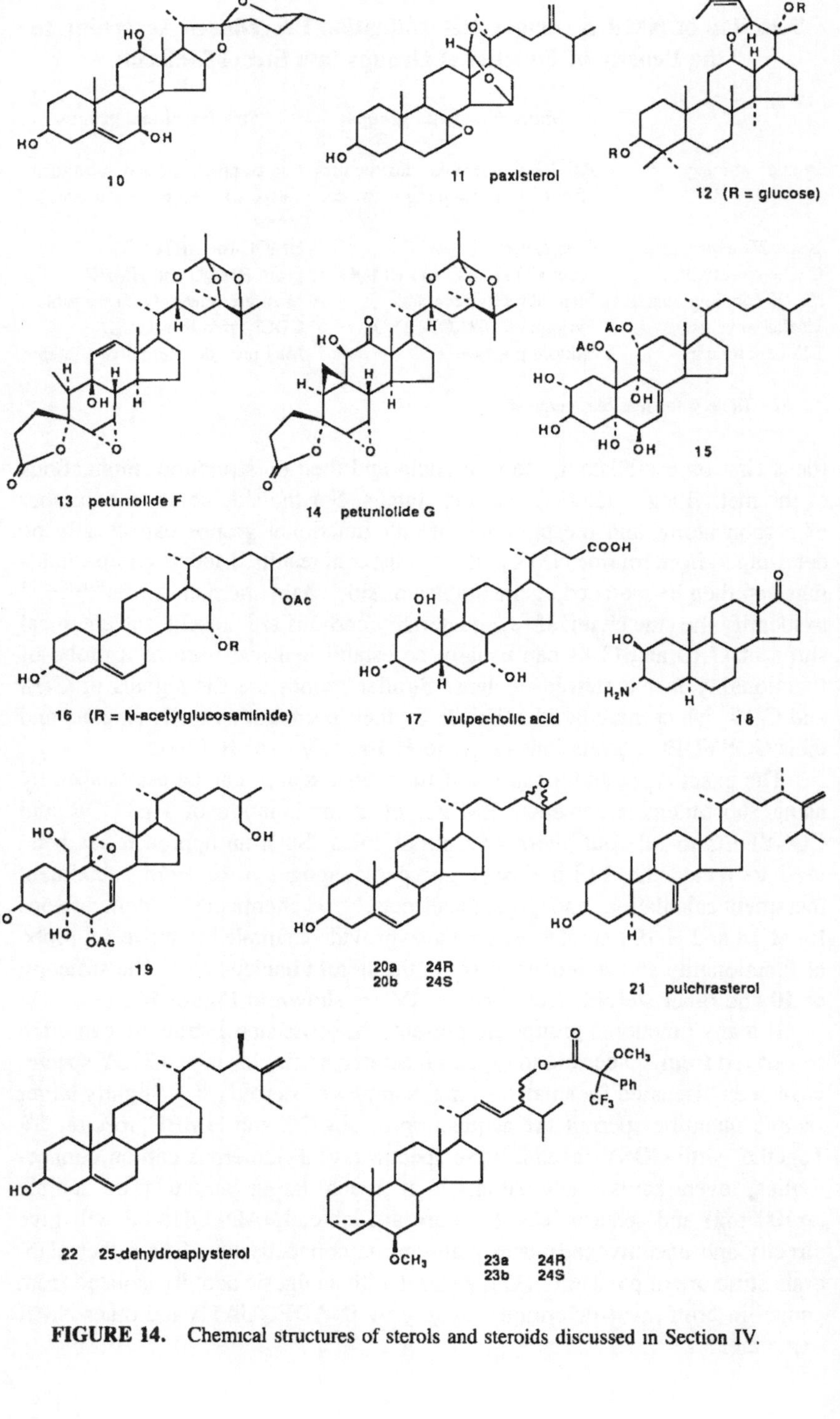

FIGURE 14. Chemical structures of sterols and steroids discussed in Section IV.

The method of building and assembling substructures from 2D correlation spectra has been applied to the structure determination of ginsenoside La (**12**), a minor triterpene saponin isolated from the leaves of ginseng.[69] Substructures in the aglycone portion were built from combinations of 1D, COSY, and ^{1}H-^{13}C shift correlation spectra with essentially no use of chemical shift comparisons or substituent increments. The substructures were put together from long-range C-H correlations in the HMBC spectrum.[69] The connections across a heteroatom, including the ether linkage between C-12 and C-23 and the positions of the sugar residues, could not have been established by INADEQUATE.

Even with HMBC and INADEQUATE, connectivities are sometimes difficult to establish at quaternary carbons and heteroatoms. Some structures containing novel arrangements of heteroatoms adjacent to quaternary carbons or chiral centers in acyclic regions cannot be completely solved by NMR techniques. If a seminal structure can be established by X-ray crystallography or chemical synthesis, then the structures of derivatives and analogs can usually be determined by NMR chemical shift comparisons. For example, X-ray crystal structures of petuniolides **13** and **14** opened the way for the structure determination of numerous analogs by NMR.[24] Petuniolides are a potent subclass of petuniasterones, a group of ergostanoid orthoesters found in petunia leaves and having deleterious effects on feeding insect larvae. Petuniolides vary greatly in potency depending on their substitution.[24]

A combination of the substructure approach and chemical shift comparisons is illustrated by the structure elucidation of a polyhydroxy sterol isolated from a deep sea sponge.[128] (The following discussion is adapted with modifications from the original report.) The ^{13}C NMR spectrum shows 31 carbon signals, of which four correspond to two acetate groups and 14 to substructure A (Figure 15), which can be established from chemical shift comparisons with cholestanol signals in ring D and the side chain. The remaining 13 carbon signals are shown in Figure 15 along with COSY, HETCOR, and ^{1}H data. The high density of functionality (7 of the 13 carbons apparently bear oxygen) permits the building of substructures through carbon connectivities emanating from the COSY spectrum. Thus, substructure B is built from HETCOR, COSY, and ^{1}H data for the signal at δ_H 4.43, which is seen to be geminal to a hydroxyl and vicinal only to an olefinic proton. Substructure B is evidently terminated by quaternary carbons. Signals at δ 2.1 and 2.6 show COSY connections to each other and to a signal at δ 5.42, whose downfield position indicates a geminal acetate. The absence of additional COSY cross peaks for resonances at δ 2.1 and 5.42 suggests that substructure C is flanked by quaternary carbons as shown. Substructure D is assembled similarly, with the complication of overlapping signals at δ 3.1 and 2.6. The remaining protonated carbon is accounted for by the substructure C–CH_2–OAc.

These substructures can be assembled in only a few ways if the final structure is to have the usual steroid nucleus. Substructure D fits only in ring

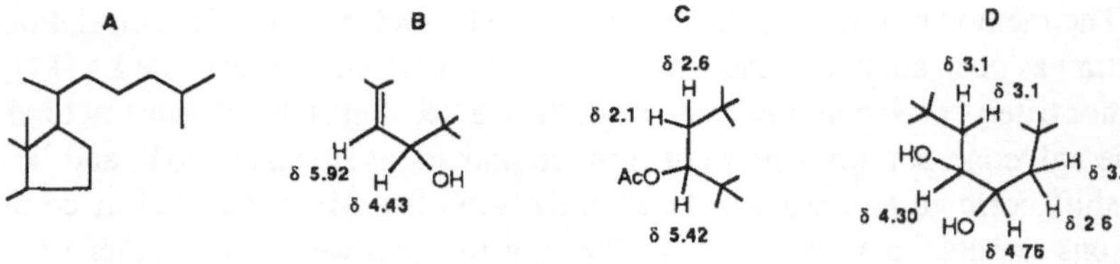

1D Proton data

5.92	d	J = 4.2
5.42	dd	J = 4.2, 11.0
5.24	d	J = 13.7
4 80	d	J = 13.6
4.76	m	
4.43	d	J = 4.2
4.30	dd	J = 4.4, 11.2
3.12	m	
3.12	m	
3.12	m	
2.92	m	(H-14)
2.6	m	
2.6	m	
2.1	d	J = 11

COSY

5.92 ⇔ 4.43	4.76 ⇔ 4.30	4.30 ⇔ 3.1*
5.42 ⇔ 2.60	4.76 ⇔ 3.1	3.1 ⇔ 2.6*
5.42 ⇔ 2.10	4.76 ⇔ 2.6	2.92 ⇔ 2.6
		2.60 ⇔ 2.1

DEPT and HETCOR

C	CH	CH_2
140.5	123.9 ⇔ 5.92	65.8 ⇔ 4.80, 5.24
77.3	73.1 ⇔ 4.30	41.1 ⇔ 2.6, 2.10
76 4	74.9 ⇔ 5.42	40.0 ⇔ 2.6, 3.1
47.9	72.3 ⇔ 4.76	32.7 ⇔ 3.1, 3.1
	71.4 ⇔ 4.43	

FIGURE 15. Determination of substructures of a polyhydroxylated sterol (**15**) from a combination of ^1H, ^{13}C, DEPT, HETCOR, and COSY data. All data are explicitly given in Reference 128 except for the anticipated COSY connections marked with an asterisk.

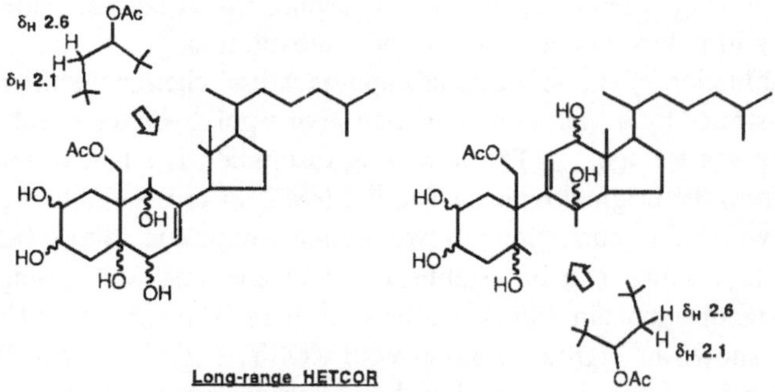

Long-range HETCOR

C-18 ⇔ δ_H 2.1, 2.6

FIGURE 16. Assembling substructures of a polyhydroxylated sterol (**15**) based on long-range HETCOR data from West, R. R. and Cardellina, J. H., II, *J. Org. Chem.*, 53, 2782, 1988. With permission

A and therefore is terminated by C-5 and C-10. Substructures B and C can be placed into rings B or C in two ways, as shown in Figure 16. Substructure C can fit in either of two orientations. The correlation of C-18 with ^1H signals at δ 2.1 and 2.6 in the long-range HETCOR spectrum indicates that these signals belong to the C-12 protons, thereby eliminating all but one of the possible structures. The stereochemistry of the hydroxyl and acetate groups was established from NOEDS experiments and ^1H–^1H coupling constants to give the final structure **15**.[128]

A combination of strategies based on substructures and comparisons is also typical for sterols with few functional groups. For example, the structure of (25*R*)-cholest-5-ene-3α,15α,26-triol (**16**), an ichthyotoxin from the class of pavoninins, was determined largely by comparisons, but the 15-hydroxyl group was located from a substructure built from the results of several decoupling difference experiments.[23]

B. STEREOCHEMISTRY IN THE STEROID NUCLEUS

Karplus relationships are equations that express vicinal J_{HH} values as a function of molecular structure; the most important term is the H–C–C–H dihedral angle θ. A very simple empirical Karplus relationship of modest accuracy is given by $^3J_{HH} = 7.76 \cos^2\theta - 1.1 \cos\theta + 1.4$ (1.2-Hz root-mean square deviation between experimental and calculated values).[129] Modifications[129,130] of this equation to include terms correcting for effects of electronegativity[131] and C–C–H valency angles markedly improve the accuracy. Such relationships provide a powerful tool for assigning configuration of substituents on the steroid nucleus, especially in saturated six-membered rings in which dihedral angles are normally either ~180° (axial-axial) or ~60° (equatorial-equatorial or equatorial-axial). In simple cases, judicious use of vicinal ^1H coupling constant comparisons (see Table 3) may be sufficient to establish substituent configuration.

The stereochemical orientation of a secondary steroidal alcohol in rings A, B, and C can be determined on simple 60–90 MHz spectrometer by means of ASIS,[59,61] Zürcher increments,[93] or the half-height width ($W_{1/2}$) of the downfield C<u>H</u>OH resonance.[90] A wide multiplet indicates an axial proton usually having two large axial-axial couplings, whereas a narrow multiplet indicates an equatorial proton with smaller couplings. Measurement of individual coupling constants at high field enhances this method, as exemplified by the structure determination of vulpecholic acid (**17**), a novel bile acid isolated from a marsupial. The hydroxyl configurations of **17** were established from the C<u>H</u>OH coupling patterns either by comparison with those of similar bile acids (H-3β and H-7β) or according to differences in coupling constants between axial and equatorial hydroxyls (H-1β, dd, J = 12.7,3.5 Hz).[132]

If the proton geminal to the substituent is buried in the methylene envelope, the coupling profile can often be extracted from the COSYDEC spectrum or by another method given in Table 4. Substituent configurations can also be determined from NOEDS experiments and from ^{13}C substituent increments, especially those reflecting the differences between γ-gauche and γ-anti effects.[126]

Use of ^1H–^1H coupling constants to make stereochemical assignments is less reliable when non-chair conformers are present. Many structural features, such as *cis* ring junctions[133,134] and certain substitution patterns,[88,123,135] are linked with conformational flexibility that may invalidate naive applications of a Karplus relationship. For example, ring A of vicinal amino-alcohol **18** is in the twist conformation,[135] thus largely precluding the use of substituent increments or ^1H–^1H coupling constants to assign substituent configurations.

Attempts to exploit small differences in predicted ^1H–^1H coupling constants can lead to erroneous conclusions. For example, although an olefinic proton in the plane of a ring might be expected to show almost no coupling to a pseudoaxial allylic proton (~90° dihedral angle) and a significant coupling to a pseudoequatorial allylic proton (~0° dihedral angle), our cursory sampling of molecular mechanics structures suggests marginal discriminating power (~4–6 Hz for pseudoequatorial couplings and ~2–3 Hz for pseudoaxial couplings). This application of ^1H–^1H coupling constants was shown to be unreliable for determining the configuration of the allylic 6-hydroxyl group of an ecdysteroid (**19**) isolated from the pupal ovaries of a silkworm, and the stereochemistry was established by ASIS.[61]

Pyridine-induced shifts at protons vicinal to hydroxyl are large for O–C–C–H dihedral angles of ~60° and small for dihedral angles of ~180°, a useful relationship for assigning substituent configuration. However, because of possible unanticipated steric or conformational effects, ASIS are best studied with model compounds for particular substitution patterns.[60,61,91] In the case of the ecdysteroid **19**, pyridine-induced shifts measured on a series of 5α,6α- and 5α,6β-dihydroxy Δ^7-sterols showed a consistent pattern in which H-4β was shifted downfield to δ ~3.0 in the 6β-hydroxysterols and H-4α shifted downfield to δ ~3.0 in the 6α-hydroxysterols. Signals for H-4α and H-4β were easily distinguished by coupling pattern differences between axial and equatorial protons. Once the pattern of pyridine-induced shifts had been established for this substitution pattern, the configuration of the 6-hydroxy group in **19** could be confidently assigned as α.[61]

Conformational flexibility makes definitive stereochemical assignments in ring D rather elusive. Ring D of sterols is usually a 13β envelope or a 13β,14α half-chair, but substitution or unsaturation in ring D may alter the conformation(s). This potential variability in the conformation of ring D reduces the precision of ^{13}C substituent increments and dictates the use of sophisticated molecular modeling in applications of a Karplus relationship. The configurations of D-ring substituents in many sterols and steroids have been established by chemical synthesis, X-ray crystallography, and other physical methods, and the NMR data recorded for these compounds can be used in chemical shift comparisons. For example, NMR assignments have been reported for various 16-methyl-17-oxygenated steroids[136] and for many 15-oxygenated sterols.[23,62,65,79,124,137] Chemical shift comparisons are also subject to the vagaries of the D-ring conformations, as are LIS, ^1H coupling constants, and NOEDS. Determinations of substituent configurations in ring D can be strengthened by molecular modeling and by collecting NMR data for both the α- and β-substituted sterols.

The ring junction stereochemistry at C-5 and C-14 is normally evident from the chemical shifts of C-19 and C-18. C-19 in 5β-sterols and C-18 in

14β-sterols each have at least two fewer γ-gauche interactions than in the corresponding 5α and 14α sterols and are consequently shifted ~10 ppm downfield.

C. SIDE-CHAIN STEREOCHEMISTRY

Proton coupling constants, ^{13}C γ-gauche effects, substituent increments, and 2D correlation experiments are of little value in establishing the configuration of chiral centers in the side chain. By far the most important NMR method for determination of stereochemistry in the side chain is chemical shift comparison. Chemical shift differences between C-22 and C-23 epimers are substantial,[122,139] but for many C-24 and C-25 epimeric pairs the largest differences are on the order of 0.01 ppm for ^{1}H and 0.2 ppm for ^{13}C.[109] Making meaningful comparisons requires chemical shifts of high precision for both the unknown sterol and each epimer of the reference sterol. The configuration of the first sterol in a new series is usually determined by chemical synthesis or degradation[140] or by X-ray crystallography,[139,141] although a few have been established by NMR with Mosher-type reagents.[23,137,142,143]

The importance of high precision in the measurement of chemical shifts is illustrated by the 24*R* and 24*S* epimers of 24,25-epoxycholesterol (**20a, 20b**),[54] sterols that have been reported to inhibit enzymes involved in cholesterol biosynthesis. Despite maximal chemical shift differences between **20a** and **20b** of only ≤0.003 ppm for resolved ^{1}H signals and 0.3 ppm for ^{13}C, the ^{13}C differences are sufficient to permit identification and quantitation of both individual epimers and epimeric mixtures.[54] Similarly, the ratio of C-25 epimers of 3β,26-dihydroxy-5α-cholest-8(14)-en-15-one (**5**) can be determined by either ^{1}H or ^{13}C NMR despite maximal chemical shift differences of 0.02 ppm for ^{1}H and 0.2 ppm for ^{13}C.[55] These 26-hydroxysterols are metabolites of 15-ketosterol **1**; the 25*R* epimer is very potent in lowering HMG-CoA reductase activity in cultured cells.[105] C-25 epimers of other 26-hydroxysterols have also been distinguished by NMR.[23,137,142,144] Once the initial assignment of C-25 stereochemistry is established by chemical synthesis,[105] X-ray crystallography,[105] physical methods,[144] or an LIS technique,[23] subsequent samples of 26-hydroxysterols can be identified by chemical shift comparisons, often even from a single epimer provided that care is taken in sample preparation and spectral measurement.

Precise ^{1}H and ^{13}C chemical shifts have been reported for the four 24-ethyl-26-hydroxycholesterol isomers both with and without a Δ^{22} double bond, and the previously accepted ^{13}C assignments of C-26 and C-27 have been reversed.[109] In some cases, distinguishing C-24 epimers by comparisons requires matching a signal to one of two chemical shifts separated by less than 0.2 ppm. The small chemical shift differences between side-chain epimers requires a much more careful approach to chemical shift comparisons than is needed for comparisons in the steroid nucleus.

Such an approach is illustrated by the assignment of C-24 configuration of pulchrasterol (**21**),[145] the major sterol component of a deep sea sponge. ^1H chemical shifts for H-28 were compared indirectly with those of 25-dehydroaplysterol (**22**), a sterol with one carbon fewer in its side chain.[145] To bridge this difference in side-chain structure, two pairs of epimers, one with the C_{11} pulchrasterol side chain and one with the C_{10} 25-dehydroaplysterol side chain were synthesized. Although the H-28 chemical shifts were only 0.006 to 0.008 ppm apart in each pair of epimers, it was possible to relate the C-24 configuration from **22** via the synthetic epimeric pairs to pulchrasterol by meticulous chemical shift comparisons. A thorough circular dichroism study confirmed the 24(R) stereochemistry for **21**.[145]

The above configuration assignment was dependent on accurate chemical shifts for 25-dehydroaplysterol. NMR data for a large number of C_7 to C_{12} sterol side chains with great diversity in position and type of functionality have been reported. Most of this work is from the laboratories of Djerassi,[139-141,145] Akihisa,[101,127] Ikekawa,[21,61] Seo,[109] Riccio,[137,142,143] and Kobayashi.[146] A few data compilations are available.[100,101,109]

Precise chemical shift comparisons tend to be difficult for highly functionalized sterols, because tenacious polar solvent impurities and other factors limit the reproducibility of chemical shifts. An alternative approach for determining stereochemistry in a side chain containing hydroxyl is based on the (R)- or (S)-α-methoxy-α-trifluoromethylphenyl acetate (MPTA) derivative. For example, the (R)-(+)-MPTA ester of a synthetic pair of 24-hydroxysterol epimers **23a** and **23b** enhanced their chemical shift differences from 0.01 to 0.08 ppm.[143] The shifts induced by MPTA derivatization were accounted for by conformational analysis, and the pattern of MPTA-induced shifts was used to assign the configuration of 24-hydroxysterols isolated from starfish.[143] The C-25 stereochemistry of 26-hydroxysterols was similarly determined for echinoderm saponins[142] and pavoninins[23] by an LIS experiment on an MTPA ester.

V. CONFORMATIONAL ANALYSIS

The ^{13}C spectrum of a sterol provides a sensitive probe of conformation.[135] Deviations between observed ^{13}C chemical shifts and those estimated from substituent increments have indicated that the conformation of ring A of the vicinal amino-alcohol **18** is not predominantly the chair form.[135] Chemical shift differences between sterols can also suggest long-range conformational effects, as illustrated in Figure 17. Whereas acetylation of dihydrolanosterol has no discernable effect on the chemical shifts in rings C and D, chemical shift comparisons between the $\Delta^{4,6}$ and $\Delta^{5,7}$ sterols **24** and **25**[105] show small but significant differences in the ^{13}C chemical shifts of ring D and some side chain carbons. (See Figure 18 for structures of **24**, **25**, and other steroids discussed in Section V.) High precision in the ^{13}C chemical shifts clearly sharpens this conformational probe. Conformational and electronic factors

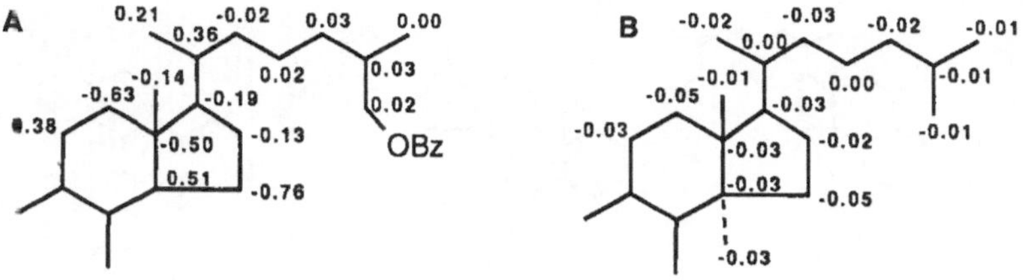

FIGURE 17. Use of chemical shift differences as conformational probes. Chemical shifts differences between: A, cholesta-4,6-diene-3β,26-diol dibenzoate and its $\Delta^{5,7}$ isomer (δ(**24**) - δ(**25**) from Ref. 105 and B, dihydrolanosterol and its acetate (δ(acetate) - δ(free sterol)) from Emmons, G. T., Wilson, W K., and Schroepfer, G. J., Jr., *Magn. Reson. Chem.*, 27, 1012, 1989)

have been used to explain similar observations in comparisons between androstanes and 17-ketoandrostanes[135] and between saturated and Δ^5 steroids.[51] These effects, which may eventually be understood by quantum mechanical calculations,[147] are minor in the latter case and lead to torsion angle deviations in ring D of $<1°$ by molecular mechanics and to observed $^3J_{HH}$ value deviations of ~ 1 Hz.[51] Deviations in ^1H chemical shifts,[88] in geminal ^1H–^1H coupling constants,[123,148] and in the appearance of the standard profiles for ^1H multiplets (see Figure 12) furnish other simple probes of conformational variation.

Although ^{13}C shieldings can reveal minor structure variations, other methods are needed to define the exact conformational changes. The most direct and definitive method of determining conformation by NMR entails a comparison of observed and calculated vicinal ^1H–^1H coupling constants. The calculated $^3J_{HH}$ values are obtained by applying a generalized Karplus relationship to dihedral angles from an energy-minimized molecular mechanics structure. A thorough conformational analysis requires as many observed $^3J_{HH}$ values as possible in the region of interest.

A. CONFORMATION IN THE STEROID NUCLEUS

Rings A, B, and C of saturated 5α,14α-steroids are typically in a chair conformation, which is easily recognized by the uniformity of vicinal coupling constants between various combinations of axial and equatorial protons. Standard ranges of such vicinal J_{HH} values are given in Table 3. The variability of $^3J_{HH}$ values reflects slight puckering or flattening of the chair conformation as well as other effects, such as electronegativities of nearby substituents. Because of the lack of conformational novelty in these rigid steroid rings, most studies of steroid conformational analysis focus on flexible rings, such as ring D, ring A of 3-keto-Δ^4 steroids, and rings adjacent to a *cis* ring junction.

Rigorous analyses of steroid conformation have been done mainly on steroids related to sex hormones, perhaps because the NMR analysis is simpler than that for sterols. Also, hypotheses relating structure and conformation to

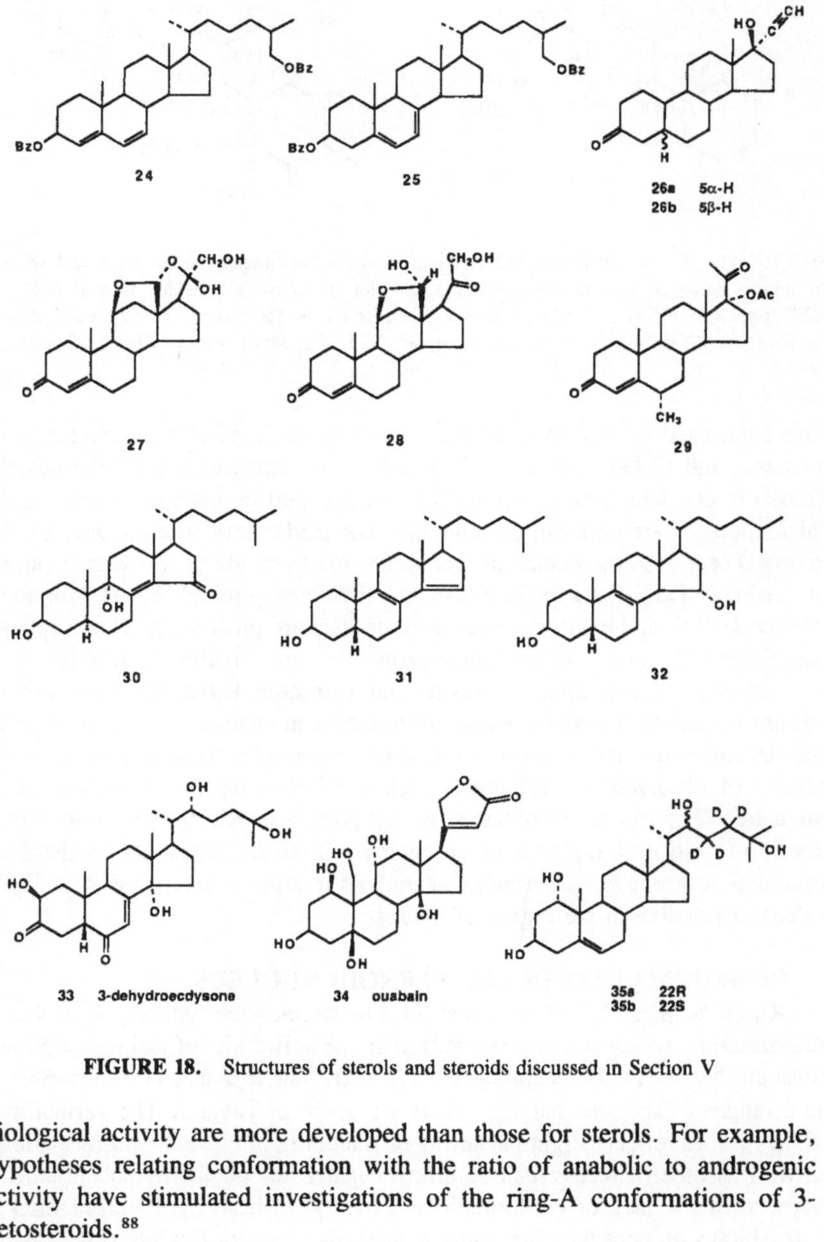

FIGURE 18. Structures of sterols and steroids discussed in Section V

biological activity are more developed than those for sterols. For example, hypotheses relating conformation with the ratio of anabolic to androgenic activity have stimulated investigations of the ring-A conformations of 3-ketosteroids.[88]

Although full sets of steroid-coupling constants first became available in 1980,[53] thorough conformational analyses were not reported until several years later.[51,85] In one of these reports,[85] 17β-hydroxy-19-nor-5α,17α-pregn-20-yn-3-one **(26a)** and its 5β-isomer **(26b)**, photodecomposition products of the contraceptive norethisterone, were exhaustively analyzed by NMR. Vicinal J_{HH} values were determined from 2D J spectra with rather high accuracies

($\pm$0.5 Hz). Even such small uncertainties in $^3J_{HH}$ values correspond to substantial inaccuracies in the calculated dihedral angles (Table 4). These uncertainties combined with the imprecision of the Karplus relationship would lead to a rather fuzzy NMR conformation. Therefore, dihedral angles obtained from molecular mechanics calculations and X-ray crystal structures were used in an extended Karplus equation to calculate expected $^3J_{HH}$ values, which were then compared with the observed coupling constants.[85] This combination of NMR and molecular modeling, an active area of computer software development, gives more reliable and refined conformations than would be possible by either method alone.

In the case of ring D, a set of equations relating the pseudorotation parameters (phase angle P and puckering amplitude ϕ_m) was established and solved by least squares fitting.[85] The resulting values of P (9.6°) and ϕ_m (40.4°) indicated a distorted 13β envelope for both the 5α and 5β isomers, in agreement with X-ray and molecular mechanics structures. However, there was a large ($\pm$1.1 Hz rms) discrepancy between observed and calculated coupling constants. The large deviations were attributed to inadequacies of a generalized Karplus equation[129] but might also be connected with deficiencies in molecular mechanics calculations for five-membered rings. This determination of the D ring conformation from pseudorotation parameters appears not to be strongly affected by an outlying $^3J_{HH}$ value.

Similar conformational analyses were reported for an extensive series of C_{19} steroids.[51] Because changes in the conformation of ring D produce only moderate changes in coupling constants, simple comparisons of observed and calculated coupling constants have little power to discriminate between conformers. Instead of the above description of ring D by pseudorotation parameters, the ring-D geometry was quantified by the torsion angle C14–C15–C16–C17. A plot of this torsion angle vs. calculated coupling constants illustrated the limited ability of vicinal J_{HH} values to differentiate similar ring D conformers.[51]

A simpler but less rigorous method of conformational analysis is based on HMBC cross-peak intensities augmented by easily accessible $^3J_{HH}$ values and results of NOESY spectra.[52] The intensity of an HMBC cross peak is a function of its three-bond C-H coupling constant, which is in turn estimated from the dihedral angle by a Karplus relationship. Geminal protons can often be distinguished as equatorial or axial based on their *syn* or *anti* orientation relative to a particular carbon atom. Anti orientation (~180° dihedral angle) gives much stronger cross peaks than syn orientations (~60° dihedral angle). This method, which appears not to suffer from the irregular relationship[48] found in long-range HETCOR between cross peak intensities and $^3J_{CH}$, has even been applied to analysis of ring D.[52]

Conformational analysis based on cross-peak intensities has similarly been described for FUCOUP, a long-range HETCOR experiment.[82] Another use of cross peak intensities in conformational analysis of ring D involves the

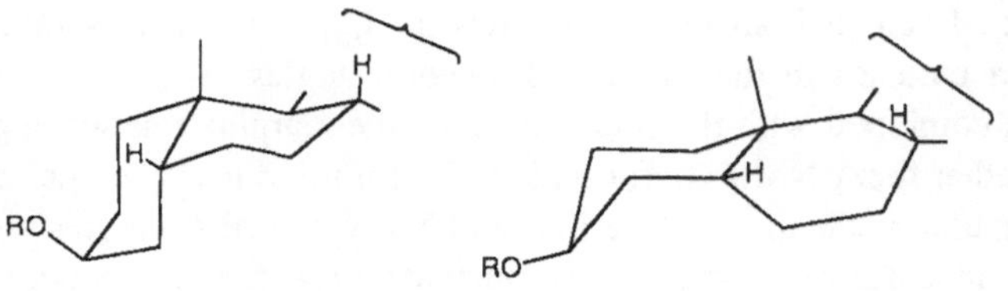

Conformation A Conformation B

FIGURE 19. Conformers A and B of 5β-sterols.

COSY spectrum of the two equilibrating forms **27** and **28** of aldosterone. The absence of COSY cross peaks for H-16β and H-15β in the hemiacetal tautomer and for H-16β and H-15α in the 20-keto form suggested different D-ring conformations for the two forms.[31]

When two interconverting conformers are present in solution, the less stable conformer is sometimes observed in the X-ray crystal structure. By contrast, NMR analysis often reveals the presence of both conformers and permits rough quantitation of their populations. For example, NMR analysis of the A ring of 17α-acetoxy-6α-methylprogesterone (**29**) showed mainly the normal 1α,2β half-chair conformer and lesser amounts of the inverted 1β,2α half chair observed by X-ray crystallography.[57] Conformational analysis of 19-nor-3-keto-Δ^4 steroids, which have normal and inverted conformers of similar energy, has been carried out by both X-ray crystallography and NMR.[123] In 2β-substituted-3-keto steroids, ring A has been shown by NMR to favor the inverted boat conformer.[88]

NMR has also been used to evaluate conformations in sterols. In a study of 3β,9α-dihydroxy-5α-cholest-8(14)-en-15-one (**30**), a potent inhibitor of sterol synthesis in cultured cells, it was recognized that the 1,3-diaxial interactions between the 9α-hydroxyl and H-12α, H-7α, H-5α, and H-1α would be partially relieved if ring B were in a boat form. However, the boat form was excluded by the large (4.7 ppm) upfield shift of C-7 attributed to the γ-gauche effect, which is expected only for the chair form.[102] This conclusion is in accord with our subsequent molecular mechanics calculations indicating that the boat form is >7 kcal higher in enthalpy than the chair.

Ring B may exist partially as a twist form in some substituted 5β-steroids. In the usual conformer (A), rings A and B are both chairs, and C-19 is axial in ring B and equatorial in ring A. In the other conformer (B), ring A is a chair with an axial C-19, and ring B is a twist or boat. Conformers A and B are shown in Figure 19. Although conformer B has apparently not been reported for saturated sterols, the two conformers are of comparable energy in 5β-Δ^8-sterols. The absence of H-8β and H-9α permits ring B to assume a half-chair conformation. Conformer B was first detected by the broadening of ^{13}C signals of 5β-cholesta-8,14-dien-3β-ol (**31**) at 75 MHz and was rig-

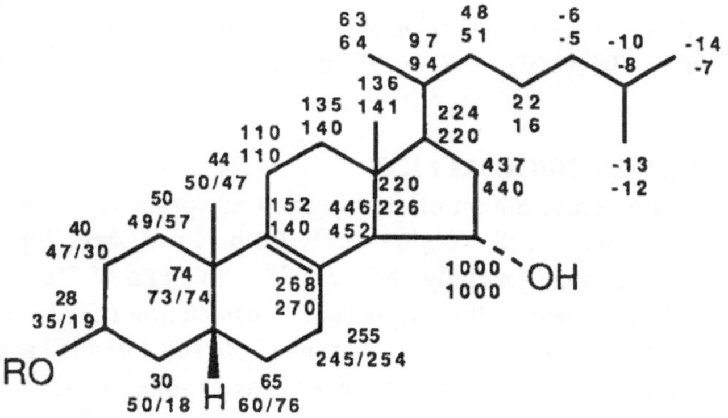

FIGURE 20. Lanthanide-induced shifts for the 3-*tert*-butyldimethylsilyl derivative of **32**: observed (upper numbers) and calculated (lower numbers) LIS. For carbons in rings A and B, two calculated LIS values are given: LIS calculated for conformer A (lower left numbers), and LIS calculated for conformer B (lower right numbers). (Data from Wilson, W. K., Pinkerton, F. D., Kirkpatrick, N. D., and Schroepfer, G. J., Jr., *J. Lipid Res.*, 30, 919, 1989.)

orously defined by an X-ray structure of a derivative of **31**.[133] NMR analysis suggested roughly equal amounts of conformers at ambient temperature. The absence of any broadening in the ^{13}C signals of **31** at 22.5 MHz illustrates the value of high magnetic fields in conformational analysis.

A related sterol, 5β-cholest-8-ene-3β,15α-diol (**32**), was also shown by the same broadening of analogous ^{13}C signals to consist of an interconverting mixture of conformers A and B.[65] Comparison of observed and calculated LIS for the 3-*tert*-butyldimethylsilyl ether of **32** showed that the observed values in ring A were generally intermediate between those calculated for conformers A and B (Figure 20). Diol **32** showed unexpectedly high potency in blocking sterol synthesis in cultured cells, a result that may be related to its ability to exist in two quite different conformations.[65] Similar conformational heterogeneity has been observed in ring C of certain 14β-sterols.[134]

The minor species in 5β-steroids may not always be conformer B. The ^{1}H spectrum of 3-dehydroecdysone[20,149] (**33**) in D_2O showed the major component to be in equilibrium with a minor component that was not observed in $CDCl_3$ solution. The ^{13}C spectrum in D_2O indicated the absence of a 3-keto signal in the minor component, which was identified as the ketone hydrate [20,149]

The cardenolide ouabain (**34**), an inhibitor of Na^+,K^+-adenosine triphosphatase, shows several broadened ^{13}C signals in DMSO-$CDCl_3$ solution.[75] Fast chemical exchange at 60° C permitted determination of vicinal J_{HH} values from HMQC and COSY spectra. Conformational analysis based on molecular modeling and the ^{1}H coupling constants indicated that the major conformer was the all-chair form. Although the minor conformer was frozen out at −20° C, signal broadening prevented measurement of ^{1}H coupling constants.[75] The presence of two *cis*-ring junctions and a 1β,3β,5β-trihydroxy system in ring A makes ouabain an intriguing subject for conformational analysis.

B. SIDE-CHAIN CONFORMATION

The side-chain conformation of sterols has been proposed to be critical to their structural role in cell membranes.[150] Both X-ray crystallography[151] and molecular mechanics generally show a 17β side chain of 20*R* configuration to be fully extended in the "right-hand" orientation (C-22 *anti* to C-13). Sterols with 12β-substituents,[152] certain C-16 substituents, and other combinations of functional groups[153] appear to favor other side-chain orientations according to our molecular mechanics calculations. Sterols having different side-chain orientations as a consequence of a $\Delta^{17,20}$ double bond or 20*S* configuration do not support the growth of yeast (*Saccharomyces cerevisiae*), a situation that has been attributed to their lack of uptake by the cells.[150] Several NMR techniques, including measurements of $^1H-^1H$ coupling constants, LIS, ASIS, and ^{13}C chemical shift comparisons, can be used to distinguish 20*R* and 20*S* epimers[91,154] and to analyze the side chain rotamers about the C17–C20 bond.

Although $^1H-^1H$ coupling constants in the side chain are difficult to interpret because they are averaged over several significantly populated conformers, conclusions are nevertheless sometimes possible. A large difference has been observed in the activities of the 22*R* and 22*S* epimers of 22-hydroxy vitamin D derivatives.[155] Conformational analysis based on a combination of molecular mechanics and $^1H-^1H$ coupling constants was carried out on the side chains of model compounds. The 23,23,24,24-tetradeuterio derivatives **35a** and **35b** were prepared to simplify NMR analysis of the H-22 resonance. Molecular mechanics indicated that the best conformer of **35b** (22*S*) had an extended (zigzag) side chain, whereas the best conformer of **35a** (22*R*) had a bent side chain (70° dihedral angle for C17–C20–C22–C23). The values of $J_{H22\text{-}H20}$ for **35a** and **35b** (3.1 and ~0 Hz) were consistent with the H22–C22–C20–H20 dihedral angles predicted by molecular mechanics (~50° and ~90°). The low activity of the 22*R* epimer was ascribed to the unfavorability of its conformer with an extended (zigzag) side chain. This conformational feature was suggested to be necessary for biological activity of vitamin D analogs.[155]

VI. CONCLUDING REMARKS

The development of mid- and high-field spectrometers and a variety of 1D and 2D multipulse experiments have provided a plethora of new tools for structure determination. Sterol structures can now usually be determined solely by NMR, although supporting information greatly simplifies the task. Strategies for structure elucidation depend on the quantity of sample and features of available NMR spectrometers. Given ≥20 mg of sample, a thorough structure determination can be done in simple cases from a routine set of mainly ^{13}C-based 1D and 2D experiments that require only ~1 h of mid-field spectrometer time. A high-field spectrometer with a probe optimized for ^{1}H greatly enhances the value of ^{1}H, COSY, HMQC, and HMBC spectra and permits full structural and conformational analysis for more difficult structural problems. By revealing carbon connectivities, these methods permit structure determination through the construction and assembly of substructures. A signal assignment strategy using only ^{1}H-based 1D and 2D experiments requiring ≤1 mg of material has been successfully implemented for numerous moderately and highly functionalized steroids.[32] This strategy can be extended and adapted for structure determination of both sterols[58] and steroids.

Chemical shift comparisons and substituent increments are powerful adjuncts to structure elucidation by substructure strategies. These methods, requiring a modest database of NMR information, can often compensate for unavailable 2D correlation experiments. Comparisons are especially important for assigning the configuration of side-chain carbons, a task for which even the highest fields and most powerful 2D experiments are generally of little value. An important test for the correctness of a proposed structure is the consistency of its ^{1}H and ^{13}C assignments with chemical shift estimates obtained from substituent effects, comparisons of closely related compounds, and other predictive methods.

Until recently, the role of NMR in conformational analysis was mainly as a probe for irregularities to be investigated more thoroughly by other analytical techniques. This situation has changed now that vicinal coupling constants can be measured for essentially all ^{1}H resonances. These vicinal proton coupling constants can be used to judge conformations suggested by molecular modeling. The observed $^{3}J_{HH}$ data can be augmented with information from NOEDS, NOESY, LIS, and HMBC experiments to enhance discrimination between low-energy conformers from molecular mechanics.

Solution conformations of numerous C_{19}–C_{21} steroids have been evaluated by this methodology, which is largely applicable to sterols as well. However, conclusions relating the NMR conformation to biological activities must be evaluated in light of several possible complications. Conformation[75,156] and even structure[20] may vary from solvent to solvent. Conformational heterogeneity involving a low energy barrier for interconversion may be misinterpreted as a single intermediate form. In particular, the flexibility of ring D usually limits the power of NMR to discriminate between its possible conformers. Finally, although minor conformers can sometimes be frozen out in low temperature spectra recorded at high field, the detection and quantitation of minor conformers (e.g., <10%) by NMR is otherwise likely to be unreliable. Minor conformers may play an important role in biological activities of sterols. Despite these caveats, the combination of NMR and molecular modeling is a powerful method for evaluating solution conformation.

The rapid development of improved NMR methodologies noted in this review greatly facilitates the determination of structures of sterols. With more general application of these new techniques, acceleration of progress in the identification of isolated sterols, studies of their biosynthesis and metabolism, and studies of their conformations can be anticipated. Such advances should provide much more rigorous evidence for assignments of structures and information critical to considerations of structure-function relationships.

ACKNOWLEDGMENTS

The research in our laboratory has been supported by grants from the National Institutes of Health (HL-15376 and HL-22532), the Robert A. Welch Foundation (C-583), the Texas Advanced Technology Program (3037), and the American Cyanamid Company.

REFERENCES

1. **Bloch, K. E.,** Sterol structure and membrane function, *CRC Crit. Rev. Biochem.*, 14, 47, 1983.
2. **Kandutsch, A. A., Chen, H. W., and Heiniger, H.-J.,** Biological activity of some oxygenated sterols, *Science*, 201, 498, 1978.
3. **Schroepfer, G. J., Jr.,** Sterol biosynthesis, *Annu. Rev. Biochem.*, 50, 585, 1981.
4. **Kandutsch, A. A. and Thompson, E. B.,** Cytosolic proteins that bind oxygenated sterols, *J. Biol. Chem.*, 255, 10813, 1980.
5. **Kandutsch, A. A. and Shown, E. P.,** Assay of oxysterol-binding protein in a mouse fibroblast, cell-free system, *J. Biol. Chem.*, 256, 13068, 1981.
6. **Taylor, F. R., Saucier, S. E., Shown, E. P., Parish, E. J., and Kandutsch, A. A.,** Correlation between oxysterol binding to a cytosolic binding protein and potency in the repression of hydroxymethylglutaryl coenzyme A reductase, *J. Biol. Chem.*, 259, 12382, 1984.
7. **Schroepfer, G. J., Jr., Parish, E. J., Chen, H. W., and Kandutsch, A. A.,** Inhibition of sterol biosynthesis in L cells and mouse liver cells by 15-oxygenated sterols, *J. Biol. Chem.*, 252, 8975, 1977.
8. **Miller, L. R., Pinkerton, F. D., and Schroepfer, G. J., Jr.,** 5α-Cholest-8(14)-en-3β-ol-15-one, a potent inhibitor of sterol synthesis, reduces the levels of activity of enzymes involved in the synthesis and reduction of 3-hydroxy-3-methylglutaryl coenzyme A in CHO-K1 cells, *Biochem. Int.*, 1, 223, 1980.
9. **Pinkerton, F. D., Izumi, A., Anderson, C. M., Miller, L. R., Kisic, A., and Schroepfer, G. J., Jr.,** 14α-Ethyl-5α-cholest-7-ene-3β,15α-diol, a potent inhibitor of sterol biosynthesis, has two sites of action in cultured mammalian cells, *J. Biol. Chem.*, 257, 1929, 1982.
10. **Pajewski, T. N., Pinkerton, F. D., Miller, L. R., and Schroepfer, G. J., Jr.,** Inhibitors of sterol synthesis. Studies of the metabolism of 5α-cholest-8(14)-en-3β-ol-15-one in Chinese hamster ovary cells and its effects on activities of early enzymes in cholesterol biosynthesis, *Chem. Phys. Lipids*, 48, 153, 1988.

11. **Miller, L. R., Needleman, D. H., Brabson, J. S., Wang, K.-S., and Schroepfer, G. J., Jr.,** 5α-Cholest-8(14)-en-3β-ol-15-one. A competitive substrate for acyl coenzyme A.cholesterol acyl transferase, *Biochem. Biophys. Res. Commun.*, 148, 934, 1987.
12. **Needleman, D. H., Strong, K., Stemke, K. A., Brabson, J. S., Kisic, A., and Schroepfer, G. J., Jr.,** Inhibitors of sterol synthesis. Effect of dietary 5α-cholest-8(14)-en-3β-ol-15-one on ACAT activity of jejunal microsomes of the rat, *Biochem. Biophys. Res. Commun.*, 148, 920, 1987.
13. **Schroepfer, G. J., Jr., Christophe, A., Needleman, D. H., Kisic, A., and Sherrill, B. C.,** Inhibitors of sterol synthesis. Dietary administration of 5α-cholest-8(14)-en-3β-ol-15-one inhibits the intestinal absorption of cholesterol in lymph-cannulated rats, *Biochem. Biophys. Res. Commun.*, 146, 1003, 1987.
14. **Schroepfer, G. J., Jr., Monger, D., Taylor, A. S., Chamberlain, J. S., Parish, E. J., Kisic, A., and Kandutsch, A. A.,** Inhibitors of sterol synthesis. Hypocholesterolemic action of dietary 5α-cholest-8(14)-en-3β-ol-15-one in rats and mice, *Biochem. Biophys. Res. Commun.*, 78, 1227, 1977.
15. **Schroepfer, G. J., Jr., Parish, E. J., Kisic, A., Jackson, E. M., Farley, C. M., and Mott, G. E.,** 5α-Cholest-8(14)-en-3β-15-one, a potent inhibitor of sterol biosynthesis, lowers serum cholesterol and alters the distribution of cholesterol in lipoproteins in baboons, *Proc. Natl. Acad. Sci. USA*, 79, 3042, 1982.
16. **Schroepfer, G. J., Jr., Sherrill, B. C., Wang, K.-S., Wilson, W. K., Kisic, A., and Clarkson, T. B.,** 5α-Cholest-8(14)-en-3β-ol-15-one lowers serum cholesterol and induces profound changes in the levels of lipoprotein cholesterol and apoproteins in monkeys fed a diet of moderate cholesterol content, *Proc. Natl. Acad. Sci. USA*, 81, 6861, 1984.
17. **Kudo, K., Emmons, G. T., Casserley, E. W., Via, D. P., Smith, L. C., St. Pyrek, J., and Schroepfer, G. J., Jr.,** Inhibitors of sterol synthesis. Chromatography of acetate derivatives of oxygenated sterols, *J. Lipid Res.*, 30, 1097, 1989.
18. **Smith, L. L.,** *Cholesterol Autoxidation*, Plenum Press, New York, 1977
19. **Smith, L. L.,** Cholesterol autoxidation 1981-1986, *Chem. Phys. Lipids*, 44, 87, 1987.
20. **Girault, J.P., Blais, C., Beydon, P., Rolando, C., and Lafont, R.,** Synthesis and nuclear magnetic resonance study of 3-dehydroecdysteroids, *Arch. Insect Biochem. Physiol.*, 10, 199, 1989.
21. **Takatsuto, S., Yazawa, N., Ikekawa, N., Takematsu, T., Takeuchi, Y., and Koguchi, M.,** Structure-activity relationship of brassinosteroids, *Phytochemistry*, 22, 2437, 1983.
22. **McMorris, T. C., Seshadri, R., Weihe, G. R., Arsenault, G. P., and Barksdale, A. W.,** Structures of oogoniol-1, -2, and -3, steroidal sex hormones of the water mold, *Achlya. J. Am. Chem. Soc.*, 97, 2544, 1975.
23. **Tachibana, K., Sakaitani, M., and Nakanishi, K.,** Pavoninins, shark-repelling and ichthyotoxic steroid N-acetylglucosaminides from the defense secretion of the sole *Pardachirus pavoninus* (Soleidae), *Tetrahedron*, 41, 1027, 1985.
24. **Elliger, C. A., Waiss, A. C., Jr., Benson, M., and Wong, R. Y.,** Ergostanoids from *Petunia parodii*, *Phytochemistry*, 29, 2853, 1990.
25. **Schroepfer, G. J., Jr.,** Sterol biosynthesis, *Annu Rev. Biochem.*, 51, 555, 1982.
26. **Schroepfer, G. J., Jr., Pascal, R. A., Jr., Shaw, R., and Kandutsch, A. A.,** Inhibition of sterol biosynthesis by 14α-hydroxymethyl sterols, *Biochem. Biophys. Res. Commun.*, 83, 1024, 1978.
27. **Schroepfer, G. J., Jr., Parish, E. J., Pascal, R. A., Jr., and Kandutsch, A. A.,** Inhibition of sterol biosynthesis in animal cells by 14α-hydroxymethyl sterols, *J. Lipid Res.*, 21, 571, 1980.
28. **Gibbons, G. F., Pullinger, C. R., Chen, H. W., Cavenee, W. K., and Kandutsch, A. A.,** Regulation of cholesterol biosynthesis in cultured cells by probable natural precursor sterols, *J. Biol. Chem.*, 255, 395, 1980.

29. **Schroepfer, G. J., Jr., Parish, E. J., Tsuda, M., and Kandutsch, A. A.,** Inhibitors of sterol synthesis. Chemical syntheses, properties and effects of 4,4-dimethyl-15-oxygenated sterols on sterol synthesis and on 3-hydroxy-3-methylglutaryl coenzyme A reductase activity in cultured mammalian cells, *Chem. Phys. Lipids,* 47, 187, 1988.
30. **Koszewski, N. J., Reinhardt, T. A., Beitz, D. C., Napoli, J. L., Baggiolini, E. G., Uskokovic, M. R., and Horst, R. L.,** Use of Fourier transform ^{1}H NMR in the identification of vitamin D_2 metabolites, *Anal. Biochem.,* 162, 446, 1987.
31. **Carter, B. G., Kirk, D. N., and Burke, P. J.,** 18-substituted steroids XIV. The high-field ^{1}H and ^{13}C nuclear magnetic resonance spectra of aldosterone; full assignments for the main equilibrating forms in solution, *J. Chem. Soc. Perkin Trans. 2,* 1247, 1987.
32. **Kirk, D. N., Toms, H. C., Douglas, C., White, K. A., Smith, K. E., Latif, S., and Hubbard, R. W. P.,** A survey of the high-field ^{1}H NMR spectra of the steroid hormones, their hydroxylated derivatives, and related compounds, *J. Chem. Soc. Perkin Trans. 2,* 1567, 1990.
33 **Smith, I. C. P. and Dufourc, E. J.,** Physicochemical determinations of sterols by deuterium and carbon-13 nuclear magnetic resonance, in *Analysis of Sterols and other Biologically Significant Steroids,* Nes, W. D. and Parish, E. J., Eds., Academic Press, New York, 1989, chap. 15.
34. **Barnes, S. and Kirk, D. N.,** Nuclear magnetic resonance spectroscopy of bile acids, in *The Bile Acids: Chemistry, Physiology, and Metabolism,* Setchell, K. D. R., Kritchevsky, D., and Nair, P. P., Eds., Plenum Press, New York, 1988, chap. 3.
35. **Bernstein, M. A.,** Strategies for applying combinations of two-dimensional NMR experiments, in *Two-dimensional NMR Spectroscopy,* Croasmun, W. R. and Carlson, R. M. K., Eds., VCH Publishers, New York, 1987, chap. 2.
36. **Morris, G. A.,** Modern NMR techniques for structure elucidation, *Magn. Reson. Chem.,* 24, 371, 1986.
37. **Sadler, I. H.,** The use of NMR spectroscopy in the structure determination of natural products: one-dimensional methods, *Nat. Prod. Reports,* 5, 101, 1988.
38. **Derome, A. E.,** The use of NMR spectroscopy in the structure determination of natural products: two-dimensional methods, *Nat. Prod. Reports,* 6, 111, 1989.
39 **Croasmun, W. R. and Carlson, R. M. K.,** Steroid structure analysis by two-dimensional NMR, in *Two Dimensional NMR Spectroscopy,* Croasmun, W. R. and Carlson, R. M. K., Eds., VCH Publishers, New York, 1987, chap. 7.
40. **Wehrli, F. W., Marchand, A. P., and Wehrli, S.,** *Interpretation of Carbon-13 NMR Spectra,* 2nd ed., John Wiley & Sons, New York, 1988, 101–114, 184–190, 190–199.
41. **Martin, G. E. and Zektzer, A. S.,** *Two-Dimensional NMR Methods for Establishing Molecular Connectivity,* VCH Publishers, New York, 1988, chap. 2–5.
42. **Kirk, D. N. and Toms, H. C.,** ω_1-Decoupled ^{1}H homonuclear shift-correlated nuclear magnetic resonance spectroscopy (COSYDEC) applied to steroids, *Steroids,* 56, 195, 1991.
43. **Bruch, M. D.,** Two-dimensional NMR spectroscopy of liquids, in *NMR Spectroscopy Techniques,* Dybowski, C. and Lichter, R. L., Eds., Marcel Dekker, New York, 1987, chap. 3.
44. **Neuhaus, D. and Williamson, M. P.,** *The Nuclear Overhauser Effect in Structural and Conformational Analysis,* VCH Publishers, New York, 1989.
45. **Andersen, N. H., Eaton, H. L., and Lai, X.,** Quantitative small molecule NOESY: a practical guide for derivation of cross-relaxation rates and internuclear distances, *Magn. Reson. Chem.,* 27, 515, 1989.
46. **Reynolds, W. F., McLean, S., Perpick-Dumont, M., and Enríquez, R. G.,** ^{13}C–^{1}H shift correlation with full ^{1}H–^{1}H decoupling. II. Further significant improvements in resolution and sensitivity, *Magn. Reson. Chem.,* 26, 1068, 1988.
47. **Bax, A. and Subramanian, S.,** Sensitivity-enhanced two-dimensional heteronuclear shift correlation NMR spectroscopy, *J. Magn. Reson.,* 67, 565, 1986.

48. **Martin, G. E. and Zektzer, A. S.,** Long-range two-dimensional heteronuclear chemical shift correlation, *Magn Reson. Chem.*, 26, 631, 1988.

49. **Bax, A., Aszalos, A., Dinya, Z., and Sudo, K.,** Structure eludication of the antibiotic desertomycin through the use of new two-dimensional NMR techniques, *J. Am. Chem. Soc.*, 108, 8056, 1986

50. **Sanders, J. K. M., and Hunter, B. K.,** *Modern NMR Spectroscopy*, Oxford University Press, New York, 1987, 111, 114–124.

51. **Schneider, H.-J., Buchheit, U., Becker, N., Schmidt, G., and Siehl, U.,** ^{1}H NMR analyses, shielding mechanisms, coupling constants, and conformations in steroids bearing halogen, hydroxy, oxo groups, and double bonds, *J. Am. Chem. Soc.*, 107, 7027, 1985.

52. **Zeng, B., Pollack, R. M., and Summers, M. F.,** Conformational analysis via heteronuclear correlation NMR spectroscopy, *J. Org. Chem.*, 55, 2534, 1990.

53. **Hall, L. D. and Sanders, J. K. M.,** Complete analysis of ^{1}H NMR spectra of complex natural products using a combination of one- and two-dimensional techniques. 1-Dehydrotestosterone, *J. Am. Chem. Soc.*, 102, 5703, 1980.

54. **Emmons, G. T., Wilson, W. K., and Schroepfer, G. J., Jr.,** 24,25-Epoxysterols: differentiation of 24R and 24S epimers by ^{13}C nuclear magnetic resonance spectroscopy, *J. Lipid. Res.*, 30, 133, 1989.

55. **St. Pyrek, J., Vermilion, J. L., Stephens, T. W., Wilson, W. K., and Schroepfer, G. J., Jr.,** Inhibitors of sterol synthesis. Characterization of side chain oxygenated derivatives formed upon incubation of 3β-hydroxy-5α-cholest-8(14)-en-15-one with rat liver mitochondria, *J. Biol. Chem.*, 264, 4536, 1989.

56. **Sanders, J. K. M. and Mersh, J. D.,** Nuclear magnetic double resonance; the use of difference spectroscopy, *Prog. Nucl. Magn. Reson. Spectrosc.*, 15, 353, 1982.

57. **Marat, K., Templeton, J. F., and Kumar, V. P. S.,** An NMR study of A-ring conformation in some 4-en-3-one steroids, *Magn. Reson. Chem.*, 25, 25, 1987.

58. **Koehn, F. E, Gunasekera, M., and Cross, S. S.,** New antiviral sterol disulfate ortho esters from the marine sponge *Petrosia weinbergi*, *J. Org. Chem.*, 56, 1322, 1991.

59. **Demarco, P. V., Farkas, E., Doddrell, D., Mylari, B. L., and Wenkert, E.,** Pyridine-induced solvent shifts in the nuclear magnetic resonance spectra of hydroxylic compounds, *J. Am. Chem. Soc.*, 90, 5480, 1968.

60. **Iida, T. and Kikuchi, M.,** Lanthanide- and aromatic solvent-induced shift effects on proton resonances in C-4-methylated steroids and tetracyclic triterpenoids, *Chem. Phys. Lipids*, 20, 157, 1977.

61. **Fujimoto, Y., Yamada, T., and Ikekawa, N.,** Pyridine-induced deshielding of 4-methylene protons for the determination of C-6 stereochemistry of sterols having a 5α,6-diol moiety. Revision of the C-6 stereochemistry of marine sterol isolated from a sponge, *Dysidea* sp., *Chem. Pharm. Bull.*, 33, 3129, 1985.

62. **Tsuda, M., Parish, E. J., and Schroepfer, G. J., Jr.,** Carbon-13 nuclear magnetic resonance studies of allylic hydroxysterols. Assignment of structure to 5α-cholest-8(14)-ene-3β,7α,15α-triol, an inhibitor of sterol synthesis, *J. Org. Chem*, 44, 1282, 1979.

63. **Wenzel, T.,** *NMR Shift Reagents*, CRC Press, Boca Raton, FL, 1987.

64. **Schneider, H.-J., Buchheit, U., Agrawal, P. K.,** Topological, stereochemical and NMR line assignments in alcohols and ketones by normalized NMR shifts induced by ytterbium, *Tetrahedron*, 40, 1017, 1984.

65. **Wilson, W. K., Pinkerton, F. D., Kirkpatrick, N. D., and Schroepfer, G. J., Jr.,** Inhibitors of sterol synthesis. Chemical synthesis of 5β-cholest-8-ene-3β,15α-diol and its effects on 3-hydroxy-3-methylglutaryl coenzyme A reductase activity in CHO-K1 cells, *J. Lipid Res.*, 30, 919, 1989.

66. **Fujiwara, H., Da, Y.-Z., Zheng, D., Sasaki, Y., Takai, Y., and Sawada, M.,** NMR relaxation studies on testosterone in solution. Computer-assisted analysis of ^{13}C T_1 data measured at different temperatures, *J. Chem. Soc. Perkin Trans. 2*, 97, 1990.

67. **Barrett, M. W., Farrant, R. D., Kirk, D. N., Mersh, J. D., Sanders, J. K. M., and Duax, W. L.**, The solution conformation of 17α-acetoxy-6α-methylprogesterone ('medroxyprogesterone acetate'): use of circular dichroism, nuclear Overhauser difference and two-dimensional J spectroscopy, *J. Chem. Soc. Perkin Trans. 2*, 105, 1982.
68. **Mersh, J. D. and Sanders, J. K. M.**, Indirect negative nuclear Overhauser effects: observation and applications, *Org. Magn Reson.*, 18, 122, 1982.
69. **Zhang, S., Yao, X., Chen, Y., Cui, C., Tezuka, Y., and Kikuchi, T.**, Ginsenoside L_A, a novel saponin from the leaves of *Panax Genseng, Chem. Pharm. Bull.*, 37, 1966, 1989.
70. **Wang, H. K., He, K., Ji, L., Tezuka, Y., Kikuchi, T., and Kitagawa, I.**, Asernestioside C, a new minor saponin from the roots of *Astragalus ernestii* COMB; first example of negative nuclear Overhauser effect in the saponins, *Chem. Pharm. Bull.*, 37, 2041, 1989.
71. **Robsinson, V. J., Bain, A. D., and Rodger, C. A.**, A complete proton and carbon-13 NMR analysis of fluocinonide, a fluorinated corticosteroid, *Steroids*, 48, 267, 1986
72. **Holland, H. L., Xu, W., and Hughes, D. W.**, Stereochemistry of reduction by the 5α-reductase enzyme of *Penicillium decumbens* and the ^1H NMR assignment of 5α-dihydrotestosterone, *J. Chem. Soc. Chem. Commun.*, 1760, 1989.
73. **Hull, W. E.**, Experimental aspects of two-dimensional NMR, in *Two-Dimensional NMR Spectroscopy*, Croasmun, W. R. and Carlson, R. M. K., Eds., VCH Publishers, New York, 1987, chap. 2.
74. **Snyder, J. K., Zhang, X., Joshi, B. S., and Pelletier, S. W.**, Complete spectral assignments of tatsidine by two-dimensional NMR techniques, *Magn. Reson. Chem.*, 27, 1057, 1989.
75. **McIntyre, D. D., Germann, M. W., and Vogel, H. J.**, Conformational analysis and complete assignment of the proton and carbon NMR spectra of ouabain and ouabagenin, *Can. J. Chem.*, 68, 1263, 1990.
76. **Schmieder, P., Zimmer, S., and Kessler, H.**, Increased resolution in proton detected heteronuclear NMR experiments by folding in the hetero-dimension, *Magn. Reson. Chem.*, 29, 375, 1991.
77. **Summers, M. F., Marzilli, L. G., and Bax, A.**, Complete ^1H and ^{13}C assignments of coenzyme B_{12} through the use of new two-dimensional NMR experiments, *J. Am. Chem. Soc.*, 108, 4285, 1986.
78. **Bax, A. and Summers, M. F.**, ^1H and ^{13}C assignments from sensitivity-enhanced detection of heteronuclear multiple-bond connectivity by 2D multiple quantum NMR, *J. Am. Chem. Soc.*, 108, 2093, 1986.
79. **Emmons, G. T., Wilson, W. K., and Schroepfer, G. J., Jr.**, ^1H and ^{13}C NMR assignments for lanostan-3β-ol derivatives: revised assignments for lanosterol, *Magn. Reson. Chem* , 27, 1012, 1989.
80. **Reynolds, W. F., McLean, S., Perpick-Dumont, M., and Enríquez, R. G.**, Improved ^{13}C–^1H shift correlation spectra for indirectly bonded carbons and hydrogens: the FLOCK sequence, *Magn. Reson. Chem.*, 27, 162, 1989.
81. **Yasuzawa, T., Yoshida, M., and Sano, H.**, Structure of paxisterol, a sterol from *Penicillium paxilli, J. Chem. Soc. Perkin Trans. 1*, 3145, 1990.
82. **Waterhouse, A. L.**, Conformational analysis via vicinal carbon-hydrogen coupling, *Magn. Reson. Chem.*, 27, 37, 1989.
83. **Schun, Y. and Cordell, G. A.**, Cytotoxic steroids of *Gelsemium sempervirens, J. Nat. Prod.*, 50, 195, 1987.
84. **Hughes, D. W.**, Application of relayed coherence transfer two-dimensional nuclear magnetic resonance spectroscopy to the assignment of ^1H chemical shifts in steroids, *Magn. Reson. Chem.*, 26, 214, 1988

85. **Sedee, A. G. J., Beijersbergen van Henegouwen, G. M. J., Guijt, W., and Haasnoot, C. A. G.,** Conformational analysis of steroids in solution: 17β-hydroxy-19-nor-5α,17α-pregn-20-yn-3-one and its 5β-isomer studied by nuclear magnetic resonance, *J. Org. Chem.*, 50, 4182, 1985.
86. **Seto, H.,** New methods in NMR spectroscopy: HOHAHA, *Kagaku (Kyoto)*, 44, 293, 1989.
87. **Hayamizu, K., Ishii, T., Yanagisawa, M., and Kamo, O.,** Complete assignments of the ^{1}H and ^{13}C NMR spectra of testosterone and 17α-methyltestosterone and the ^{1}H parameters obtained from 600 MHz spectra, *Magn. Reson. Chem.*, 28, 250, 1990.
88. **Marat, K., Templeton, J. F., Gupta, R. K., and Kumar, V. P. S.,** Ring A conformation in steroids. II. NMR study of C-2 monomethyl- and dimethyl-substituted 5α-androstan-3-ones, *Magn. Reson. Chem.*, 25, 730, 1987.
89. **Cui, Y.-X., Hua, S.-M., Cheng, J.-L., Yue, J. M., and Chen, Y.-Z.,** Two-dimensional NMR study of ganwuweizic acid, *Magn. Reson. Chem.*, 27, 998, 1989.
90. **Bhacca, N. S. and Williams, D. H.,** *Applications of NMR Spectroscopy in Organic Chemistry*, Holden-Day, San Francisco, 1967, 79–82 and 135–150.
91. **Cooley, G., Kirk, D. N., Morgan, R. E., and Sae Melo, M. L.,** Pyridine-induced shifts in the ^{1}H nuclear magnetic resonance spectra of 20-hydroxypregnane derivatives, *J. Chem. Soc Perkin Trans. 1*, 1390, 1977.
92. **STN Karlsruhe,** C-13 NMR/IR file, available through STN International, Columbus, OH.
93. **Bridgeman, J. E., Cherry, P. C., Clegg, A. S., Evans, J. M., Jones, E. R. H., Kasal, A., Kumar, V., Meakins, G. D., Morisawa, Y., Richards, E. E., and Woodgate, P. D.,** Microbiological hydroxylation of steroids. I. Proton magnetic resonance spectra of ketones, alcohols, and acetates in the androstane, pregnane, and œstrane series, *J. Chem. Soc. (C)*, 250, 1970.
94. **Blunt, J. W. and Stothers, J. B.,** ^{13}C N.m.r. spectra of steroids — a survey and commentary, *Org. Magn. Reson.*, 9, 439, 1977.
95. **Smith, W. B.,** Carbon-13 NMR spectroscopy of steroids, *Annu. Rep. NMR Spectrosc.*, 8, 199, 1978.
96. **Wehrli, F. W. and Nishida, T.,** The use of carbon-13 nuclear magnetic resonance spectroscopy in natural products chemistry, *Fortschr. Chemie Org. Naturst* , 36, 1, 1979.
97. **Agrawal, P. K., Jain, D. C., Gupta, R. K., and Thakur, R. S.,** Carbon-13 NMR spectroscopy of steroidal sapogenins and steroidal saponins, *Phytochemistry*, 24, 2479, 1985.
98. **Breitmaier, E. and Voelter, W.,** *Carbon-13 NMR Spectroscopy*, VCH Publishers, New York, 1987, 337.
99. **Robien, W., Kopp, B., Schabl, D., and Schwarz, H.,** Carbon-13 NMR spectroscopy of cardenolides and bufadienolides, *Prog. NMR Spectroscopy*, 19, 131, 1987.
100. **Akihisa, T., and Matsumoto, T.,** ^{13}C NMR spectra of sterols and triterpene alcohols, *Yukagaku*, 36, 301, 1987.
101. **Akihisa, T.,** ^{13}C-NMR spectral identification of sterols, in *Analysis of Sterols and other Biologically Significant Steroids*, Nes, W. D. and Parish, E. J., Eds., Academic Press, New York, 1989, chap. 12
102. **Tsuda, M. and Schroepfer, G. J., Jr.,** Inhibitors of sterol biosynthesis. Carbon-13 nuclear magnetic resonance studies of 9α-fluoro-5α-cholest-8(14)-en-3β-ol-15-one and related compounds, *J. Lipid Res.*, 22, 1188, 1981.
103. **Tsuda, M. and Schroepfer, G. J., Jr.,** Carbon-13 nuclear magnetic resonance studies of C_{27} sterol precursors of cholesterol, *J. Org. Chem.*, 44, 1290, 1979.
104. **Wilson, W. K. and Schroepfer, G. J., Jr.,** Acid-catalyzed isomerization of 7-dehydrocholesterol benzoate. A revised mechanism and an improved synthetic procedure, *J. Org. Chem.*, 53, 1713, 1988.

105. **Kim, H.-S., Wilson, W. K., Needleman, D. H., Pinkerton, F. D., Wilson, D. K., Quiocho, F. A., and Schroepfer, G. J., Jr.**, Inhibitors of sterol synthesis. Chemical synthesis, structure, and biological activities of (25R)-3β,26-dihydroxy-5α-cholest-8(14)-en-15-one, a metabolite of 3β-hydroxy-5α-cholest-8(14)-en-15-one, *J. Lipid Res.*, 30, 247, 1989.
106. **Iida, T., Tamura, T., Matsumoto, T.**, Carbon-13 NMR spectra of some C-6 methylated steroids, *Magn. Reson. Chem.*, 25, 558, 1987.
107. **Martynow, J. and Paryzek, Z.**, Epoxidation-induced shifts in the carbon-13 NMR spectra of steroid: lanostane derivatives, *Magn. Reson Chem.*, 27, 258, 1989.
108. **Lin, L.-J., Shiao, M.-S., and Lee, K. R.**, ^{13}C-NMR correlation of stereochemistry in lanostanoid triterpenes, *J. Nat. Prod.*, 52, 595, 1989.
109. **Horibe, I., Nakai, H., Sato, T., Seo, S., Takeda, K., and Takatsuto, S.**, Stereoselective synthesis of the C-24 and C-25 stereoisomeric pairs of 24-ethyl-[26-^{2}H)]sterols and their Δ^{22} derivatives: reassignment of ^{13}C N.m.r. signals of the *pro*-R and the *pro*-S methyl groups at C-25 of 24-ethylsterols, *J. Chem. Soc. Perkin Trans. 1*, 1957, 1989.
110. **Duddeck, H.**, Substituent effects on ^{13}C chemical shifts in aliphatic molecular systems. Dependence on constitution and stereochemistry, *Topics Stereochem.*, 16, 219, 1986.
111. **Römer, J., Scheller, D., and Grossmann, G.**, Carbon-13 nuclear magnetic resonance study of the bromo derivatives of 5α-cholestan-3-one and cholest-4-en-3-one, *Magn. Reson. Chem.*, 25, 135, 1987.
112. **Schnedier, H.-J. and Jung, M.**, Comparison of substituent-induced proton NMR shifts in steroids and decalins, *Magn. Reson. Chem.*, 26, 679, 1988.
113. **Dalling, D. K. and Grant, D. M.**, Carbon-13 magnetic resonance. The perhydroanthracenes and perhydrophenanthrenes, *J. Am. Chem. Soc.*, 96, 1827, 1974.
114. **Beierbeck, H., Saunders, J. K., and ApSimon, J. W.**, The semiempirical derivation of ^{13}C nuclear magnetic resonance chemical shifts. Hydrocarbons, alcohols, amines, ketones, and olefins, *Can. J. Chem.*, 55, 2813, 1977.
115. **Eggert, H. and Djerassi, C.**, Carbon-13 nuclear magnetic resonance spectra of monounsaturated steroids. Evaluation of rules for predicting their chemical shifts, *J. Org. Chem.*, 46, 5399, 1981.
116 **Schneider, H.-J., Buchheit, U., Hoppen, V., and Schmidt, G.**, ^{13}C- and ^{1}H-NMR shielding effects in aliphatic *gauche/trans* fragments, *Chem. Ber.*, 122, 321, 1989.
117. **Sutton, G. P., Anker, L. S., and Jurs, P. C.**, Evaluation of automated methods for the selection of models for simulation of carbon-13 nuclear magnetic resonance spectra of keto steroids, *Anal. Chem.*, 63, 443, 1991.
118 **Bastard, J., Bernassau, J. M., Bertranne, M., and Maia, E. R.**, Computer predictions of ^{13}C spectra – alcohols and ketones, *Magn. Reson. Chem.*, 26, 992, 1988.
119. **Tsuda, M. and Schroepfer, G. J., Jr.**, Nuclear magnetic resonance spectroscopy of sterols: ^{13}C olefinic carbon shieldings in sterols and related cyclic compounds, *Chem. Phys. Lipids*, 25, 49, 1979.
120. **Whitesell, J. K. and Minton, M. A.**, Anti, vicinal hydrogen-hydrogen interactions. A fundamental shift effect in ^{13}C NMR spectroscopy, *J. Am. Chem. Soc.*, 109, 225, 1987.
121. **Kelecom, A.**, Use of calculated carbon chemical shift tables of saturated C-7 to C-11 sterol side chains for structure elucidation, *Bull Soc. Chim. Belg.*, 89, 343, 1980.
122. **Zielinski, J., Li, H. T., and Djerassi, C.**, Stereoselective synthesis of (22R)- and (22S)-22-methylcholesterol, *J. Org. Chem.*, 47, 620, 1982.
123. **Wong, T. C., Guo, W., Bohl, M., Hübner, M., Luck, G., Steiger, T., and Reck, G.**, Solution conformation of three steroid 19-nor-4-en-3-ones determined from two-dimensional nuclear magnetic resonance spectroscopy, coupling constant calculations, and circular dichroism, *J. Chem. Soc. Perkin Trans. 2*, 765, 1988.
124. **Parish, E. J.**, Analysis of oxysterols, in *Analysis of Sterols and other Biologically Significant Steroids*, Nes, W. D. and Parish, E. J., Eds., Academic Press, New York, 1989, chap. 8.

125. **Christie, B. D. and Munk, M. E.,** The role of two-dimensional nuclear magnetic resonance spectroscopy in computer-enhanced structure elucidation, *J. Am. Chem. Soc.*, 113, 3750, 1991.
126. **Blunden, G., Patel, A. V., and Crabb, T. A.,** Microbiological transformations of hecogenin and diosgenin by *Cunninghamella elegans*, *Phytochemistry*, 29, 1771, 1990.
127. **Akihisa, T., Tamura, T., Matsumoto, T., Kokke, W. C. M. C., Ghosh, P., and Thakur, S.,** (22*Z*,24*S*)-stigmasta-5,22,25-trien-3β-ol and other novel sterols from *Clerodendrum scandens:* first report of the isolation of a *cis*-Δ^{22}-unsaturated sterol from a higher plant, *J. Chem. Soc. Perkin Trans. 2*, 2213, 1990.
128. **West, R. R. and Cardellina, J. H., II,** Isolation and identification of eight new polyhydroxylated sterols from the sponge *Dysidea etheria*, *J. Org Chem.*, 53, 2782, 1988.
129. **Haasnoot, C. A. G., de Leeuw, F. A. A. M., and Altona, C.,** The relationship between proton-proton NMR coupling constants and substituent electronegativities. I. An empirical generalization of the Karplus equation, *Tetrahedron*, 36, 2783, 1980.
130. **Imai, K. and Osawa, E.,** An empirical generalization of the Karplus equation, *Magn. Reson. Chem.*, 28, 668, 1990.
131. **Altona, C., Ippel, J. H., Hoekzema, A. J. A. W., Erkelens, C., Groesbeek, M., and Donders, L. A.,** Relationship between proton-proton NMR coupling constants and substituent electronegativities, *Magn. Reson. Chem.*, 27, 546, 1989
132. **Lee, S. P., Lester, R., and St. Pyrek, J.,** Vulpecholic acid (1α,3α,7α-Trihydroxy-5-cholan-24-oic acid): a novel bile acid from a marsupial *Trichosurus vulpecula* (Lesson), *J. Lipid Res.*, 28, 19, 1987.
133. **Wilson, D. K., Wilson, W. K., Quiocho, F. A., and Schroepfer, G. J., Jr.,** Concerning the structure of 3β-benzoyloxy-5β-cholesta-8,14-diene, a major byproduct in the chemical synthesis of 5α-cholest-8(14)-en-3β-ol-15-one, *Chem. Phys. Lipids*, 47, 273, 1988.
134. **Bull, J. R., Dillen, J. L. M., and Sefton, M. A.,** Conformational deformation of ring C in 14β-estra-1,3,5(10),15-tetraen-17-ones, *Tetrahedron*, 46, 8143, 1990.
135 **Schneider, H.-J., Buchheit, U., Geschwendtner, W., and Lonsdorfer, M.,** Steric distortions and polar effects in steroids: molecular mechanics calculations and ^{13}C-NMR-investigations, in *Molecular Structure and Biological Activity*, Griffin, J P. and Duax, W. L., Eds., Elsevier, New York, 1982, 165.
136. **Schnedier, G., Meskó, E., Hackler, L., Dombi, G., Aeigan, D., and Engelhardt, G.,** Steroids. XXXII. Configurational analysis of 16-methyltestosterone derivatives, *J. Chem. Soc. Perkin Trans. 1*, 1597, 1985.
137 **Finamore, E., Minale, L., Riccio, R., Rinaldo, G., and Zollo, F.,** Novel marine polyhydroxylated steroids from the starfish *Myxoderma platyacanthum*, *J. Org. Chem.*, 56, 1146, 1991.
138 **VanAntwerp, C. L., Eggert, H., Meakins, G. D., Miners, J. O., and Djerassi, C.,** Additivity relationships in carbon-13 nuclear magnetic resonance spectra of dihydroxy steroids, *J. Org. Chem.*, 42, 789, 1977.
139. **Li, H.-T., Massey, I. J., Swenson, D. C., Duax, W. L., and Djerassi, C.,** Synthesis of (23R)- and (23S)- methylcholesterol, *J. Org. Chem.*, 48, 48, 1983.
140. **Cho, J.-H. and Djerassi, C.,** Minor and trace sterols from marine invertebrates 58. Stereostructure and synthesis of new sponge sterols jaspisterol and isojaspisterol, *J. Org. Chem.*, 52, 4517, 1987.
141. **Kerr, R. G., Kerr, S. L., Pettit, G. R., Herald, D. L., Groy, T. L., and Djerassi, C.,** Sterols of marine invertebrates 63. Isolation and structure elucidation of sutinasterol, the major sterol of the marine sponge *Xestospongia sp.*, *J. Org. Chem.*, 56, 58, 1991.
142. **D'Aurai, M. V., De Riccardis, F., Minale, L., and Riccio, R.,** Synthesis of 24-methyl-26-hydroxysteroid side-chains: models for stereochemical assignments in polyhydroxylated marine steroids, *J. Chem. Soc. Perkin Trans. 1*, 2889, 1990.

143. **Riccio, R., Finamore, E., Santaniello, M., and Zollo, F.**, Stereoselective synthesis of (24*S*)- and (24*R*)-24-(hydroxymethyl)cholesta-5,22(*E*)-dien 3β-ol. model compounds for stereochemical assignments of polyhydroxylated marine steroids, *J. Org. Chem.*, 55, 2548, 1990.
144 **Batta, A. K., Williams, T. H., Salen, G., Greeley, D. N., and Shefer, S.**, Differentiation between the 25*R*- and 25*S*-isomers of 5β-cholestane-3α,7α,26-triol by ^{13}C NMR spectroscopy, *J. Lipid Res*, 21, 130, 1980.
145. **Crist, B. V., Li, X., Bergquist, P. R., and Djerassi, C.**, Sterols of marine invertebrates. 44. Isolation, structure elucidation, partial synthesis, and determination of absolute configuration of pulchrasterol. The first example of double bioalkylation of the sterol side chain at position 26, *J Org. Chem.*, 48, 4472, 1983.
146. **Kobayashi, M., Kanda, F., Rao, C. V. L., Kumar, S. M. D., Trimurtulu, G., and Rao, C. B.**, Marine sterols. XVI. Polyhydroxysterols of the soft corals of the Andaman and Nicobar coasts. 1. Isolation of (24*S*)-24-methylcholest-5-ene-3β,25ζ,26-triol and (24*S*)-24-methylcholestane-3β,5β,6α,25-tetrol, *Chem. Pharm. Bull.*, 38, 1724, 1990.
147. **Klasinc, L., Tolic, L. P., Spiegl, H., Knop, J. V., and McGlynn, S. P.**, Electronic structure and long-range interaction in androstanes, *Comput. Chem.*, 14, 287, 1990.
148. **Wong, T. C. and Clark, G. R.**, Measurement of proton geminal coupling constants *via* selective two dimensional indirect J spectroscopy. Application to the study of the conformation of steroids, *J. Chem. Soc. Chem. Commun.*, 1518, 1984.
149. **Howarth, O. W., Thompson, M. J., and Rees, H. H.**, Reaction of 3-dehydroecdysone with certain n.m.r. solvents, *Biochem. J.*, 259, 299, 1989.
150. **Nes, W. R.**, Structure-function relationships for sterols in *Saccharomyces cerevisiae*, in *Ecology and Metabolism of Plant Lipids*, Fuller, G. and Nes, W. R., Eds., American Chemical Society, Washington, D.C., 1987, 252.
151. **Duax, W. L., Griffin, J. F., Rohrer, D. C., and Weeks, C. M.**, Conformational analysis of sterols: comparison of X-ray crystallographic observations with data from other sources, *Lipids*, 15, 783, 1980.
152. **Lindeman, S. V., Timofeeva, T. V., Chernov, S. V., Reshetova, I. G., and Struchkov, Y. T.**, The molecular structure of 5β-chola-7,14-diene-3α,12β,24-triol, *Bioorg. Khim.*, 14, 397, 1988.
153. **Wilson, W. K., Wheeler, M. E., Pinkerton, F. D., Pyrek, J. S., and Schroepfer, G. J., Jr.**, Inhibitors of sterol synthesis. Characterization of β,γ-unsaturated analogs of 3β-hydroxy-5α-cholest-8(14)-en-15-one and their effects on 3-hydroxy-3-methylglutaryl coenzyme A reductase activity in CHO-K1 cells, *J. Lipid Res.*, 32, 1215, 1991.
154. **Ibuka, T., Taga, T., Shingu, T., Saito, M., Nishii, S., and Yamamoto, Y.**, New stereoselective synthesis of 20S and 20R steroidal side chains. Remarkable stereoselectivity differences between saturated and α,β-unsaturated steroidal esters, *J. Org. Chem.*, 53, 3947, 1988.
155. **Eguchi, T., Yoshida, M., and Ikekawa, N.**, Synthesis and biological activities of 22-hydroxy and 22-methoxy derivatives of 1α,25-dihydroxyvitamin D_3: importance of side chain conformation for biological activities, *Bioorg. Chem.*, 17, 294, 1989.
156. **Helmer, B., Schnoes, H. K., and DeLuca, H. F.**, ^{1}H nuclear magnetic resonance studies of the conformations of vitamin D compounds in various solvents, *Arch. Biochem. Biophys.*, 241, 608, 1985.

Chapter 3

THEORETICAL INVESTIGATIONS ON STEROID STRUCTURE AND QUANTITATIVE STRUCTURE-ACTIVITY RELATIONSHIPS

Martin Bohl

TABLE OF CONTENTS

I. INTRODUCTION

In the past two decades, several theoretical techniques have been developed and established in modern drug research:

- Quantum-chemical calculations can be performed on different levels of approximation and computational effort. *Ab initio* computations using various standard basis sets will be applied if very accurate results are needed, transition or electronic excited states are involved, or no semiempirical method with an appropriate parametrization is available. Several approximate molecular orbital (MO) methods can be used to determine ground-state properties (molecular geometry, charge distribution, dipole moments, coupling constants, etc.) of a wide range of organic molecules.
- Various empirical force fields for organic compounds have been parameterized and tested. A few of them are widely accepted and prove to provide reliable geometrical and energetical information. They are used to optimize bond distances, valence, and torsion angles, to investigate the conformational space, to describe drug-protein or drug-solvent interactions, or to run dynamic simulations.
- Basic and more sophisticated approaches have been developed and applied to establish quantitative relationships between chemical structure and observed biological activity data. Free-Wilson and Hansch analyses, principal component, discriminant, or pattern recognition analyses are also used in steroid research.
- The rapid development of computer hardware has supported all these methodological enhancements. The drastically increased speed, memory, and storage capacity of modern computers allows one to calculate more integrals in MO methods, to include more atoms in molecular mechanics problems, to simulate larger time intervals in molecular dynamics (MD), to keep and search in large data bases, or to process time-consuming statistical validations on large data sets. Moreover, molecular graphic systems have been developed based on high-quality display technologies. Several molecular modeling software packages are available that combine all of the above-mentioned techniques with powerful capabilities of visualization of the geometrical structure and other molecular properties.

At this stage, theoretical methods have become a new tool, not to replace but to complement the existing methods in drug research. This tool should contribute to finding better drugs in a shorter time, which is highly desired since the cost to develop one new drug is estimated at over 200 million of German marks and requires the synthesis of 10,000 to 20,000 new compounds. The ways in which these techniques could help are manifold:

- To see the molecular architecture in a visual way rather than looking at purely numerical data could, in the easiest way, induce a new way of thinking about the problem.
- Because all personal and material resources are limited and many research programs overtaxed, theoretical models could establish a ranking order in which substances are to be synthesized and/or tested.
- Validated structure-activity correlations can be used for a more rational design of new compounds, in the best case, to derive new lead structures.
- QSAR results focus attention on molecular features that are important to the biological potency of interest. In this way, they contribute to an understanding of the molecular mode of action or help in confirming or contradicting proposed mechanisms. QSAR findings may be helpful in future drug-protein modeling.
- Modeling studies can suggest further experiments or guide and direct further investigations, e.g., by identifying outliers in QSAR correlations.

Since none of the theoretical methods can provide totally exact and comprehensive prediction, healthy scepticism is highly recommended. The approximate character is given in both the inaccuracy of the theoretical method applied and the model that simulates the biological phenomenon. Additionally, biological tests used to describe individual subprocesses of steroid behavior or a certain sum of these subprocesses, given in the following, have a different applicability to *in vivo* conditions of human beings: drug uptake, binding to carrier proteins in the blood and transport, binding to active sites of metabolizing enzymes as well as type and rate of metabolism, bioavailability due to distribution, degradation or excretion, interference with other drugs, cell-membrane penetration, high-affinity and nonspecific binding to receptor proteins, mechanism of interaction with chromatin, etc.

It is likely that these processes are influenced differently by a particular steroid structural modification. The contribution of theoretical investigations toward the understanding of at least a part of these interrelations and toward the explanation of at least a part of the existing biological data shall be reviewed in this paper. Previous reviews covering some of the aspects considered here have been given by Hüttenrauch,[1] Hansch,[2] Wolff,[3] Zeelen,[4] Hanč and Urbánková,[5] and Kubinyi.[6]

II. QSAR ANALYSES

A. ACTIVITY CONTRIBUTIONS OF SUBSTITUENTS (FREE-WILSON TYPE ANALYSES)

For a series of compounds that vary only in the substituents at one or more substitution sites, it is an obvious aim to assign to each substituent an individual contribution to the biological activity. A given substituent, e.g., a methyl group, attached at different substitution sites and, therefore, located in different areas of drug-environment interactions, is expected to have varying amounts of power to influence the overall activity. Even a given substituent at a given substitution site could contribute to a different extent depending on the particular combination of other substituents in the molecule. The reason for this can be found in direct substituent-substituent interactions (e.g., intramolecular hydrogen bonding, hydrophobic interference, modified side-chain orientation due to steric repulsion), or indirect effects like distortions of the three-dimensional structure of the chemical core common to all compounds. With respect to the latter effect, the steroid skeleton is, compared to other drugs, considered a relatively rigid body. Thus, if in addition the direct intramolecular interaction can be neglected because of, for example, the spatial separation between the substituents, the biological potency A can be split into independent contributions a_{ij} of the substituent i in position j and the contribution c of the core:

$$\log A = \sum_{ij} a_{ij} b_{ij} + c \tag{1}$$

The elements b_{ij} of the structure matrix have a value of one when the substituent i is present in position j and otherwise have a value of zero. The determination of the activity contributions of all substituents (also called enhancement factors,[3,7] *de novo* substituent constants,[3,18] or functional group factors[4,16]) can be achieved (1) simply by comparing two compounds differing in only one substituent and with subsequent calculation of the mean value of different a_{ij} determinations or (2) by applying a regression analysis of the Free-Wilson or Fujita-Ban type.[6,21,22] A precondition for the first method is that the data set includes all single-substituent combinations needed to compute all parameters a_{ij} at least once. The standard deviation for a particular activity contribution a_{ij} obtained by method (1) and the confidence interval of a particular regression coefficient by method (2) give an indication of how independent the contribution of that substituent really is and, likewise, whether the additivity concept in this approach is justified or not.

Such activity descriptors of various substituents have been reported for several steroid types and biological effects. Summaries of these investigations and of the results obtained are given in Tables 1 and 2 to 5, respectively.

Needless to say, data entries in these tables should only be expected to have comparable values if the scale of activity measurements and the conditions of biological testing (type of biological effect, animal species, route of administration) are the same. However, the signs of the data are synchronized (positive values indicate an enhanced activity in all cases) and the tendency within a given column can be compared to those in other columns. Activity predictions for unknown compounds are limited in these types of analyses to the range of substituents under study.

TABLE 1
Free-Wilson Type Structure-Activity Investigations

Biological effect	Species	Number of steroids in data set	Number of activity contributions[a]	Type of approach	Statistics F	Statistics s	Ref.	Year
Glycogen deposition, anti-inflammatory	Rat	nd	13, 9	Direct comparison	nd	nd	7	1958
Anti-inflammatory	Rat, man	nd	13, 10	Direct comparison	nd	nd	8	1963
Glycogen deposition	Rat	44	17	Fujita-Ban	4.7	0.868	9	1978
Glucocorticoid receptor binding	Rat	24	15	Direct comparison	nd	nd	10	1978
Affinity to corticoid-binding globulin	Guinea pig	39	24	Direct comparison	nd	nd	11	1980
Progestational potency	Rabbit	64, 69	13, 14	nd	nd	0.87, 0.91	12	1977
Progesterone receptor binding	Rabbit	60	59	Direct comparison	—	—	13	1982
Affinity to progesterone-binding globulin	Guinea pig	103	nd	Direct comparison	nd	nd	14	1978
Uterotropic activity	Mouse	28	22	Fujita-Ban	5.7	0.98	15	1980
Estrogen receptor binding	Woman	15	10	Direct comparison	nd	nd	16	1979
Activity of androgens and anabolics	nd	nd	nd	Fujita-Ban	nd	nd	17	1984
Inhibition of Na^+,K^+-ATPase	Human	95	57	Fujita-Ban	141.9	0.997[b]	18	1988

Note: nd = no data.

[a] Without core contribution c.
[b] For subset of 42 compounds and 21 substituents.

Normally, substitutions with high-acitivity contributions from each substitution site will be combined to design new compounds unless these compounds are part of the original data set. Once the numerical contribution of an individual substituent to the overall activity is known, a second step might be an attempt to identify the partial process or processes in which this substitution exerts its greatest influence.

For corticoids, Fried and Borman[7] showed that the 9α-fluorine enhances the rat liver glycogen deposition and the antiinflammatory potency considerably and to a very similar extent when introduced in cortisone, cortisol, prednisolone, corticosterone, 11β-hydroxyprogesterone, and 11-ketoprogesterone derivatives. The logarithms of the enhancement factors they found in the glycogen deposition assay result for selected structural features in the following order (see also Table 2):

$$9\alpha\text{-F} > 21\text{-OH} > 1\text{-ene} > 6\alpha\text{-CH}_3 > 17\alpha\text{-OH} > 16\alpha\text{-OH}$$

When compared with the free-energy data provided by Wolff et al.[10] for rat glucocorticoid receptor binding, a similar influence can be seen at the receptor level:

$$6\alpha\text{-CH}_3 > 9\alpha\text{-F} \approx 21\text{-OH} > 1\text{-ene} > 17\alpha\text{-OH} > 16\alpha\text{-OH}$$

Contrary to the overall activity, the 17α-hydroxyl group has no stabilizing effect in the interaction with the glucocorticoid receptor and the influencing power of the 6α-methyl group is, with respect to other substituents, strengthened at the receptor level. A conformity in lowering both glucocorticoid activity and receptor affinity is found for the 9α-bromo substituent while 9α-F and 9α-Cl lead to pronounced activity in these biological tests. It should be noted that the glucocorticoid receptor can only just interact with 11-oxo steroids. This is an indication for metabolite activity in the corresponding *in vivo* contribution. It was noticed by Wolff et al.[10] that their increments are not fully independent of one another.

Justice[9] reported Fujita-Ban regression results significant at the 0.001 level for rat glycogen deposition and subcutaneous administration. These data correspond to a great extent with those of Fried and Borman. The free-energy contributions for binding to the guinea-pig corticosteroid-binding globulin (CBG) provided by Mickelson and Westphal[11] indicate that only certain key steroid sites like 11β-OH and 21-OH are important for optimal binding to the transport protein.

In the case of progestational steroids, Seeley et al.[13] provided receptor affinity data for a wide range of substituents and substitution sites. However, all values represent a binding difference between two single compounds, one of them being constantly progesterone as reference. Only very few of the

TABLE 2
Additive logA Contributions of Individual Substituents to Different Biological Effects of Corticoids

Substituent	Glycogen deposition (Rat)[7,a]	Glycogen deposition (Rat)[9]	Anti-inflammatory effect (Rat)[7,a]	Anti-inflammatory effect (Rat)[8,a]	Anti-inflammatory effect (Man)[8,a]	Glucocorticoid receptor binding (rat)[10,b]	Affinity to CBG (guinea pig)[11,b]
1-Dehydro	+0.48 to +0.60	+0.564	+0.48 to +0.60	+0.48	+0.60	+0.29	−0.9 to −1.5
2α-Methyl	+0.48 to +0.78	+0.611	0.00 to +0.60	+0.65	−0 15	nd	−0.6
6-Dehydro	−0.30 to −0.15	+0.108	nd	−0.30	—[c]	nd	−0.2
6-Dehydro-6-methyl	nd	+1 129	nd	nd	nd	nd	nd
6α-Fluoro	nd	+0.793	nd	+0.90	+0.40	+0.36	nd
6α-Methyl	+0.30 to +0.48	+0.782	0.00 to +0.30	+0.30	+0.11	+1.09	−1.6
9α-Bromo	−0.40[d]	nd	nd	nd	nd	−1.89	nd
9α-Chloro	+0.48 to +0.70	+0.312	+0.48	nd	nd	+0.71	−2.9
9α-Fluoro	+1.00	+0.652	+0.85 to +1.00	+0.90	+1.00	+0.57	−3.0 to −3.1
9α-Methoxy	nd	nd	nd	nd	nd	−2.21	nd
11-Oxo	nd	+1.920	nd	nd	nd	−1.67	−0.5
11α-Hydroxy	nd	nd	nd	nd	nd	+0.82	−0.9 to −1.0
11β-Hydroxy	nd	+2.010	nd	nd	nd	+0.89	+1.0 to +1.2
12α-Chloro	+0.60[e]	nd	nd	nd	nd	nd	nd
12α-Fluoro	+0.78 to +0.90[e]	nd	nd	nd	nd	nd	nd
16-Methylene	nd	nd	nd	+0.18	nd	nd	nd
16α-Hydroxy	−0.40 to −0.30	+0.043	−1.00 to −0.70	−0.70	−0.30	−0.86	nd
16α-Methyl	nd	+0.702	nd	+0.18	+0.30	+0.11	−0.6
16β-Methyl	nd	nd	nd	+0.26	+0.40	+0.21	nd
17α-Acetoxy	nd	+0.162	nd	nd	nd	nd	−2.6
17α-Hydroxy	0.00 to +0.30	+0.536	+0.60	nd	nd	−0.49	0.0 to +0.3
20-Oxo	nd	nd	nd	nd	nd	nd	+1.8

TABLE 2 (continued)
Additive logA Contributions of Individual Substituents to Different Biological Effects of Corticoids

Substituent	Glycogen deposition		Anti-inflammatory effect			Glucocorticoid receptor binding	Affinity to CBG
	(Rat)[7,a]	(Rat)[9]	(Rat)[7,a]	(Rat)[8,a]	(Man)[8,a]	(rat)[10,b]	(guinea pig)[11,b]
21-Acetoxy	nd	+0.630	nd	nd	nd	nd	nd
21-Fluoro	+0.30	+0.022	+0.30	+0.30	nd	nd	nd
21-Hydroxy	+0.60 to +0.85	+0.754	+1.40	+1.00	>+1.0	+0.56	+0.2 to +0.4

Note: nd = no data.

[a] Logarithm of enhancement factors (positive values indicate activity increase).
[b] Negative free-energy contributions ($-\Delta G$) in kcal/mol (positive values indicate increase in binding affinity).
[c] Variable, usually <0.00.
[d] In 1-dehydrosteroids this value is +0.60.
[e] In the presence of a 17α-hydroxyl group this value is < −2.00.

considered modifications to progesterone can increase the affinity to the rabbit progesterone receptor: introduction of a 17α-methyl or 6α-fluoro substituent, removal of the 10-methyl group (see also data in Table 3).

Both carbonyl oxygens of progesterone contribute significantly to the total receptor binding energy: 3-oxo (1.2 kcal/mol) and 20-oxo (1.7 kcal/mol).

With exception of the 10-methyl group, hydroxyl and methyl substituents at several positions of the progesterone backbone were also found by Blanford et al.[14] to decrease the affinity for the guinea-pig progesterone-binding globulin (PBG).

Zeelen[12] reported enhancement factors for the progestational potency (Clauberg assay) starting with a data set of 70 steroids. The correlation statistics are satisfactory in view of the complexity of the biological processes involved and of the fact that data from other sources have been taken into account. The activity contributions of several substituents (e.g., 6-ene, 16α-methyl, 17α-alkyl, 21-acetoxy) differ depending on the route of application. The receptor binding increase from the 6α-fluorine and of the 19-nor progesterone derivatives is also reflected in these activity data, whereas the effect caused by 6α-methyl and 17α-acetoxy substitutions is contrary in potency and receptor binding.

To interpret estrogenicity in mice, Tang et al.[15] performed a Fujita-Ban analysis including 28 steroids. Fanchenko et al.[16] extracted group contributions from human estrogen receptor binding measurements. Besides the description of the heteroatom functions of estradiol at C3 and C17, an agreement between both investigations is found for the activity-reducing influence of the 16α-hydroxyl group (see Table 4).

A Fujita-Ban analysis for androgenic and anabolic activity[17] is discussed in more detail in Chapter 10.

Schönfeld and Repke[18] carried out a Fujita-Ban analysis for the inhibition of human-heart Na^+,K^+-ATPase taking 95 steroids into account in the whole investigation. For a subset of 42 steroids, they derived 21 binding contributions of individual substituents with a correlation coefficient of $r = 0.997$, a standard deviation of $s = 0.94$ kJ/mol and an F-test value of 141.9. The contributions of the remaining 36 substituents occurring only once in the data set were determined by a second step. The additivity concept was proved for the derivatives of the 5β,14β-androstane backbone. This observation leads the authors to conclude that a common inhibition mechanism is valid for all derivatives. It is interesting to see that, in most cases, substituents not in 3β and 17β positions diminish the binding affinity (see data in Table 5). This is true even for the 12β-hydroxyl group which is present, e.g., in digoxigenin derivatives.

Free-Wilson type analyses generally provide a single activity value for each chemical substructure but no information about the nature of molecular interactions forming the integral activity contribution. One piece of infor-

TABLE 3
Additive logA Contributions of Individual Substituents to Different Biological Effects of Progestins

Substituent	Clauberg test (rabbit)[12] s.c.[a]	Clauberg test (rabbit)[12] Oral[a]	Progesterone receptor binding (rabbit)[13,b]	Affinity to PBG (guinea pig)[14,b]
1-Dehydro	nd	nd	−0.99	−0.5
1α,2α-Methylene	+0.46	0 00	nd	nd
2α-Fluoro	nd	nd	−0.68	nd
2α-Hydroxy	nd	nd	<−2.31	−4.6
2α-Iodo	nd	nd	−0.63	nd
2α-Methyl	nd	nd	−1.37	−1.2
2β-Methyl	nd	nd	<−2.57	nd
3-Oxo	nd	nd	+1.21	nd
4-Acetoxy	nd	nd	−0.60	nd
4-Hydroxy	nd	nd	<−2.84	nd
4-Methyl	nd	nd	−0.80	nd
6-Dehydro	−0.52	+0.60	−0.69	+0.1
6-Dehydro-6-chloro	+0.52	+1.79	nd	nd
6-Dehydro-6-fluoro	nd	+1.63	nd	nd
6-Dehydro-6-methyl	0.00	+1.74	nd	nd
6α-Chloro	nd	nd	−0.16	nd
6α-Fluoro	+0.43	+0.61	+0.06	nd
6α-Hydroxy	nd	nd	−1.99	−1.4
6α-Methyl	+0.38	+0.63	−0.36	nd
6β-Chloro	nd	nd	−1.45	nd
6β-Fluoro	nd	nd	−0.65	nd
6β-Hydroxy	nd	nd	<−2.46	−1.0
6β-Methyl	nd	nd	−0.12	nd
7α-Hydroxy	nd	nd	<−2.68	nd
7α-Methyl	nd	nd	−0.72	nd
7β-Acetoxy	nd	nd	<−3.00	nd
7β-Methyl	nd	nd	−1.26	nd
9α-Hydroxy	nd	nd	<−2.71	nd
10-Methyl	−0.93	−1.07	−0.41	−0.3
11α-Hydroxy	nd	nd	−2.08	−1.1
11β-Hydroxy	nd	nd	−0.86	−1.4
12α-Acetoxy	nd	nd	<−3.00	nd
14α-Hydroxy	nd	nd	−1.61	nd
15α-Methyl	nd	nd	−0.34	nd
15β-Hydroxy	nd	nd	−1.25	nd
15β-Methyl	nd	nd	−0.19	nd
16α-Hydroxy	nd	nd	<−2.70	nd
16α-Methyl	0.00	−0.52	−0.74	−0.8
16β-Hydroxy	nd	nd	−2.15	nd
16β-Methyl	nd	nd	−1.85	−0.7
17α-Acetoxy	−0.84	+1.12	−0.54	−1.3
17α-Bromo	nd	nd	−0.28	nd
17α-Ethyl	0.00	+1.04	nd	nd
17α-Fluoro	nd	nd	−0.18	nd
17α-Hydroxy	nd	nd	<−2.31	−1.5
17α-Methoxy	nd	nd	−0.25	nd

TABLE 3 (continued)
Additive logA Contributions of Individual Substituents to Different Biological Effects of Progestins

Substituent	Clauberg test (rabbit)[12] s.c.[a]	Clauberg test (rabbit)[12] Oral[a]	Progesterone receptor binding (rabbit)[13,b]	Affinity to PBG (guinea pig)[14,b]
17α-Methyl	0.00	+0.49	+0.84	nd
18-Hydroxy	nd	nd	<−2.66	nd
19-Hydroxy	nd	nd	<−2.44	nd
20-Oxo	nd	nd	+1.70	nd
21-Acetoxy	−1.00	0.00	nd	−0.7
21-Fluoro	+0.30	0.00	−0.54	nd
21-Hydroxy	nd	nd	−1.07	−0.4
21-Iodo	nd	nd	−0.60	nd

Note: nd = no data.

[a] Logarithm of enhancement factors (positive values indicate activity increase).
[b] Reference compound is progesterone; negative free-energy contributions (−ΔG) in kcal/mol (positive values indicate binding-affinity increase).

TABLE 4
Additive logA Contributions of Individual Substituents to Different Biological Effects of Estrogens

Substituent	Uterotropic activity (mouse)[15]	Estrogen receptor binding (woman)[16,a]
2-Bromo-3-hydroxy	nd	−0.61
2-Methyl	−2.22	nd
3-Benzoate	+2.00	−0.08
3-Cyclopentyloxy	+1.42	nd
3-Hydroxy	+1 04	+1.21
3-Methoxy	+0.82	−0.35
4-Bromo-3-hydroxy	nd	−0.61
6α-Hydroxy	−1.97	nd
15β-Methoxy	−1.97	nd
16α-Cyclopentyloxy	−1.79	nd
16α-Hydroxy	−0.85	−0.54
16β-Hydroxy	−1.15	nd
16β-Hydroxymethyl	−2.32	nd
17-Oxo	+1.23	−0.21
17α-Ethyl	−0.71	nd
17α-Ethynyl	+0.96	+0.01
17α-Hydroxy	−1.20	nd
17β-Hydroxy	+1.13	+0.24

Note nd = no data.

[a] Logarithm of association constant.

TABLE 5
Selected Free-Energy Contributions of Individual Substituents ($-\Delta G$) in kJ/mol for Inhibition of Human-Heart Na^+,K^+-ATPase

Substituent	$-\Delta G$ contribution[18]
3α-Methyl	−5.87
3β-Acetoxy	+0.40
3β-Deoxy-3α-hydroxy	−9.97
3β-Digitoxosyloxy	+6.73
3β-Rhamnosyloxy	+8.14
3β-Tridigitoxosyloxy	+5.83
11α-Hydroxy	−1.05
12β-Hydroxy	−2.36
16α-Acetoxy	−0.37
16α-Hydroxy	−11.54
16β-Acetoxy	−3.90
16β-Hydroxy	−4.50
17-Oxo	−5.08
17β-Amino	+7.09
17β-(2′,5′-Dihydro-2′-oxo-fur-4′-yl)	+19.79
17β-(2′,5′-Dihydro-3′-hydroxy-2′-oxo-fur-4′-yl)	+15.32
17β-(2′,5′-Dihydro-3′-methoxy-2′-oxo-fur-4′-yl)	+8.62
17β-(2′,5′-Dihydro-3′-methyl-2′-oxo-fur-4′-yl)	+16.32
17β(Pyridazin-4′-yl)	+18.83
17β-(1′,2′-Pyron-5′-yl)	+26.54

Note: Positive values indicate an increased inhibitory activity.

mation in this respect can be taken from the study of the temperature dependence of binding affinities and subsequently calculating thermodynamic ΔH and ΔS data.[10,11] The arguments will be reviewed in the following section within the context of discussing the underlying physicochemical forces in steroid activity.

B. PHYSICOCHEMICAL PARAMETERS (HANSCH TYPE ANALYSES)

Each substituent can be considered by a range of hydrophobic (x_h), steric (x_s) and electrostatic (x_e) parameters or any combination of these descriptors (e.g., linear combination).[20-26] If the activity is directly related to these physicochemical parameters

$$\log A = a_h x_h + a_s x_s + a_e x_e \tag{2}$$

with x_i being the parameters themselves or a function of thereof, the a_i coefficients can be determined by use of (1) a set of observed activity data for the whole compounds or determined activity contributions of individual substituents (see Section II.A), (2) the known physicochemical descriptor data (preferably of linear free-energy type) corresponding to the chemical structure under study, and (3) a presumed analytical expression between both. In general, the a_i coefficients will be different for each substitution site since they reflect the nature of interactions involved in this specific molecular area. Therefore, Equation 2 is solved separately for each substitution site by multiple regression. Due to the pioneer work of Corwin Hansch, this method is called Hansch analysis as long as the classical hydrophobic, steric, and electrostatic parameters are applied. Due to the fact that there are no predefined selections of parameters and mathematical functions for the set-up of Equation 2, certain consequences follow:

- The optimized coefficients and the derived interpretation of the activity data can change with another choice for the right side of Equation 2.
- Some effort should be made in selecting an appropriate mathematical expression for Equation 2, including, e.g., use of knowledge from independent sources about the nature of interactions and systematic variation in Equation 2 with statistical evaluation and/or validation of the obtained model by checking predicted phenomena.
- There are methodical developments for new types of parameters, e.g., the application of three-dimensional field data instead of scalar quantities for spatial extension or atomic charges (see Section II.D).

The first item can be best illustrated with a series of investigations on progesterone analogues.

1. Progestins

Teutsch et al.[27] measured the Clauberg potency in a series of fifteen 17α-(acetoxy-16-methylene-4,6-pregnadiene-3,20-dione analogues differing in the 6-substituent. They gave a ranking of the substituents and discussed qualitatively the activity variation on the basis of steric (volume requirements and distance of extension from C6 carbon) and electronic influences. On several occasions[12,28-31] this data set was quantitatively reexamined. The best correlations found in each study are included in Table 6. Wolff et al.[28,29] used hydrophobic π, electronic F and R, and steric MR parameters to demonstrate that hydrophobicity is of major importance (accounting for over half the variance) while both resonance and steric effects play either no role or only a limited one (see Equations 3 and 4 in Table 6). For the methyl substituent which was excluded from the data set, a different mechanism of action was suggested.

Topliss and Shapiro[30] found Equation 5 in Table 6 by additional use of the steric extension C. This correlation accounts for 73% of the variance, includes the methyl compound and confirms the unimportance of electronic terms. The relationship favors small, hydrophobic substituents at C6. Equation

TABLE 6
Hansch-Type QSAR Results for Progestational Steroids

Equation no.	Steroid core	Substitution site	Activity type	Correlation equation[a]	Statistics[b] n	r	s	F	Ref.
3	—[c]	6	—[d]	$\log A = -1.05\,\pi^2 + 1.06\,\pi + 3.46\,F - 0.17$	13	0.890	0.492	11.4	28,29
4	—[c]	6	—[d]	$\log A = 0.97\,\pi + 2.84\,F - 0.07\,MR + 0.26$	13	0.892	0.489	11.6	28,29
5	—[c]	6	—[d]	$\log A = 1.00\,\pi - 0.05\,C + 1.32$	14	0.86	0.54	nd	30
6	—[c]	6	—[d]	$\log A = 1.05\,\pi - 0.03\,MV$	12	0.85	0.60	nd	12
7	—[c]	6	—[d]	$\log A = 0.89\,\pi - 0.08\,\Delta V + 1.37$	14	0.911	0.431	26.8	31
8	—[c]	6	—[d]	$\log A = 2.84\,LUMO + 19.90$	4	0.977	nd	nd	32
9	—[f]	11β	—[d]	$\log A = -1.46\,E_s^x - 0.14\,MR + 1.97$	8	0.91	0.34	nd	33,34
10	—[g]	13β	—[d]	$\log A = 0.40\,f - 0.87\,L + 2.04$	5	0.87	nd	nd	36
11	—[h]	16α,17α	—[d]	$\log A = -1.57\,f^2 + 4.22\,f - 1.82\,B_1 + 0.79$	9	0.93	0.25	10.1	37
12	—[i]	16α,17α	—[d]	$\log A = -0.58\,f^2 + 0.86\,f - 0.30\,L + 1.10\,d + 1.08$	12	0.94	0.32	12.3	38
13	—[k]	17β	—[j]	$\log A = -0.668\,(\log V_R)^2 + 6.40\,\log V_R - 13.5$	12	0.924	0.257	32.9	39
14	—[m]	Various	—[l]	$\log RBA = 0.23\,\pi_a + 1\,29\,\pi_b + 0.011\,SA_t + -0.020\,SA_o - 1\,34\,MK - 0.22\,cc + 1.77$	56	0.92	0.45	46	40
15	—[m]	Various	—[n]	$\log RBA = 0.18\,\pi_a + 1.45\,\pi_b + 0.010\,SA_t + -0.012\,SA_o - 0.99\,MK - 0.33\,cc + 1.79$	55	0.88	0.54	29	40
16	—[o]	Various	—[n]	$\log RBA = 0.854\,f - 0.725\,MTD + 3.209$	34	0.962	0.262	194.6	41
17	—[p]	Various	—[n]	$\log RBA = 0.636\,f - 1.092\,MTD + 3.213$	25	0.955	0.294	114.0	41

Note: nd = no data.

[a] Hydrophobicity, π; inductive effect, F; molar refractivity, MR; steric extension in one direction, C; molecular volume, MV; energy of lowest unoccupied molecular orbital, LUMO; difference in van der Waals' volume, ΔV; steric Taft-Charton parameter, E_s^x; Rekker's hydrophobic fragmental constant, f; steric Verloop parameters, L, B_1; Bondi's surface area, SA; minimal topological difference, MTD; lipophilicity from HPLC, V_R.

[b] n = number of data points (substituents or steroids); r = correlation coefficient; s = standard deviation; F = F-test variance ratio.

[c] 17α-Acetoxy-16-methylene-4,6-pregnadiene-3,20-dione.

[d] Progestational Clauberg potency (rabbit).
[f] 17α-Ethynyl-4-estren-17β-ol.
[g] 18-Norprogesterone.
[h] 16α,17α-Methyleneprogesterone.
[i] Acetal of 16α,17α-dihydroxyprogesterone.
[j] Contraceptive potency.
[k] Norethisterone, levonorgestrel
[l] Progesterone receptor binding affinity (human).
[m] Pregnane, androstane.
[n] Progesterone receptor binding affinity (rabbit).
[o] Progesterone.
[p] Testosterone.

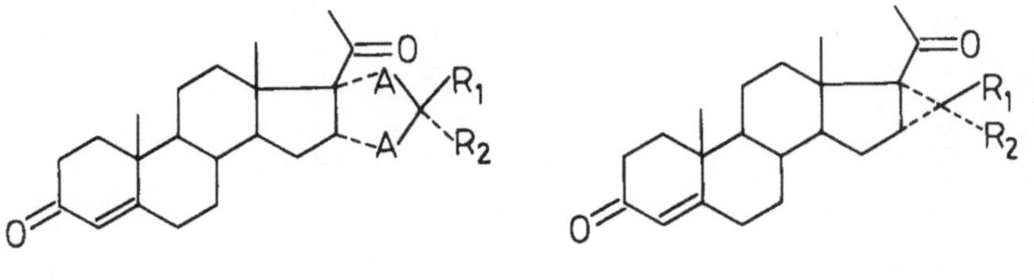

FIGURE 1.

6 (Table 6) represents the best correlation found by Zeelen.[12] Coburn and Solo[31] tested in a systematic way the inclusion of a variety of more conventional steric terms, among them Taft's constant E_s, Exner's molecular volume MV, van der Waals volume V, and parachor Pa. Equation 7 (Table 6) accounts for 83% of the variance and was considered by Coburn and Solo to indicate that the steric factor is of some importance. A correlation with an electronic descriptor (energy of the lowest unoccupied molecular orbital) was found by Hartog et al.[32] (Equation 8 in Table 6). However, their biological data set consists of only four compounds and differs, for example, in the activity ranking of the methyl compound H(13) < CH_3(30) < F(50) < Cl(75) compared to above-used data,[27] H(1) < F(55) < Cl(77) < CH_3(91).

The influence of substituents in the 11β position on the oral Clauberg activity was studied by Zeelen et al.[33,34] using 17α-ethynyl-4-estren-17β-ol (lynestrenol) as basic core. According to Equation 9 in Table 6, the progestational activity can be enhanced by groups giving rise to a strong 1,3-diaxial interaction (E_s^x term) with the angular methyl group at C13, thus inducing a bowing of the steroid backbone toward the α-side. The second steric term (molar refractivity) controls the total bulk of the 11β-substituent. No evidence was found in this study for a hydrophobic contribution in the 11β region.

In agreement with these results, Kloosterboer and Bergink[35] claimed an 11-methylene and 18-methyl group for optimal binding to the human progesterone receptor.

In the 13β position, the progestational potency (s.c. Clauberg test) is influenced by hydrophobic and steric parameters[36] (Equation 10 in Table 6). However, the significance of this result is limited by the small number of compounds employed.

Zeelen[37,38] reported Equations 11 and 12 (Table 6) for two series of progesterone derivatives with modifications in the 16α,17α region (see Figure 1). The s.c. Clauberg activity is shown to depend in a non-linear function on hydrophobicity in that α-region of ring D and linearly on the steric extension in one direction (Verloop parameter B_1 or L for the R_1 substituent). The indicator variable d has the value of one if R_2 is an α-furyl ring and otherwise has the value of zero. To explain the exceptionally high progestational activity of the furyl analogues, an additional hydrogen bond to the receptor is used.

Karunanithy et al.[39] found a non-linear relationship between lipophilicity expressed as high-performance liquid chromatographic (HPLC) retention data

V_R and the contraceptive potency for 17β-esters of norethisterone and levonorgestrel (see Equation 13 in Table 6). Hydrolysis of these long-acting agents is assumed to determine the rate of release of the active parent steroids.

An overall description of the requirements in steroid binding to the progesterone receptor was given in two QSAR investigations.[40,41] Lee et al.[40] studied binding affinity data of 56 steroids for human, rabbit, sheep, and guinea-pig receptor preparations. Equations 14 and 15 (Table 6) represent some results where π_b is the sum of Hansch's hydrophobic coefficients for all polar groups in 17α, 20α, and 20β positions and π_a for polar groups in other positions. The SA_i and SA_o are the Bondi's surface areas in and out of the hydrophobic pockets, respectively. The MK indicator variable takes the value of one if there is a ketone at any position other than C3 or for compounds lacking the oxo-O3 oxygen (misplaced ketone term) and of zero otherwise. The cc (conformational changes) parameter counts carbon atoms with changes in hybridization from testosterone as reference. It is evident that the calculation of SA values requires a knowledge of the hydrophobic pocket locations. This information was derived by Lee et al. by comparing binding affinities of pairs of compounds differing in only one nonhydroxyl substituent. In addition, the allowable size of these pockets was estimated while investigating substituents with excessive size and with the help of QSAR correlations found in an initial step.

These pocket sizes are compared in Table 7 to the rabbit-progesterone receptor binding-energy differences of hydroxyl- and methyl-substituted analogues taken from Seeley et al.[13] and to the results of a minimal topological difference (MTD) study,[41,42] which uses the sterical MTD variable and hydrophobicity in all regions but 3- and 17β-positions, which are regarded as hydrogen bonding areas. Equations 16 and 17 in Table 6 and the vertex characteristics obtained by the MTD study suggest that progesterone and testosterone derivatives are bound to the progesterone receptor in a slightly different orientation, mainly with respect to the D-ring region. The data column in Table 7 referring to our MTD approach describes the steric requirements in addition to hydrophobic forces. Thus, a vertex characteristic of zero suggests no steric restrictions on putting a hydrophobic atom next to the steroid attachment atom. The data in Table 7, in particular those of Seeley et al., and QSAR relationships summarized in Table 6 indicate the presence of hydrophobic regions in the vicinity of 2α, 6α, 6β, 11β, 13β, 15β, 16α, 17α, and 21 positions of progesterone. There are conflicting data for the 4 and 7α region. In the 10β direction, the steroid-receptor contact seems to be very close. The exceptionally small binding-energy difference between the 16β-hydroxyl and the 16β-methyl group may be due to a direct intramolecular contact with the 17β-acetyl side chain in both cases. This speculation is supported by the low affinity of the 16β-methylprogesterone and by X-ray structural data which will be further discussed in Section III.E together with results of modeling studies.

TABLE 7
Description of Hydrophobic Pockets in the Rabbit Progesterone Receptor Due to Different Investigations[13,40,41]

Steroid site	Binding-energy difference[13,a]	Maximum allowable substituent size (Å)[40]	Vertex characteristics[41,b]
2α	>0.94	nd	nd
4	>2.04	nd	+1
6α	>1.63	22	0[c]
6β	>2.34	nd	nd
7α	>1.96	0	+1[c]; 0[d]
10β	nd	0	+1
11β	nd	22	0
15β	1.06	nd	nd
16α	>1.96	45	+1[c]
16β	0.30	nd	nd
17α	>3.15	nd	0[c]; −1[d]
18	nd	22[c]; 0[d]	+1
21	nd	63	nd

Note: nd = no data.

[a] Difference in binding energy (kcal/mol) between hydroxy- and methyl-substituted progesterone derivatives.
[b] −1 (cavity vertex, binding-contact area); 0 (sterically irrelevant vertex); +1 (wall vertex, repulsive-contact area).
[c] Only for progesterone derivatives.
[d] Only for testosterone derivatives.

Hoppen and Hammann[43] reported correlations between the ^{13}C-NMR resonance signal of the C17 carbon and the binding affinity to the progesterone receptor and with the progestin/androgen receptor-binding ratio for eight steroids.

2. Corticoids

Fried and Borman[7] discussed electronic effects of 9α-substituted cortisol derivatives on the promoted glucocorticoid potency. The first QSAR study on a series of 9α-substituted cortisol 21-acetate analogues was performed by Wolff and Hansch.[29,44,45] Equation 18 in Table 8 accounts for 92% of the variance in the activity data. It is a three-parameter correlation based on only seven compounds, but a further three compounds could be explained with this equation. The glucocorticoid activity can be increased according to Equation 18 by electron-withdrawing, hydrophobic, and/or small 9α substituents. In this case, the hydrophobic effect is not the dominating one.

Moreover, Coburn and Solo[31] demonstrated that the inclusion of the hydrophobic parameter is no longer justified when the 9α-hydroxy compound is omitted from the data set (see Equation 19 in Table 8).

TABLE 8
Hansch-Type QSAR Results for Corticoids

Equation no.	Steroid core	Substitution site	Activity type	Correlation equation[a]	Statistics[b] n	r	s	F	Ref.
18	—[c]	9α	—[d]	$\log A = 2.779\ \sigma_I - 0.220\ MR + 0.764\ \pi + 0.073$	7	0.959	0.327	11 3	29, 44, 45
19	—[c]	9α	—[d]	$\log A = 2.519\ \sigma_I - 0.152\ MR + 0.030$	6	0.955	0.327	15.6	31
20	—[c]	9α	—[d]	$\log A = 3.025\ \sigma_I - 0.0405\ Pa + 0.823\ \pi + 0.722$	7	0.96	0.34	10.6	46
21	—[e]	—[e]	—[d]	$A = 7814.960 \log Pa - 22674.348$	20	0.84	85.8	44.2	46
22	—[e]	—[e]	—[f]	$\log A = 0.006\ Pa - 4.422$	19	0.67	0.71	13.5	46
23	—[e]	—[e]	—[g]	$\log K_D = -0.022\ SA - 0.59\ PG + 1.50\ \text{tilt} + 6.10\ X - 6.52$	29	0.97	0.26	106	10
24	—[e]	—[e]	—[h]	$\log A = -0.672\ R_m + 2.290$	9	0.964	0.094	nd	50
25	—[e]	—[e]	—[j]	$\log A = 0.110 \log P + 1.626$	15	0.729	0.056	34.9	51[i]
26	—[e]	—[e]	—[k]	$\log RBA = -0.823\ D + 3.13$	6	0.885	0.207	nd	53
27	—[e]	—[e]	—[k]	$\log RBA = -0.164\ \delta + 29.7$	6	0.829	0.250	nd	53

Note: nd = no data.

[a] Inductive effect, σ_I; molar refractivity, MR; hydrophobicity, π; parachor, Pa; Bondi's surface area, SA; hydrophobicity from r-TLC, R_m; octanol-water partition coefficient, P.
[b] n = number of data points (substituents or steroids); r = correlation coefficient; s = standard deviation; F = F-test variance ratio.
[c] Cortisol 21-acetate.
[d] Rat-liver glycogen deposition.
[e] Various.
[f] Antigranuloma activity.
[g] Rat glucocorticoid receptor binding affinity.
[h] Albumin binding.
[i] Equation was recalculated using the published data.[51]
[j] Human plasma protein binding.
[k] Rat mineralocorticoid receptor binding affinity.

Ahmad and Mellors[46] showed that Quale's parachor Pa of the whole steroid can be used as a steric parameter instead of molar refractivity to obtain a correlation with similar significance (see Equation 20 vs. 18 in Table 8). The authors pointed out that parachor could be a useful tool for the diagnosis of non-specific interactions based mainly on hydrophobicity and molar volume, in contrast to interactions being specific to polar groups or stereochemical configurations. While corticoids have high parachor values (above 770), sex hormonal steroids have low parachor values (below 770). With Equations 21 and 22, they were able to explain 71 and 45%, respectively, of the variance in glucocorticoid potency of a larger series containing substituents at different sites of the steroid skeleton.

With the use of 12 steroids, Dahlberg et al.[47] were able to show a highly significant correlation ($r = 0.985$) between the logarithm of the binding affinity to the rat cytosolic glucocorticoid receptor and the logarithmic *in vivo* topical glucocorticoid activity. Furthermore, they assumed that hydrophobicity in the 16α,17α region significantly influenced receptor binding.

Investigating the binding affinity for the human cytosolic glucocorticoid receptor of 35 steroids, Ponec et al.[48,49] found that lipophilicity as a single parameter could only be related to receptor binding for certain substitution sites. They discussed qualitatively the role of hydrophobic and steric effects for 17α- and 21-esters of cortisol and betamethasone. Longer ester chains in both positions (increased lipophilicity) lead to enhanced affinity.

Wolff et al.[3,10] measured the temperature dependency of corticosterone binding to the rat glucocorticoid receptor and calculated the thermodynamics of this binding. From the negative change in heat capacity, positive ΔS and negative ΔH values upon association, they concluded that hydrophobicity is the main driving force for receptor binding. For the entropy of activation being also positive, hydrophobic effects are also considered to play a key role in the formation of the transition state. This conclusion was critized by Mickleson and Westphal,[11] who pointed out thatΔS is negative at physiological temperatures and the principal driving force could therefore not be hydrophobic in nature. In the QSAR correlation (Equation 23 in Table 8) that Wolff et al.[3,10] obtained using 29 glucocorticoids, the hydrophobic parameter SA (Bondi's surface area in excess to progesterone) accounts for only 22% of the variance in the binding affinity (logarithm of equilibrium dissociation constant). Another 35% is explained by the size limitation of the hydrophobic pocket in 9α position (parameter X), 29% by the polar interaction term PG counting oxygen-containing substituents at carbons C11, C16, C17, and C21 in a special way, and 4% by the overall shape descriptor, tilt, parameterized through the C3-C17 interatomic distance. Hydrophobicity, however, was shown by Gandolfi et al.[50] to be the primary factor controlling binding of glucocorticoids and androgens to albumin binding. The correlation (Equation 24, Table 8) is linear with the R_m parameter from reversed-phase thin layer chromatography (TLC).

A linear correlation was also found by Ebling et al.[51] between the log octanol-water partition coefficient and nonspecific binding to human plasma proteins. Equation 25 in Table 8 was calculated for the purpose of this review on the basis of the numerical values published in this paper.

The logarithm of the apparent association constant for human serum albumin (HSA) binding of several unsaturated pregnane analogues was demonstrated[52] to be linearly related to the logarithm of the isobutanol-water partition coefficient.

As far as the binding to the rat mineralocorticoid receptor is concerned, lipophilicity seems to play no essential role. Yamakawa et al.[53] used measured R_m and calculated log P values to characterize hydrophobic properties, X-ray diffraction, and NMR data to study steric and electronic effects. For six steroids they found a linear relationship between the logarithm of the binding affinity and the distance of the O3 oxygen from the least-squares plane through atoms C5 to C14 (D in Å, see Equation 26, Table 8) and also with the ^{13}C-NMR chemical shift of atom C5 (δ, see Equation 27 in Table 8). Accordingly, the flatness of the steroid skeleton must be considered as an important prerequisite for optimal binding to the mineralocorticoid receptor.

3. Estrogens

The free energy of binding to the human myometrical estrogen receptor amounts for estradiol at 20°C to −13.5 kcal/mol where the entropy change is relatively large and positive ($\Delta H = -7.1$ kcal/mol, $T\Delta S = +6.4$ kcal/mol).[54] Once again there is speculation about protein conformational changes and hydrophobic effects contributing to this entropy difference.

On the basis of a decreasing order in the relative binding affinities for the estrogen receptor (1) of 7α-substituents: $-H(100\%) \approx -CH_3(104\%) > -CH_2OCH_3(28\%) > -OH(0.9\%)$ and (2) of 11β-substituents: $-H(100\%) > -CH_2CH_3(78\%) > -CH_3\ (65\%) > -CH_2OCH_3(37\%) > -OH(7\%)$, Zeelen and Bergink[55] argued that the active site of the estradiol receptor is hydrophobic. They explained the fact that hydrophobic substituents do not enhance receptor binding with the absence of preformed hydrophobic cavities in the receptor protein. Accordingly, the formation of the steroid-receptor complex should be accompanied by a conformational change of the active-site receptor structure and by a significant hydrophobic bonding contribution.

For the D-ring region at least, two investigations reveal strong evidence for an essential hydrophobic influence. We[56] examined 3-methoxy derivatives of estradiol substituted at C14, C15 and found significant correlations between the logarithm of the uterotropic activity in mice and hydrophobic parameters (Hansch's π and Rekker's f), e.g., for 14α-configurated steroids:

$$\log A = -0.996\,\pi - 0.392 \qquad (28)$$

$$n = 9,\ r = 0.943,\ s = 0.235,\ t = 7.5$$

and for 14β-configurated steroids:

$$\log A = -1.071\ \pi + 0.349 \tag{29}$$

$$n = 7,\ r = 0.946,\ s = 0.281,\ t = 6.5$$

As the slopes are almost equal in both cases, the hydrophobicity has the same influencing power for both types of steroids. According to Franke,[22] the regression coefficient for the π parameter of approximately 1.0 ± 0.2 indicates a mechanism of two-side binding in a hydrophobic pocket rather than a case of adsorption on a protein surface (π coefficient of 0.5 ± 0.2). Furthermore, a few rat uterine estrogen receptor binding data of 14,15-methylene and 14,15-epoxy analogues in both α- and β-configuration are consistent with the observed *in vivo* potency trend.[56]

A direct relationship of the binding affinity for the rat uterine estrogen receptor and the hydrophobicity was found for 16α-substituted estradiol analogues by Ponsold et al.:[57]

$$\log A = -0.721\ \Delta R_m^2 + 0.126\ \Delta R_m + 1.842 \tag{30}$$

$$n = 7,\ r = 0.985,\ s = 0.058$$

$$\log A = -0.346\ \pi^2 + 0.187\ \pi + 1.831 \tag{31}$$

$$n = 7,\ r = 0.980,\ s = 0.072$$

While Hansch's π and ΔR_m data from reversed-phase TLC gave significant correlations, no influence was seen of sterical and inductive parameters for the small 16α-substituents considered. Fevig et al.[58] studied 26 estradiol derivatives substituted in the 16α position but were not able to establish a quantitative relationship. The estradiol receptor from rat and lamb uterus cytosol seems to prefer small, non-polar groups, has a variable affinity for side chains of moderate size, and appears to respond to geometry or degree of unsaturation in the 16α region.

Peters et al.[59] tried to correlate the antifertility potency to the size and partition coefficient of 17α-(trialkylsilyl)ethynyl derivatives of estradiol. They could only see that the maximum desired effect is achieved if the van der Waals' volume of the three alkyl groups on silicon is between 51 and 72 cm^3/mol.

The MTD method was used by Popoviciu et al.[60] to describe the estrogenicity in rats of 21 steroids and nonsteroidal compounds:

$$\log A = -0.857\ \text{MTD} + 1.126\ \pi + 5.536 \tag{32}$$

$$n = 21,\ r = 0.971,\ s = 0.394,\ F = 180$$

Besides the steric MTD variable, hydrophobicity was found to contribute to the activity variance. The optimized receptor map was used to predict activity data for 25 compounds not included to derive Equation 32.

4. Androgens

Hansch-type structure-activity correlations for androgens were mainly dealing with 17β-esters of testosterone or its derivatives. Biagi et al.[61] found a parabolic relationship between the hemolytic activity and the chromatographic R_m (Equation 33 in Table 9) or Hansch's π parameter as an example for penetration through biological membranes. The linear Equations 34 and 35 (Table 9) describe the dominating influence of the lipophilicity on the androgenic activity of these testosterone esters.[62-64] The fit between R_m and activity data could be improved by selecting an appropriate stationary and mobile phase for thin-layer chromatography in the R_m determination. While log A in Equations 34 and 35 is expressed by the time of maximum biological effect, Equation 36 refers to the overall effect as area under time-response curves.[65] Furthermore, James et al.[65] found relationships between the hydrophobicity and the ester hydrolysis rate (Equation 37) and between hydrolysis rate and the overall androgenic activity ($r = 0.997$) for five compounds. The latter correlation is considered by the authors to be unfortunate because the duration of the biological potency could not be dependent on a slow rate of hydrolysis. The partition coefficient of saturated 17β-esters of testosterone is also found to be the major determinant in the extent of lymphatic transport in rats. Noguchi et al.[66] observed a linear correlation for five steroids with $r = 0.982$.

The anabolic effect (weight of *levator ani* muscles) of nandrolone esters was shown to be related to the logarithm of the oleate-water distribution coefficient by Chaudry and James[67] (Equation 38 in Table 9).

The results illustrate that a single molecular property like lipophilicity can have an impact on the complex biological system in different ways, but remains throughout all individual processes the primary factor controlling the overall potency in a series of analogs. Further discussion of the influence of steroid structure on androgenic and anabolic activity is given by Hüttenrauch,[1] Vida,[68] Kochakian,[69] and by Urbánková et al. in Chapter 10.

5. Cardiotonic Steroids

Digitalis steroids are among the most often prescribed drugs but belong also to the most dangerous ones. To develop new cardioactive steroids with a lower risk of toxicity has therefore always been an area of active research in which knowledge of structure-activity relationships is indispensable. Several attempts were made to identify hydrophobic, electronic, and steric influences.

Biagi[70] found a parabolic relationship between the chromatographic R_m parameter and the toxicity in cats. Using nine compounds, the correlation

TABLE 9
Hansch-Type QSAR Results for Androgens

Equation no.	Steroid core	Substitution site	Activity type	Correlation equation[a]	Statistics[b] n	r	s	F	Ref.
33	—[c]	17β	—[d]	$\log A = -1.723 R_m^2 + 1.561 R_m + 1.502$	14	0.954	0.173	nd	61
34	—[c]	17β	—[e]	$\log A = 0.416 R_m + 0.295$	7	0.934	0.085	nd	62
35	—[c]	17β	—[f]	$\log A = -0.438 R_m + 0.554$	7	0.993	0.037	nd	63
36	—[c]	17β	—[f]	$\log A = -2.394 R_m^2 - 0.671 R_m + 2.837$	5	0.998	0.035	1024	65
37	—[c]	17β	—[g]	$\log k = -1.643 R_m^2 - 0.446 R_m + 2.059$	5	0.999	0.027	805.3	65
38	—[h]	17β	—[i]	$\log A = -0.636 (\log P)^2 + 7.33 \log P - 17.8$	7	0.970	0.064	112.9	67

Note: nd = no data.

[a] Hydrophobicity from r-TLC, R_m; ethyl oleate-water partition coefficient, P.
[b] n = number of data points (substituents or steroids); r = correlation coefficient; s = standard deviation; F = F-test variance ratio.
[c] Testosterone.
[d] Hemolytic activity (rat).
[e] Androgenic activity (capon's comb test).
[f] Androgenic activity (rat).
[g] Hydrolysis rate.
[h] Nandrolone.
[i] Anabolic activity (rat).

coefficient was only 0.619. Davydov[71] obtained a linear correlation for 17 glycosides between the lethal dose and the logarithm of the retention volume in reversed-phase HPLC (Equation 39 in Table 10). It should be noted that the retention volume could be split in additive contributions of individual functional groups.

Klaus et al.[72] tried to relate toxicity, inotropic potency, and Na^+,K^+-ATPase inhibition to lipophilicity but found that lipophilic parameters derived from reversed-phase HPLC, TLC, and octanol-water partition coefficient only account for at most 19% of the variation in ATPase inhibitory activity. There was also no correlation seen between lipophilicity of digitalis glycosides and protein binding.[73]

The steric requirements of the Na^+,K^+-ATPase binding site, considered digitalis receptor or a good model for the receptor site, were studied by means of the minimal topological difference (MTD) method.[74] In addition to the sterical MTD variable, Equation 40 in Table 10 contains an indicator variable s which refers to the strong inhibitory effect of C15 substituents. Wall vertices denoting tight steroid-protein contacts are also found in positions 11β, 12β, 16β, and 22 of the cardenolides. Cavity vertices for both lactone-ring carbonyl and 3β oxygens are in agreement with widely accepted hydrogen bonding mechanisms in these regions. In MTD investigations by Simon et al.[75] (Equation 41) and Motoc[76] (Equation 42) to explain toxicity data for cats, the steroid ring A, however, was found not to be very relevant.

Mori et al.[77] provided a correlation between ^{13}C-NMR chemical shifts (δ) of the carbonyl C24 atom in bufadienolide analogues and cardioactivity data using isolated guinea pig papillary muscles. Although only six compounds were analyzed (see Equation 43 in Table 10) this finding draws some attention to the importance of the charge distribution in the lactone ring. Electronic effects are also considered in the model of Dittrich et al.[78-80]

Guinea pig Na^+,K^+-ATPase inhibition data were used to derive linear Equations 44 to 46 (Table 10) on the basis of a dipole-dipole interaction model. Equation 44 refers to the effective dipole moment μ of digitoxigenin derivatives in the (C21,O14) conformation. This 17β-side-chain orientation places the C21 carbon atom adjacent to the O14 oxygen. Equations 45 and 46 are part of a larger set of correlations found for all, and for highly active cardenolides, respectively. Köpke et al.[80] recommended a multistep procedure using several equations in sequence to estimate the inhibitory activity of an unknown compound. The parameter W is a selected projection of the dipole-dipole interaction force and ϕ is an angle between a projected dipole-moment vector and a particular axis in the model. It is important to know that all results obtained by Köpke et al.[80] are based on the assumption that the (C22,O14) conformation of digitoxigenin is retained in the receptor interaction.

The same assumption is made in the model developed by Rohrer et al.[81,82] Inhibition of rat brain Na^+,K^+-ATPase (Equation 47 in Table 10) was found to correlate linearly with the relative carbonyl-oxygen separation D (in Å)

TABLE 10
Hansch-Type QSAR Results for Cardiotonic Steroids

Equation no.	Steroid core	Substitution site	Activity type	Correlation equation[a]	Statistics[b] n	r	s	F	Ref.
39	—[c,d]	—[e]	—[f]	$LD = 0.104 \ln V_R - 0.044$	17	0.964	nd	nd	71
40	—[c]	—[e]	—[g]	$\log I = -0.753\ MTD - 0.923\ s + 7.383$	20	0.916	0.340	44.2	74
41	—[c,d]	—[e]	—[h]	$\log A = -1.15\ MTD + 0.70$	30	0.894	0.38	nd	75
42	—[c,d]	—[e]	—[h]	$\log A = -0.870\ MTD + 1.462$	30	0.892	nd	nd	76
43	—[i]	17β	—[j]	$A = -0.25\ \delta + 48.35$	6	0.95	nd	nd	77
44	—[c]	—[e]	—[g]	$\Delta G^\circ = 0.46\ \mu - 6.73$	20	0.968	nd	nd	79
45	—[c]	—[e]	—[g]	$\ln I = -0.35\ \Phi - 0.83\ W - 9.13$	32	0.622	1.46	9.2	80
46	—[c]	—[e]	—[g]	$\ln I = 0.75\ \mu_z - 10.85$	13	0.878	0.44	37.0	80
47	—[c]	—[e]	—[k]	$\log I = 0.459\ D - 6.471$	9	0 996	nd	nd	81,82
48	—[c]	—[e]	—[l]	$\log A = 0.54\ D - 6.85$	4	0.99	nd	nd	84
49	—[c,d]	—[e]	—[k]	$\log I = 0.46\ D - 6.37$	16	0.99	0.14	nd	85
50	—[c,d]	—[e]	—[g]	$\log I = 0.1085\ U - 2.885$	6	0.982	0.265	106	86
51	—[c,d]	—[e]	—[g]	$\log I = 0.1015\ U - 3.078$	6	0.978	0.294	86.5	86

Note: nd = no data.

[a] Lipophilicity from HPLC, V_R; minimal topological difference, MTD; NMR chemical shift, δ; effective dipole moment, μ; lactone-ring carbonyl oxygen separation, D; interaction energy, U.
[b] n = number of data points (substituents or steroids); r = correlation coefficient; s = standard deviation, F = F-test variance ratio.
[c] Cardenolide.
[d] Bufadienolide.
[e] Various.
[f] Toxicity.
[g] Na^+,K^+-ATPase inhibition (guinea pig).
[h] Toxicity (cat).
[i] Proscillaridin.
[j] Cardioactivity (guinea pig).
[k] Na^+,K^+-ATPase inhibition (rat).
[l] Inotropic activity (cat).

compared to digitoxigenin. For all the nine steroids under study, X-ray crystallographic data were taken to measure the O23 oxygen separation. Ahmed et al.[83] demonstrated by using five steroids that this relationship also holds with Na^+,K^+-ATPase preparations from hog kidney and cat heart. The ATPase inhibition and oxygen displacement distance also were shown to closely correlate with the inotropic potency in intact cat cardiac tissue[84] (see Equation 48 in Table 10). Fullerton et al.[85] reviewed the whole work and extended the number of data points in the correlation to 16 (see Equation 49), including bufalin, for which a negative distance is used. The results provide strong evidence that a steric complementariness is a major requirement in the receptor interaction.

The idea of using well-defined crystallographic steroid structures and the hydrogen bonding mechanism for the receptor-lactone ring contact, but without missing electronic effects, leads us[86] to develop an approach based on empirical potentials and a simple receptor hydrogen atom model. Equations 50 and 51 reveal that correlations between the guinea-pig heart Na^+,K^+-ATPase inhibitory activity and the calculated interaction energy U (in kJ/mol) can be found for both the (C21,O14) and (C22,O14) digitoxigenin conformations, respectively. However, the significance levels for the two equations do not allow a decision as to which 17β-side-chain orientation is adopted in the bioactive state.

A detailed theoretical study including molecular-mechanics geometry computations for all possible conformations of 46 cardenolide and bufadienolide derivatives as well as structure-activity investigations was performed by Shamovsky et al.[87] From the resulting coordinate maps viewed under hydrogen bonding requirements, the authors concluded that the (C21,O14) conformation for cardenolides and the (C22,O14) conformation for bufadienolides are preferred in the receptor protein environment. There is in addition strong evidence from NMR interpretations[19,88] of structure and activity data, as well as from modeling results discussed in Section III.E, that digitoxigenin derivatives are bound to the Na^+,K^+-ATPase in the (C21,O14) conformation.

Cardiotonic steroids are a good example that the conformation frequently observed in X-ray crystal structures and also used in statistically well-established models (Equations 47 to 49) must not necessarily be the one actually producing the biological effect.

With respect to the sugar moiety of cardiac glycosides, it was concluded that only one sugar, i.e., the one closest to the steroid ring A, takes part in the interaction with the Na^+,K^+-ATPase binding site.[19,85,86] Takiura et al.[90] discussed the cardiotoxicity being influenced by the hydrophobicity of the sugars.

An additional binding site at 16β-substituents, like esters, was suggested.[85,89,91]

Based on NMR data for the C20 atom, Thomas et al.[92,93] proposed a separate charge interaction between this atom and the receptor site. Schönfeld et al.[19] identified the 5β,14β-androstane-3β,14-diol as the minimum structural requirement for specific receptor recognition and as the key structure in inducing the receptor-protein conformational change.

6. Other Steroid Activities Related to Hydrophobicity

As already given in Equation 33 (Table 9), the lipophilicity of testosterone esters mainly controls the hemolytic activity. Biagi et al.[94] showed that this relationship also holds when glucocorticoids, estradiol and progesterone analogs are included in the data set:

$$\log A = -0.514\, R_m^2 + 0.516\, R_m + 3.933 \qquad (52)$$
$$n = 17,\ r = 0.898,\ s = 0.281$$

With the same data set, a similar type of correlation was found for the binding to the erythrocyte membrane closely related to hemolysis.[94]

$$\log A = -0.239\, R_m^2 + 0.782\, R_m + 1.280 \qquad (53)$$
$$n = 17,\ r = 0.951,\ s = 0.137$$

Hydrophobicity can generally be expressed by various parameters being determined theoretically, taken from solubility or partitioning experiments, or from TLC or HPLC chromatography etc. The cross-correlation between different quantities is normally checked and a large amount of data for steroids or their substituents is available.[20,21,23,26,71,95] The octanol-water partition system has not proved to be the best system to optimally describe all biological situations, as was shown, for example, in modeling a *stratum corneum* environment for diffusion through intact skin.[96] For the permeability of *stratum corneum*, a good correlation with the isotropic molecular surface, being the nonpolar part of the solvent-accessible surface area, was detected by Dunn et al.[97]

Lundberg[98] studied the transfer of several steroids from water to octanol at 293 K. The entropy change is found to determine the free energy of transfer for progesterone ($\Delta G = -17.9$ kJ/mol; $\Delta H = +4.2$ kJ/mol; $T\Delta S = +22.1$ kJ/mol) and for testosterone ($\Delta G = -21.5$ kJ/mol; $\Delta H = +1.4$ kJ/mol; $T\Delta S = +22.9$ kJ/mol). Introduction of substituents, like in 17α-hydroxyprogesterone ($\Delta G = -18.9$ kJ/mol; $\Delta H = -10.8$ kJ/mol; $T\Delta S = +8.1$ kJ/mol), can considerably modify the thermodynamic contributions to the free energy although the ΔG value remains almost unchanged. Such a study of the free-energy partitions might also be useful for the steroid-receptor binding to gain further insight in the molecular mode of interactions.

For example, Repke and Dittrich[99] found that for cardioactive steroids at 37°C the formation of both the steroid-Na^+,K^+-ATPase activated and inhibitory complex is an endothermic process which is driven by an entropy increase. The entropy change was argued to be related to protein conformational rearrangements, removal of hydrogen-bonded water, and formations of intrapeptide salt bridges.

The overall ΔS and ΔH data for binding to the glucocorticoid and estrogen receptor have been mentioned in Sections II.B.2 and II.B.3, respectively. The entropy seems to play a role in the binding process for the steroid hormone receptors as well. Unfortunately, the influence of individual substituents on enthalpy and entropy has as yet not been investigated.

C. PATTERN RECOGNITION, DISCRIMINANT, AND FACTOR ANALYSES

Factorial analyses and principal component analyses are used to filter out basic relationships in large data matrices and to systematically reduce the dimensionality of the data space. Such methods can be applied to recognize the main explanatory physicochemical descriptors from a variety of similar parameters, e.g., molecular field data, or to simultaneously analyze data from different biological tests, e.g., different receptor types.

Sokolov et al.[100] calculated net atomic charges by the extended Hückel method (EHT) for 8-azasteroids and found a correlation between the activity coefficients for erythrocyte hemolysis and two principal components derived from the EHT charges:

$$\log A = 0.268\, y_8 - 0.088\, y_7 + 0.914 \qquad (54)$$

$$n \text{ not given},\ r = 0.836,\ s = 0.017,\ F = 9.3$$

where the component y_8 is related to the electron densities on atoms 1, 2, and 8.

Two correspondence analyses based on χ^2 metrics were performed by Doré et al.,[101,102] including 48 and 34 steroids and binding affinity data for the progestin, androgen, glucocorticoid, and mineralocorticoid receptors. As a result of the correspondence analysis, each steroid can be positioned in a space of principal factorial axes. The new set of axes has to be determined by a matrix diagonalization procedure such that each successive axis contains a successively smaller portion of the variance in the data, and much of the variance is projected into a lower dimensional space. In the first analysis,[101] three factorial axes account for 59%, 31%, and 10% of the total variance. The first factorial axis is characterized to be 65% androgenic and 35% progestational in nature. Glucocorticoid (49%) and mineralocorticoid (38%) binding contribute to the second factorial axis. One aim of such an analysis is to identify compounds being most specific for a particular receptor and to establish the structural features they have in common. Therefore, the effect of individual structural modifications, such as the side-chain lengthening at C13 or increasing unsaturation from 4-ene, 4,9-diene to 4,9,11-triene, was investigated by following the locations of the corresponding steroids in the distribution maps, i.e., the two-dimensional projections along selected factorial axes. Introduction of two double bonds, the presence of a 13β-ethyl or 2-oxo group, or the elimination of the 3-keto oxygen decreases the specificity of binding. Besides the normally observed high-specificity high-affinity compounds, the analysis reveals high-specificity low-affinity binders in the case of the androgen receptor. In the second analysis, promegestone and demegestone are shown to define the progestin-receptor pole, while progesterone and 16α-methylprogesterone display some deviation from that pole. The mineralocorticoid pole is, for example, represented by aldosterone, 11-deoxycorticosterone, and cortexolone. A more detailed description of the results can also be found in the chapter of Ojasoo et al. in this book. The correspondence analyses and also a subsequent hierarchical cluster analysis[102] was found useful to highlight the distinctive nature of each receptor class and the biological impact of individual steroid-substructure modifications. This type of factor analysis does not depend on the choice of preselected physicochemical parameters, but it can only be used for steroids or steroid structural features for which activity data are known.

Bodor et al.[103] classified 122 anti-inflammatory steroids into two categories (potent, nonpotent) and applied a pattern-recognition method to describe activity variations in two tests: McKenzie-Stoughton vasoconstrictor test (human) and granuloma cotton pellet test (rat). Potency classification has the advantage that normally more data points can be used, i.e., also data with lower accuracy and from different laboratories. However, all results must be seen in the light of a limited activity information content. Two major problems in pattern recognition are the appropriate choice of structural descriptors and the determination of the discriminant function that defines a hyperplane separating the data space of each class. As descriptors, Bodor et al. used 33 and 17 steroid substructural features (double bonds, substituents) for the two data sets. The number of descriptors was reduced to 22 and 13, respectively, by the weight-sign change feature selection technique that identifies nonessential descriptors. It should be noted that these descriptors, when considered separately, are not responsible for switching activity between the two classes, but they affect activity to some extent and this influence is negated with their elimination.

The discriminant functions of the two data sets were determined by means of the linear learning machine and found to classify all steroids in the training set correctly. In 88 to 89% of the cases, new discriminant functions were able to predict the correct class of the steroid left out by applying a leave-one-compound-out procedure. Since this pattern recognition study has some similarities with the Free-Wilson analysis in dealing with substructures, the ranking of the weight-vector elements of the discriminant function might be compared to the activity contributions from Table 2. The order of some activity-enhancing substructures in the pattern-recognition results for the human test is

$$6\alpha\text{-F}(+0.68) > 16\alpha\text{-Methyl}(+0.38) \approx 9\alpha\text{-F}(+0.37)$$
$$> 1\text{-ene}(+0.25) > 21\text{-OH}(+0.17)$$

whereas the data in Table 2 give the following ranking:

$$21\text{-OH}(>1.0) > 9\alpha\text{-F}(+1.00) > 1\text{-ene}(+0.60)$$
$$> 6\alpha\text{-F}(+0.40) > 16\alpha\text{-Methyl}(+0.30)$$

The similarity is restricted to the fact that all substructures mentioned are described as increasing potency in both schemes. Due to the activity classification, the weight-vector elements are not activity contributions but are contributions to classification. However, Bodor et al. found numerical activity data calculated from weight vectors to be correlated with observed log A data with correlation coefficients of 0.830 and 0.853.

The human vasoconstrictor data set was reexamined by Stouch and Jurs[104,105] who also applied the pattern-recognition method. Instead of substructure descriptors they used the following 10 physicochemical descriptors: Hansch's π hydrophobicity for substituents R in C6-R, C9-R, C17-R, C17-OCOR, C20-R, C20-OR, C20-CH_2OCOR, and three indicator variables for the presence/absence of 1-ene, 16,17-acetonide, and 16ξ-methyl. For the hydrophobicity in 17- and 21-positions, a nonlinear term was chosen. The discriminant function obtained by use of a linear learning machine could correctly classify all 88 steroids of the training set. Other techniques used to check the calculated discriminant function were the K nearest-neighbor (KNN) method and data projections in the space of the first two principal components; 93% of the steroids could be correctly classified by using class information of the first-nearest neighbors. The predictive power of the discriminant function found is underlined by complementary use of various classification criterions (first nearest-neighbor distance, KNN class, multiple discriminants, principal components plots) applied to a new set of steroids. Out of 14 prediction steroids of unambiguous activity, only one was consistently misclassified. All these classification results provide some evidence that the ten above-given physicochemical descriptors are able to form a ten-dimensional space in which both activity classes are properly separated. These molecular properties therefore account for the variance in the antiinflammatory potency.

Pattern recognition methods as well as linear and quadratic discriminant analyses were used by Henry and Block[106,107] to classify 46 compounds as estrogens, androgens, progestins, corticoids, or cardiotonic steroids. Out of four different types of fragment molecular connectivity indices, the first-order molecular connectivity MC_1 was found to give the best results.[106] MC_1 values for the steroid carbon atoms C3, C5,C10, C17, C20 (or atoms occupying topologically equivalent positions) were used as optimal descriptors to differentiate between steroid classes. The discriminant functions determined by means of various methods were able to classify the following numbers out of 46 steroids correctly:

Linear discriminant analysis:	38 (83%)[106]
Quadratic discriminant analysis:	41 (89%)[106]
Single Andrews discriminant function:	34 (74%)[107]
K nearest-neighbor analyses (K = 3,4,5):	41 (89%)[107]
Multicategory linear learning machine analysis:	39 (85%)[107]
Statistical isolinear multicomponent analysis:	39 (85%)[107]
Linear discriminant analysis:	41 (89%)[108]
Adaptive least-squares analysis:	44 (96%)[108]

An inspection of the misclassified steroids reveals that particularly progestins and androgens are not easily separated by the connectivity descriptors chosen. It must be estimated that the predictive value of the described discriminant functions is rather limited:

- 15 to 45% of steroids in a nine membered test set were misclassified.[106-108]
- A manual classification by a steroid researcher would probably give better results.
- For steroids it is known that relatively small variations in the steroid structure may cause profound changes in biological activity (e.g., androgenic testosterone vs. anabolic 19-nortestosterone vs. gestagenic 17α-ethynyltestosterone).

It is difficult to believe that topological indices of a few atoms can account for a large variety of compounds with a certain reliability. On the other hand, the identification of structural descriptors with activity-discriminatory power is a first step on the way to establishing new QSAR correlations using multivariate statistical methods.

D. MOLECULAR FIELD STUDIES

The use of physicochemical parameters in linear free-energy QSAR relationships (Hansch-type analyses) have been demonstrated in Section II.B to describe properties of steroids or their substituents. In the previous section, physicochemical descriptors are used in pattern recognition or discriminant analyses to define the space in which activity information is projected and interpreted. In all these approaches, hydrophobic data of fragments, net charges on atoms, topological information as MTD, etc. were taken as single variables but not as information describing the three-dimensional space around a molecule in a systematic way. Such a three-dimensional description of molecular properties can take into account electron distribution, conformational details, the nature of atoms and bond contained in a molecule, and other relevant information with respect to its actual location in space. Some methodical work has been done in this area, e.g., the distance geometry approach with emphasis on conformational properties,[109] the GRID program with emphasis on interaction of a molecule with a probe,[110] or development of hydrophobic fields.[111-113]

A general three-dimensional QSAR description can be given if three-dimensional field data are introduced in the right side of Equation 2 in Section II.B. This type of QSAR investigation is covered by the comparative molecular field analysis (CoMFA)[114] which, in the present SYBYL program implementation,[115] takes the steric and eletrostatic fields into consideration. In this method the three-dimensional molecular structure of all molecules in the data set is aligned, steric and electrostatic interactions with a probe are computed for each point of a three-dimensional grid around the molecules, and the variance in these field data is used to explain the variance in the activity data by means of partial least-squares (PLS) and cross-validation statistics.

This method was introduced with an application on steroids by Cramer et al.[114] The affinity of 21 steroids for corticosteroid-binding (CBG) and testosterone-binding globulins (TBG) was studied. Structures were minimized using the standard TRIPOS force field and superimposed by fitting C3, C5, C6, C13, C14, and C17 carbons. The influence of the probe van der Waals' radius, of the separation between grid points, and of the relative grid-origin position were investigated. Individual variations in all these methodical parameters are shown to improve the correlation but only to a limited extent. The correlations with two PLS components (two degrees of freedom in indepdendent variables) by using the standard (not the best) parameters are found to have conventional correlation coefficients of $r = 0.947$ (CBG) and $r = 0.934$ (TBG) and predictive cross-validated r^2 of 0.662 and 0.555, respectively. The CBG activity of seven steroids that were not part of the training set could be predicted with $r^2 = 0.81$ whereas predictions for three steroids with new structural features (1-ene, 2α-methyl, 9α-fluorine) were pocrer.

In addition to the predictive ability of CoMFA, the regions in the steric and electrostatic field around the molecules that contribute most to the activity variations can be identified. With the SYBYL molecular modeling software,[115] color-coded isopotential maps for each field visualize regions in which appropriate substituents increase or decrease the activity.

There are other successful CoMFA studies on steroids, which have not yet been published.

A method bearing some resemblance to CoMFA has been developed by Allèra and Klein:[116] the automated receptor modeling (ARM) method. Atoms of the unknown active site of the protein are treated as pseudo-atoms. Locations, van der Waals' radii, and charges of these pseudo-atoms are optimized to calculate ligand-protein interaction energies that quantitatively reproduce the variations in experimental binding affinities. Using this ARM method and a data set of 15 steroids, it was possible to explain binding affinities for a rat liver plasma membrane-inserted carrier.[116] The role of 21-OH, 11-oxo, and 11β-OH functions was discussed. The validity of the carrier model was demonstrated by correct prediction of affinities of more than 15 steroids which were not used to generate the model.

III. MOLECULAR MODELING AND CONFORMATIONAL ANALYSES

As briefly outlined in the previous section, the CoMFA method can be regarded as a link between the classical Hansch-type QSAR technique and molecular modeling approaches.[117] One prerequisite for the use of CoMFA is the knowledge of the molecular structures and their alignments, where molecular structure refers to the geometrical, including conformational, and electronic structure. The modeling papers published in the steroid field are mainly dealing with the isolated steroid molecule. This is, obviously, the first step but structures may be different in natural crystal packing, solvent, or protein active-site environment.

Interestingly, approaches have been reported to model the binding domain of the human androgen receptor and of the human corticosteroid binding globulin[118] (see also chapter by Ojasoo et al. in this book and articles cited therein).

In the following section, the results in computing the structure of steroid molecules will be summarized.

A. STEROID BACKBONE GEOMETRY

Caillet and Pullman[119] used the all-valence electron quantum-chemical method of pertubative configuration interaction using localized orbitals (PCILO) and standard valence geometry to calculate the relative stability of the sterane ring system with different ring fusions. They found those natural ring junctions most stable where the B/C and C/D rings are *trans* but where A/B rings may be either *trans* or *cis* with somewhat greater probability for the latter arrangement of being also *trans* (see Table 11). On the contrary, the experimental and force field data presented by Allinger and Wu[120] indicate a higher thermodynamical stability of the C/D *cis* fusion in androstanes. The relative stability of the A/B *trans* rings junction was also confirmed by MINDO/2 calculations.[121] More stability data for steroid-relevant systems including oxygen-substituted compounds are provided by Allinger and Tribble.[122] The data of Table 12 give an indication of the overall accuracy of force-field calculations. The year in which the data were published is included in Table 12 since all force-field parameterizations are in continuous development. In 1974 Romers et al.[123] gave an overview about X-ray and molecular mechanics structure data known at that time. With respect to biological activity, they speculated about the role of the overall bent shape of steroids. This shape

TABLE 11
Relative Stabilities of Steroid Skeleton Isomers in kcal/mol

Isomer	Relative PCILO energy[119,a]	Relative force-field energy[120,b]	Relative MINDO/2 heat of formation[121] —[c]	—[d]	Experimental ΔG at 573 K[120,b]
5α,14β	4	0.00	nd	nd	0.00
5β,14β	7	0.73	nd	nd	1.48 ± 0.35
5α,14α	0	1.18	0.00	0.00	1.77 ± 0.45
5β,14α	2	1.87	0.34	0.33	2.66 ± 0.22

Note: nd = no data.

[a] Gonane skeleton.
[b] Androstane skeleton.
[c] Androstan-17-one skeleton.
[d] Androstane-3,17-dione skeleton.

controls the separation of the oxygen atoms attached to the A and D rings and the appropriate directions of the corresponding C-O vectors. The flexibility of the A, B, and D rings is considered to be a necessary condition for the formation of hydrogen bonds with the receptor at both steroid ends.

Already Allinger's 1973 force field was shown[129] to describe the structure of androsterone that 90% of the bond lengths and bond angles in the calculated structure are within the reported experimental errors.

Shamovsky et al.[130] developed a specific force field to investigate steroids and compared the calculated structures with experimental data by use of, for example, a skeleton bowing index, pseudorotational parameters, and the O3-O17 oxygen separation.

Cox and Turner[131] found only small differences in geometry between molecular mechanics and X-ray data of 5α-androst-16-en-3-one. The only exception is the C16=C17 double bond where the X-ray study gave 1.302 Å and the calculation 1.336 Å, which is close to the strain-free value and fits into the generally observed range of 1.327 to 1.354 Å for 16-ene steroids substituted at C17. A good agreement between experimental and MM2(77) theoretical structure data was reported by Castells et al.[132] for sp-nitrogen containing compounds including the progestational steroid STS-557 when the cyano-group parameters were readjusted.

For 4,4-dimethyl-5α-androstane-3-one, an *ab initio* geometry optimization using the 4-21G basis set was performed by van Alsenoy et al.,[133] who pointed out that bond lengths and angles are reliable at a 0.001 Å and 0.1° level, respectively.

When applying several molecular mechanics methods to compute the structure of 20-methyl-5-pregnene-3β,20-diol for which very accurate low-temperature neutron diffraction data are available, Duax et al.[126,134,135] found

TABLE 12
Accuracy of Internal Coordinates Determined by Molecular Mechanics (RMS Deviations Between Diffractional and Theoretical Data)

				Nonhydrogen atom geometry			
Year	Diffraction method	Steroid	Force field	Bond length (Å)	Bond angle (°)	Torsion angle (°)	Ref.
1972	X-ray	—[a,b]	Allinger	0.010	1.31	nd	122
1974	X-ray	—[a,c]	Allinger	0.006	nd	nd	123
			Altona	0.008	1.38	nd	
			Warshel-Lifson	0.011	0.62	nd	
			Boyd	0.012	0.74	nd	
1974	X-ray	—[a,d]	Allinger	0.007	nd	nd	124
			Altona	0.006	nd	nd	
			Warshel-Lifson	0.012	nd	0.7	
			Boyd	0.011	nd	1.5	
1977	X-ray	—[e]	GEMO	0.01	0.9	2.0	125
1983	Neutron	—[f]	MM2p	0.02	2.5	nd	126
			QCFF/MCA	0.04	5	nd	
			PROPHET	0.12	8	nd	
1985	X-ray	—[g]	MM2	0.0086	0.79	nd	127
1988	X-ray	—[h]	MMP2	0.014	1.4	3.7	128
		—[i]	MMP2	0.018	1.4	5.2	
		—[j]	MMP2	0.020	1.4	2.6	

Note: nd = no data.

[a] Only C-C bonds considered.
[b] 3α-Hydroxy-5α-androstan-17-one.
[c] 20-methyl-5α-pregnane and other 5α-steroids.
[d] 17β-Isopropyl-5α-androstane.
[e] 17α-Hydroxyprogesterone.
[f] 20-Methyl-5-pregnene-3β,20-diol.
[g] 17β-Hydroxy-19-nor-5α,17α-pregn-20-yn-3-one.
[h] Gestodene.
[i] Mifepristone.
[j] 17α-Ethynylestradiol.

1. That PROPHET, QCFF/MCA, and MMPI programs underestimate the bond lengths involving hydrogen atoms and PROPHET also overestimates the bond lengths involving nonhydrogen atoms
2. That the MM2p program generates the most reliable bond lengths and angles (see Table 12)
3. A tendency for MM2p calculations on fused ring systems with a double bond to enforce planarity to a greater degree than is appropriate
4. A systematic difference in the 17β-side chain orientation between X-ray and MM2p data most probably due to inadequate force-field parametrization

Osawa et al.[136] tested the MM2(77) force field with respect to barrier heights of internal rotations about C–C bonds in cholestane, 20-isocholestane, and other compounds. This force field reveals satisfactory results for simple hydrocarbons but systematically underestimates barrier heights for highly congested molecules.

The effect of a substituent on a remote portion of the steroid skeleton, also known as conformational transmission, is discussed by several authors.[123,133,134,137-142] Schmit and Rousseau,[142] for instance, examined the influence of individual structural modifications in corticoids on backbone characteristics. 1-Ene unsaturation and 11β-hydroxy substitution was found to increase the steroid skeleton convexity.

B. ELECTRONIC STRUCTURE

Besides conformational steric transmission, conformational electronic transmission was considered by Kollman et al.[143] for cortisol derivatives. They calculated the charge distribution of 6α- and 9α-fluorocortisol by the semiempirical CNDO/2 method and discussed long-range effects relevant to biological activity between C9 and C4, between C6 and C11 carbons, as well as an increased acidity of the 11β-hydroxyl hydrogen caused by the 9α-fluoro substitution. X-ray crystal structure data of these corticosteroids suggest a relationship between the antiinflammatory potency and the bowing of the steroid backbone towards the α-face.[144]

Schneider et al.[140,141] also described long-range influences, e.g., of the 3β-substituent on the ^{13}C NMR chemical shift at C6 correlating with the electric field effect of the 3β-substituent.

The electronic structure in terms of molecular orbital energies, charge distribution, Mulliken population, and dipole moments was calculated and compared for several representatives of sex hormones by the extended Hückel theory (EHT),[145] of corticoids by EHT,[146] and of cardiotonic steroids by EHT and CNDO/2.[147,148] We[148] also used the CNDO/2 wave functions of molecular fragments (C and D rings and 17β substituent) to compute molecular electrostatic potential (MEP) contour maps of some Na^+,K^+-ATPase inhibitors. The location of the potential minima and the observed inhibitory activities support the concept of a flexible geometric adaptation in the binding process. Chaudhary et al.[149] performed PCILO and CNDO/2 calculations on bisquarternary azasteroids. A high degree of delocalization of the positive charge was seen.

Gale and Saunders[150] used dipole moments from CNDO/2 computations of A- and D-ring fragments to explain the influence of the 17α-ethynyl group on the solubility of testosterone and estradiol analogues.

Gassiot et al.[151] determined net atomic charges and the total electronic energy of 59 steroids by means of the Del Re method. They found highly significant linear correlations of the charges at 12 different steroid carbon atoms, the electronic energy, and the molecular volume with chromatographic retention parameters for various stationary phases. The multiple regression coefficients are in the range of 0.985 to 0.995.

C. STRUCTURE CALCULATIONS USED IN NMR INTERPRETATIONS

NMR experiments only implicitly provide information about charge distribution being related to NMR chemical shifts and conformational details encoded in the coupling constants. Therefore, structure data from calculations or diffraction studies are taken to aid the interpretation of NMR results. This has to be done with caution since there may be significant conformational differences due to crystal packing forces, solvent interactions, and/or methodical errors. In general, the dependence of coupling constants on torsion angles is given by the Karplus equation.

Reviews about NMR investigations on steroid structure are given in Chapter 2, by Croasmun and Carlson,[152] and by Kirk.[153]

Altona and Hirschmann[154] were able to explain the observed J(17α-H,20-H) of 7.5 Hz for 5α-pregnane-3β,20α-diol by a time-averaged coupling of the two energetically preferred conformations determined by molecular mechanics: (1) 75% conformer E (20-H anti to 17α-H) with J_{calc} = 9.2 Hz and (2) 25% conformer F (20-O anti to 17α-H) with J_{calc} = 0.9 Hz. The experimental J(17α-H,20-H) value of 9.0 Hz for the 20β-epimer conforms with the theoretical result showing that the conformer with J_{calc} = 9.2 Hz predominates in the equilibrium to 98%.

Sedee et al.[127] predicted vicinal coupling constants on the basis of MM2(77)-optimized structures of 17β-hydroxy-19-nor-17α-pregn-20-yn-3-one isomers. The RMS deviations between observed and calculated couplings in the range of 0.54 to 0.67 Hz compare to the estimated experimental error of 0.5 Hz. In this comparison of spectroscopical and theoretical ^3J values and in the analysis of pseudorotational parameters of the D ring, it was found that the 15α-H/16α-H coupling was not well described by the Karplus equation. A selective reduction of that coupling by the influence of the C13 carbon atom in the observed 13β-envelope D-ring conformation was discussed. The conformational features observed in solution by NMR are found to be in close agreement with the crystal and force-field optimized structures.

In the interpretation of geminal coupling constants for a series of 19-nor-4-en-3-one steroids, we[155] were able to assign the observed ^2J(6α-H,6β-H) values of 14.5 to 15.0 Hz to structures having a normal A-ring conformation (see Section III.D). This conclusion is based (1) on a linear correlation between experimental and calculated J constants, and (2) on a significant difference in the coupling at C6 theoretically determined for the normal and inverted A-ring conformers (ΔJ = 2 Hz) by use of molecular mechanics and INDO molecular orbital computations.

The conclusion that the normal A-ring conformation is predominant in solution was also drawn by van Geerestein[128] by comparing observed and computed vicinal coupling constants based on MMP2-optimized structures of 10-methyl (testosterone) and 19-nor-4-en-3-one steroids (norgestrel, 3-keto-desogestrel).

Schneider et al.[156] discussed the D-ring conformations of 17-oxo and 17-acetyl steroids as well as A-ring conformations of 10-methyl-4-en-3-one steroids by comparing vicinal coupling constants from MM2 calculations and NMR. It was concluded that the A-ring sofa and the D-ring half-chair/envelope intermediate conformations are preferred in solution.

In addition, Schneider et al.[140,141,156,157] used structure data generated by molecular mechanics and electric-field parameters of substituents to distinguish between the steric and electronic nature of observed substituent-induced NMR shifts.

^{13}C-NMR chemical shifts of cholesterol analogues with differently substituted 17β side chains were calculated by Gonzalez-Sierra et al.,[158] taking a predetermined conformer population into account. A generalization of this approach might be helpful in the validation of theoretical predictions of side-chain orientations in solution. Other sources of information are provided by intramolecular distances taken from NOE (nuclear Overhauser effect) measurements or barriers to internal rotation determined by NMR relaxation studies.

The latter type of NMR investigations was applied by Bernassau et al.[159] to the barriers of 10β-and 13β-methyl rotation in 5α-androstane, 5α-androst-2-ene, androst-4-ene, androst-5-ene, and 5α-androst-7-ene. The general trend in the spectroscopically observed rotational barriers were found to be well reproduced by MM2(77) energy profiles. The molecular mechanics energy contributions were further used to identify the nature of underlying interactions. Differences mainly in torsional energies and van der Waals' contacts between the minimum- and maximum-energy conformations are responsible for the variations in barrier heights.

D. GEOMETRY IN THE A,B-RING REGION

The conformational mobility of the first sugar at the C3 position of cardiac glycosides was studied by Fullerton et al.[85,160,161] Potential-energy contour maps were calculated by rigid-body rotation about the C3-O3 and O3-C1′ bonds of digitoxigenin β-D-digitoxoside and digitoxigenin α-L-rhamnoside (see Figure 2) using the CAMSEQ program. The calculations indicate very limited conformational freedom in agreement with crystal structure data.[161] In addition to nonbonded interactions, electronic effects[85,161] are mentioned that may further restrict the conformational space. Chiu and Watson[162] confirmed the potential-energy results by use of the COMOL program. Superimposing the preferred conformations of the α-L-rhamnose and β-D-digitoxose orientation in the 5β-H cardenolides with the structure of the conformationally rigid and highly potent 5α-configurated gomphoside (see Figure 2), they developed a model for the sugar-binding site of the cardiac receptor. This

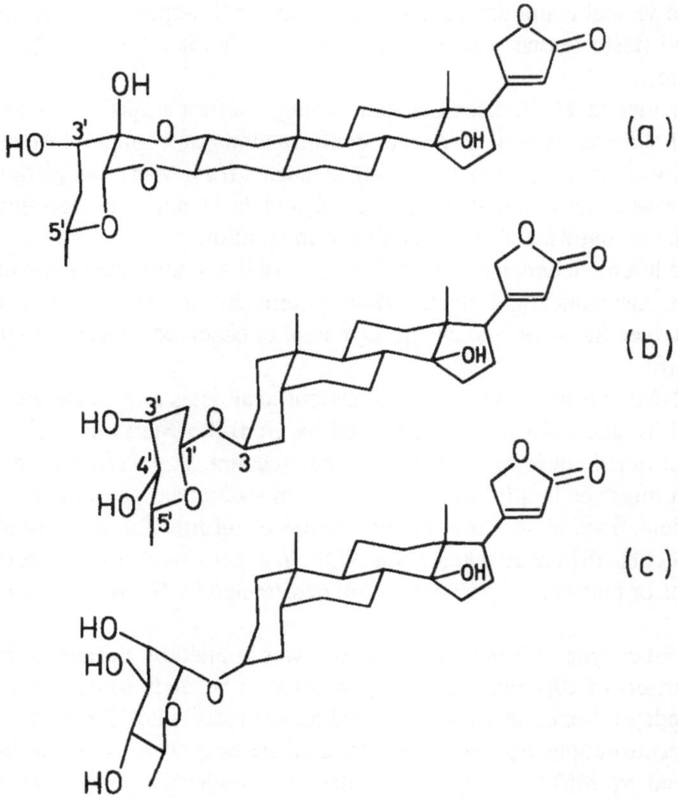

FIGURE 2. (a) Gomphoside, (b) digitoxigenin β-D-digitoxoside, and (c) digitoxigenin α-L-rhamnoside.

model that also takes data of inotropic activity at isolated guinea-pig atria for gomphoside analogues into account consists of a hydrogen bond to the axial 3′-hydroxyl group and a broad hydrophobic region around the 5′-methyl group. In the hypothetical binding conformation of digitoxigenin α-L-rhamnoside (torsion angle C2-C3-O3-C1′ of 150 to 160° and torsion angle C3-O3-C1′-C2′ of 170 to 180°), the 3′-hydroxyl is separated from that in gomphoside by about 3 to 3.5 Å, the 4′-hydroxyl by about 2 to 2.5 Å, and the 5′-methyl by less than 1 Å. These substituents were also found in other investigations[93,163] to play an important role in receptor binding.

Another example for mapping the hydrogen bonding zone around the A-ring was presented by Delettré[164] and Raynaud et al.[165,167] They used superpositions of crystal and GEMO molecular-mechanics optimized structures as well as progesterone-receptor-binding affinities of progesterone and three A-nor steroids to propose locations for hydrogen bond donors to the O3 carbonyl oxygen.

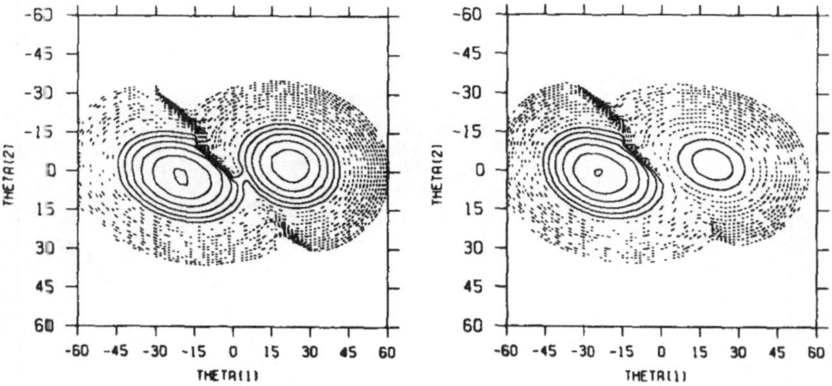

FIGURE 3. MMP2(85) Potential-energy surfaces describing A-ring conformations of norgestrel (left) and 10-methyl substituted 3-keto-desogestrel (right), contour lines are drawn at 2 kJ/mol intervals, broken lines start 10 kJ/mol higher than global minimum energies, theta(1) = C4-C5-C10-C1 and theta(2) = C2-C3-C4-C5. (From van Geerestein, V. J., Conformational Aspects of Synthetic Sex Steroids, Thesis, Utrecht, 1988.)

In an extensive comparison of crystallographically observed conformations of 10-methyl-4-ene, 19-nor-4-ene, 4,9-diene, and 4,9,11-triene steroids, Delettré et al.[168] found the dienes and trienes to be more mobile and flexible than the monoenes. This conclusion was supported by GEMO molecular mechanics energy calculations of the minimum-energy structure deformation. About 1 kcal/mol is required to allow the different backbone bowing observed for 4,9-dienes and 4,9,11-trienes.

In addition to the overall bowing of the steroid skeleton, 4-ene steroids can adopt two basic A-ring conformations (normal: 1α-sofa, 1α,2β-half chair, 2β-sofa inverted: 1β-sofa, 1β,2α-half chair, 2α-sofa) while 4,9-dienes have alternatives for both the A- and B-ring conformation (normal: 6β-sofa,6β, 7α-half chair, 7α-sofa inverted: 6α-sofa, 6α,7β-half chair, 7β-sofa).[128,168-171] In accordance with crystal-structure observations, GEMO optimizations indicate a preference for the normal A-ring conformation for 10-methyl-4-en-3-one steroids[170,172] and for the inverted A-ring conformation for 4,9-dien-3-ones.[170] In the same calculations, a removal of the 10-methyl group was found to lower the relative stability of the normal A-ring conformation by 4 to 7 kJ/mol. Accordingly, 19-nor-4-en-3-one steroids reveal a high A-ring conformational flexibility that is also experimentally proved.[155,173-175] GEMO potential-energy curves[170] and two-dimensional MMP2 energy contour maps[128] for selected 19-nor steroids characterize the interconversion between the normal and inverted A-ring conformers (see Figure 3). Molecular mechanics calculations were also used to investigate the influence of individual substituents on the relative A-ring conformational stability: 1α-methyl and 2α-methyl substitutions in 3-keto-desogestrel[128] lead to a preference of the normal A-ring conformer while a 1β-methyl group[128] increases the stability

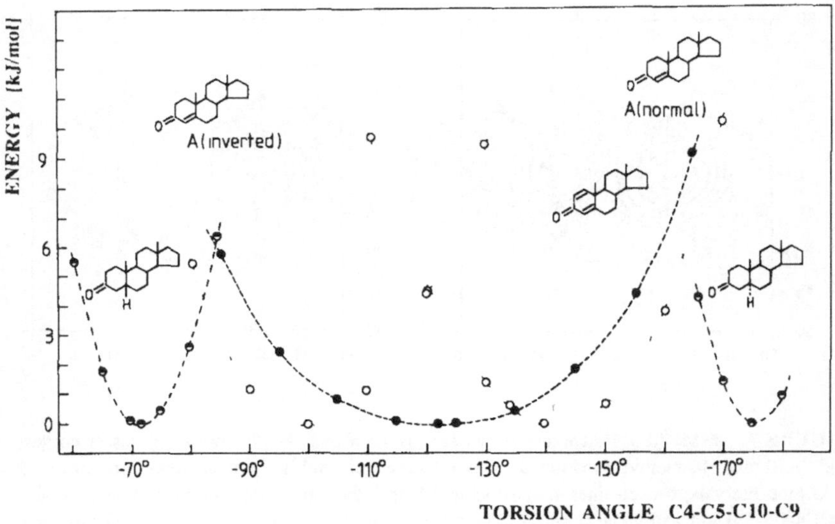

FIGURE 4. MM2p Potential-energy curves for variation in the torsion angle C4-C5-C10-C9 of 5β-androstan-3-one, 4-androsten-3-one (normal and inverted A-ring conformation), 1,4-androstadien-3-one, and 5α-androstan-3-one (all individual energy-minimum conformations are set to zero energy).

of the inverted conformation; 11β-phenyl in 13β-configurated 4,9-dien-3-one steroids tends to stabilize the normal A-ring conformation[171] so that A(inverted)B(normal) and A(normal)B(normal) conformations are energetically preferred[128,171] while the 1α-methyl-11β-phenyl dienone derivative favors the normal A-ring conformers by about 25 kJ/mol compared to A(inverted)B(inverted) and A(inverted)B(normal) conformations.

Figure 4 illustrates the different flexibilities of 3-one, 4-en-3-one, and 1,4-dien-3-one 10β-methyl steroids in bending the A ring with respect to the remaining steroid skeleton. For each point of these potential curves, all internal coordinates with exception of the driving torsion angle C4–C5–C10–C9 were optimized by using MM2p molecular mechanics. As can be seen from the broad potential well, the 1,4-unsaturation significantly increases the conformational mobility. This is confirmed by an analysis of 26 relevant crystal structures of 1,4-dien-3-one steroids. The separation of the O3 oxygen atom from the least-squares plane through C,D-ring carbon atoms ranges from 197 to 351 pm in these X-ray determinations. The MM2p-optimized structure of the 1,4-dienone steroid is found to fit well into the crystal structural region and takes an intermediate A-ring location between the normal and inverted A-ring conformations of the 4-en-3-one steroid.[176]

In the receptor binding process, each steroid will take advantage of its individual conformational freedom to meet the steric and non-geometrical requirements of the active site and, thus, to optimize the interaction energy.

TABLE 13
Relative Binding Affinity to the Progesterone Receptor and Location of the O3 Oxygen Atom

Steroid structure type	Relative BE[a]	Steroid CSD code[b]	Distance of O3 to the C,D-ring plane (pm)
10-Methyl-4-en-3-one (normal ring A conformation)	0.0	BPROGE	128
5α-H-3-one	0.7	OXANTS	141
1,4-Dien-3-one	4.2	CUFXOQ	197
10-Methyl-4-en-3-one (inverted ring A conformation)	nd[c]	MACXPR10	308
5β-H-3-one	8.2	(MM2 optimized)	340

[a] Relative binding energies (in kJ/mol) for 10-methyl-17β-acetyl steroids (progesterone derivatives) reported by Seeley et al.[13]
[b] CSD = Cambridge Structural Database.
[c] nd = no data.

It is therefore not the energy-minimum conformation but the most planar observed structure of each type of 10-methyl steroids which is taken in Table 13 for a comparison with the relative binding affinities for the progesterone receptor. No binding data for the 10-methyl compound with an inverted A-ring structure are known although a value of more than 10 kJ/mol was reported for 2β-methyl-4-pregnene-3,20-dione[13] which probably adopts an inverted A-ring conformation.

The close relationship between the binding energy and the displacement of the O3 oxygen from the steroid-backbone plane gives evidence that (1) highly planar progestins are preferably bound by the progesterone receptor and (2) the normal A-ring conformation of 4-en-3-one steroids seems to be the bioactive structure.

This is in agreement with the proposed location of the hydrogen bonding zone mapped by Delettré et al.[164-167] approximately within the plane of a rather flat progestin. It is also consistent with the finding that the 10-methyl group may cause steric hindrance in the receptor contact[41,170] possibly by forcing the O3 oxygen to occupy a more α-side directed position. For 4,9-dien-3-ones, the relative O3 oxygen locations of normal and inverted A-ring conformers are found to be very close.[171] This enables the inverted A-ring conformation of the dienones to occupy an optimal orientation with respect to the receptor as well.

On the contrary, Duax et al.[169,177-179] claimed that the inverted 1β,2α-half-chair conformation of 4-en-3-ones is optimal for high-affinity binding to the progesterone receptor (see also Chapter 1). This hypothesis has already been questioned by van Geerestein[128] and by us[170] when comparing receptor affinity data and preferred conformations in the 4-ene, 4,9-diene, and 4,9,11-triene series.

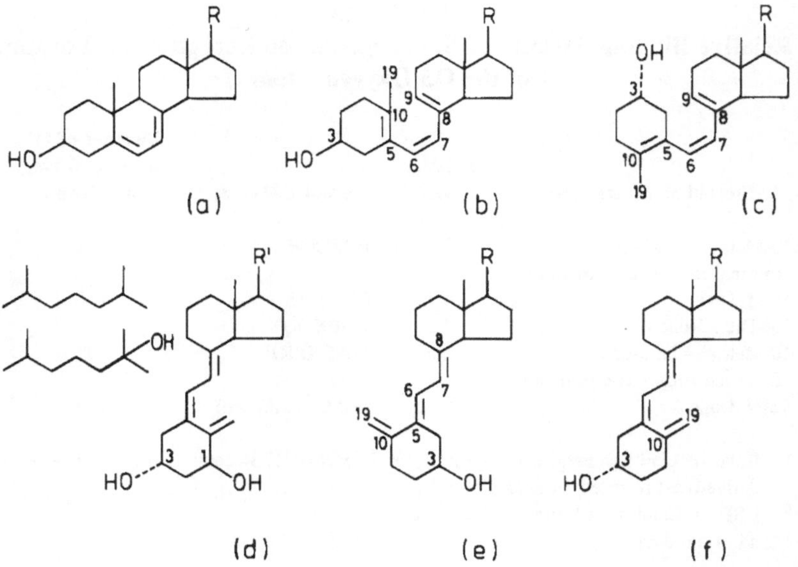

FIGURE 5. (a) Provitamin D_3, (b) previtamin D_3 (cZc), (c) previtamin D_3 (tZc), (d) 1α,25-dihydroxyvitamin D_3, (e) vitamin D_3 (5*E*), (f) vitamin D_3 (5*Z*).

For the discussion of the enzymatic hydroxylation of aromatase substrates, MM2 and PCILO calculations were performed to study the 10β side-chain orientations of steroids bearing 19-hydroxyl groups.[180,181] Only small differences were found between the energetically preferred conformations and barriers to internal rotations of 19-hydroxy-5α-androstane-3,17-dione, 19-hydroxyandrost-4-ene-3,17-dione, and 19-hydroxyandrosta-4,6-diene-3,17-dione. For the 4-ene and 4,6-diene, the global energy-minimum orientation around the C10–C19 bond corresponds to the enzymatically preferred 19-pro*R* hydrogen location in out-of-ring position.

With regard to compounds of the vitamin D family, ground-state conformational details were studied using molecular mechanics for previtamin D_3[182] and for vitamin D_3 derivatives[183,184] including the natural hormone 1α,25-dihydroxy-vitamin D_3 (see Figure 5). Dauben and Funhoff[182] calculated MMP2 3D-plots and energy-contour maps by varying the C6–C7–C8–C9 and C10–C5–C6–C7 torsion angles of previtamin D_3 in both 2,3-half chair conformations of ring A (corresponding to pseudoaxial and pseudoequatorial 3-OH). They found a distribution at 298K of 46% cZc and 54% tZc forms while the global minimum is given by the (+)tZ(−)c geometry with a pseudo-equatorial 3-OH. Mosquera et al.[183,184] applied the MMPI force filed to determine, for example, the equilibrium populations at 298K for the equatorial/axial 3-OH group of 68%/32% (5Z-vitamin D_3), of 64%/36% (5*E*-vitamin D_3), and of 46%/54% (1α,25-dihydroxy-vitamin D_3). These ratios agree with those deduced from NMR data within the limits of the experimental error.

The dienic region B was found to be distorted from planarity in maximum by 20 to 30°C, depending on the A-ring substitution. A-ring chair conformations were calculated to have lower energies than half-chair conformations in most cases of vitamin D_3 derivatives. The authors of these investigations think that selective alterations in geometrical features could help in discovering potentially more powerful and selective analogues.

E. GEOMETRY IN THE C,D-RING REGION

Based on a careful analysis of a bulk of X-ray crystal structure data, Duax et al.[177,178,185] concluded that the tight steroid-receptor contact involves only the A ring of estrogens, gestagens, and corticoids, while the B, C, and D rings of those steroids play no or a minor role in receptor binding but are the most important factor in determining agonist vs. antagonist behavior (see also Chapter 1). Differences in van der Waals' volumes between superimposed crystal structures of *trans*-tamoxifen vs. estradiol and a selected progestin vs. antiprogestin[178,185] highlight some similarities in the spatial regions specifically occupied by those antagonists. Powerful antiprogestins and antiglucocorticoids were found by substitutions at the 11β position.[186-189] Structure calculations and modeling approaches on this type of steroids were performed by using the SCRIPT,[187] SYBYL,[189] MMP2(85),[128] and MM2p[171] programs.

The influence of the removal of the 13β-methyl group in C/D *trans* and *cis* fused steroid on the D-ring conformation and 17β-side chain orientation was studied by Shamovsky et al.[130] using molecular mechanics. MM2 computations revealed in accordance with X-ray observations that the 13β-ethyl conformation is affected by the presence of the C15=C16 double bond in 14α-H steroids.[128,190] D-ring saturated steroids adopt an orientation of the terminal 18-methyl group being *anti*-periplanar to C14 (energy stabilization by 6 kJ/mol in 3-keto-desogestrel[128] and by 8 kJ/mol in levonorgestrel[190]), whereas this methyl in the 15-ene steroid gestoden is energectically favored by 2kJ/mol in *anti*-periplanar location with respect to C12.[128] The energy situation for gestoden was confirmed by Eckle et al.[190] using the semiempirical MNDO method.

Because of its importance to biological activity, conformational aspects of side chains attached to C17 are studied in a series of papers. Schmit et al.[191] used GEMO geometry optimizations to evaluate the influence of the 17α-methyl group on the D-ring structure of androgenic and anabolic steroids, to describe hydrogen bonding directions for the 17β-hydroxyl group, and the potential-energy curve for rotation about the C17–O17 bond of the 17β-methoxy derivative of testosterone. We[192] studied potential interferences between 17α-ethyl and 17β-hydroxyl conformations by computing an MM2 potential-energy surface and found some evidence for an IR and NMR interpretation suggesting that the 17α-ethyl substituent modifies the geometrical capabilities of the 17β-hydroxyl in 17α-ethylestradiol to form hydrogen bonds to the receptor. Osawa et al.[136] performed MM2 calculations for side-chain rotation about the C17–C20 bond of cholestane and 20-isocholestane for which a mixture of several rotamers should be expected at normal temperatures.

The orientation of the 17β-substituent of cardioactive and potentially cardioactive steroids were determined by CNDO/2[148,193] and several force-field computations.[81,85-87,194-196] Most force-field investigations[81,86,194,195] involved only rigid-body rotations. In accordance with X-ray observations, traditional cardenolides were found to favor two conformations, namely the (C21,O14) and (C22,O14) conformation (see also Section II.B.5) having nearly equal energy in digitoxigenin according to MM2p full geometry optimizations.[85] Introduction of the C8=C14 double bond was shown to influence the D-ring conformation, the directionality of the C17–C20 bond relative to the constant steroid-skeleton portion, and to give rise to broad potential-energy wells.[81,86,193] The lactone-ring carbonyl oxygen of the 8(14)-unsaturated digitoxigenin analogue is, therefore, located in regions not well suited to form receptor hydrogen bonds (see Section II.B.5). Consequently, these compounds display weak Na^+,K^+-ATPase inhibition. In our modeling study,[86] a 17β-furan-2′-carbaldehyde substituent was designed to counterbalance the effects of the 8(14)-unsaturation. The resulting steroid was calculated to be only capable of forming hydrogen bonds to that receptor hydrogen corresponding to the (C21,O14)-conformation of cardenolides. The compound was prepared as 19-nor analogue and results of the guinea-pig Na^+,K^+-ATPase inhibition test[197] are compared in Figure 6 to those of the 8(14)-saturated cardenolide equivalent, 19-nordigitoxigenin. Three conclusions can be drawn from the strong inhibitory activity observed for the furan-2-carbaldehyde derivative:

1. Evidence is given for the reliability of our modeling results.
2. The necessity of the presence of the 14β-hydroxyl group and of the C/D *cis* ring fusion claimed, e.g., with the lead structure proposed by Schönfeld et al.[19] is questioned.
3. This is an additional indication that the (C21,O14)-conformation of cardenolides is involved in binding to the Na^+,K^+-ATPase (other arguments are mentioned in Section II.B.5).

In further modeling approaches on 14β-configurated steroid, the conformational flexibility in the steroid-skeleton D-ring region should be taken into account.[196]

Much work has been done to clarify the role of the 17β-acetyl side chain conformation in progesterone receptor binding and progestational activity. From more than 80 crystal structure analyses, it is known that 16β-nonsubstituted progesterone derivatives exclusively adopt a conformation characterized by a C16–C17–C20–O20 torsion angle range between 0 and $-46°$ (mean value of $-21°$).[198] This orientation is also referred to as (C16,O20)-conformation denoting that the O20 oxygen is pointing towards the C16 carbon. 16β-Substituents have the power to alter this 17β-acetyl directionality[198] as

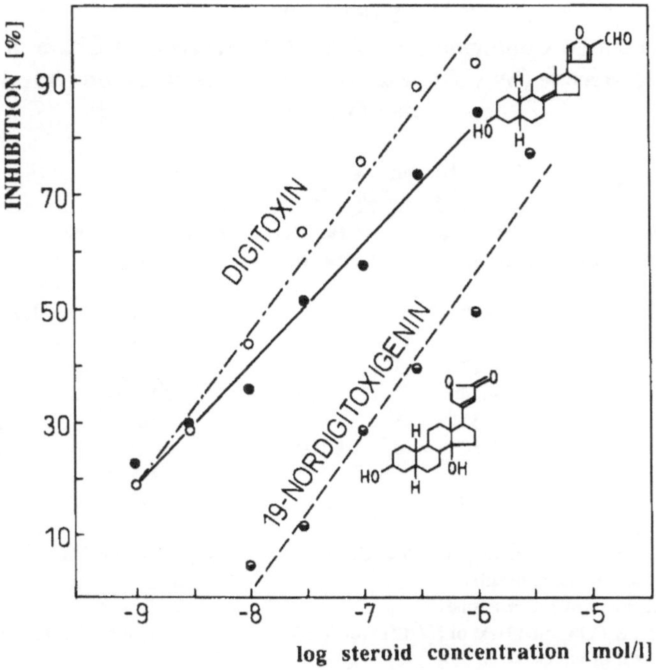

FIGURE 6. Inhibition of guinea-pig heart Na^+, K^+-ATPase vs. logarithm of steroid concentration for digitoxin, 19-nordigitoxigenin, and 4-(3'β-hydroxy-5'β-estr-8'(14')-en-17'β-yl)-furan-2-carbaldehyde. (From Hübler, D., Schubert, G., and Bohl, M., unpublished data.)

also found by MM2 computations.[199] Table 14 summarizes the results of theoretically predicted conformations. Although the agreement with the crystallographically observed torsion angle is not very satisfactory for all the calculations (see, for example, results of the sophisticated quantum-chemical MNDO method), computed relative stabilities revealed the preference of the (C16,O20)-conformation also being deduced from circular dichroism measurements.[206] Contradicting theoretical data were mainly provided for the exact carbonyl directionality and the relative stability of the second energy-minimum orientation, namely the (C16,C21)-conformation. MM2,[199,200] SYBYL,[200] and AM1[200] (see Figure 7) potential energy curves obtained with complete geometry optimizations show a second local minimum in the torsion range of −126° to −157° and an additional shoulder between +140° and +170° (MM2: +160°; SYBYL: +144°; AM1: +172°). Assuming a Boltzmann distribution of the two energy-minimum conformations at 298 K, and using energies and dipole moments determined by AM1 (16% (C16,C21)-conformation with μ = 4.60 D; 84% (C16,O20)-conformation with μ = 2.25 D, leads to a calculated dipole moment of 2.6 D in very close agreement with the experimental value of 2.7 D.[199]

TABLE 14
Calculated Conformations of the 17β-Acetyl Side Chain of Progesterone (by Using Full Geometry Optimization Unless Noted Otherwise)

		Torsion angle C16–C17–C20–O20			
Method	Steroid or fragment	(C16,O20)-conformation	(C16,C21)-conformation	Energy difference (kJ/mol)[a]	Ref.
MM2	—[b]	−4°	−133°	1.3	126,199,200
MM2	—[c]	−8°	−134°	0.8	201
SYBYL[d]	—[b]	−2°	−126°	2.4	200
PCILO[e]	—[c]	−14°	+170°	13.4	203
MNDO	—[b]	−35°	nd[f]	nd	204
MNDO	—[c]	−50°	+160°	8	205
AM1[g]	—[b]	−29°	−157°	4.1	200
X-ray	—[b]	−10°	nd	nd	126

[a] Positive energy values indicate that the (C16,O20)-conformation is more stable.
[b] Whole progesterone molecule.
[c] D-ring fragment of progesterone.
[d] Using the TRIPOS force field of SYBYL version 5.4[115] with termination criterion: maximum displacement of atoms = 10^{-5} Å.
[e] Coordinates of D-ring atoms were fixed whereas the 17β-acetyl structure was optimized.
[f] nd = no data.
[g] Using the MOPAC version 5.0 with PRECISE(100) option from SYBYL version 5.4.[115]

Some published calculations[206-209] have not been included in Table 14 for different reasons (too large torsion-angle increments, no geometry optimization, known limitations, nondocumented parametrization, etc.).

The effect of several substitutions on conformational properties were calculated as well. Molecular-mechanics[142,199,201,202,208] and PCILO investigations[203] were carried out for structural modifications leading to corticoids while CNDO/2,[210] INDO,[211] and force-field calculations[209,212-215] were applied to evaluate charge distribution, 17β-acetyl and D-ring conformations of progesterone derivatives bearing an additional cycle fused to 16α,17α positions.

In the following, those results are summarized that give, considered as a whole, evidence for the (C16,O20)-conformation being the bioactive acetyl orientation in progestational steroids (not necessarily also for antiprogestins):

1. Surcouf et al.[216] found for 21-phenylseleno-substituted progesterone a torsion angle C16–C17–C20–O20 of −31° by X-ray diffraction and a progesterone-receptor binding affinity of 59%, while the same side group in 17α position alters the torsion to +13° and the binding affinity to 2%.
2. As described by Raynaud et al.,[166,188,217] 21-hydroxy-21-methylpreg-

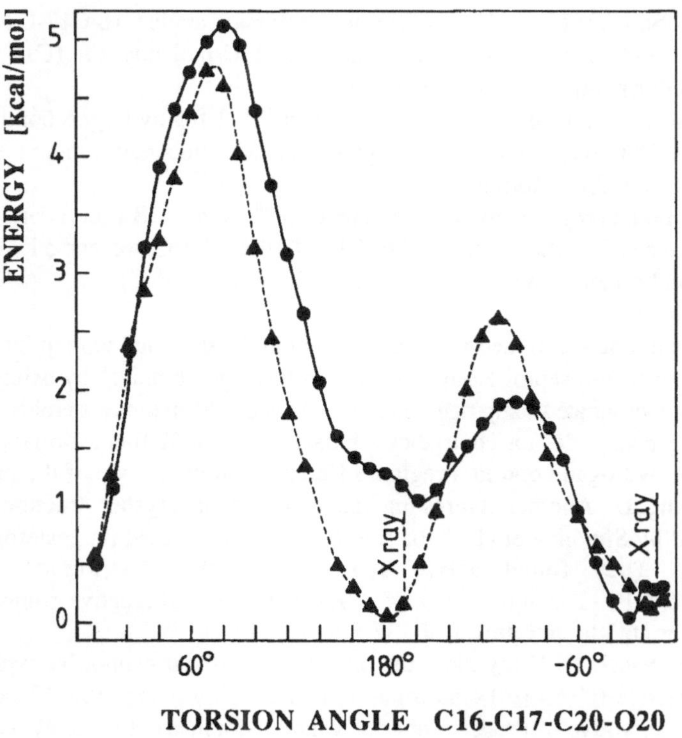

FIGURE 7. Potential-energy profiles calculated by the AM1 molecular orbital method (MOPAC version 5.0, SYBYL version 5.4 [115]) for 17β-acetyl side chain rotation of progesterone (solid line) and 16α,17α-methyleneprogesterone (broken line).[200]

nadienone derivatives adopt different 17β-side chain orientations depending on the configuration at C21. The (21*S*) compound RU 27987 having the normal (C16,O20)-conformation displays a higher activity in the Clauberg test than the (21*R*) isomer RU 988 for which the unusual (C16,C21)-orientation was observed in the crystal structure.

3. The introduction of the 16β-methyl group into progesterone leads to a receptor-binding affinity reduction from 100% to approximately 30%[13] (see also Table 3) and to a loss of progestational potency.[215] The stabilization of unusual acetyl conformations by 16β-substituents was already mentioned above.
4. In accordance with an enhanced binding affinity and favorable progestin activity, 16α,17α-cyclohexano steroids were shown to have an even stronger preference for the (C16,O20)-conformation[214] (MM2 relative stability of 7.5 kJ/mol compared with 1.3 kJ/mol for progesterone).
5. An additional methyl group attached to the 16α,17α-hexano ring and located in *syn*-periplanar orientation to the C20 carbon was found to strongly reduce the progestational potency and to have a preference for

adopting the (C16,C21)-conformation (torsion angle C16-C17-C20-O20 of $-143°$), which is more stable by 5.4 kJ/mol than the (C16,O20)-conformation in MM2 calculations.[215]

6. The zone postulated by Raynaud et al.[164-167] for hydrogen bonding to the O20 oxygen is in full agreement with receptor binding to the (C16,O20)-conformation.
7. A model proposed by Schönemann et al.[36] taking 13β-acetyl derivatives into consideration requires the (C16,O20)-conformation and a hydrogen bond donor location over ring D in the vicinity of C15.

It should be mentioned that Herrmann et al.[189] also tried to map locations of progesterone-receptor atoms involved in hydrogen bonding by using 13α- and 13β-configurated antiprogestational 17α- and 17β-hydroxy steroids. Their modeling results do not contradict a bioactive (C16,O20)-conformation but place the hydrogen bond in a region different from the β-side of the progesterone ring D. Another interesting model based on crystal structures was proposed by Simonov et al.[218] for five 16α,17α-substituted progesterone analogues. They found active steroids to have the torsion angle C15–C16–C17–C20 in the range of 149.1 to 151.3° and inactive compounds to be restricted to an interval of 132.8 to 133.5°.

With respect to X-ray data, the only exception that cannot be explained by a bioactive (C16,O20)-conformation is the highly potent 16α,17α-cyclopropano derivative of progesterone for which the unusual 17β-acetyl conformation is seen in the crystal.[37,212,219] However, the AM1 potential-energy profile of the acetyl side chain rotation shown in Figure 7 reveals that the energy difference between both basic conformations is very small (0.3 kJ/mol) for this steroid.

F. CONFORMATIONS OF UNUSUALLY CONFIGURATED STEROIDS

In order to understand the different alcohol proportions in the reduction of estr-5(10)-en-3-one and estra-5(10),9(11)-dien-3-one, Cohen[220] calculated the relative stabilities of A-ring conformers of both steroid reactants by means of GEMO molecular mechanics. He found the 2α,3β-half chair to be preferred to the 2β,3α-half chair by 0.9 kcal/mol in case of the 5(10)-monoene, whereas the 5(10),9(11)-diene favors the 2β,3α-half chair ring A conformation by only 0.3 kcal/mol, which suggests a greater conformational mobility for the latter steroid. That the 5(10)-enone is very flexible can already be derived from X-ray diffraction studies:[221] three structures were observed to have a 2α-envelope and, additionally, 2β-envelope, 2α,2β-half chair, and 2β,2α-half chair conformations were found once each in the solid state.

As already mentioned in Section III.A (see also 14α- vs. 14β-steroids in Table 11), 5α,14β-configurated androstane was evaluated to be thermodynamically favored. Another example is the estra-1,3,5(10)-trien-11-one system[222,223] for which the unnatural 9β-configuration has a lower MM2 strain energy than the 9α-isomer. This situation is reversed when a 14α-methyl group is present.[223] Much work on unnaturally configurated steroids, especially 9β-steroids, has been done by Bull et al.[223-228] When a chair-like conformation is retained for the B and C rings, the B/C *cis* ring fusion of 9β-steroids results in a rather bent structure which has some intrinsic strain energy due to the more congested state. This strain can be relieved by adopting alternative ring conformations (ring deformations). Moreover, the stability of these conformations is strongly influenced by appropriate substitutions. In this ways, Bull et al. were able to selectively generate molecules of different overall bent shape useful in future mapping the steric demands of receptor binding. They expect to find substantial differences in binding properties and in the pattern of metabolic changes.[224]

Molecular mechanics computations together with X-ray diffraction, NMR, and CD spectroscopy were applied to determine the relative conformational stabilities of various compounds; in the 14α-methylestra-1,3,5(10)-trien-17-one series,[223,224] 8β,9α (Figure 8a) and 8α,9α-isomers (Figure 8c) were found to favor a HC conformation (H: half chair-like ring B; C: chair-like ring C), whereas four conformers of the 8β,9β-isomer (Figure 8b) have comparable MM2 steric energies within 0.79 kcal/mol: BT (B: boat-like ring B; T: twist-boat like ring C), HC, BC, HT with the two latter ones having the lowest energy. With introduction of a 11β-hydroxyl group into the 8β,9β-isomer (Figure 8d), the relatively flat BT conformation is stabilized as a consequence of the 1,3-diaxial interaction between the 11β-OH and 13β-methyl in the strongly bent BC conformation.[223] Other substitutions were also shown to modify the degree of conformational flexibility and the ratio of preferred conformers.[223-225] In the 4,4,9,14-tetramethyl-19-nor-5β,9β,10α,14α-pregnane series,[225] MM1 calculations predict flat CBB conformations (C: chair-like ring A; B: boat-like ring B; B: boat-like ring C) for both the 6α,11β- (Figure 8e) and 6α,11α-diols (Figure 8h), a bent CCC conformation for the 6β,11α-diol (Figure 8g), and a mixture of CBB and CCC forms for the 6β,11β-diol (Figure 8f). The results were confirmed by three crystal structures but showed some deviations from NMR interpretations possibly ascribed to solvent effects. The dominating influence of the 14α-methyl group in flattening the B,C-ring structure of the 6α-OH compounds by α-face steric repulsion was also demonstrated.[225] For the 5α-steroid given in Figure 8i, MM2 predicted the following order of conformer stability[226] in accordance with crystallographic data: CBB (0.00 kcal/mol) > CBC (0.60 kcal/mol) > CCC (2.47 kcal/mol). The contributions of various structural features to the energy-minimum CBB conformation were discussed.[226] Differences in the ring-A conformation of steroids given in Figure 8j to l were also studied by MM2 calculations using CNDO net atomic charges.[227]

The importance of taking these unnaturally configurated steroids into account to derive more general structure-activity relationships is underlined

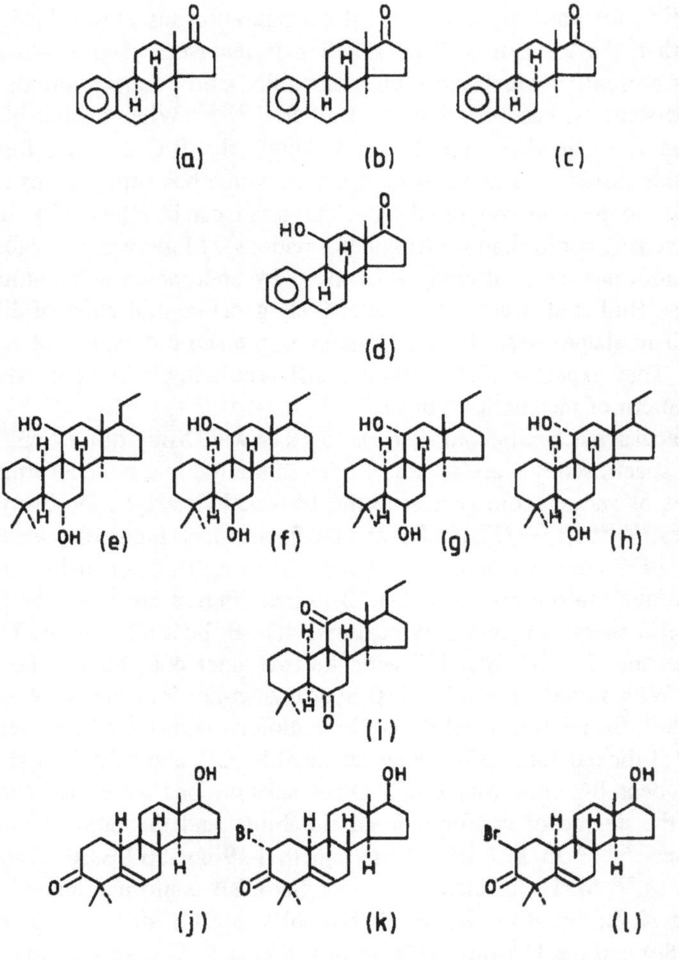

FIGURE 8. Unnaturally configurated steroids.

by the promising activity data reported: for example 9β,10α-progesterone is approximately 25 times as active as progesterone in the s.c. McPhail test,[229] 3β-hydroxy-9β-estra-1,3,5(10)-triene-11,17-dione with a strongly bent X-ray structure exhibits a 10 times more enhanced estrogenic potency than the flat-structured 9α-configurated steroid.[230] All the structure and potency data of unusually configurated steroids can be expected to permit improvement in future modeling attempts and should be an integral part in such investigations in which interactions with protein binding sites[118] or DNA[231,232] are modeled.

ACKNOWLEDGMENTS

I wish to thank Dr. Richard Cramer for reading the manuscript and Miss Sarah Whiting for technical assistance.

REFERENCES

1. **Hüttenrauch, R.**, Chemische und physikalisch-chemische Faktoren für die Wirksamkeit von Steroiden, *Pharmazie*, 28, 84, 1973.
2. **Hansch, C.**, Recent advances in biochemical QSAR, in *Correlation Analysis in Chemistry, Recent Advances*, Chapman, N. B. and Shorter, J., Eds., Plenum Press, New York, 1978, 397.
3. **Wolff, M. E.**, Steroids and other hormones, in *Quantitative Structure-Activity Relationships of Drugs*, Topliss, J. G., Ed., Academic Press, New York, 1983, 351.
4. **Zeelen, F. J.**, QSAR of steroids, *Quant. Struct.-Act. Relat.*, 5, 131, 1986.
5. **Hanč, O. and Urbánková, I.**, Steroidni hormony a QSAR, *Československ. Farm.*, 37, 90, 1988.
6. **Kubinyi, H.**, Free Wilson analysis. Theory, applications and its relationship to Hansch analysis, *Quant. Struct.-Act. Relat.*, 7, 121, 1988.
7. **Fried, J. and Borman, A.**, Synthetic derivatives of cortical hormones, in *Vitamins and Hormones*, Vol. 16, Harris, R. S., Marrian, G. F., and Thimann, K. V., Eds., Academic Press, New York, 1958, 303.
8. **Sarett, L. H., Patchett, A. A., and Steelman, S. L.**, The effects of structural alteration on the anti-inflammatory properties of hydrocortisone, in *Progress in Drug Research*, Vol. 5, Jucker, E., Ed., Birkhäuser, Basel, 1963, 11.
9. **Justice, J. B., Jr.**, A Fujita Ban structure-activity analysis of 44 steroids, *J. Med. Chem.*, 21, 465, 1978.
10. **Wolff, M. E., Baxter, J. D., Kollman, P. A., Lee, D. L., Kuntz, I. D., Bloom, E., Matulich, D. T., and Morris, J.**, Nature of steroid-glucocorticoid receptor interactions: thermodynamic analysis of the binding reaction, *Biochemistry*, 17, 3201, 1978.
11. **Mickelson, K. E. and Westphal, U.**, Steroid-protein interactions. Influence of steroid structure and temperature on the binding of steroids to guinea pig corticosteroid-binding globulin, *Biochemistry*, 19, 585, 1980.
12. **Zeelen, F. J.**, Interactions of steroids with receptors, in *Biological Activity and Chemical Structure*, Keverling Buisman, J. A., Ed., Elsevier, Amsterdam, 1977, 147.
13. **Seeley, D. H., Wang, W.-Y., and Salhanick, H. A.**, Molecular interactions of progesterone analogues with rabbit uterine cytoplasmic receptor, *J. Biol. Chem.*, 257, 13, 359, 1982.
14. **Blanford, A. T., Wittman, W., Stroupe, S. D., and Westphal, U.**, Steroid-protein interactions. XXXVIII. Influence of steroid structure on affinity to the progesterone-binding globulin, *J. Steroid Biochem.*, 9, 187, 1978.
15. **Tang, Z., Wu, J., Mao, X., Chen, M., and Li, Y.**, Quantitative studies on the structure-activity relationships of 1,3,5(10)-estratriene derivatives, *Yao Hsueh Hsueh Pao*, 15, 410, 1980.
16. **Fanchenko, N. D., Sturchak, S. V., Shchedrina, R. N., Pivnitsky, K. K., Novikov, E. A., and Ishkov, V. L.**, The specificity of the human uterine receptor, *Acta Endocrinol.*, 90, 167, 1979.
17. **Urbánková, I.**, Effect of substitution on the activity of androgens and anabolics, *Československ. Farm.*, 33, 331, 1984.
18. **Schönfeld, W. and Repke, K. R. H.**, A Free-Wilson analysis of 5β,14β-androstane derivatives inhibiting the Na/K-ATPase from human heart, *Quant. Struct.-Act. Relat.*, 7, 160, 1988.
19. **Schönfeld, W., Weiland, J., Lindig, C., Masnyk, M., Kabat, M. M., Kurek, A., Wicha, J., and Repke, K. R. H.**, The lead structure in cardiac glycosides is 5β,14β-androstane-3β,14-diol, *Naunyn-Schmiedeberg's Arch. Pharmacol.*, 329, 414, 1985.
20. **Hansch, C. and Leo, A.**, *Substituent Constants for Correlation Analysis in Chemistry and Biology*, John Wiley & Sons, New York, 1979.

21. **Seydel, J. K. and Schaper, K.-J.,** *Chemische Struktur und biologische Aktivitat von Wirkstoffen*, Verlag Chemie, Weinheim, 1979.
22. **Franke, R.,** *Theoretical Drug Design Methods*, Elsevier, Amsterdam, 1984.
23. **Hansch, C., Leo, A., Unger, S. H., Kim, K. H., Nikaitani, D., and Lien, E. J.,** "Aromatic" substituent constants for structure-activity correlations, *J Med. Chem.*, 16, 1207, 1973.
24. **Moriguchi, I.,** Quantitative structure-activity studies. I. Parameters relating to hydrophobicity, *Chem. Pharm. Bull.*, 23, 247, 1975.
25. **Testa, B. and Seiler, P.,** Steric and lipophobic components of the hydrophobic fragmental constant, *Drug Res.*, 31, 1053, 1981.
26. **Van de Waterbeemd, H., Tayar, N. E., Carrupt, P.-A., and Testa, B.,** Pattern recognition study of QSAR substituent descriptors, *J. Comp.-Aided Mol. Design*, 3, 111, 1989.
27. **Teutsch, G., Weber, L., Page, G., Shapiro, E. L., Herzog, H. L., Neri, R., and Collins, E. J.,** Influence of 6-azido and 6-thiocyanato substitution on progestational and corticoid activities and a structure-activity correlation in the Δ6-6-substituted progestational series, *J. Med. Chem.*, 16, 1370, 1973
28. **Wolff, M. E. and Hansch, C.,** A quantitative reexamination of structure-activity relationships in the Δ6-6-substituted progesterone series, *J. Med. Chem.*, 17, 898, 1974.
29. **Wolff, M. E., Hansch, C., Giannini, D. D., Kollman, P. A., Duax, W. L., and Baxter, J.,** Correlation of glucocorticoid and progestational activity with steric, electronic and hydrophobic parameters, *J. Steroid Biochem.*, 6, 211, 1975.
30. **Topliss, J. G. and Shapiro, E. L.,** Quantitative structure-activity relationships in the Δ6-6-substituted progesterone series. A reappraisal, *J. Med. Chem.*, 18, 621, 1975.
31. **Coburn, R. A. and Solo, A. J.,** Quantitative structure-activity relationships among steroids. Investigations of the use of steric parameters, *J. Med. Chem.*, 19, 748, 1976.
32. **Hartog, J., Halkes, S. J., Morsink, L., de Wachter, A. M., and Schlatmann, J. L. M. A.,** Structure-activity relationships of a series of substituted 17α-hydroxy-9β,10α-progesterones, *J. Steroid Biochem.*, 6, 577, 1975.
33 **v.d. Broek, A. J., Broess, A. I. A., v.d. Heuvel, M. J., de Jongh, H. P., Leemhuis, J., Schönemann, K. H., Smits, J., de Visser, J., van Vliet, N. P., and Zeelen, F. J.,** Strategy in drug research Synthesis and study of the progestational and ovulation inhibitory activity of a series of 11β-substituted-17α-ethynyl-4-estren-17β-ols, *Steroids*, 30, 481, 1977.
34 **Zeelen, F. J.,** Strategy in drug research, structure-activity relationship of 11-substituted progestational steroids, in *Advances in Pharmacological Research and Practise*, Vol. 3, Darvas, F., Ed., Pergamon Press, Oxford, 1980, 43.
35. **Kloosterboer, H. J. and Bergink, E. W.,** Structural requirements for optimal receptor binding of contraceptive progestogens, *Pharmaceut. Weekbl. Sci. Ed.*, 7, 295, 1985.
36. **Schönemann, K. H., van Vliet, N. P., and Zeelen, F. J.,** Structure and progestational activity of 13-substituted-18-norpregn-4-ene-3,20-diones, a pilot study, *Steroids*, 45, 297, 1985.
37. **Zeelen, F. J.,** Structure and activity of progesterone derivatives in *QSAR and Strategies in the Design of Bioactive Compounds*, Seydel, J. K., Ed., VCH, Weinheim, 1985, 135.
38. **Zeelen, F. J.,** Drug receptor interactions, construction of a map for the ring-D binding part of the rabbit progesterone receptor, in *Proceedings of the F.E.C.S. Third International Conference on Chemistry and Biotechnology of Biologically Active Natural Products*, Vol. 3, VCH, Weinheim, 1987, 196.
39. **Karunanithy, R., Leung, S. L., and Yeo, S. H.,** Use of high-performance liquid chromatography for quantitative structure-activity relationship studies of a series of potential contraceptive steroidal esters, *J. Pharm. Biomed. Anal.*, 5, 597, 1987.
40. **Lee, D. L., Kollman, P. A., Marsh, F. J., and Wolff, M. E.,** Quantitative relationships between steroid structure and binding to putative progesterone receptors, *J. Med. Chem.*, 20, 1139, 1977.

41. **Bohl, M., Simon, Z., Vlad, A., Kaufmann, G., and Ponsold, K.**, MTD calculations on quantitative structure-activity relationships of steroids binding to the progesterone receptor, *Z. Naturforsch. C*, 42, 935, 1987.
42. **Gergen, I., Bohl, M., Simon, H., and Simon, Z.**, Structure-activity relations for steroids by the MTD method. Superposition procedure for molecules with different condensed cycles, *Rev. Roumaine Chim.*, 34, 995, 1989.
43. **Hoppen, H.-O. and Hammann, P.**, The influence of structural modification on progesterone and androgen receptor binding of norethisterone. Correlation with nuclear magnetic resonance signals, *Acta Endocrinol.*, 115, 406, 1987.
44. **Wolff, M. E. and Hansch, C.**, Correlation of physicochemical parameters and biological activity in steroids. 9α-Substituted cortisol derivatives, *Experientia*, 29, 1111, 1973.
45. **Wolff, M. E., Hansch, C., Kollman, P. A., Giannini, D. G., and Duax, W. L.**, Quantitation of chemical and biological effects in steroids by Hansch-type and molecular orbital calculations, in *Proceedings of the Conference on Chemical Structure-Biological Activity Relationships: Quantitative Approaches*, Tichy, M., Ed., Birkhäuser, Basel, 1976, 31.
46. **Ahmad, P. and Mellors, A.**, Glucocorticoid potency and parachor, *J. Steroid Biochem.*, 7, 19, 1976.
47. **Dahlberg, E., Thalén, A., Brattsand, R., Gustafsson, J-Å., Johansson, U., Roempke, K., and Saartok, T.**, Correlation between chemical structure, receptor binding, and biological activity of some novel, highly active, 16α,17α-acetal-substituted glucocorticoids, *Mol. Pharmacol.*, 25, 70, 1984.
48. **Shroot, B., Caron, J. C., and Ponec, M.**, Glucocorticoid specific binding: structure-activity relationships, *Br. J. Dermatol.*, 107, 30, 1982.
49. **Ponec, M., Kempenaar, J., Shroot, B., and Caron, J.-C.**, Glucocorticoids: binding affinity and lipophilicity, *J. Pharm. Sci.*, 75, 973, 1986.
50. **Gandolfi, O., Barbaro, A. M., and Biagi, G. L.**, Relationship between R_m values and protein binding of corticosteroids and androgens, *Experientia*, 29, 689, 1973.
51. **Ebling, W. F., Milsap, R. L., Szefler, S. J., and Jusko, W. J.**, 6α-Methylprednisolone and 6α-methylprednisone plasma protein binding in humans and rabbits, *J. Pharm. Sci.*, 75, 760, 1986.
52. **Westphal, U.**, *Steroid-Protein Interactions*, Springer, Berlin, 1971, 1986.
53. **Yamakawa, M., Ezumi, K., Shiro, M., Nakai, H., Kamata, S., Matsui, T., and Haga, N.**, Relationships of the molecular structure of aldosterone derivatives with their binding affinity for mineralocorticoid receptor, *Mol. Pharmacol.*, 30, 585, 1986.
54. **Lövgren, T., Pettersson, K., Kouvonen, I., and Punnonen, R.**, Binding characteristics of estrone, estradiol and estriol to the human myometrial estrogen receptor, *J. Steroid Biochem.*, 9, 803, 1978.
55. **Zeelen, F. J. and Bergink, E. W.**, Structure-activity relationships of steroid estrogens in *Cytotoxic Estrogens in Hormone Receptive Tumors*, Raus, J., Martens, H., and Leclercq, G., Eds., Academic Press, London, 1980, 39.
56. **Bohl, M., Schubert, G., Koch, M., Reck, G., Strecke, J., Wunderwald, M., Prousa, R., and Ponsold, K.**, Quantitative structure-activity relationships of estrogenic steroids substituted at C14, C15, *J. Steroid Biochem.*, 26, 589, 1987.
57. **Ponsold, K., Draffehn, J., and Schönecker, B.**, Quantitative Struktur-Wirkungs-Beziehungen zwischen physikochemischen Parametern 16α-heterosubstituierter Ostradiole und dem Bindungsvermögen an Rezeptoren des Rattenuterus *in vitro*, *Pharmazie*, 32, 596, 1977.
58. **Fevig, T. L., Mao, M. K., and Katzenellenbogen, J. A.**, Estrogen receptor binding tolerance of 16α-substituted estradiol derivatives, *Steroids*, 51, 471, 1988.
59. **Peters, R. H., Crowe, D. F., Tanabe, M., Avery, M. A., and Chong, W. K. M.**, Steroidal silicon side-chain analogues as potential antifertility agents, *J. Med. Chem.*, 30, 646, 1987.

60. **Popoviciu, V., Holban, S., Bădilescu, I. I., and Simon, Z.**, Steric mapping of oestrogenic receptor site by a minimal steric difference method, *Studia Biophys* , 69, 75, 1978.
61. **Biagi, G. L., Guerra, M. C., and Barbaro, A. M.**, Relationship between lipophilic character and hemolytic activity of testosterone and testosterone esters, *J. Med. Chem.*, 13, 944, 1970.
62. **Biagi, G. L., Barbaro, A. M., and Guerra, M. C.**, Relationship between R_m values and biological activity of steroid esters, *Experientia*, 27, 918, 1971.
63. **James, K. C.**, R_m Values and biological action of testosterone esters, *Experientia*, 28, 479, 1972.
64. **James, K. C., Richards, G. T., and Turner, T. D.**, A comparison of chromatographic methods for the assessment of the distribution coefficients of androgen esters, *J. Chromatogr.*, 69, 141, 1972.
65. **James, K. C., Nicholls, P. J., and Richards, G. T.**, Correlation of androgenic activities of the lower testosterone esters in rat, with R_m values and hydrolysis rates, *Eur. J. Med. Chem.*, 10, 55, 1975.
66. **Noguchi, T., Charman, W. N. A., and Stella, V. J.**, The effect of drug lipophilicity and lipid vehicles on the lymphatic absorption of various testosterone esters, *Int. J. Pharm.*, 24, 173, 1985.
67. **Chaudry, M. A. Q. and James, K. C.**, A Hansch analysis of the anabolic activities of some nandrolone esters, *J. Med. Chem.*, 17, 157, 1974.
68. **Vida, J. A.**, *Androgens and Anabolic Agents*, Academic Press, New York, 1969.
69. **Kochakian, C. D., Ed.**, *Anabolic-Androgenic Steroids*, Springer, Berlin, 1976.
70. **Biagi, G. L.**, Correlazione fra coefficiente di ripartizione, Olio di silicone/aqua di glicosidi cardioattivi e loro attività biologica, *Fitoterapia*, 38, 110, 1967.
71. **Davydov, V. Ya.**, Correlation between the molecular structure of cardiac glycosides, steroid hormones and carbohydrates and their retention in high-performance liquid chromatography, *J. Chromatogr.*, 365, 123, 1986.
72. **Klaus, W., Rogatti, M., and Fricke, U.**, Physikochemische Eigenschaften und Wirkung herzwirksamer Glykoside an Biomembranen — Beziehung zur biologischen Antwort, in *Wandlungen in der Therapie der Herzinsuffizienz*, Rietbrock, N., Schnieders, B., and Schuster, J., Eds., Vieweg & Sohn, Braunschweig, 1983, 55.
73. **Hempelmann, F. W. and Heinz, N.**, Lipophilie und Proteinbindung bei Cardenoliden, *Drug Res.*, 28, 2185, 1978.
74. **Bohl, M. and Simon, Z.**, MTD Approach to quantitative structure-activity relationships for cardiotonic steroids, *Z. Naturforsch. C*, 40, 858, 1985.
75. **Simon, Z., Dragomir, N., Plauchithiu, M. G., Holban, S., Glatt, H., and Kerek, F.**, Receptor site mapping for cardiotoxic aglycones by the minimal steric difference method, *Eur. J. Med. Chem.*, 15, 521, 1980.
76. **Motoc, I.**, Molecular shape descriptors, in *Steric Effects in Drug Design*, (Topics in Current Chemistry, Vol. 114), Charton, M. and Motoc, I., Eds., Springer, Berlin, 1983, 93.
77. **Mori, J., Nagai, S.-I., Sakakibara, J., Takeya, K., Hotta, Y., and Ando, H.**, Studies on cardiac ingredients of plants. III. Structural confirmation and biological activity of reduced proscillaridins, *Chem. Pharm. Bull.*, 35, 1839, 1987.
78. **Dittrich, F. and Repke, K. R. H.**, Correlations between dipole moments and biological activities of cardiotonic steroids, in *Proceedings of the Conference on Chemical Structure-Biological Activity Relationships: Quantitative Approaches*, Tichy, M., Ed., Birkhäuser, Basel, 1976, 249.
79. **Dittrich, F., Berlin, P., Köpke, K., and Repke, K. R. H.**, Stereo-electronic interaction between cardiotonic steroids and Na,K-ATPase: molecular mechanism of digitalis action, *Curr. Topics Membr. Transp.*, 19, 251, 1983.
80. **Köpke, K., Dittrich, F., and Berlin, P.**, Abschätzung der Wirkstärke von Cardenoliden, *Pharmazie*, 39, 256, 1984.

81. **Rohrer, D. C., Fullerton, D. S., Yoshioka, K., From, A. H. L., and Ahmed, K.,** Functional receptor mapping for modified cardenolides: use of the PROPHET system, in *Computer Assisted Drug Design,* (ACS Symposium Series, Vol. 112), Olson, E. C. and Christofferson, R. E., Eds., American Chemical Society, Washington, DC, 1979, 259.
82. **Fullerton, D. S., Yoshioka, K., Rohrer, D. C., From, A. H. L., and Ahmed, K.,** Digitalis genin activity: side-group carbonyl oxygen position is a major determinant, *Science,* 205, 917, 1979.
83. **Ahmed, K., Rohrer, D. C., Fullerton, D. S., Deffo, T., Kitatsuji, E., and From, A. H. L.,** Interaction of (Na^+,K^+)-ATPases and digitalis genins, *J. Biol. Chem.,* 258, 8092, 1983.
84. **From, A. H. L., Fullerton, D. S., Deffo, T., Kitatsuji, E., Rohrer, D. C., and Ahmed, K.,** The inotropic activity of digitalis genins is dependent upon C(17) side-group carbonyl oxygen position, *J. Mol. Cell Cardiol.,* 16, 835, 1984.
85. **Fullerton, D. S., Ahmed, K., From, A. H. L., McParland, R. H., Rohrer, D. C., and Griffin, J. F.,** Modelling the cardiac steroid receptor, in *Molecular Graphics and Drug Design,* Burgen, A. S. V., Roberts, G. C. K., and Tute, M. S., Eds., Elsevier, Amsterdam, 1986, 257.
86. **Bohl, M., Ponsold, K., and Reck, G.,** Quantitative structure-activity relationships of cardiotonic steroids using empirical molecular eletrostatic potentials and semiempirical molecular orbital calculations, *J. Steroid Biochem.,* 21, 373, 1984.
87. **Ovchinnikov, A. A., Shamovsky, I. L., and Barenboim, G. M.,** Conformational analysis of the structure-activity relationship of cardiac glycosides, in *Proceedings of the F. E. C. S. Third International Conference on Chemistry and Biotechnology of Biologically Active Natural Products,* Vol. 3, VCH, Weinheim, 1987, 20.
88. **Hintsche, R., Megges, R., Pfeiffer, D., Portius, H. J., Schönfeld, W., and Repke, K. R. H.,** Biological potency and solution conformation of cardenolides determined by a new ^{1}H NMR method, *Eur. J. Med. Chem.,* 20, 9, 1985.
89. **De Pover, A. and Godfraind, T.,** Influence of 16β formylation on Na,K-ATPase inhibition by cardiac glycosides, *Naunyn-Schmiedeberg's Arch. Pharmacol.,* 321, 135, 1982.
90. **Takiura, K., Yuki, H., Okamoto, Y., Takai, H., and Honda, S.,** Studies of oligosaccharides. XIV. Structure-activity relationship of variations in the sugar moiety of digitoxin, *Chem. Pharm. Bull.,* 22, 2263, 1974.
91. **Griffin, J. F., Rohrer, D. C., Ahmed, K., From, A. H. L., Hashimoto, T., Rathore, H., and Fullerton, D. S.,** The effect of 16β-substitution on the structure and activity of digitoxigenin: is there an additional binding interaction with Na^+,K^+-ATPase?, *Mol. Pharmacol.,* 29, 270, 1986.
92. **Thomas, R., Boutagy, J., and Gelbart, A.,** Synthesis and biological activity of semisynthetic digitalis analogs, *J. Pharm. Sci.,* 63, 1649, 1974.
93. **Thomas, R., Gray, P., and Andrews, J.,** Digitalis: its mode of action, receptor and structure-activity relationships, in *Advances in Drug Research,* Vol. 19, Testa, B., Ed., Academic Press, London, 1990, 310.
94. **Biagi, G. L., Barbaro, A. M., Gandolfi, O., Guerra, M. C., and Cantelli-Forti, G.,** R_m values of steroids as an expression of their lipophilic character in structure-activity studies, *J. Med. Chem.,* 18, 873, 1975.
95. **Caron, J. C. and Shroot, B.,** Determination of partition coefficients of glucocorticosteroids by high-performance liquid chromatography, *J. Pharm. Sci.,* 73, 1703, 1984.
96. **Saket, M. M., James, K. C., and Kellaway, I. W.,** The partitioning of some 21-alkyl steroid esters between human stratum corneum and water, *Int. J. Pharm.,* 27, 287, 1985.
97. **Dunn, W. J., III, Koehler, M. G. and Grigoras, S.,** The role of solvent-accessible surface area in determining partition coefficients, *J. Med. Chem.,* 30, 1121, 1987.
98. **Lundberg, B.,** Temperature effect on the water solubility and water-octanol partition of some steroids, *Acta Pharm. Suecica,* 16, 151, 1979.

99. **Repke, K. R. H. and Dittrich, F.,** Thermodynamics of information transfer from cardiotonic steroids to receptor transport ATPase, *Trends Pharmacol. Sci.,* 1, 398, 1980.
100. **Sokolov, Yu.A., Golubovich, V. P., Gurskii, S. G., and Ahrem, A. A.,** Prediction of the biological activity of some 8-azasteroids, *Dokl. Akadem. Nauk BSSR,* 31, 662, 1987.
101. **Doré, J.-C., Gilbert, J., Ojasoo, T., and Raynaud, J.-P.,** Correspondence analysis applied to steroid receptor binding, *J. Med. Chem.,* 29, 54, 1986.
102. **Ojasoo, T., Doré, J.-C., Gilbert, J., and Raynaud, J.-P.,** Binding of steroids to the progestin and glucocorticoid receptors analyzed by correspondence analysis, *J Med. Chem.,* 31, 1160, 1988.
103. **Bodor, N., Harget, A. J., and Phillips, E. W.,** Structure-activity relationships in the antiinflammatory steroids: a pattern-recognition approach, *J. Med. Chem.,* 26, 318, 1983.
104. **Stouch, T. R. and Jurs, P. C.,** Computer-aided studies of the structure-activity relationships between the structure of some steroids and their antiinflammatory activity, *J. Med. Chem.,* 29, 2125, 1986.
105. **Jurs, P. C., Stouch, T. R., Czerwinski, M., and Narvaez, J. N.,** Computer-assisted studies of molecular structure-biological activity relationships, *J. Chem. Inf. Comput. Sci.,* 25, 296, 1985.
106. **Henry, D. R. and Block, J. H.,** Steroid classification by discriminant analysis using fragment molecular connectivity, *Eur J. Med. Chem.,* 15, 133, 1980.
107. **Henry, D. R. and Block, J. H.,** Pattern recognition of steroids using fragment molecular connectivity, *J. Pharm. Sci.,* 69, 1030, 1980.
108. **Moriguchi, I.,** Pattern discrimination in structure-activity studies: reexamination of the set of 46 steroidal compounds, in *Multivariate Chemometrics in QSAR: a Dialogue,* Mager, P. P., Ed., Research Studies Press, Letchworth, 1988, 282.
109. **Crippen, G. M.,** *Distance Geometry and Conformational Calculations,* Research Studies Press, Letchworth, 1981.
110. **Goodford, P. J.,** A computational procedure for determining energetically favorable binding sites on biologically important macromolecules, *J. Med. Chem.,* 28, 849, 1985.
111. **Ghose, A. K., Pritchett, A., and Crippen, G. M.,** Atomic physicochemical parameters for three dimensional structure directed quantitative structure-activity relationships. III. Modeling hydrophobic interactions, *J. Comput. Chem.,* 9, 80, 1988.
112. **Furet, P., Sele, A., and Cohen, N. C.,** 3D Molecular lipophilicity potential profiles: a new tool in molecular modeling, *J. Mol. Graphics,* 6, 182, 1988.
113. **Fauchère, J.-L., Quarendon, P., and Kaetterer, L.,** Estimating and representing hydrophobicity potential, *J. Mol. Graphics,* 6, 203, 1988.
114. **Cramer, R. D., III, Patterson, D. E., and Bunce, J. D.,** Comparative molecular field analysis (CoMFA). I. Effect of shape on binding of steroids to carrier proteins, *J. Am. Chem. Soc.,* 110, 5959, 1988.
115. **Molecular** Modelling Program SYBYL, TRIPOS Associates, 1699 South Hanley Road, St. Louis, Missouri 63144.
116. **Alléra, A. and Klein, R.,** A computer-generated model of a glucocorticoid-carrier in the rat liver plasma membrane, *Biol. Chem. Hoppe-Seyler,* 370, 870, 1989, and personal communication.
117. **Vinter, J. G.,** Molecular modelling systems, in *Molecular Graphics and Drug Design,* Burgen, A. S. V., Roberts, G. C. K., and Tute, M. S., Eds., Elsevier, Amsterdam, 1986, 15.
118. **(a) Bissery, V., Gaboriaud, C., Thomas, A., Varloot, L., and Mornon, J. P.,** Molecular modeling of the hormone binding domain of human androgen receptor, 9th Annual Meeting of the Molecular Graphics Society, Uppsala, July 2 to 6, 1990. (b) **Thomas, A., Bissery, V., and Mornon, J. P.,** Three dimensional modelling of human corticosteroid binding globulin, 9th Annual Meeting of the Molecular Graphics Society, Uppsala, July 2 to 6, 1990.

119. **Caillet, J. and Pullman, B.,** Quantum-mechanical studies on the conformational and electronic properties of steroids. I. The sterane (perhydrocyclopentanophenanthrene) ring system and its constituent subunits, *Theor. Chim. Acta,* 17, 377, 1970.
120. **Allinger, N. L. and Wu, F.,** Conformational analysis. LXXVII. The relative stabilities of the isomeric androstane ring systems, *Tetrahedron,* 27, 5093, 1971.
121. **Bodor, N. and Dewar, M. J. S.,** Ground states of σ-bonded molecules. XI. Conformational analyses by MINDO/2 method, *J. Am. Chem. Soc.,* 92, 4270, 1970.
122. **Allinger, N. L. and Tribble, M. T.,** Conformational analysis. LXXX. The hydrindanone ring system, *Tetrahedron,* 28, 1191, 1972.
123. **Romers, C., Altona, C., Jacobs, H. J. C., and de Graaff, R. A. G.,** Steroid conformations from X-ray analysis data, *Terpenoids and Steroids,* 4, 531, 1974.
124. **Altona, C. and Faber, D. H.,** Empirical force field calculations, in *Topics in Current Chemistry,* Vol. 45, Springer, Berlin, 1974, 1.
125. **Schmit, J.-P. and Rousseau, G. G.,** Structure-activity relationships for glucocorticoids. II. Theoretical approach of molecular structures based on energy optimisation of a Westheimer model, *J. Steroid Biochem.,* 8, 921, 1977.
126. **Duax, W. L., Griffin, J. F., and Rohrer, D. C.,** Steroid conformation, receptor binding, and hormone action, in *Steroid Conference in X-Ray Crystallography and Drug Action,* Horn, A. J., Ed., Oxford University Press, Oxford, 1983, 405.
127. **Sedee, A. G. J., Beijersbergen van Henegouwen, G. M. J., Guijt, W., and Haasnoot, C. A. G.,** Conformational analysis of steroids in solution: 17β-hydroxy-19-nor-5α,17α-pregn-20-yn-3-one and its 5β-isomer studied by nuclear magnetic resonance, *J. Org. Chem.,* 50, 4182, 1985.
128. **Van Geerestein, V. J.,** Conformational Aspects of Synthetic Sex Steroids, Thesis, Utrecht, 1988.
129. **Allinger, N. L., Tribble, M. T., and Yuh, Y.,** Androsterone. The structure by force-field calculations, *Steroids,* 26, 398, 1975.
130. **Shamovsky, I. L., Barenboim, G. M., and Ovchinnikov, A. A.,** Estimation of the optimum parameters of atom-atom potential functions of steroids from calculating their conformations, *Bioorg. Chim.,* 8, 1677, 1982.
131. **Cox, P. J. and Turner, A. B.,** Synthesis, X-ray structure and molecular mechanics studies of the boar taint steroid (5α-androst-16-en-3-one), *Tetrahedron,* 40, 3153, 1984.
132. **Castells, J., Jaime, C., López-Calahorra, F., Santaló, N., and Velasco, D.,** Parametrization of cyano group MM2 constants in peracetylated aldononitriles, *J. Org. Chem.,* 53, 5363, 1988.
133. **van Alsenoy, C., Wolinski, K., Siam, K., and Schäfer, L.,** Ab initio studies of structural features not easily amenable to experiment, *J. Mol. Struct. (Theochem),* 180, 343, 1988.
134. **Duax, W. L., Fronckowiak, M. D., Griffin, J. F., and Rohrer, D. C.,** A comparison between crystallographic data and molecular mechanics calculations on the side chain and backbone conformations of steroids, in *Intramolecular Dynamics,* Jortner, J. and Pullman, B., Eds., D. Reidel, Dordrecht, 1982, 505.
135. **Duax, W. L., Griffin, J. F., and Rohrer, D. C.,** A comparison of crystallographic observations and molecular mechanics calculations of the conformations of steroids and steroid analogues, in *Natural Products Chemistry,* Zalewski, R. I. and Skolik, J. J., Eds., Elsevier, Amsterdam, 1985, 385.
136. **Osawa, E., Shirahama, H., and Matsumoto, T.,** Application of force field calculations to organic chemistry. VIII. Internal rotation in simple to congested hydrocarbons including 2,3-dimethylbutane, 1,1,2,2-tetra-*tert*-butylethane, 2,2,4,4,5,5,7,7,-octamethyloctane, and cholestane, *J. Am. Chem. Soc.,* 101, 4824, 1979.
137. **Bucourt, R.,** L'emploi des angles dièdres en analyse conformationnelle. Corrélations géométriques et énergétiques relatives aux problèmes des monocycles et des structures stéroides, *Bull. Soc. Chim. France,* 2080, 1964.

138. **Bucourt, R. and Hainaut, D.**, L'emploi des angles diedres en analyse conformationnelle Calcul des géométries et des énergies conformationnelles du cyclohexane, du cyclohexène et de quelques systèmes bicycliques, *Bull. Soc. Chim. France*, 1366, 1965.
139. **Bucourt, R.**, Conformational transmission in steroids, in *Conformational Analysis*, Chiurdoglu, G., Ed., Academic Press, New York, 1971, 59.
140. **Schneider, H.-J. and Gschwendtner, W.**, Polar and steric long-range transmission effects in cholestanes, cholestenes, and 9-substituted decalins. Carbon-13 nuclear magnetic resonance shifts and force field and linear electric field calculations, *J. Org. Chem.*, 47, 4216, 1982.
141. **Schneider, H.-J., Buchheit, U., Gschwendtner, W., and Lonsdorfer, M.**, Steric distortions and polar effects in steroids: molecular mechanics calculations and ^{13}C-NMR-investigations, in *Molecular Structure and Biological Activity*, Griffin, J. F. and Duax, W. L., Eds., Elsevier, Amsterdam, 1982, 165.
142 **Schmit, J.-P. and Rousseau, G. G.**, Structure-activity relationships for glucocorticoids. IV. Effects of substituents on the overall shape of steroids which bind to the glucocorticoid receptor, *J. Steroid Biochem.*, 9, 921, 1978.
143. **Kollman, P. A., Giannini, D. D., Duax, W. L., Rothenberg, S., and Wolff, M. E.**, Quantitation of long-range effects in steroids by molecular orbital calculations, *J. Am. Chem. Soc.*, 95, 2869, 1973.
144. **Weeks, C. M., Duax, W. L., and Wolff, M. E.**, A comparison of the molecular structures of six corticosteroids, *J. Am. Chem. Soc.*, 95, 2865, 1973.
145. **Repmann, H.**, Die Elektronenstruktur von Steroidhormonen. I. Die Sexualhormone, *Theor. Chim. Acta*, 17, 396, 1970.
146. **Repmann, H.**, Die Elektronenstruktur von Steroidhormonen. II. Die Nebennierenrindenhormone, *Theor. Chim. Acta*, 28, 141, 1973.
147. **Bohl, M. and Ponsold, K.**, Electronic structure of cardiotonic steroids, Wiss. Z. Friedrich-Schiller-Univ. Jena, *Naturwiss. R.*, 36, 559, 1987.
148. **Bohl, M. and Süssmilch, R.**, Calculations on molecular structure and electrostatic potentials of cardiotonic steroids, *Eur. J. Med. Chem.*, 21, 193, 1986.
149. **Chaudhary, A. K., Gombar, V. K., and Singh, H.**, Steroids and related studies. LXIII. Molecular orbital calculations on some neuromuscular blocking azasteroids, *Indian J. Chem.*, 24B, 2, 1985.
150. **Gale, M. M. and Saunders, L.**, The solubilisation of steroid by lysophosphatidylcholine. Testosterone, estradiol and their 17α-ethinyl derivations, *Biochim. Biophys. Acta*, 248, 466, 1971.
151. **Gassiot, M., Fernandez, E., Firpo, G., Carbó, R., and Martin, M.**, Empirical quantum chemical approach to structure-gas chromatographic retention index relationships, *J. Chromatogr.*, 108, 337, 1975.
152. **Croasmun, W. R. and Carlson, R. M. K.**, Steroid structural analysis by two-dimensional NMR, in *Two-Dimensional NMR Spectroscopy, Applications for Chemists and Biochemists*, Methods in Stereochemical Analysis, Vol. 9, Croasmun, W. R. and Carlson, R. M. K., Eds., VCH, New York, 1987, 387.
153. **Kirk, D. N.**, Steroids: physical methods, *Natur. Prod. Reports*, 3, 505, 1986.
154. **Altona, C. and Hirschmann, H.**, Empirical valence-force calculations of steroids. I. Conformations and relative energies of the 20-epimers of 5α-pregnane-3β,20-diol, *Tetrahedron*, 26, 2173, 1970.
155. **Wong, T. C., Guo, W., Bohl, M., Hübner, M., Luck, G., Steiger, T., and Reck, G.**, Solution conformation of three steroid 19-nor-4-en-3-ones determined from two-dimensional nuclear magnetic resonance spectroscopy, coupling constant calculations, and circular dichroism, *J. Chem. Soc., Perkin Trans. 2*, 765, 1988.
156. **Schneider, H.-J., Buchheit, U., Becker, N., Schmidt, G., and Siehl, U.**, ^{1}H NMR analyses, shielding mechanisms, coupling constants, and conformations in steroids bearing halogen, hydroxy, oxo groups, and double bonds, *J Am. Chem. Soc.*, 107, 7027, 1985.

157. **Schneider, H.-J., Gschwendtner, W., and Weigand, E. F.,** Conformational relaxation as limitation of chemical models. Empirical force field calculations and ^{13}C NMR shielding effects for some cyclohexanes, bicyclo[2.2.1]heptanes, bicyclo[3.3.1]nonane, and 11β-substituted estrenes, *J. Am. Chem. Soc.*, 101, 7195, 1979.
158. **Gonzalez-Sierra, M., Bustos, D. A., Ruveda, E. A., Olivieri, A. C., and Grasselli, M.,** Semiempirical calculation of ^{13}C nuclear magnetic resonance chemical shifts of acyclic hydrocarbons. Application to the stereochemical analysis of steroidal side chains, *Can J. Chem.*, 66, 71, 1988.
159. **Bernassau, J. M., Fetizon, M., Hanna, I., and Pinheiro, J. A.,** Axial hydrogen interactions and methyl group rotation in steroids, *J. Phys. Chem.*, 90, 3941, 1986.
160. **Fullerton, D. S., Kitatsuji, E., Deffo, T., Rohrer, D. C., Ahmed, K., and From, A. H. L.,** Use of PROPHET and MMS-X computer graphics in the study of the cardiac steroid receptor site of Na,K-ATPase, in *Current Topics in Membranes and Transport: Structure, Mechanism and Function of the Na/K Pump,* Vol. 19, Hoffman, J. F. and Forbush, B., Eds., Academic Press, New York, 1983, 257.
161. **Rohrer, D. C., Kihara, M., Deffo, T., Rathore, H., Ahmed, K., From, A. H. L., and Fullerton, D. S.,** Cardiac glycosides. IV. A structural and biological analysis of β-D-digitoxosides, β-D-digitoxose acetonides, and their genins, *J. Am. Chem. Soc.*, 106, 8269, 1984.
162. **Chiu, F. C. K. and Watson, T. R.,** Conformational factors in cardiac glycoside activity, *J. Med. Chem.*, 28, 509, 1985.
163. **Rathore, H., From, A. H. L., Ahmed, K., and Fullerton, D. S.,** Cardiac glycosides. VII. Sugar stereochemistry and cardiac glycoside activity, *J. Med. Chem.*, 29, 1945, 1986.
164. **Delettré, J., Mornon, J. P., Ojasoo, T., and Raynaud, J. P.,** An approach to the mapping of the steroid hormone receptor, in *Perspectives in Steroid Receptor Research,* Bresciani, F, Ed., Raven Press, New York, 1980, 1.
165. **Raynaud, J. P., Delettré, J., Ojasoo, T., Lepicard, G., and Mornon, J. P.,** Steps towards mapping of steroid hormone receptors, in *Physiopathology of Endocrine Diseases and Mechanisms of Hormone Action,* Soto, R. J., de Nicola, A., and Blaquier, J., Eds., Alan R. Liss, New York, 1981, 461.
166. **Raynaud, J. P. and Ojasoo, T.,** The relevance of structure-affinity relationships in the study of steroid hormone action, in *Steroid Hormone Receptors: Structure and Function,* Eriksson, H. and Gustafsson, J.-Å., Eds., Elsevier, 1983, 141.
167. **Ojasoo, T., Delettré, J., Mornon, J. P., Turpin-VanDycke, C., and Raynaud, J. P.,** Towards the mapping of the progesterone and androgen receptors, *J. Steroid Biochem.*, 27, 255, 1987.
168. **Delettré, J., Mornon, J. P., Lepicard, G., Ojasoo, T., and Raynaud, J. P.,** Steroid flexibility and receptor specificity, *J. Steroid Biochem.*, 13, 45, 1980.
169. **Duax, W. L., Cody, V., Griffin, J. F., Rohrer, D. C., and Weeks, C. M.,** Molecular conformation and protein binding affinity of progestins, *J. Toxicol. Environ. Health,* 4, 205, 1978.
170. **Bohl, M., Kaufmann, G., Hübner, M., Reck, G., and Kretschmer, R.-G.,** A-ring conformational stability and progesterone-receptor binding affinity of 4-en-3-one steroids, *J. Steroid Biochem.*, 23, 895, 1985.
171. **Bohl, M., Schubert, G., Ponsold, K., Reck, G., Höhne, E., and Simon, K.,** Molecular mechanics and X-ray crystal structure investigations on conformations of 11β substituted 4,9-dien-3-one steroids, *J. Mol. Graph.*, 7, 122, 1989.
172. **Bucourt, R., Cohen, N. C., and Lemoine, G.,** Nature quasi-*trans* ou quasi-*cis* de la fusion de deux cycles comportant un atome trigonal à la jonction, *Bull. Soc. Chim. France,* 903, 1975.
173. **Reck, G., Hübner, M., Bohl, M., Bannier, G., and Kretschmer, R.-G.,** Crystal and molecular structures of modified progestagens. I. 17α-Cyanomethyl-17β-hydroxy-estr-4-en-3-one, $C_{20}H_{27}NO_2$, *Cryst. Res. Technol.*, 21, 1035, 1986.

174. **Reck, G., Hübner, M., Bohl, M., and Leibnitz, P.,** Crystal and molecular structures of modified progestagens. II. 17α-Azidomethyl-17β-hydroxy-estr-4-en-3-one, $C_{19}H_{27}N_3O_2$, *Cryst. Res Technol.*, 21, 1041, 1986.
175. **Reck, G., Hübner, M., and Pfeiffer, D.,** Crystal and molecular structures of modified progestagens III. 17α-Chloromethyl-17β-hydroxy-estr-4-en-3-one, $C_{19}H_{27}ClO_2$, *Cryst. Res. Technol.*, 21, 1185, 1986.
176. **Scharfenberg-Pfeiffer, D., Megges, R., Bohl, M., Simon, K., and Stopsack, H.,** Crystal and molecular structure of 17α-acetoxy-21-hydroxy-1,4-pregnadiene-3,20-dione, *Cryst. Res. Technol.*, 25, 151, 1990.
177. **Duax, W. L., Griffin, J. F., Rohrer, D. C., Weeks, C. M., and Ebright, R. H.,** Steroid hormone action interpreted from X-ray crystallographic studies, in *Biochemical Actions of Hormones*, Vol. 11, Litwack, G., Ed., Academic Press, New York, 1984, 187.
178. **Duax, W. L., Griffin, J. F., and Weeks, C. M.,** Models for steroid-receptor interactions derived from X-ray crystallographic investigations, in *Interaction of Steroid Hormone Receptors with DNA*, Sluyser, M., Ed., VCH, Weinheim, 1985, 83.
179. **Duax, W. L., Griffin, J. F., Weeks, C. M., and Wawrzak, Z.,** The mechanism of action of steroid antagonists: insights from crystallographic studies, *J Steroid Biochem.*, 31, 481, 1988.
180. **Beusen, D. D., Kalman, B. L., and Covey, D. F.,** Role of substrate conformational features in the stereospecificity of aromatase, *Biochemistry*, 25, 662, 1986.
181. **Bohl, M., Simon, Z., and Lochmann, J. R.,** Theoretical investigations on the role of steroid-skeleton C4═C5 unsaturation in competitive aromatase inhibition, *Z. Naturforsch. C*, 44, 217, 1989.
182. **Dauben, W. G. and Funhoff, D. J. H.,** Theoretical evaluation of the conformations of previtamin D_3, *J. Org. Chem.*, 53, 5070, 1988.
183. **Mosquera, R. A., Rios, M. A., and Tovar, C. A.,** Conformational analysis of 5*Z*- and 5*E*-vitamin D_3 dihydroderivatives by molecular mechanics, *J. Mol. Struct. (Theochem)*, 168, 125, 1988.
184. **Mosquera, R. A., Rios, M. A., and Tovar, C. A.,** Conformational analysis of vitamin D_3 derivatives by molecular mechanics. II. 1α,25-Dihydroxyvitamin D_3 and analogues, *J. Mol. Struct.*, 213, 297, 1989
185. **Duax, W. L. and Griffin, J. F.,** Structural differences distinguishing steroid antagonists from agonists, in *Receptor Mediated Antisteroid Action*, Agarwal, M. K., Ed., Walter de Gruyter, Berlin, 1987, 1
186. **Teutsch, G.,** 11Beta-substituted 19-norsteroids: at the crossroads between hormone agonists and antagonists, in *Adrenal Steroid Antagonism*, Agarwal, M. K., Ed., Walter de Gruyter, Berlin, 1984, 43.
187. **Teutsch, G.,** Analogues of RU 486 for the mapping of the progestin receptor: synthetic and structural aspects, in *The Antiprogestin Steroid RU 486 and Human Fertility Control*, Baulieu, E.-E. and Segal, S. J., Eds., Plenum Press, New York, 1985, 27.
188. **Teutsch, G., Ojasoo, T., and Raynaud, J. P.,** 11β-Substituted steroids, an original pathway to antihormones, *J. Steroid Biochem.*, 31, 549, 1988.
189. **Herrmann, C., Hoyer, G.-A., Neef, G., Schomburg, D., and Wiechert, R.,** Röntgenstrukturanalysen und Struktur-Wirkungs-Beziehungen von Antigestagenen, *Liebigs Ann. Chem.*, 1317, 1986.
190. **Eckle, E., Hoyer, G-A., Hofmeister, H., Laurent, H., Schomburg, D., and Wiechert, R.,** Die Kristall- und Molekülstruktur von Gestoden, *Liebigs Ann. Chem.*, 199, 1988.
191. **Schmit, J.-P., Quivy, J. I., and Rousseau, G. G.,** The androgen receptor binding site: a conformational study of steroids in relation with their affinity. I. Interaction with the D-ring, *J. Steroid Biochem.*, 13, 1387, 1980.
192. **Bohl, M.,** Molecular mechanics investigation on side-chain conformations of a 17α-ethyl-17β-hydroxy steroid with regard to receptor binding, *Z. Naturforsch. C*, 42, 221, 1987.

193. **Scharfenberg, P. and Sklenar, H.,** Bestimmung von Molekülstrukturen mittels quantenchemischer Methoden, in *Quantitative Structure-Activity Analysis: Proceedings of the Second Symposium on Chemical Structure-Biological Activity Relationships: Quantitative Approaches,* Franke, R. and Oehme, P., Eds., Akademie-Verlag, Berlin, 1978, 175.

194. **Berlin, P.,** Konformationsberechnungen von Effektoren der (Na,K)-ATPase, *Ergebn Exp. Med.,* 24, 181, 1977.

195. **Rohrer, D. C., Fullerton, D. S., Yoshioka, K., Kitatsuji, E., Ahmed, K., and From, A. H. L.,** Structures of modified cardenolides. IV. [20 *S*]20(22)-Dihydrodigitoxigenin analogues, *Acta Cryst. B,* 39, 272, 1983.

196. **Bohl, M. and Wunderwald, M.,** Molecular mechanics investigation on conformational flexibility of 14β steroid in drug-receptor interactions, *Z. Naturforsch. C,* 41, 297, 1986.

197. **Hübler, D., Schubert, G., and Bohl, M.,** unpublished data.

198. **Duax, W. L., Griffin, J. F., and Rohrer, D. C.,** Conformation of progesterone side chain: conflict between X-ray data and force-field calculations, *J. Am. Chem. Soc.,* 103, 6705, 1981.

199. **Profeta, S., Jr., Kollman, P. A., and Wolff, M. E.,** Conformation of the progesterone side chain: resolution of the apparent conflict between X-ray data and force-field calculations using MM2, *J. Am. Chem. Soc.,* 104, 3745, 1982.

200. **Bohl, M.,** unpublished data.

201. **Carballeira, L. and Rios, M. A.,** Estudio por mecanica molecular de la conformacion de la cadena lateral de hormonas tipo 20-oxopregnano, *Anal. Quim. Real Soc. Espan. Quim. A,* 81, 191, 1985.

202. **Carballeira L. and Rios, M. A.,** Influencia de la sustitucion sobre la conformacion de la cadena lateral de hormonas tipo 20-oxo-pregnano, *Anal. Quim. Real Soc. Espan. Quim.,* 82, 46, 1986.

203. **Viana, L. M. and Takahata, Y.,** PCILO Study of the conformation of 20-oxopregnane hormones, *Int. J. Quant. Chem.,* 22, 265, 1982.

204. **Adelantado, J. M. and Richards, W. G.,** The geometry of a steroid from quantum mechanical calculation: progesterone, *J. Chem. Soc., Perkin Trans. 2,* 1253, 1986

205. **Lindeman, S. V., Timofeeva, T. V., and Struchkov, Yu. T.,** Capabilities of conformational changes in molecules of biologically active steroids, in *Proceedings of the 7th All-Union Symposium on Intramolecular Interactions and Molecular Conformations,* Pushino, USSR, 1986, 131, and personal communication.

206. **Watanabe, S., Saito, T., and Ikeda, M.,** Conformational study of the acetyl group and cyclohexenone in progesterone interacting with phosphatidyllipid by means of circular dichroism, *J Steroid Biochem.,* 31, 853, 1988.

207. **Kier, L. B.,** The conformation of 20-oxopregnane hormones from molecular orbital calculations and a consideration of the cortisol receptor, *J. Med. Chem.,* 11, 915, 1968.

208. **Schmit, J.-P. and Rousseau, G. G.,** Structure-activity relationships for glucocorticoids. III. Structural and conformational study of the rings and side-chain of steroid which bind to the glucocorticoid receptor, *J. Steroid Biochem.,* 9, 909, 1978.

209. **Lindeman, S. V., Tomofeeva, T. V., Shklover, V. E., Struchkov, Yu. T., Turuta, A. M., and Kamernitzky, A. V.,** The X-ray investigation of 16α,17α-thiazolidine derivatives of Δ^4- and Δ^5-pregnanes and conformational analysis of their oxathiolane analog, *Steroids,* 43, 125, 1984.

210. **Kamernitzky, A. V., Turuta, A. M., Fundieler, I. N., Pavlov, V. A., Antipin, M. Yu., Struchkov, Yu. T., Polishchuk, A. P., and Snatzke, G.,** α-Epithioketosteroids: conformation and reactivity, *Tetrahedron,* 38, 165, 1982.

211. **Lindeman, S. V., Struchkov, Yu. T., Turuta, A. M., Kamernitzky, A. V., Korshunov, V. A., Chuvylkin, N. D., and Snatzke, G.,** The structure of 16α,17α-epiminopregnenolone 3β-acetate in crystal and solution *Bioorg. Chim.,* 13, 252, 1987.

212. **Shamovsky, I. L., Ovchinnikov, A. A., Kamernitzky, A. V., Khazheeva, Z. I., and Simonov, V. I.,** Conformational flexibility of β-acetyl chain and biological activity of 16α,17α-cycloalkanoprogesterones, *Bioorg Chim.,* 8, 261, 1982.

213. **Kamernitzky, A.**, Structural-energetic aspect of steroid-receptor interaction, in *Proceedings of the F.E.C.S. Third International Conference on Chemistry and Biotechnology of Biologically Active Natural Products*, Vol. 2, VCH, Weinheim, 1987, 295.
214. **Bohl, M., Kasch, H., and Levina, I. S.**, MM2-Kraftfeld-Untersuchungen zur 17β-Acetylseitenketten-Konformation bei 16α,17α-Cyclohexanosteroiden, *Z. Chem*, 26, 378, 1986.
215. **Kamernitzky, A. V., Levina, I. S., Kulikova, L. E., Korkhov, V. V., Nikitina, G. V., and Bohl, M.**, Relationship between progestagenic action and conformation of the 17β-acetyl side chain in the pregna-D_6'-pentarane series, *Bioorg. Chim.*, 14, 828, 1988.
216. **Surcouf, E., Lepicard, G., Mornon, J.-P., Ojasoo, T., and Raynaud, J.-P.**, Additional conformational data for the mapping of the progestin binding site: crystal structures of 21-(phenylseleno)progesterone and 17α-(phenylseleno)progesterone, *J. Med. Chem.*, 26, 1320, 1983.
217. **Raynaud, J. P., Ojasoo, T., Bignon, E., Doré, J. C., and Surcouf, E.**, Receptor binding as a tool in the development of selective new bioactive steroids and nonsteroids, in *Innovative Approaches in Drug Research*, Harms, A. F., Ed., Elsevier, Amsterdam, 1986, 47.
218. **Simonov, V. I., Tseikinsky, V. M., Kamernitzky, A. V., Kulikova, L. E., and Levina, I. S.**, The structure and biological function of 16α,17α-cycloalkanoprogesterones (pregna-D'-pentaranes), *Bioorg. Chim.*, 7, 920, 1981.
219. **Tseikinsky, V. M., Rybakov, V. B., Simonov, V. I., Kamernitzky, A. V., Ignatov, V. N., and Levina, I. S.**, The molecular and crystal structure of 16α,17α-cyclopropanoprogesterone, *Bioorg. Chim.*, 6, 752, 1980.
220. **Cohen, N. C.**, GEMO, a computer program for the calculation of the preferred conformations of organic molecules, *Tetrahedron*, 27, 789, 1971.
221. **Declercq, J.-P., van Meerssche, M., and Zeelen, F. J.**, Conformational analysis of 3-oxo 5(10)-unsaturated steroids. Single-crystal X-ray structure analysis of 17-hydroxy-7α-methyl-19-nor-17α-pregn-5(10)-en-20-yn-3-one (Org OD 14), *Recl. Trav. Chim. Pays-Bas*, 103, 145, 1984.
222. **Liang, C. D., Baran, J. S., Allinger, N. L., and Yuh, Y.**, Synthesis and conformational stabilities of 11-oxo-9α- and 9β-estradiol 3-benzyl ether, *Tetrahedron*, 32, 2067, 1976.
223. **Bischofberger, K., Bull, J. R., Dillen, J. L. M., and van Rooyen, P. H.**, 14-Methyl steroids. VI. Synthesis, conformational analysis, and X-ray crystallography of 14α-methylestra-1,3,5(10)-triene derivatives: stereoselective hydroboration of *rac*-14α-methylestra α-1,3,5(10),9(11)-tetraenes, *S. Afr. J. Chem.*, 40, 123, 1987.
224. **Bull, J. R.**, Steroidal hormone analogues of unnatural configuration: models for structure-activity studies, *S. Afr. J. Chem.*, 38, 155, 1985.
225. **Boeyens, J. C. A., Bull, J. R., Tuinman, A., and van Rooyen, P. H.**, Steroidal analogues of unnatural configuration. XIV. Conformational analysis of 4,4,14α-trimethyl-19(10→9β)*abeo*-5β,10α-pregnane-6,11-diols by X-ray crystallography and force-field calculations, *J. Chem. Soc., Perkin Trans 2*, 1279, 1979.
226. **Bull, J. R., Levendis, D. C., and Tuinman, A.**, Steroidal analogues of unnatural configuration. XVII. Conformational analysis of 4,4,14α-trimethyl-19(10→9β)*abeo*-5α,10α-pregnane-6,11-dione, *Tetrahedron*, 38, 347, 1982.
227. **Bull, J. R., Pachter, R., Tuinman, A., and van Rooyen, P. H.**, Steroidal analogues of unnatural configuration. XVIII. Conformational analysis of 17β-hydroxy-4,4,9-trimethyl-9β,10α-estr-5-en-3-one and the derived 2α- and 2β-bromo compounds, *Tetrahedron*, 40, 777, 1984.
228. **Bull, J. R., Dillen, J. L. M., and Steer, L. M.**, Pentacyclic steroids. II. Synthesis and conformational analysis of isomeric hexahydrobenzo-[4.5.6]cholestanes, and approaches toward 4α-alkylation of 3-substituted cholest-4-en-6-ones, *S. Afr. J. Chem.*, 40, 155, 1987.

229. **Reerink, E. H., Schöler, H. F. L., Westerhof, P., Querido, A., Kassenaar, A. A. H., Diczfalusy, E., and Tillinger, K. C.**, A new class of hormonally active steroids, *Nature*, 186, 168, 1960.
230. **Segaloff, A., Gabbard, R. B., Flores, A., Borne, R. F., Baker, J. K., Duax, W. L., Strong, P. D., and Rohrer, D. C.**, Steroid structure and function. VII. Remarkable estrogenicity of 3-hydroxy-9β-estra-1,3,5(10)-triene-11,17-dione, *Steroids*, 35, 335, 1980.
231. **Uberoi, N. K., Hendry, L. B., Muldoon, T. G., Myers, R. B., Segaloff, A., Bransome, E. D., and Mahesh, V. B.**, Structure-activity relationships of some unique estrogens related to estradiol are predicted by fit into DNA, *Steroids*, 45, 325, 1985.
232. **Hendry, L. B., Bransome, E. D., Jr., Lehner, A. F., Muldoon, T. G., Hutson, M. S., and Mahesh, V. B.**, The stereochemical complementarity of DNA and reproductive steroid hormones correlates with biological activity, *J. Steroid Biochem.*, 24, 843, 1986.

Chapter 4

TWO APPROACHES TO STRUCTURE-ACTIVITY RELATIONSHIPS IN THE FIELD OF SEX-STEROIDS AND THEIR ANALOGS

Tiiu Ojasoo, Jean-Christophe Doré, Jean-Paul Mornon, and Jean-Pierre Raynaud

TABLE OF CONTENTS

I. INTRODUCTION

Rational drug design in the 1990s is increasingly based on a knowledge of the 3D-structure of the active sites of proteins that enables the chemist to imagine ligands with the appropriate fit and reactivity to induce or block biological action. In the field of steroid hormones, such an approach is not yet feasible since no integral steroid hormone receptor (SHR) has been sufficiently purified to be crystallized. Presently, several teams are producing appreciable quantities of SHR fragments by creating fusion proteins that are overexpressed in *Escherichia coli*[1-3] or *Saccharomyces cerevisiae*[4] or of full-length SHR in a *baculovirus* system.[5,6] The solution 3D-structure of the Zn^{++}-finger DNA-binding domain (DBD) of the estrogen and glucocorticoid receptors has been determined by 2D ^{1}H NMR spectroscopy[7,8] and, very recently, the X-ray crystal structure of a DBD-fragment complexed to a DNA-response element has been published.[9] However, the nature of the steroid-binding site is still much of a mystery as well as the way in which SHR functions are coordinated. The steroid hormone apparently leads to the partial dissociation of a large complex in which SHR is only one of many elements; it increases SHR phosphorylation, promotes DNA binding, and induces gene transcription. It may not be essential to each of these functions and probably has other as-yet poorly understood properties.

In the absence of a SHR crystal structure, two convergent approaches can be adopted to gain insight into the nature of the steroid binding site: on the one hand, structure prediction methods can be used to create a 3D-model of the SHR and of the steroid binding site; on the other, binding and activity data on large numbers of SHR ligands can be analyzed in order to deduce an imprint of the active site. In this chapter, we shall briefly address both of these complementary approaches in an attempt to evaluate and bridge the gap between them.

3D-modeling of the SHR is a relatively novel venture.[10-14] We shall recall the basic principles of an easy-to-grasp prediction technique, hydrophobic cluster analysis (HCA), and update the interpretations of our model of the steroid hormone-binding domain (HBD). Since drug design on the basis of screening data has been widely discussed in many reviews,[15-23] our emphasis will be on data analysis by multivariate statistical methods.[19,20,23]

II. CHARACTERIZATION OF THE STEROID HORMONE RECEPTOR

A. DOMAIN STRUCTURE OF STEROID HORMONE RECEPTORS

SHRs,* including the receptors for vitamin D_3 (VDR)[24] and ecdysone, form part of an expanding superfamily of about 30 proteins identified so far (for reviews see References 25 to 28). Prominent members of the family are the thyroid hormone receptors,[29,30] the product of the retroviral oncogene v-*erb*A,[31] the receptors for the morphogen retinoic acid,[32,33] for aryl hydrocarbons such as dioxin,[34] and for rodent hepatocarcinogens that cause proliferation of peroxisomes.[35] Less well-known proteins often with unknown ligands, i.e., orphan proteins, are COUP-TF (chicken ovalbumin upstream promoter-transcription factor),[36] the apolipoprotein A1 regulatory protein (ARP-1),[37] the *Drosophila* proteins encoded by the *seven-up* (*Svp*),[38] *knirps*,[39,40] *ultraspiracle*,[41] *fushi tarazu*,[42] and other genes. A very recent report suggests that dopamine may be an activator for COUP-TF.[43]

Sequence analysis combined with functional studies have led to the identification of three distinct SHR domains:[44-46] (1) the hydrophobic carboxy domain of approximately 240 amino acids binds steroid, (2) the middle domain (≈65 amino acids), which is most highly conserved among SHRs of all hormone classes, contains Zn^{++}-coordinated fingers characteristic of DNA-binding proteins. The steroid- and DNA-binding domains are connected by a hydrophilic stretch of 40 to 100 amino acids (hinge region) and do not require the presence of the third highly variable N-terminal domain for ligand-dependent activation of gene transcription. (3) The N-terminal domain is highly immunogenic, of low sequence identity among SHRs, and of very variable length (24 amino acids for VDR and 603 for MR.* Although not essential to transcriptional activation, its deletion results in a decrease in induction.

The highly hydrophobic steroid-hormone binding domain (HBD) is essential to SHR function. Its truncation generates a constitutive hormone-independent mutant that retains only 1 to 5% of the transcriptional activity of the corresponding wild-type receptor.[47] Our attention will be focused on modeling its 3D-structure using, as explained below, the known 3D-crystal structure of human α_1-antitrypsin (hα_1-AT)[48] as a reference.

* Estrogen receptor (ER), progesterone receptor (PR), androgen receptor (AR), mineralocorticoid receptor (MR), glucocorticoid receptor (GR).

B. MODELING OF THE HORMONE-BINDING DOMAIN OF THE RECEPTOR BY HYDROPHOBIC CLUSTER ANALYSIS (HCA)

1. Basic Principles and Methodology

Lim[49] postulated that the nascent polypeptide produced by a ribosome is a fluctuating α-helix which then folds into the native protein. HCA[50-52] can be considered to be based on this premise, i.e., a polypeptide chain is an α-helix unless other forces overwhelm this situation. Since hydrophobic interactions are a potentially strong folding force, the distribution of nonpolar sidechains can be examined on a flattened helical presentation to find potential interactions (i.e., hydrophobic clusters) that will influence the folding pattern. By extension, one can infer that proteins with similar hydrophobic cores might have similar patterns of hydrophobic clusters.[53-55]

The stratagem of writing each amino acid sequence on a classical α-helix efficiently condenses information and enables the easy visual detection and analysis of hydrophobic clusters of amino acids on a 2D plot. The sequence is written on an α-helix of 3.6 amino acids per turn that has been smoothed on a cylinder so that, after 5 turns, residues i and i + 18 are on the same directrix. The cylinder is then cut parallel to the axis, unrolled, and duplicated in order to obtain a better visual impression of the environment of each amino acid (Plate 1).* The adjacent and adjacent-but-one amino acids cover a 17 amino-acid stretch that is considered necessary for cogent secondary structure predictions by classic methods. Sets of hydrophobic amino-acids (V, I, L, F, W, M, Y) are encircled into clusters. Ala (A) and Cys (C) are considered hydrophobic in a hydrophobic environment only.

The 2D-topology (size, shape, orientation) and distribution of the hydrophobic clusters of the DBD and HBD of human SHRs[31,56-63] are compared in Plate 1[10-12] and confronted with the HCA plot of human corticosteroid binding globulin (hCBG), a plasma protein that transports endogenous glucocorticoids and progesterone. hCBG belongs to the family of serine protease inhibitors (SERPINs)[64] and has a 45% sequence identity with hα_1-AT[65] whose HCA plot is also given in Plate 1. The crystal structures of hα_1-AT, after cleavage by elastase,[48] and of two other members of the SERPIN family, antithrombin III[66] and ovalbumin,[67] have been determined to a resolution of ≈3Å or less.

Highly similar clusters, that are often located within identical sequence patterns, are used as anchor points for the manual alignment of the HCA plots, although presently this procedure is greatly facilitated by a new program MANSEK.[52] Structural segments are delimited by the positions of Pro (*) or Gly (♦) that induce disruptions in structure or by hydrophilic areas that correspond to loops. Insertions or deletions tend to occur in these areas rather than in the hydrophobic core of the protein that maintains stability.[68]

* Plate 1 appears after page 184.

2. Visual Comparison of HCA Plots

The DBD sequences of SHRs are extremely well conserved, as are the resulting small hydrophobic clusters (Plate 1). The DBD is separated from the HBD by a hinge region (HR) that has been divided into two segments VL_1 and VL_2. Segment VL_1 of hMR is particularly rich in Pro (*) but it is within segment VL_2 that the greatest variability is observed among SHRs. That of VDR (not shown)[13] is exceptionally rich in Ser (⊡). The HBD has been divided into 15 segments. It is clear that several structural segments (isolated or combined hydrophobic clusters) of the HBD vary little among SHRs and can easily be distinguished *de visu*, e.g., the highly conserved segments S_2, S_5, S_6, S_{10}-S_{12}, S_{14}, S_{15} of hPR, hGR, hMR, hAR. Whereas the sequences of these SHRs end at S_{15}, hER possesses an additional hydrophilic C-terminal segment and an additional subsegment between S_{10} and S_{11} centered around Ser 463.

The HCA plots of hCBG and hα_1-AT can be aligned with the plots for the HBDs of SHRs without insertion nor deletion if due account is taken of proteolysis of SERPINs by elastase which releases the highly hydrophobic peptide S_{16}-S_{17}. Surprisingly, SHRs except for hER end at this cleavage point. Could this observation signify that sequenced SHRs are all proteolytic cleavage products? Neutrophil elastase can produce 52- and 30-kDA GR fragments in the cytosol of human leukemia cells;[69] SHRs appear to be able to bind serine and thiol-serine protease inhibitors[70] such as phenylmethylsulfonylfluoride (PMSF),[71] aprotinin,[72] and leupeptin[73] and incidentally also protease substrates[74] thus begging the question whether they have protease and/or protease inhibitor activity.[75]

A marked similarity in HCA plot patterns is observed in spite of the low sequence identity (6 to 10%) between the HBD of SHRs and hCBG. Conserved hydrophobic clusters are located in the segment C-terminal to finger F_2 in the DBD, in segment VL_1, and in segments S_1-S_4, S_6, S_8, S_{10}-S_{12}, S_{14}, and S_{15} of the HBD. Segment S_6, characterized by a large highly distinctive hydrophobic cluster which is present in all members of the SHR and SERPIN families, was one of the major anchors for the alignment. The tail end of this cluster has the consensus sequence 111X111FXXXW (1 = hydrophobic residue; X = any residue). A recent data bank search of 40,470 sequences limited the number of protein classes other than SERPINs, SHRs, and cytochromes with this sequence to less than 50 among which were ATPases (in particular actin and Ca^{++}-transporting ATPase), the acetylcholine receptor, and proteins of the photosystem (opsin).

HCA homology scores describing the topological correspondence of the hydrophobic residues, whatever their precise nature, within the clusters in segments S_1 to S_{15} are a rough estimate of the conservation of the hydrophobic protein core. Experience shows that a $\leq$50% score indicates unrelated 3D structures and a $\geq$75% score very similar 3D structures. According to these scores

```
hPR   727 L S V V K W S K S L P G F R N L H I D D Q I T L I Q Y S W M S L M V F G L G W R S Y    776 L Y F A P D L I L
hGR   572 I A A V K W A K A I P G F R N L H L D D Q M T L L Q Y S W M F L M A F A L G W R S Y    620 L C F A P D L I I
hMR   778 I Q V V K W A K V L P G F K N L P L E D Q I T L I Q Y S W M C L S S F A L S W R S Y    827 L Y F A P D L V F
hAR   712 V H V V K W A K A L P G F R N L N V D D Q M A V I Q Y S W M G L M V F A M G W R S F    761 L Y F A P D L V F
hER   355 V H M I N W A K R V P G F V D L T L H D Q V H L L E C A W L E I L M I G L V W R S M    402 L L F A P N L L L
hCBG  136 M N F Q D W A T A S R Q I                 P A I L V L V N Y I F F K G T W T Q P    199 F Y V D E T T V V

              ← S4        S5 →                         S6                                     ← S7   S8 →
          A, β-strand    helix f                   A3 β-strand

hPR   816 E E F L C M K V L L L L N T I P L E G L R S Q T Q F E E M R S S Y I R E L    869 R F Y Q L T K L L D N L H D
hGR   661 E E Y L C M K T L L L L S S V P D K G L K S Q E L F D E I R M T Y I K E L    714 R F Y Q L T K L L D S M H E
hMR   867 E E Y T I M K V L L L L S T I P K D G L K S Q A A F E E M R T N Y I K E L    920 R F Y Q L T K L L D S M H D
hAR   801 Q E F L C M K A L L L F S I I P V D G L K N Q K F F D E L R M N Y I K E L    854 R F Y Q L T K L L D S V Q P
hER   443 E E F V C L K S I I L L N S     *     L K S L E / I H R V L D K I T D T L    503 R L A Q L L L I L S H I R H
hCBG  241                 G T V F F I L P D K G             M N T V I A A L       I    286 G V Y D L G D V V E E M G
Cyt   346         D R N R L V L L E A T I R E V L R I R P V S P T

                                 S10 - S11                                                 S12
                   B2, B3 β-strands helices g, h                                A6 β-strand, helix i
```

FIGURE 1. Regions of high contiguous sequence identity within hSHRs and hCBG as deduced from HCA plots.[11] The asterisk denotes the longer subsegment centered on S463 in hER (between segments S_{10} and S_{11}). The 17 amino acids incorporating the steroid-binding consensus sequence of bovine cytochrome P450c17[78] have been aligned with the SHRs by HCA plot analysis and underlined.

- hCBG belongs structurally to the SERPIN family (HCA score between hGBG and hα_1-AT = 83%). No other hormone binding protein (BP), apart from human thyroxine (T_4) binding globulin (hTBG)[76] which has been identified as apolipoprotein A-1,[77] showed any significant HCA homology with SERPINs nor with SHRs (i.e., α-glycoprotein, rat prostatic BP, human sex-steroid BP, rabbit uteroglobin, rat androgen BP, human vitamin D BP, cellular retinoic acid BP, β-lactoglobulin, plasma retinol BP, transthyretin, murine α-fetoprotein).
- The HBDs of SHRs display structural homology with the SERPINs hCBG, hTBG, and hα_1-AT (HCA score = 60 to 70% when excluding the variable segment S_5).
- The HBDs of SHRs that bind 3-keto-4-ene steroids (i.e., hGR, hPR, hMR, hAR) are structurally interrelated to >80% but only related to ≈75% with the HBD of hER that binds A-ring phenolic steroids.
- SHRs (e.g., hER and hGR) have an ≈70% HCA homology score with vitamin D_3 receptor (hVDR), retinoic acid receptor (hRARα), and thyroid hormone receptor (hTRβ) which are structurally interrelated to ≥80% (HCA plots not shown).[13]

3. Sequence Alignment

By matching analogous clusters, amino acid sequences are aligned on the basis of presumed 3D topology rather than maximum sequence identity. The most precise sequence alignment is deduced by proceeding outwards from the hydrophobic center of each cluster toward the boundary of a structural segment. This alignment method has confirmed the presence of several regions of high contiguous sequence identity within SHRs (e.g., segments S_4, S_6, $S_{7/8}$, S_{10} and S_{12}) (Figure 1), some of which, because of this high degree of conservation, could partake in key functions of SHRs.

The 17 identical amino acids of cytochromes P450c17 and P450c21 that are considered part of a steroid-binding site[78] can be aligned by HCA plot analysis with the corresponding amino acids in SHRs and hCBG as shown in Figure 1, i.e., with 451 to 471 of hER and not 80 amino acids downstream as has been previously proposed.[79]

Few of the amino acids in the zones of high contiguous sequence identity are common to SHRs and hCBG. They are, however, highly reactive low-polarity amino acids: Trp (W) in segments S_4 and S_6, Tyr (Y) in segments S_7 and S_{12}, Phe (F) in segment S_6, and Leu (L) in segment S_{12}. A variety of chemical studies has suggested the presence of Trp and Tyr residues in the steroid-binding site of hCBG.[80,81] For instance, excess cortisol can protect hCBG from nitration of Tyr residues by tetranitromethane, suggesting the presence of a Tyr residue. Fluorescence quenching studies and analysis of UV difference spectra of hCBG-steroid interaction complexes have led to the conclusion that a Trp (W) residue is present. Trp-modifying agents affected first an accessible then a less-accessible Trp, with concomitant inactivation of steroid-binding sites. Two of the three Trp within the first 358 residues of hCBG are found in the SHRs (segments $S_{4/5}$ and S_6). A Tyr residue, possibly equivalent to Tyr 849 of hPR, is considered to be involved in the progesterone-binding site of uteroglobin[82] and of progesterone-binding globulin[83] and also in estradiol binding to ER[84] and to 17β-dehydrogenase.[85] An Asn instead of a Tyr at position 770 of mGR increases the dissociation rate of ligand.[86] His and Met residues have been covalently labeled in hCBG and also in hPR,[87] but only one Met in the above segments (S_6 and $S_{7/8}$) is common to both proteins.

4. Structure Prediction

Several amino acid sequences can define a single structural motif and the prime objective of HCA is to discern structural analogies rather than sequence identities. Indeed, many hydrophobic cluster shapes can be related to identifiable 3D-structural features.[50] Thus, the large highly distinctive hydrophobic cluster in segment S_6, present in all members of the SERPIN and SHR families and used as an anchor to align HCA plots, is typical of a long β-strand. The cluster of segment S_{10} is also typical of a β-strand whereas segment S_{12} is a β-strand followed by an α-helix. The 2D-structural features of hα_1-AT, as determined from its X-ray crystal structure, cap its HCA plot in Plate 1.

The relatively high HCA score (66 to 70%) corresponding to the comparison between the HBD sequences of SHRs that bind 3-keto-4-ene steroids and the sequence preceding the elastase cleavage point of hα_1-AT justifies the use of the crystal structure of cleaved hα_1-AT as a first approximation of the 3D structure of the HBD of a SHR.[12] This crystal structure, however, is probably not a valid model for the hinge region in view of the poor HCA correspondence for segments VL_1 and VL_2. Even for the HBD the hα_1-AT model has to be viewed with circumspection and constantly reassessed in the light of new structural or biochemical data. The construction of a more precise model for the HBD of each SHR by atomic substitution and energy minimization should enable confrontation with available biochemical data on the corresponding SHR[52] and consequently yield a better gauge of the validity of the model. Furthermore, the recent availability of the X-ray crystal structure of uncleaved ovalbumin,[67] another SERPIN, should help to put the hα_1-AT model into perspective.

As a first verification, in view of the role of intramolecular disulfide bonds in maintaining the architecture of proteins, we examined whether the location of Cys within the HBD of SHRs is compatible with the hα_1-AT model, although, under the reducing conditions prevalent in the cell or *in vitro* in the presence of thiol-reducing agents such as dithiothreitol (DTT), the S-S bridges may be reduced to enable steroid binding.[88-90] Two Cys residues are relatively well conserved in the HBD of SHRs: C798/C820 of hPR, C643/C665 of hGR, C783/C805 of hAR, C417/C447 of hER (the second Cys is missing in hMR) and correspond to locations within a hairpin structure of hα_1-AT. If due allowance is made for the expected variation between hα_1-AT and SHRs, the distance between these residues is compatible with the formation of a disulfide bond. GR has an additional Cys tandem (C622/C638 of hGR) whose corresponding hα_1-AT residues are also close enough to each other within the 3D model to be compatible with disulfide bond formation.

C. VISUALIZATION OF THE FUNCTIONS OF THE HORMONE-BINDING DOMAIN USING α_1-ANTITRYPSIN AS A MODEL PROTEIN

At any moment during their lifespan, SHRs may be complexed with different components necessary to their stability and/or activity (see Reference 12). These may be proteins of analogous or lower molecular weight (approximate range 29 to 90 kDa) with possibly ATPase, kinase, or protease activity, low molecular weight (inhibitory) factors, nucleic acids (RNA and, in particular, tRNA) and nucleotides (ATP, cAMP), ions, unsaturated long-chain fatty acids and phospholipids, sugars, steroid ligand, solvent (i.e., structured water). The environment thus governs SHR conformation and supplies the cofactors and energy that ensure proper folding and binding for signal transmission to DNA.

What is the SHR environment? The vast majority of experiments on SHRs have been performed on "cytosol", which is a high-speed supernatant of tissue homogenates prepared in low-salt buffers, in the presence of diverse agents such as molybdate (a phosphatase inhibitor), sucrose, EGTA (a chelating agent), dithiothreitol (DTT, a thiol-reducing agent). SHRs have also been detected in or near rough endoplasmic reticulum, in association with the cytoskeletal matrix, free ribosomes, microsomes, and membranes. However, SHRs trigger gene transcription within the nucleus. Sheridan et al.[91] have proposed that the distribution of unoccupied SHR molecules between the cytoplasmic and nuclear compartments reflects water structure and repartition, the cytoplasm being a crowded, semicrystalline (highly structured) environment, whereas the nucleus behaves as a much more aqueous compartment. Ligand binding would change the SHR to a more hydrophilic nuclear form. Later enucleation and immunocytochemical experiments have supported the contention that for most SHR classes the unfilled SHR is essentially nuclear.[92-94]

Nascent polypeptide chains emerging from the ribosome would appear to have an α-helical conformation and a similar extended conformation is observed when proteins traverse mitochondrial membranes.[95] It has been proposed that the hydrophobic regions exposed in newly synthesized incompletely folded polypeptides interact with ubiquitous heat-shock proteins (HSPs), also designated stress-induced proteins or molecular chaperones, whose role is to stabilize the protein, to prevent improper interactions resulting from transient exposure of hydrophobic or charged surfaces, and to promote proper folding. To fulfil this role, HSPs would use the energy from ATP hydrolysis since several HSPs have demonstrable ATPase activity.

SHRs in the cytosol are associated with several HSPs.[96] However, unfolded HSP-free SHR, represented by the linear succession of the secondary structural features of $h\alpha_1$-AT deduced by X-ray but also HCA analysis, would be a particularly useful model to indicate schematically the localization of the functions inherent to SHR subdomains.[12,14] Figure 2 uses this device to represent the correspondence between structural motifs and the amino acid sequences involved in several SHR functions (HSP_{90} binding,[96-101] ATP-binding,[102,103] phosphorylation,[104-106] transcription activation,[107,108] dimerization,[109] and nuclear localization[110,111]) irrespective of hormone class, tissue, and species. Folding this linear $h\alpha_1$-AT structure into the known crystal structure (Figure 3) gives a visual impression of the possible spatial distribution of these functions. For further details, the reader is referred to References 12 to 14.

III. STEROID BINDING TO STEROID HORMONE RECEPTORS

Steroid binding is the primary function of the SHR but, in order to be effective in inducing the biological end point, it has to set into motion or enhance some of the other functions described above. Whether steroid is actually required for SHR binding to DNA *in vitro* and *in vivo* has been a controversial issue.[112-114]

Ligands such as steroids that are involved in signaling pathways tend to be small lipophilic molecules that fit specifically into hydrophobic pockets within receptor proteins. Their specificity as substrates is largely determined by hydrogen bonds (H-bonds) while other forces stabilize the ligand-mac-

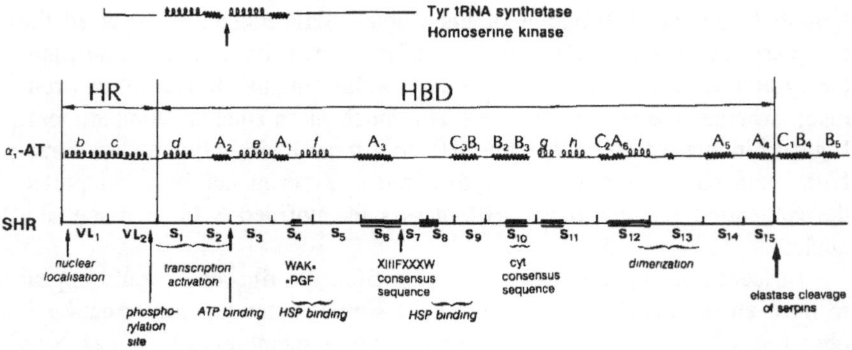

FIGURE 2. Schematical and hypothetical representation of the hinge region (HR) (segments VL_1 and VL_2) and hormone-binding domain (HBD) (S_1 to S_{15}) of a SHR. The baseline is capped by the secondary structures deduced from the 3D-crystal structure of hα_1-AT. ıııııı : α-helix, ∿: β-strand, —: loop or other. The presumed roles of several segments are given below the baseline. The underlined bold sections refer to the regions of high sequence identity among SHRs (see Figure 1). The Rossmann fold of two ATP-binding proteins (tyrosine tRNA-synthetase and homoserine kinase) with some sequence identity with a SHR or related receptor (retinoic acid)[102,103] are shown at the top of the figure.

romolecule complex (van der Waals interactions, aromatic-aromatic interactions, electrostatic interactions, etc.). Van der Waals interactions act over longer distances than H-bonds, are individually of weaker energy, but far more numerous.

Binding of most biologically active steroids to SHRs is associated with an energy of about -12 to -15 kcal/mol, corresponding to a dissociation constant of 10^{-9} to 10^{-11} M. A third to one half of this energy is supplied by the H-bonds at the C-3 and C-17 ends of the steroid (for all classes of steroid hormones) and at C-11 (for gluco- and mineralocorticoids), the remainder by van der Waals interactions. For instance, the mean intermolecular van der Waals interaction energy is, respectively, -3 and -3.5 kcal/mol for appropriately located methyl and ethynyl groups.

Removal of the C-3 and/or C-17 substituents of the natural hormones or replacement by groups unable to form the same directional H-bonds results in a drastic loss in affinity for the SHR unless, for instance, another strong interaction compensates for part of the normal H-bond interaction involving the 3-keto group (~3 kcal/mol).

In many instances, the effects of substituents on the steroid (perhydrocyclopentanophenanthrene) or an analogous (e.g., triphenylethylene) skeleton are additive with regard to binding affinity.[115-118] However, depending upon the substituent and position, transmission effects can occur that influence charge distribution, substituent orientation, and the twist of the skeleton. This has been illustrated in the past using C-7,17-methylated and C-17 ethynylated steroid 4-enes and 4,9,11 trienes as examples[119,120] which form complexes of

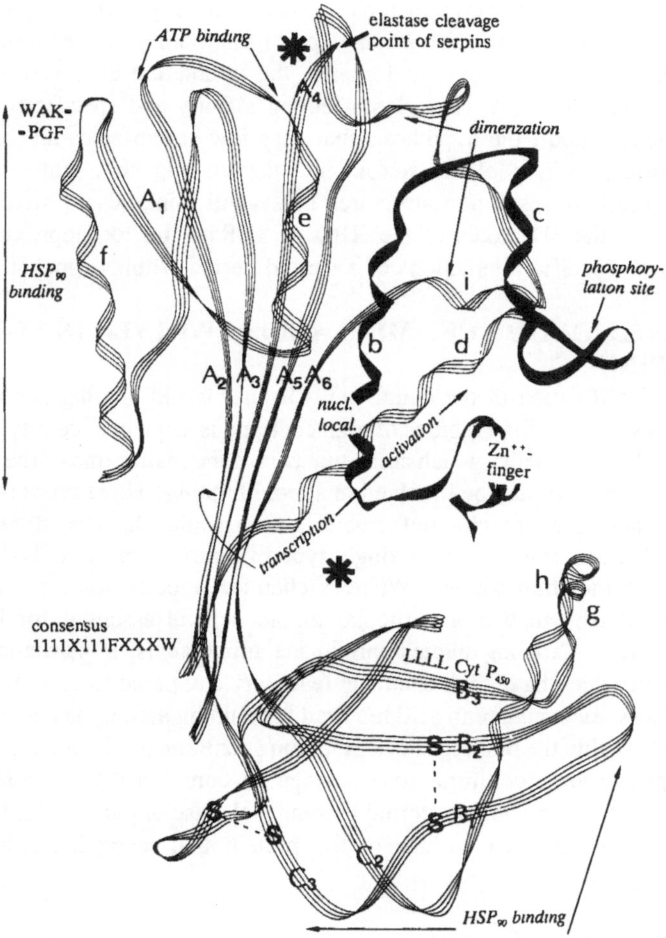

FIGURE 3. Ribbon representation of the crystal structure of proteolyzed hα_1-AT (residues 20 to 358)[48] without cleaved peptide. β-strands ($A_1 \rightarrow A_6$, $B_1 \rightarrow B_3$, $C_2 \rightarrow C_3$) and α-helices (b → i) have been labeled as in Figure 2. The dark zones (DBD and HR) are not an appropriate model for SHRs. The approximate locations of the following are indicated: regions of association with HSP_{90} (574-596 hGR;[97] 632-659 mouse GR[98]), possible ATP-binding site within an α-β-α-β Rossmann fold,[12,14] putative phosphorylation site,[105] and transcription activation sequence,[108] dimerization domain[109] with heptad repeat hydrophobic amino acids,[12] nuclear localization signal;[110] several consensus sequences (WAK··PGF, FXXXW,[12] LLLL[78]); possible disulfide bridges (S—-S), and two hydrophobic channels that could accommodate steroid (*). The lower cavity is formed by removal of the α_1-AT peptide fragment cleaved by elastase, the upper cavity by unfolding of the A_4-strand (segment S_{15}) as observed in the crystal structure of ovalbumin.[67]

different stability with PR. The trienes are flat and flexible molecules[121] that might easily enter a hydrophobic groove interacting with a number of residues in the process. This could account for their low specificity. Their most distant conformations are 3Å apart and about 1 kcal from the minimum energy conformation. Such an energy increment could be readily supplied by binding. The additional 9,11-ene double bonds, apart from modifying electronic properties, can twist the skeleton and change the orientation of a 17α-ethynyl, when present, by ~1 Å, thus also affecting affinity and specificity. These observations support the hypothesis that very few amino acids are involved in the correct locking of the steroid into the binding site. Unfortunately, although the high-resolution structures of several hundreds of steroids are known,[122,123] the 3D model of the HBD of SHRs is far too approximate to consider the specific interactions of a steroid nested within a pocket.

A. IDENTIFICATION OF AMINO ACIDS INVOLVED IN STEROID BINDING

The identification of the amino acids in the steroid-binding site is at its early stages. Some of the regions of high contiguous sequence identity (Figure 1) among different SHRs which constitute part of the mainframe of the protein structure (Figure 3) may be involved in steroid binding. Three types of methods have attempted to pinpoint crucial amino acids: the use of chemical reagents that are selective for a single type of amino acid, affinity-labeling studies, and mutation studies. Whereas chemical modification or mutation studies establish whether a particular amino acid is essential for binding affinity, affinity-labeling studies only locate amino acids, in particular those such as Cys which have high chemical reactivity compared to other residues. Nevertheless, even an amino acid required for binding affinity is by no means necessarily within the binding site; a mutation at one site can influence affinity at another site via a conformational change. A complementary method that can also yield information on steroid binding is the use of panels of antibodies to peptides at the outer surface of the protein that recognize either free and/or occupied SHR.

1. Use of Chemical Reagents

Experiments with diethylpyrocarbonate, a reagent which under certain conditions of pH is considered selective of His, have suggested that chick PR and rat and lamb ER contain a His residue in their steroid-binding sites, although this might be an unusually reactive Tyr residue.[124-126] The participation of a Tyr in estradiol binding to rat uterus ER is supported by the greater inhibition of binding by tetranitromethane (TNM) at pH 8 than pH 6 (TNM has increasing affinity for Cys over Tyr at pH 6).[84]

2. Affinity Labeling[127]

Several Met and Cys residues identified by affinity labeling with dexamethasone 21-mesylate, triamcinolone acetonide, or R5020 (promegestone), are reported to be present within the hormone-binding site of GR and PR: M622, C638, C656, and C754 in rat GR, C644 in murine GR, M759 and M909 in hPR.[128-131] C644 of murine GR is equivalent to C656 of rat GR and

M759 of hPR to M622 of rat GR. The hPR counterparts of C638 and C656 of rat GR are a Tyr and Ser, respectively. The hα_1-AT residue corresponding to M622 of rat GR and M759 of hPR is I188 (segment S_6), and that corresponding to M909 of hPR is near position 333 (segment S_{14}) (see Plate 1). The positions of these residues within the crystal structure of hα_1-AT are shown in Plate 2.* The hα-$_1$AT residues that are the counterparts of C656 and C754 in rat GR are situated at the outer surface of the protein and are not shown.

Binding of organometallic estradiol derivatives to ER[132] has suggested that vicinal space positioning of a Cys (C) and Lys (K) or Arg (R) is necessary for the binding of their D ring. These could be amino acids 447/449 of hER, that in our alignment precede the cytochrome consensus sequence in Figure 1, or amino acids 529 to 531. Cys 530 of hER has recently been identified as a site of reaction with the affinity labels ketononestrol aziridine, an estrogen, and tamoxifen aziridine, an antiestrogen.[133]

3. Studies with Mutated and Mutant Receptors

Two types of studies fall under this heading: on the one hand, experiments with artificially mutated receptors designed to demonstrate the significance of an amino acid or a sequence in binding and SHR activation and, on the other hand, experiments with naturally mutant or variant receptors that have been detected in pathological states, e.g., in cancer or in disorders such as the complete or partial androgen insensitivity syndrome (AIS) where resistance to hormone action (undervirilization and infertility) can be due to structural mutations of the AR gene. Mutations of the GR and VDR genes have also been reported in cases of glucocorticoid[134] and vitamin D_3[135] resistance. To be able to associate a pathological syndrome with a defective protein is one of the most enlightening ways of establishing relationships between cell proteins and response. However, of the mutations detected in variant proteins so far, only a fraction occur in the HBD of the protein and not necessarily within the steroid-binding site. Furthermore, many of the mutations are premature stop codons giving rise to truncated SHRs.

a. Glucocorticoid Receptor Mutations

The mutation Glu to Gly in position 546 of mouse GR results in a poor steroid-binding protein.[86] Asn instead of a Tyr at position 770 of mGR increases the dissociation rate of ligand[86] and deletion of the last 29 amino acids from the carboxy terminus of rat GR reduces steroid binding to 1%.[107] The approximate locations of these mutations within the hα_1-AT model are shown in Plate 2. Paradoxically, a 16-kDa fragment of rat GR (Thr 537-Arg 673) retains affinity for dexamethasone although this is considerably lower than for the holoreceptor.[136] Even if the true protein (HBD) is more compact than

* Plate 2 appears after page 184.

FIGURE 4. hα_1-AT correspondence of the rat GR fragment[136] that can bind dexamethasone in the presence of HSP_{90}.

in our model, it is difficult to imagine how a fragment ending between strands B_2 and B_3 (Figure 4) can remain conformationally stable enough to bind steroid unless perhaps the structure is maintained by HSP_{90} and by an appropriate disulfide bridge. Sequential point mutagenesis of each of the three Cys of this 16-kDa core (Cys 640, 656, and 661) to Ser or, in the case of Cys 656, to Gly showed that each of the mutants could bind dexamethasone, with decreased affinity for the 640 and 661 mutants but increased affinity for the 656 mutant.[137] Any two of the Cys could apparently form a disulfide bridge under appropriate conditions as determined by their behavior towards the thiol reagent methylmethanethiosulfonate.

The binding affinity for dexamethasone of an Asp 641 Val mutant receptor identified in a case of familial glucocorticoid resistance was threefold lower than that of the wild-type GR.[134]

b. Progesterone Receptor Mutations

A comparison of the sequences of the HBDs of chick PR,[138,139] that does not respond to the antiprogestin RU38486,[140] and hPR[58] that does, has revealed the importance of Gly 722 in RU38486 action.[141] Glycines, like prolines, provoke a disruption in secondary structure. The corresponding residue of chick PR is a Cys that when mutated to a Gly gives rise to a protein that responds to RU38486 in cotransfection assays. This Gly residue is located within the ATP-binding site of the HBD (Plate 2). The observation is intriguing since it could imply that the antihormonal efficiency of RU38486 is a form of selective ATP-depletion on the SHR, ATP being doubtless necessary for hormone-binding activity[142] and also for other functions including phosphorylation and conformational changes. In particular, the conformational change needed for transcriptional activation may not occur since, in the presence of RU38486, the activation sequence τ_2 (helix d and strand A_2 in Figure 3) is inactive.[143] The identification of the corresponding amino acid in hamster PR that, like chick PR, does not respond to RU38486[144] would be instructive.

*c. **Estrogen Receptor Mutations***

The substitution of a Val for a Gly at position 400 of hER from MCF_7 cells destabilizes the SHR structure and decreases estradiol affinity at 25°C but not 4°C.[145] Mutations of charged and polar amino acids near Cys 530 of hER (Glu 523 to Gln, Lys 529 and Lys 531 to Gln, Asn 532 to Asp, and Asp 538 to Asn),[146] a site that would be labeled by organometallic[132] and aziridine[133] estrogen or antiestrogen derivatives (see above), have revealed that, whereas E_2 has a three- to fourfold increased K_d for the KKN→QQD mutant (Lys 529 Gln, Lys 531 Gln, Asn 532 Asp) at 22°C, its affinity for Glu 523 Gln and Asp 538 Asn was unchanged (Plate 2). The structurally varied antiestrogens 4-hydroxytamoxifen, LY 117018 and ICI 164,384, a steroid with a 7α-amido alkyl aliphatic side chain,[147] were 5 to 50 times more potent competitors of E_2 binding to the KKN→QQD mutant as compared to wild-type ER but unchanged competitors of the Glu 523 Gln and Asp 538 Asn mutants.

Mutations of each Cys (Cys 381, 417, 447, and 530) of the HBD of hER (Gly 400) to Ala gave mutant receptors that, at 22°C, had a similar K_d for E_2 as wild-type ER.[148] All bound labeled tamoxifen aziridine and ketononestrol aziridine covalently, thus suggesting that in the Ala 530 mutant an alternate residue is covalently labeled.

Variant ER transcripts exist in human breast tumors and these may be involved in variable response to therapies directed at ER.[149-151] Among the variations noted in the C-terminal domain are a Lys to Arg substitution in the hinge region, a Leu 296 Pro, a silent polymorphism at Pro 325, a truncation at 370, alternate splicing, i.e., transcripts missing either exon 5 (transcriptionally active in the absence of estrogen) or exon 7 (inactive but able to prevent wild-type ER function).[152-154]

*d. **Androgen Receptor Mutations***

There is a high degree of genetic heterogeneity in complete AIS (cAIS). In cultured skin fibroblasts from subjects with cAIS AR binding can be either apparently absent or present at normal levels. Brown et al.[155] have identified two point mutations (Arg 774 Cys and Arg 831 Gln) that abolish androgen binding and one (Met 866 Val) that reduces the apparent binding affinity sixfold compared to wild-type hAR. Ris-Stalpers et al.[156] reported a Gly to Thr mutation in the splice donor site of intron 4 of the hAR of a patient with cAIS which abolished normal RNA splicing at the exon 4/intron 4 boundary and resulted in the activation of a cryptic splice donor site in exon 4 leading to the deletion of 123 nucleotides from the mRNA. This AR does not bind androgen. Marcelli et al.[157] have detected a premature termination codon instead of the expected Lys codon at amino acid position 588 which results in a decrease in the amount of mRNA encoding hAR and in the synthesis of a truncated hAR that is unable to bind ligand and unable to activate the long-terminal repeat of the mouse mammary tumor virus (MMTV) in cotransfection assays. The following point mutations have been identified by De Bellis et al.[158] in cAIS and are shown in Plate 2: Ser 759 Phe, Arg 774 His (see Arg 774 Cys above), Arg 831 stop (see Arg 831 Gln above), Arg 855 Cys, and Asp 864 Gly (see adjacent Met 866 Val above and Thr 868 Ala below). An Arg 840 His mutation was the cause of partial resistance. Of the mutated amino acids, only Arg 855 and Asp 864 are conserved among the SHRs that bind 3-keto-4-enes.

AIS is also observed in testicular feminized *Tfm* rodents. Yarbrough et al.[159] found a single point mutation (Arg 734 Gln) within the HBD of rat AR that gave rise to a mutant AR which had only 10 to 15% of the androgen-binding capacity of wild-type AR and reduced capacity to stimulate transcriptional activity in cotransfection assays in CV1 cells. Gaspar et al.[160] detected an AR mRNA in *Tfm*/Y mutant mice that covers the entire coding region but that is present at 10- to 20-fold lower levels than in normal animals. The expression of this mutant AR would be affected at a posttranscriptional level.

Not all of these studies have addressed the hormone-binding specificity of the mutant ARs. Having compared two potent synthetic androgens (methyltrienolone[161-163] and mibolerone[164]) with the natural hormones testosterone (T) and dihydrotestosterone (DHT), Kaufman et al.[165] suggest that, in subjects with AR-defective androgen resistance, the different ligand-AR complexes reach different dissociative states by undergoing sequential transitions along a common pathway, and that these transitions are co-regulated both by the chemical characteristics of the bound androgen and by other nonreceptor factors. Distinct specificity differences have been noted in the human prostate tumor cell line LNCaP. According to Veldscholte et al.,[166-168] the AR of these cells contains a single point mutation (Thr 868 Ala) in the HBD that, like the above Met 866 Val mutation, is located in the dimerization region (Plate 2). This mutant hAR binds androgens, progestogens, estrogens, and antiandrogens which can all activate the expression of an androgen-regulated reporter gene construct.

As shown in Plate 2, with the help of HCA, the hα_1-AT model can be used to pinpoint several of the above-listed SHR mutations that reduce or abolish ligand binding. For GR, PR, and AR, these mutations are located around the region occupied by the cleaved peptide within hα_1-AT; this is less evident for ER. However, whatever the hormone class, the deletion of A_4 and A_5 strands drastically decreases steroid-binding affinity. Establishing the effect of the deletion of the A_4 strand alone might improve our understanding of the hα_1-AT model which has a five-stranded backbone β-sheet compared to the crystallized ovalbumin model with a four-stranded sheet. In ovalbumin, the thermal stability of which is considerably lower than that of α_1-AT, the A_4 strand is unwound and forms an external loop.

B. POTENTIAL HYDROPHOBIC STEROID-BINDING CHANNELS

The hα_1-AT model has two hydrophobic channels that could accommodate a steroid: (1) a highly hydrophobic channel about 20 by 10 Å is formed by clearing the cleaved peptide (S359-Q393) which fills the core of proteolyzed hα_1-AT. Its surface appearance is shown in Figure 5 (left-hand side). Many of the residues protruding into the channel belong to the well-conserved segments S_6, S_{10}-S_{12}, S_{14}, and S_{15}. Although the size of this channel is much larger than that of a steroid, it could vary since the α/β motif opposite the β-sheet (strands A_1 to A_6) is apparently flexible. The X-ray structures of retinol-binding protein,[169] delta-5-3 ketosteroid isomerase,[170] and cholesterol oxidase[171] complexed with hormone have revealed a binding site sandwiched between a large β-sheet and an α/β motif. (2) The position of the cleavage point of hCBG suggests that the main part of strand A4 (segment S_{15}) is not buried in native hCBG but can unwind to form an external loop as found in the recently published X-ray structure of uncleaved ovalbumin. This creates a vicinal hydrophobic channel 17 by 6Å that could feasibly constitute an alternative hormone-binding site as shown in Figure 5 (right-hand side). The two proposed sites communicate.

Further data are necessary to assess the potential significance of these two hypothetical sites. Nevertheless, it is noteworthy that several lines of evidence have already suggested the possible existence of two steroid-binding sites on the SHR. Progestins accelerate the dissociation of glucocorticoids from GR by binding to a second hydrophobic site.[172-176] More recently, truncated receptor forms obtained in fusion experiments with β-galactosidase or ubiquitin have yielded curvilinear Scatchard plots for ligand binding.[2] If there is no second independent binding component that could account for these results, it is feasible that a region of the SHR such as the hinge region may articulate the DNA-binding and N-terminal domains in such a way as to allow or forbid access to one of the sites. Furthermore, the proximity of the one site to the dimerization and ATP-binding regions is intriguing. RU38486 which, like progestins, binds with high affinity to hPR[177,178] but, unlike progestins, does not accelerate labeled dexamethasone dissociation from GR,[176]

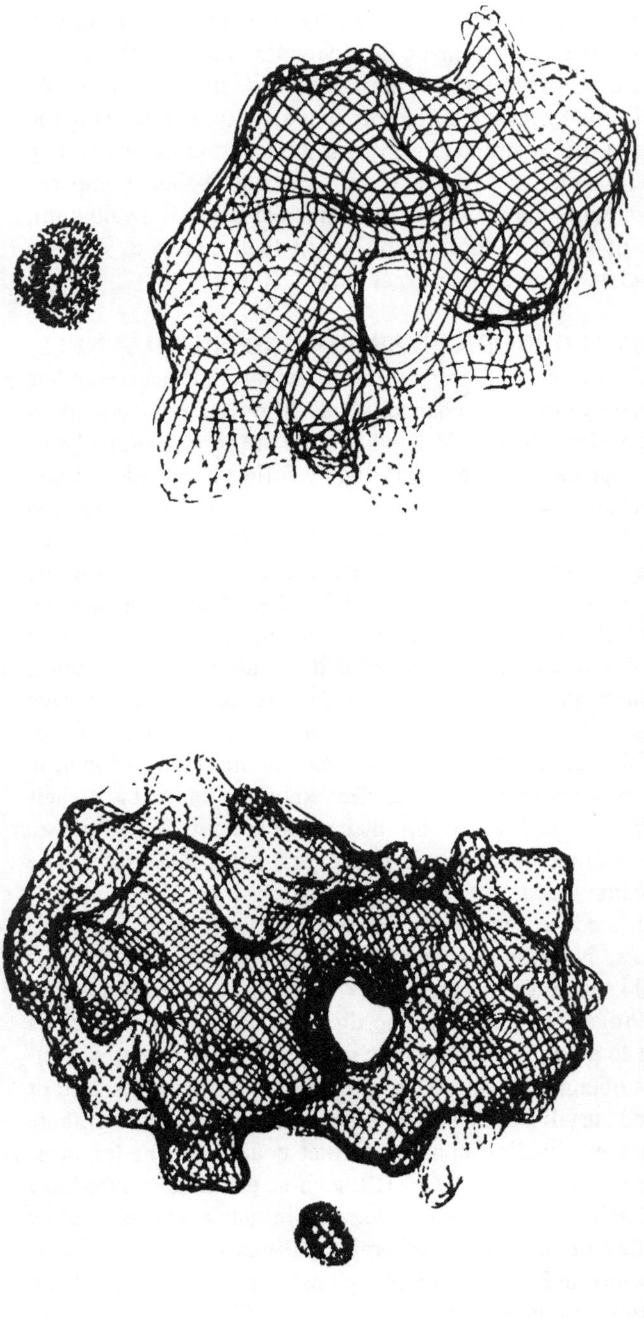

FIGURE 5. Smoothed β-spline surface of cleaved hα_1-AT showing channels formed by the cleaved peptide S359-Q393 (left-hand side) and by the mold of the A_4-strand (right-hand side). In the left panel, the orientation shown in Figures 3 and 4 and Plate 2 has been rotated anticlockwise around the vertical axis. In the right panel, the protein is beheld from the top of Figures 3 and 4 and Plate 2 downwards. A progesterone molecule, viewed along its major axis, is represented to scale (grid square ~ 1 Å^2).

does not bind to a SHR with a mutation within the ATP-binding site. This questions whether antihormones with reactive side chains often used as tools in studies on mechanism of action interact with functional sites other than the binding site of the cognate natural steroid ligand.

When opting for two sites or for a one-site allosteric model, several questions need to be addressed: Is only one site normally functional and the other masked or just an evolutionary relic? Are there different sites according to hormone class, for phenolic steroids vs. 3-keto-4-ene steroids? Unlike in the case of 3-keto-4-ene steroids, a positively charged environment as created by Lys and Arg is needed for phenol binding. Recent studies on hydroxylated diarylindenes and on a homologous series of parahydroxylated triphenylacrylonitrile (TPE) derivatives has surprisingly suggested that the orientation of the molecule within the binding site is always the same whichever ring is hydroxylated. A hydroxyl group in a given para position of the TPE always interacts with the same location within the site.[118,179] Is the HBD more compact than in our model and are the two sites in fact a single site?

IV. STRUCTURE-ACTIVITY RELATIONSHIPS ANALYZED BY MULTIVARIATE METHODS

From the above analysis of SHRs, it is clear that steroid ligands and their analogs could interfere with several functions of SHRs. They might also interfere with analogous functions on other protein targets, e.g., with similar ATP-binding sites. How their actions with regard to their multiple targets are related remains unelucidated.

In order to establish which structural determinants are involved in a particular biological function, the life scientist often relies on screening data on one or more targets using, if possible, molecules that differ by systematic single substitutions. The approach is laudable but, in the interpretation of his data, he is often faced with several problems: the data are so voluminous that he cannot grasp them in their entirety; he cannot classify the chemical structures without the assistance of a theoretical chemist; he can only infer with difficulty whether the biological variables are related to each other in any way; he cannot tell which information is redundant. Faced with these issues, he has two options: either sift the data and retain those that seem to directly confirm or refute his working hypothesis or maintain an open mind with no preconceptions on causal links between the study parameters and test compounds. In the latter instance, a multivariate analysis can draw different levels of correlation among the measured variables whether these are biological parameters or structural features.

In the present section, we shall briefly describe two types of multivariate analysis, the minimum spanning tree (MST) method[180-183] and correspondence factorial analysis (CFA),[184,185] which we have already applied in the past to our screening data, whether to the SHR-binding profiles and activity of steroids[19,20,186,187] or the ER binding and activities of nonsteroid analogs of the triphenylethylene series.[23,188-190] In each case, the crude data matrix is first converted into a Chi^2-distance table. Those who dislike and avoid statistics should simply bear in mind that the sole object is to establish proximities between points in an n-dimensional space. Proximity implies affiliation, distance implies diversity. MST and CFA analyze this distance table differently, MST being an automatic classification procedure and CFA a factorial procedure. One of the major differences is that, whereas the MST treats test compounds (rows) and biological variables (columns) separately, CFA enables the simultaneous representation on a single graph of both test compounds and biological variables. These procedures can be used for purely descriptive or analytical purposes (structural models), but also as a decision base to predict active structures (simulation models).

A. MINIMUM SPANNING TREE (MST) METHOD

1. Basic Principles

The MST method calculates the shortest route within the n-dimensional space that links the molecules into a network on the basis of their responses on the chosen battery of tests (or the biological parameters into another network on the basis of their reactivity towards the molecules). The distance separating two molecules, as measured along the pathways generated, is an indication of the closeness of the relationship between them with regard to all the test variables measured. Within the network, no ringroads nor backtracking is permitted. A frivolous comparison would be the best sales round for a commercial traveler. The criterion that the sum of the segments linking the molecules be as small as possible is far from trivial since there are nearly 500 million ways in which 12 items can be arranged linearly. The MST is thus not just a logical way of disposing the items but a true description of the inherent structure of the system under study.

A major advantage of MST is the clear visual interpretation of the data it provides.

1. Position of a molecule — Is it part of a trunk or a branch, a main branch or an atypical twig, at a bifurcation or within a long stem? The position thus tells us whether the molecule is part of a cluster, an intermediate within a dynamic progression, a new lead compound. Its structural affiliation with other molecules can be deduced.
2. Specificity/selectivity profiles — Pathways along the branches illustrate how specific properties evolve in relation to others.
3. Introduction of a new molecule — The full activity profile of a new molecule can be introduced into the system to establish its position. The new molecule may be located at a branch extremity (enhanced selectivity), between two adjacent nodes (an intermediary), or between two nonadjacent nodes (creation of a new pathway).

For the life scientist there are basically two approaches to the interpretation of an MST in chemical terms: from his knowledge of the biological targets, e.g., active sites on proteins, he can take a guess at the molecular forces that come into play: steric hindrance, charge, hydrophobicity, etc. Such new variables can be introduced into the MST; or he can now, with his chemical classification already in hand, seek the assistance of a theoretical chemist who will describe the molecules in terms of

- 3D structure (Dreiding models, van der Waals spheres translated, like any other solid, into contours, vectors, pixel densities, skeletons etc., crystalline coordinates)
- Chemical formulas (compared in their entirety by fitting procedures or as fragments by topological methods such as autocorrelation vectors or DARC-PELCO)
- Spectral signatures that can be discontinuous such as mass and NMR spectra or continuous such as UV, IR spectra and need to be transformed (e.g., sampling, Fourier transform)
- Quantum chemistry variables (isopotential map, charge density, free valency index, π electron density, etc.).

These attributes of the molecules can be used to construct another MST to reveal their purely physicochemical affiliations.[183]

2. Application of MST to Steroid Progestins: Analysis of Structural Determinants Involved in Specificity Profiles

Relative binding affinities (RBAs) of test compounds for several cytosol SHRs under standardized conditions are routinely measured in large-scale screening programs.[15-22,191] In the present example, the published RBAs of 51 compounds[19,20] (Table 1) will be submitted to a MST analysis. Initially, the compounds were chosen to demonstrate the AR and GR binding components of most progestins,[192,193] to establish whether the secondary AR and GR binding exhibited fast or slow dissociation kinetics,[192-194] and to highlight the lack of binding specificity of available glucocorticoids.[187] The published data, obtained under identical experimental conditions (with a few exceptions), were assembled into a single table and the MST analysis was performed on six biochemical tests: competition for labeled R5020 binding to PR in rabbit uterus cytosol at 0°C after 2 and 24 h incubation, competition for labeled testosterone binding to AR in rat prostate cytosol at 0°C after 30 min and 2 h incubation, and competition for labeled dexamethasone binding to GR in rat thymus cytosol at 0°C after 1 and 24 h incubation. The data in each column were scaled from 1 to 100 prior to constitution of the χ^2-distance matrix.

The shortest pathway is shown in Figure 6. The degree of analogy between the specificity profiles of the compounds are indicated by broad dark lines

TABLE 1
Relative Binding Affinities (RBAs) for Cytosol SHRs at 0°C and at Two Incubation Times in a Routine Screening System

	PR		AR		GR	
	2 h	24 h	30 min	2 h	1 h	24 h
Testosterone (T)	1	1	100	100	3	0.5
5α-DHT	1.5	1	95	120	1	1
Danazol	9	3	35	8	<0.1	<0.1
17α-MethylT	3	2	90	45	2	1
19-NorT	20	15	230	155	4	0.5
Trenbolone	75	15	250	190	9	2.5
Norgestrienone (R2010)	65	45	95	70	55	15
Gestrinone (R2323)	75	50	95	85	265	150
Metribolone (R1881)	210	190	160	205	60	20
Ethynylnortestosterone (ENT)	155	265	75	45	20	5.5
17α-EthynylT (ET)	35	15	7	0.1	2	0.5
Cyproterone ac. (CPA)	80	60	50	15	9*	6
Megestrol ac. (MA)	150	120	65	20	90+	45
Chlormadinone ac. (CA)	175	320	80	20	215	35
Progesterone (P)	100	100	20	5.5	115	75
RU25051	330	870	40	15	155	80
RU23747	225	380	3.5	2	55	25
RU22779	205	335	7.5	3.5	35	15
Demegestone (R2453)	230	420	7.5	1	40	10
Promegestone (R5020)	220	535	10	1.5	100	30
Nomegestrol ac. (NA)	170	450	20	1	20*	15
16α-MethylP	65	60	10	1	160	55
6α, 16α-Dimethyl P	50	155	5	8	395	220
RU25253	180	530	40	6	770	330

Medroxyprogesterone ac. (MPA)	125	305	40	50	470	215
RU4841	230	675	60	135	380	245
Norgestrel	170	905	110	85	210	40
RU2999	260	305	205	160	85	55
RU2420	280	330	175	180	85	25
RU25593	40	35	6	5	390	180
RU25055	70	85	20	6	550	245
Corticosterone (B)	5	3	1.2	0.5	100	100
Cortisone (E)	<0.1	—	—	<0.1	1.5	1.0
16α-Methylprednisolone	<0.1	—	—	<0.1	135	150
21-Desoxydexamethasone	<0.1	—	—	<0.1	225	190
Cortexolone (S)	0.8	0.5	<0.1	0.3	40	15
Hydrocortisone (F)	<0.1	—	—	<0.1	45	60
Cortivazol	<0.1	—	—	<0.1	300	475
Cortivazol-21OH	<0.1	—	—	<0.1	310	530
Prednisolone	<0.1	—	—	<0.1	60	115
Betametasone	<0.1	—	—	<0.1	185	380
Dexamethasone (DXM)	0.4	<0.1	0.3	<0.1	165	455
Desoximethasone	25	15	2.5	1	345	990
RU23739	35	65	1.5	0.3	450	1410
Triamcinolone acetonide	15	12	0.2	<0.1	210	790
Triamcinolone	<0.1	—	—	<0.1	55	290
Fluocinolone acetonide	38	50	0.8	0.2	275	1250
RU38140	45	85	35	15+	360*	1300
RU38486	80	530	10	25+	465*	1365
RU28289	230	440	—	10+	450*	1365
Deoxycorticosterone (DOC)	26	18	10	2.5	65	110

Note: All results are comparable except when marked with an asterisk: *4h, or plus sign, +24 h. The RBAs of testosterone, progesterone, and corticosterone for their respective receptors were set at 100. (— : not determined but considered equal to 0.1 in the multivariate analysis).

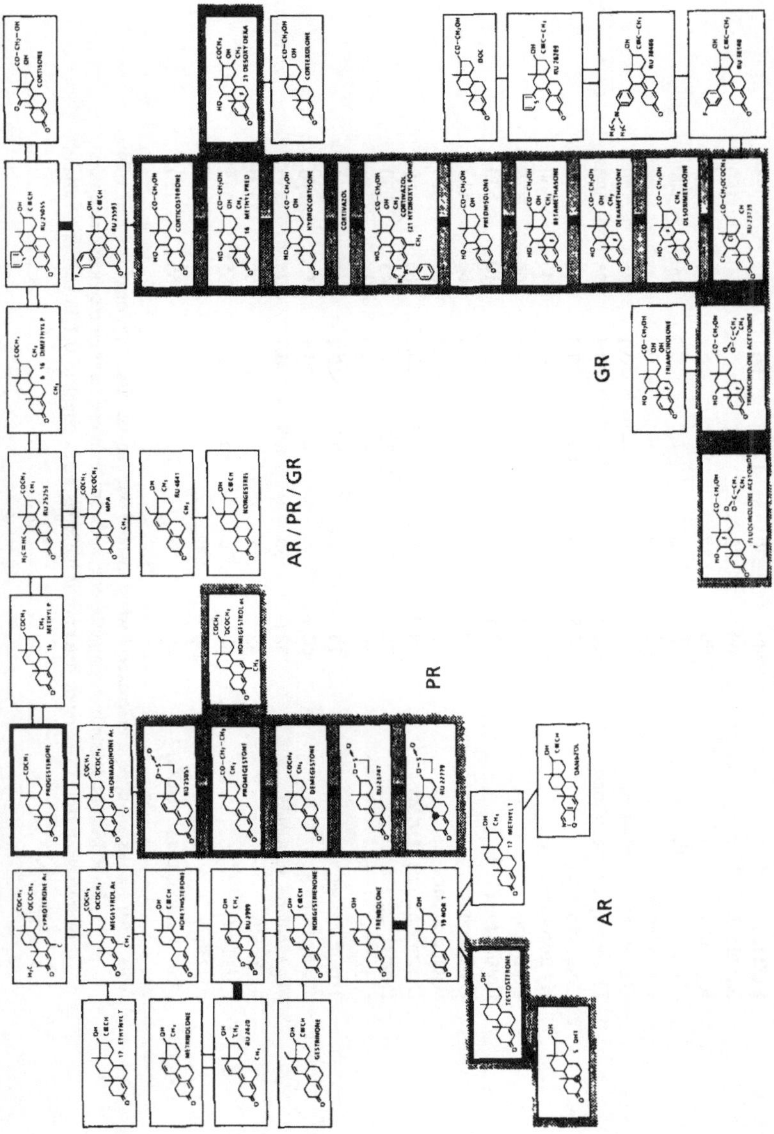

FIGURE 6. Minimum spanning tree for steroids constructed on the basis of their RBAs for PR, AR, and GR at two incubation times (see Table 1).[19,20] The relative proximity of two steroids is indicated by the nature of the line linking them. Dark broad line: close, double line: intermediate, single fine line: distant. Molecules with highly similar profiles are surrounded by stippling.

(close relationship), double lines (intermediate relationship), and fine lines (distant relationship). The MST should be viewed as a mobile; only distances along the links have any meaning. Thus, for instance, in Figure 6, the relationship between RU22779 and 19-nortestosterone is very distant and transits via chlormadinone and megestrol acetates.

This MST describes the specificity of binding *only* and does not illustrate the absolute binding levels. Within each branch, the compounds are ranked by stepwise progression towards the most distinctive profile. On the unproven assumption that, in the tissues and species under study, all the compounds are competitive inhibitors of the sites defined by the radioligand, it can be seen from the tip-of-the-branch molecules that only 5α-dihydrotestosterone is more specific than T, whereas several compounds are more specific than promegestone-R5020[195] (e.g., nomegestrol acetate,[196] demegestone, RU23747, and RU22779). The profile of nomegestrol acetate would appear to differ from those of demegestone and its sultine neighbors. Similarly, for this population of molecules under the prevalent experimental conditions, one can consider that dexamethasone or rather its neighbor-once-removed RU23739 is a precursor of two distinct types of profiles, one close to that of dexamethasone and leading to fluocinolone acetonide and triamcinolone, the other more diverse and leading to deoxycorticosterone (DOC). Overall, the compounds can be classified into two opposing groups according to their inherent androgenic (left-hand branch) and glucocorticoid (right-hand branch) nature. Between these two branches lies the progesterone molecule from which two alternate pathways emanate, one towards pure progestins, the other towards aspecific molecules of the norgestrel type.

The central and in some ways key position of progesterone is intriguing. It appears to reflect a multifaceted capacity of the molecule in which differential aspects can be enhanced by appropriate substitution. This is in accord with its known biological (antiandrogenic and antiglucocorticoid) properties. 16α and 6α methyls,[197] or a C11-vinyl substituent (RU25253)[192,198] initiate the pathway towards more prominent GR binding. This trend is continued by the C-11 substituted ethynyl nortestosterone derivatives (RU25055 and RU25593) which are weakly connected to corticosterone. From corticosterone onwards, one enters the field of closely related GR binders, except for the diverticulum at RU23739 towards RU38140 which has a particularity that will be discussed below (see Section IV.B.2).

Returning to our departure point, progesterone, one notes that an offshoot from the main trunk of RU25253 leads to norgestrel via medroxyprogesterone acetate (MPA) and RU4841. The AR component of these three molecules precludes their insertion in the GR branch; on the other hand, their marked GR component excludes them from the AR and PR branches. The GR component of MPA could be partly accounted for by its 6α-methyl and that of RU4841 by the introduction of a 7α-methyl into a C17,18-dimethyl substituted triene. All other trienes[199] are in a cluster in the AR branch. Also departing from progesterone is a branch characterized by 19-norpregnane or sultine structures that ensure high PR specificity. These molecules are highly similar. The AR and PR branches are bridged by the acetoxy progesterone derivatives, except for MPA.

Although this MST teaches us little that is new compared to the information already published on these compounds, it illustrates how a multivariate method can be used to classify information not only for a limited population of molecules as shown here but for up to several hundred molecules on a hundred or so tests. The representation is a true description of the organization of the system; the use of the crude RBA data (i.e., without scaling from 1 to 100 to distribute the weight evenly among columns) gave a highly similar structure with only a few differences in branch points. However, it should be borne in mind that the conclusions drawn only apply to the population of molecules under study, under the experimental conditions used, and on the basis of the chosen yardstick to express the results.

The above data were also submitted to an alternative multivariate analysis, correspondence factorial analysis (CFA), as described below.

B. CORRESPONDENCE FACTORIAL ANALYSIS (CFA)

1. Basic Principles

As mentioned above, statistical analysis is based on the measurement of distances by an appropriate metric such as the Euclidean metric used to establish correlations among quantitative variables (e.g., as in analysis of variance) and the χ^2-metric used to compare and contrast profiles of qualitative variables. The aims of a multidimensional analysis are to define relationships (correlations) within a family of variables measured on a number of items (i.e., among observations). It engenders clouds of points providing a visual description of the variables and observations. A first step in representing p variables in a p-dimensional space is to define a reduced space that best accounts for the total variance and that represents the initial data with minimum deformation. This procedure assumes that the inherent information is partially redundant as a result of strong correlations between certain variables. The reduced space is defined by a new system of orthogonal or *factorial* axes that are ranked by their order of importance in accounting for the total variance, i.e., according to their contribution to the inertia of the clouds of points. Establishing factorial axes implies, in fact, creating new variables that are linear combinations of the initial variables. From a mathematical viewpoint, multidimensional analysis consists in treating a data table of rows and columns as follows: (1) transforming the data according to the chosen metric to obtain a working data-matrix, (2) calculating the eigenvalues and eigenvectors of this matrix which yield the rank order of the factorial axes, (3) calculating the relative (RC) and absolute (AC) contributions of each initial variable with respect to the most significative factorial axes in order to interpret these axes (ΣAC = 100% and ΣRCs of each item to all factorial axes = 1), and

(4) selecting one or more pertinent planes for the graphical representation of the analysis.

2. Application of CFA to the Receptor-Binding Profiles of Steroid Progestins

CFA was used to convert the multidimensional table of data for the 51 steroids screened for binding (RBAs) to the 3 receptors under 2 incubation conditions (Table 1) into 2D-factorial maps which *simultaneously* represent test compounds and biochemical variables. Proximity within the maps highlights correlations among the tests, among the compounds, and between tests and compounds. The $\Phi_1\Phi_2$ factorial map (Figure 7) was defined by the molecules and biochemical tests that were the most discriminatory and incorporated a very high proportion of the total variance of the system (56.69 + 30.48 = 87.17%). The remaining axes ($\Phi_3 \rightarrow \Phi_5$) accounted for only 12.82% (7.87 + 3.09 + 1.86) of the variance (Figure 8).

According to Table 2, the most striking feature of the chosen population of test molecules is the dichotomy between GR and AR binding. The absolute contributions (ACs) to the Φ_1 factorial axis show that these properties of the molecules are opposed to the following extents: [+]57.7% (18.5 + 39.2) for GR vs. [−]39.8% (19.2 + 20.6) for AR. (The signs in square brackets refer to the direction of the projection on the factorial axes). The ACs of AR are similar for short and long incubation times, but the AC of GR is double for long over short incubation times. This observation is further qualified (see Φ_2 factorial axis) by an opposition of PR binding ([−]57.7% = 27.0 + 30.7) with primarily AR binding ([+]31.0% = 13.1 + 17.9) and to some degree with GR binding at long ([+]10.5%) but not short incubation times. Thus, binding specificity differences predominate in the $\Phi_1\Phi_2$ factorial map, whereas differences between incubation times, i.e., between ligand-receptor interaction kinetics,[200,201] are highlighted by the lower order axes. These axes ($\Phi_3 \rightarrow \Phi_5$) occult interference from specificity differences and basically oppose long and short incubation times for GR ([−]62.1% vs. [+]35.0%), PR and AR, respectively.

The clusters of molecules that bring to light the contrasted biochemical properties are situated around each pole in the $\Phi_1\Phi_2$ factorial map and confirm the tip-of-the-branch clusters of the MST, i.e., highly specific progestins around the PR pole, androgens at the AR pole, and glucocorticoids around GR. Of greater interest is the location of the intermediate molecules. Those with the least variance in the system, i.e., nearest to the origin, are MPA and RU4841 which interact to considerable extents with all three receptors. The androgenic trienes are in the top left-hand quadrant, the less androgenic progestins in the lower left-hand quadrant. Compared to the results of the MST, only the position of norgestrel seems somewhat unexpected being closer to the acetoxyprogestins than to RU4841. However, a closer examination reveals that its ACs to the $\Phi_1\Phi_2$ axes of this map are very low compared to

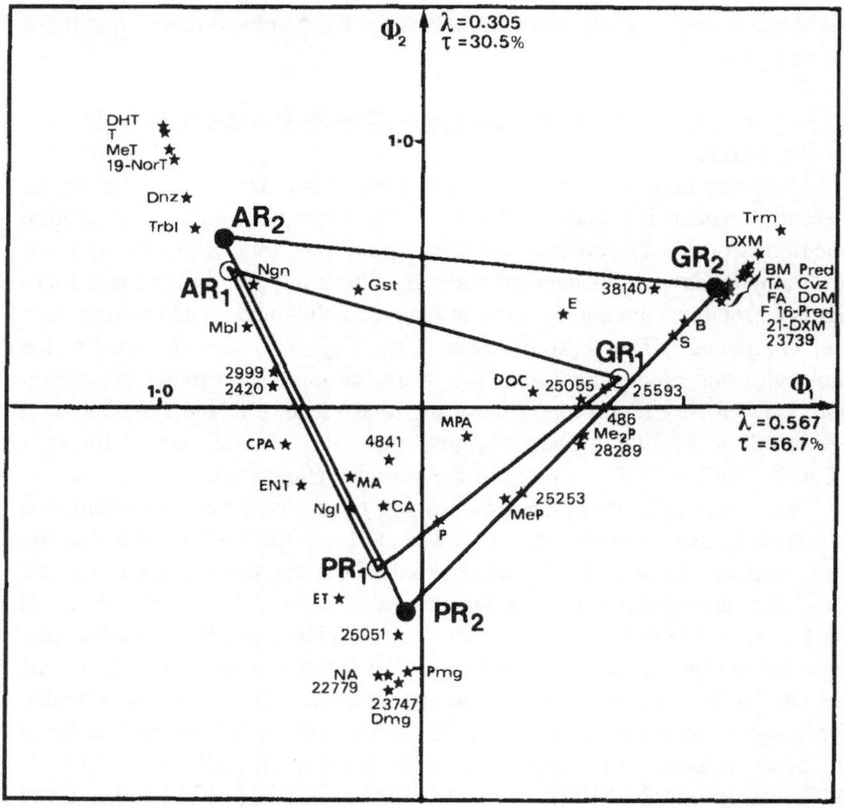

FIGURE 7. $\Phi_1\Phi_2$ distribution map derived by CFA of the data in Table 1[19,20] showing the relative locations of the biochemical variables (AR,PR,GR) at short (○) and long (●) incubation times and of the test steroids (*). The abbreviations used are as in Table 1 and as follows: MeT = 17α-methyltestosterone, Dnz = danazol, Trbl = trenbolone, Ngn = norgestrienone, Mbl = metribolone, Gst = gestrinone, Ngl = norgestrel, Pmg = promegestone, Dmg = demegestone, MeP = 16α-methylprogesterone, Me_2P = 6α,16α-dimethylprogesterone, 21-DXM = 21-deoxydexamethasone, 16-Pred = 16α-methylprednisolone, FA = fluocinolone acetonide, DoM = desoximethasone, Cvz = cortivazol, TA = triamcinolone acetonide, Pred = prednisolone, BM = betamethasone, and Trm = triamcinolone.

its AC to the Φ_4 axis which discriminates between PR (RC = 22%) and accessorily AR (RC = 5%) binding kinetics (Figure 8). Indeed, as Table 1 shows, the RBA of norgestrel for PR increases over five-fold on lengthening incubation time. It thus forms an extremely stable complex with this receptor. Similarly, the Φ_4 axis reveals the stability of the RU38486 and RU4841 complexes with PR.

Inspection of the glucocorticoid-like molecules provides some interesting comments. Intermediate molecules between PR and GR are, as expected from the MST, 16α-methylprogesterone and RU25253. Somewhat closer to the

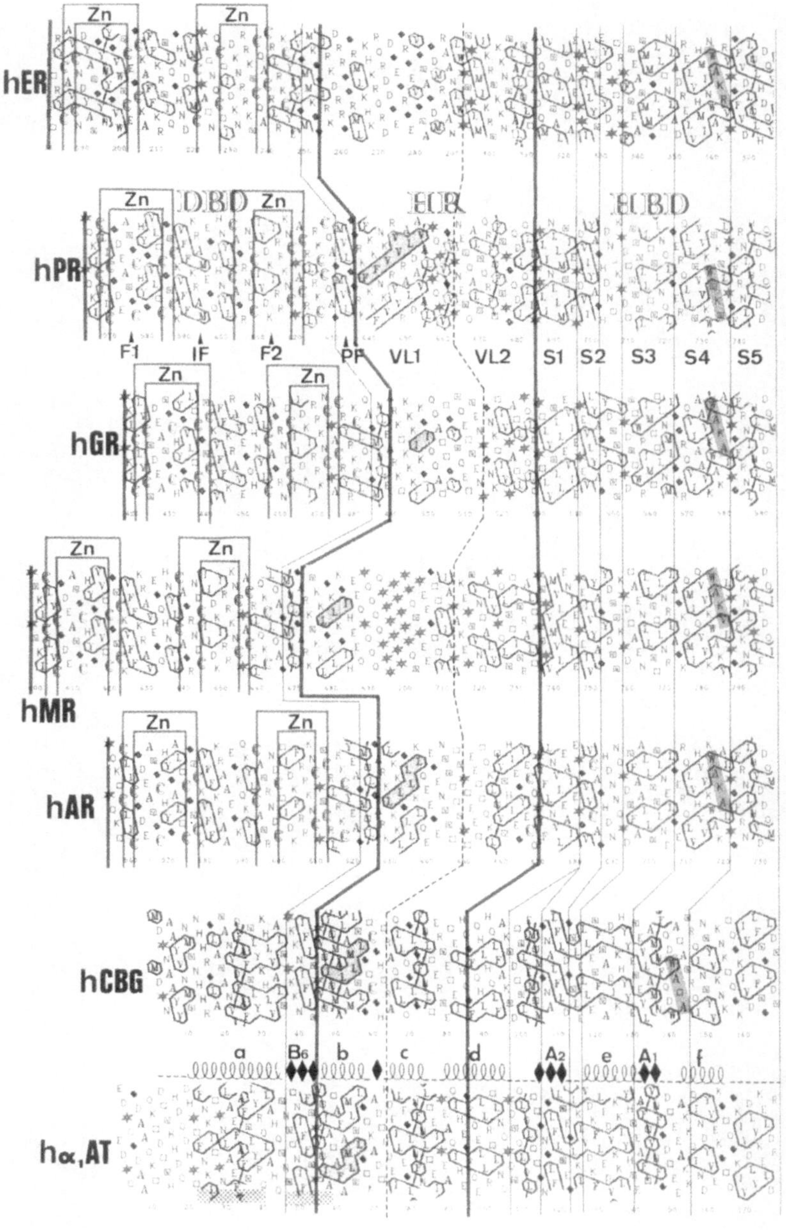

PLATE 1. Alignment of the HCA plots of the DNA-binding domain (DBD), hinge region (HR) and hormone-binding domain (HBD) of SHRs with the HCA plots of two SERPINs ($h\alpha_1$-AT and hCBG) The DNA-binding SHRs are distinguished from the SERPINs by the presence of Cys-coordinated Zn^{++}-fingers (F_1 and F_2) that are schematically represented in blue (IF and PF interfinger and postfinger units, respectively) In the HBD and corresponding regions, alterations in 3-D structure induced by the presence of Pro, Gly, or hydrophilic amino-acids are denoted by vertical lines which, when joined, show the correspondence between the major segments (S_1 and S_{15}) of all the proteins The SERPINs have an additional segment (S_{16}-S_{17}) which is cleaved by neutrophil elastase, the red arrow indicates the cleavage point hER differs from the other SHRs by a highly hydrophilic C-terminal extension and a longer loop between S10 and S11 (S456-K472) Examples of char-

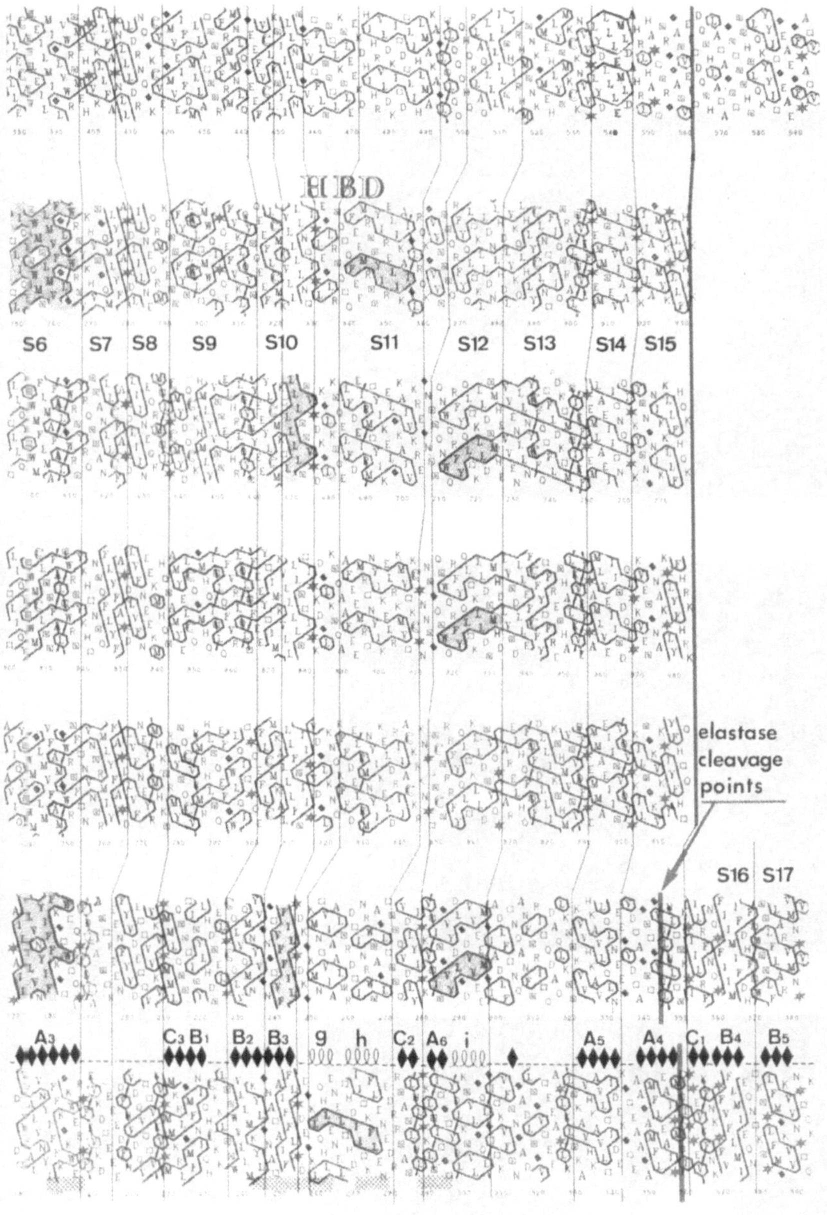

PLATE 1 (continued).

acteristic hydrophobic clusters are stippled in black The HCA plot of hα_1-AT is capped by a schematic representation of its secondary structure deduced from X-ray crystallography studies (helices a to i, β-strands A to C. See Figure 3) Green stippled areas below hα_1-AT indicate the components of the channel that would be created by elastase cleavage of S16 and S17 and clearing of the C-terminal peptide (see lower star in Figure 3). The four amino-acids in orange (WAK·)(See Figure 1) are an example of high conservation of sequence among the proteins Abbreviations for the amino-acid residues are: A, Ala; C, Cys, D, Asp; E, Glu; F, Phe; ♦, Gly; H, His, I, Ile, K, Lys, L, Leu; M, Met, N, Asn, *, Pro, Q, Gln, R, Arg, ⊡, Ser, □, Thr, V, Val; W, Trp, and Y, Tyr

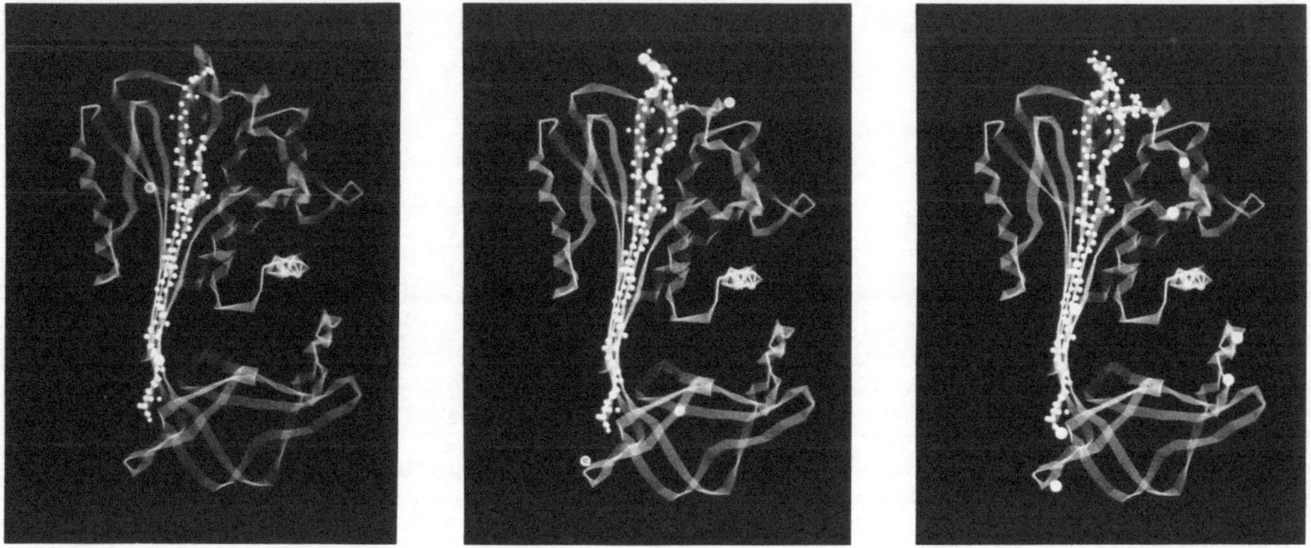

A B C

PLATE 2. Approximate location of selected deletions and mutations within the HBD of SHRs as represented by the crystal structure of cleaved hα_1-AT (residues 20 to 358) (A) **Glucocorticoid and progesterone receptors** The main chain carbon atoms of the last 29 amino acids of the carboxy terminus of rat GR whose deletion reduces glucocorticoid binding to 1%[107] are indicated by small white dots Amino acids affinity-labelled by R5020 and triamcinolone acetonide, *i e* , Met 909 of hPR and Met 622 of rat GR,[128–131] are shown by yellow van der Waals spheres The Tyr 770 Asn and Glu 546 Gly mutants of mouse GR with decreased steroid binding affinity[86] are represented by pale blue van der Waals spheres The Gly 722 Cys hPR mutant that does not bind RU38486[141] is in red (B) **Estrogen receptor** hER has a C-terminal extension of 33 amino acids which has no equivalent in hα_1-AT Deletion of a further 27 amino acids in mouse ER (small white dots) gives rise to a non-DNA-binding mutant [79] The Val 400 Gly mutation that decreases estradiol binding is represented by an orange van der Waals sphere (lower left-hand corner) Possible cysteines affinity-labelled with organometallic[132] or aziridine[133] derivatives are shown in red Cys 530 (top red), Cys 448 (bottom red) The Lys 529 Gln, Lys 531 Gln, Asn 532 Asp mutants with decreased affinity for estradiol and increased affinity for anti-estrogens[148] encompass Cys 530, whereas the Glu 523 Gln and Asp 538 Asn mutants with unchanged affinity for estradiol[148] are at a distance (C) **Androgen receptor** The deletion of the last 41 amino acids of hAR giving rise to a non-steroid binding protein[156] are shown by white dots as in panels A and B The Thr 868 Ala mutant from LNCaP cells with modified specificity[166–168] is in red Mutations giving rise to the partial or complete androgen insensitivity syndrome (Ser 759 Phe, Arg 774 His, Arg 831 stop signal, Arg 840 His, Arg 855 Cys, Asp 864 Gly)[158] are shown in yellow

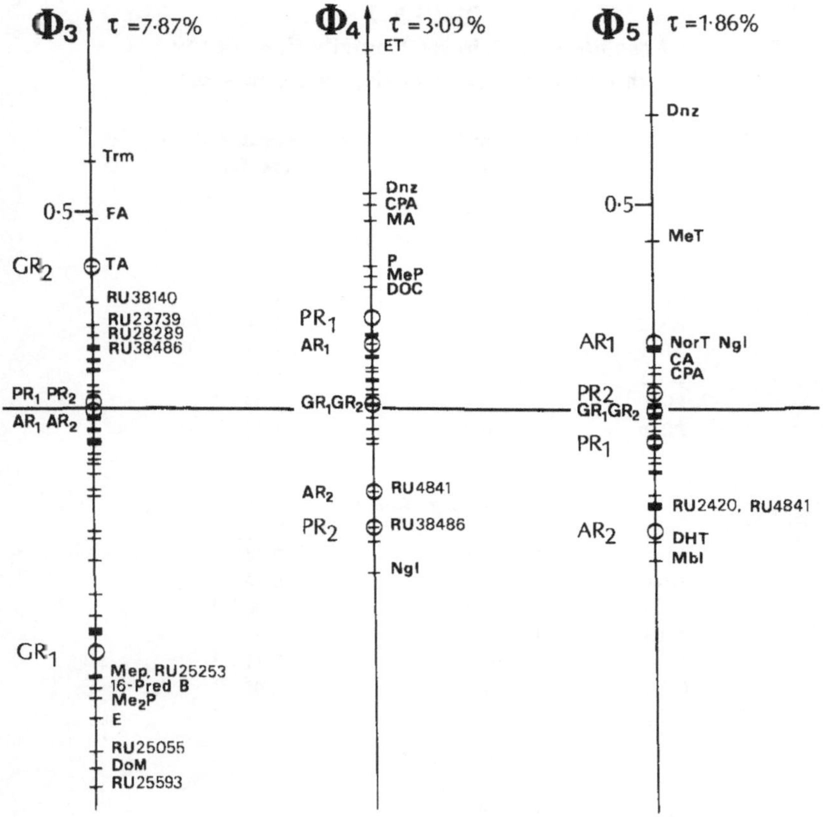

FIGURE 8. Distribution of test steroids and of biochemical parameters along the $\Phi_3 \rightarrow \Phi_5$ axes. Only those test compounds that contribute significantly to these axes are specified. Biochemical variables in large characters contribute significantly more than those in small characters (see Table 2).

GR_1 pole are 6α,16α-dimethylprogesterone, RU25055, and RU25593, but also located in this area are the molecules of the diverticulum at the bottom of the GR branch in the MST. These are associated with particular high RBA values at short incubation times. This is not at all surprising since, as indicated in Table 1, these measurements were performed at 4 h and not at 1 h as for the other compounds. The CFA method has thus highlighted the difference arising from a discrepancy in experimental conditions and that cannot, in the absence of further information, be attributed to a difference in structure of the test compounds. Similar arguments concerning experimental conditions might also explain the singular positions of cyproterone and nomegestrol acetates within the MST. These results therefore emphasize the absolute need for standard conditions for performing comparisons and puts a great value on data generated by routine screening programs.

TABLE 2
Absolute and Relative Contributions of the Binding Parameters to the Factorial Axes

	Absolute contributions (%)	Relative contributions ($\cos^2 \Theta$)
	$\Phi_1\ \tau_1 = 56.69\%$	
GR-2	[+]39.2	0.79
AR-2	[−]20.6	0.64
AR-1	[−]19.2	0.68
GR-1	[+]18.5	0.67
PR-1	[−] 2.1	0.11
PR-2	[−] 0.4	0.02
	$\Phi_2\ \tau_2 = 30.48\%$	
PR-2	[−]30.7	0.84
PR-1	[−]27.0	0 76
AR-2	[+]17.9	0.30
AR-1	[+]13.1	0.25
GR-2	[+]10.5	0.11
GR-1	[+] 0.7	0.01
	$\Phi_3\ \tau_3 = 7.87\%$	
GR-1	[−]62.1	0.31
GR-2	[+]35.0	0.10
PR-1	[+] 1.6	0.12
PR-2	[+] 1.3	0.01
AR-2	0	0
AR-1	0	0
	$\Phi_4\ \tau_4 = 3.09\%$	
PR-2	[−]40.0	0.11
PR-1	[+]33.5	0.10
AR-2	[−]16.0	0.03
AR-1	[+]10.5	0.02
GR-1	0	0
GR-2	0	0
	$\Phi_5\ \tau_5 = 1.87\%$	
AR-1	[+]42.5	0.05
AR-2	[−]31.5	0.03
PR-1	[−]13.9	0.02

TABLE 2 (continued)
Absolute and Relative Contributions of the Binding Parameters to the Factorial Axes

	Absolute contributions (%)	Relative contributions ($\cos^2 \Theta$)
PR-2	[+]12.0	0.02
GR-1	0	0
GR-2	0	0

Note: RBAs for PR, AR, GR after short (1) and long (2) incubation times. τ: Percent variance associated with each factor (total = 100% for all items). Absolute contribution (AC): extent to which an item contributes to the variance explained by a factor (total = 100% for all items of each field). $\cos^2\Phi$ [relative contribution (RC)]: extent, expressed as a percentage/100, to which an item is dispersed among factors (total = 1 for all factors). +, −: signs of the projection coordinates on the factorial axes.

3. Use of CFA as a Model System to Examine the Relationships between Growth Inhibition by Triphenylethylene Anti-Estrogens and their Action on Protein Kinase C (PKC) Activity

CFA maps can also be used as models to introduce individual data *either* on new test compounds *or* new tests. This procedure first establishes, on the basis of the RCs, whether the new data are relevant to the previous analysis and, second, helps set aside any data that constitutes redundant information for any future analysis. An example of the integration of new data on steroid binding and activity has already been published.[186] Here we shall describe an unreported example relating to the biological activities of a series of homologous hydroxylated triphenylacrylonitrile derivatives.

Over the last few years considerable interest has focused on the estrogen regulation of cell growth in human breast cancer[202] and on the design of so-called anti-estrogens[147,203-206] more effective than tamoxifen to halt this growth. With this in mind, our triphenylacrylonitriles were screened on several tests:[207] ER binding (RBAs), MCF_7 cell proliferation (AGONISM), inhibition of such proliferation (ANTAGONISM), and cytotoxicity (CYTO). The total variance of the system can be divided by CFA into subsystems displaying different degrees of correlation among these variables. The largest proportion of the variance is inherent in the $\Phi_1\Phi_2$ factorial map shown in Figure 9a which illustrates the relative positions of the tested variables. (The positions of the molecules are not indicated). Several variables have two independent locations within the chart. These describe different facets of activity derived either from assessments of concentrations of test compound giving 50% of maximal activity ($1/EC_{50}$ or $1/IC_{50}$) (diamonds) or from percentage response levels achieved with the test compound with respect to a control (%) (squares).

This CFA illustrates how the effective nanomolar concentrations of TPEs (EC_{50}s) required to promote MCF_7 cell growth (AGONISM) can be fairly

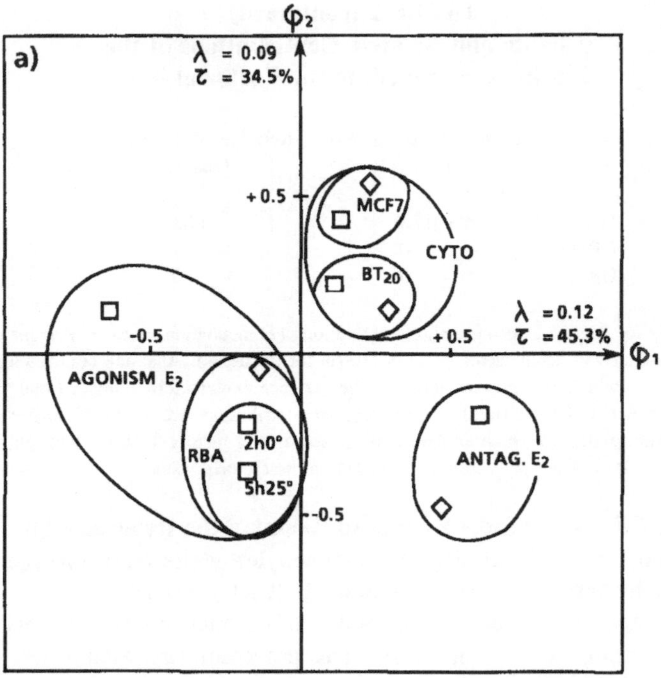

FIGURE 9. (a) Published $\Phi_1\Phi_2$ distribution map derived by CFA from data on a population of TPEs. The map shows proximities between the following biological variables: RBAs for ER under two sets of incubation conditions (2 h 0°C, 5 h 25°C), MCF_7 cell proliferation (AGONISM), inhibition of E_2-promoted MCF_7 cell proliferation (ANTAGonism), inhibition by high TPE concentrations (CYTO) of MCF_7, and BT_{20} cell growth. Activities have been measured either as concentrations that inhibit maximal activity by 50 or 30% (diamonds) or as percentages of a control response at a 10^{-5} *M* TPE concentration (squares). The smaller the contour, the closer the relationship. (From Bignon, E., Pons, M., Crastes de Paulet, A., Doré, J. C., Gilbert, J., Abecassis, J., Miquel, J. F., Ojasoo, T., and Raynaud, J. P., *J. Med. Chem.*, 32, 2092, 1989. With permission.) (b) Use of the above $\Phi_1\Phi_2$ distribution map (open symbols) as a model to position the data for TPE action on PKC subspecies I, II, III activity (solid symbols) under different activation conditions. (A: Ca^{++}- and PS-dependent phosphorylation of H_1 histone in the absence of diolein; A′: in the presence of diolein; B: Ca^{++}- and PS-independent phosphorylation of protamine sulfate). A 10^{-4} *M* TPE concentration was used to measure percent control response.

closely related to the RBAs of these TPEs for ER. The maximum growth responses induced by the TPEs and their RBAs for ER were not similarly correlated. Some of the tested TPEs could also inhibit estradiol (E_2)-stimulated cell proliferation within the nanomolar concentration range (ANTAG.E_2), an effect that was reversed by estrogen, that was not observed in an ER-negative cell line (BT_{20}) but that, in the context of this particular analysis, was not found to be correlated with RBAs for ER. On the other hand, part of the estrogen-irreversible growth inhibition of both ER-positive and -negative cells

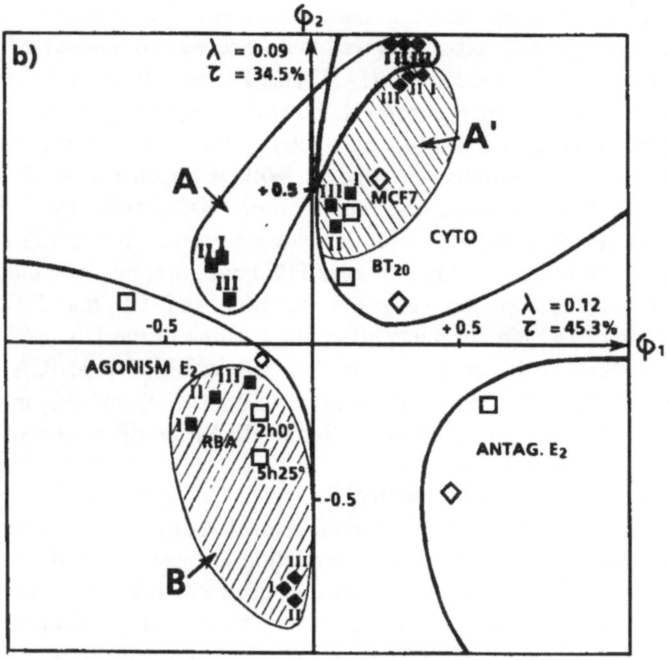

FIGURE 9b.

(CYTO) leading to a cytolytic effect at micromolar TPE concentrations appeared anticorrelated with ER binding.

Another molecular target that has been implicated in the growth inhibition of TPEs is protein kinase C (PKC), a Ca^{++}- and phospholipid-dependent enzyme activated by diacylglycerol that is involved in cell surface transduction and in the control of cell proliferation and tumor promotion.[208-211] Its expression in human breast tumor biopsies is elevated relative to normal breast tissue.[212] Tamoxifen can bind to PKC;[210] four 1,1-diphenylethylenes (DPEs) and two TPEs of our test series can influence PKC activity by different molecular mechanisms.[213,214]

The effect of the TPEs on the activities of three PKC isoforms [I(γ), II(β), and III(α)] were screened under different experimental conditions:[189] on the one hand, using histone as a substrate in the presence of phosphatidylserine (PS) and Ca^{++} in which case maximal enzyme activity is obtained in the presence of the physiological activator, diacylglycerol (diolein = DO); on the other hand, using protamine sulfate as a substrate in which case the enzyme activity is Ca^{++} and phospholipid independent.

The question thus arose whether our previous screening data on the ER binding and cell proliferation of TPEs (Figure 9a) and the new data on their PKC activities were in any way related. Using the above CFA analysis as a model, we introduced the individual data on the new tests to establish their relevance. The procedure was found to be statistically valid as indicated by the RCs of PKC to these factorial axes (not given). The positions taken up within the map by PKC activity showed that there was no marked difference among the three PKC subspecies (I, II, and III) as intimated by previous results (Figure 9b).[213] They depended, however, largely upon the mode of PKC activation and upon the classic parameters used to describe dose-response curves (% maximum activity and $1/IC_{50}$). With H_1 histone as substrate and PS, Ca^{++}, and DO as cofactors, inhibition of PKC activity by TPEs was clearly most closely correlated with the estrogen-irreversible cell growth inhibition (CYTO) observed at micromolar TPE concentrations (whether $1/IC_{30}$ or %) (CA'). The correlation was somewhat closer in MCF_7 than BT_{20} cells. The omission of DO did not much affect the location of the $1/IC_{50}$ PKC pole but shifted the % PKC pole towards the %-AGONISM pole (CA). With protamine sulfate as substrate, in the absence of PS, Ca^{++}, and DO, inhibition of PKC activity could be essentially allied to RBAs for ER as shown by the proximity of the poles (CB).

The introduction of each column of PKC data one by one into the published analysis used as a model thus indicated the relevance of this variable measured under different activation conditions in relation to those that had been measured previously. In order to substantiate this result and to be able to accurately position the test molecules with respect to the biochemical variables, a full-scale CFA was performed on the data for all available molecules — old and new — but restricted to the most essential and informative parameters only. This CFA analysis (not shown) revealed a close correlation between cytotoxicity and inhibition of DO-activated PKC which was closest for TPEs with a basic amino side chain.[215] Since these two activities occurred within the same concentration ranges, the link may be causal suggesting that the cytotoxic action of high TPE concentrations could be partly explained by inhibition of PKC in a situation of maximal activation, i.e., of diacylglycerol-activated and membrane-inserted PKC *in vivo*. These TPEs could by virtue of their amphiphilic characteristics inhibit PKC activity by interfering with the lipid-protein interface within the membrane.[216] Indeed, tamoxifen inhibits PKC-mediated signal transduction in neutrophils[217] and can also modify membrane fluidity of human breast cancer cells.[218] The correlation also suggests that cytotoxicity may be fostered by the presence of DO which increases the affinity of PKC for Ca^{++} and PS[219] and halves the amount of TPE required to inhibit PKC.[189]

The cytotoxicity of this population of TPEs in MCF_7 cells was correlated to a lesser extent (20%) to ER binding. There may be no causal link between these two effects since *in vitro* they occur at different concentrations. Nevertheless, TPEs might, by virtue of their antiestrogenic action, render ER-positive cells more fragile and susceptible to a cytotoxic action. This interpretation is supported by the observation that TPEs are cytotoxic at lower concentrations when no estrogenic activity is manifest in the cells. Finally, TPE cytotoxicity was accounted for by a major independent element that could not be related to any of the measured biological variables.

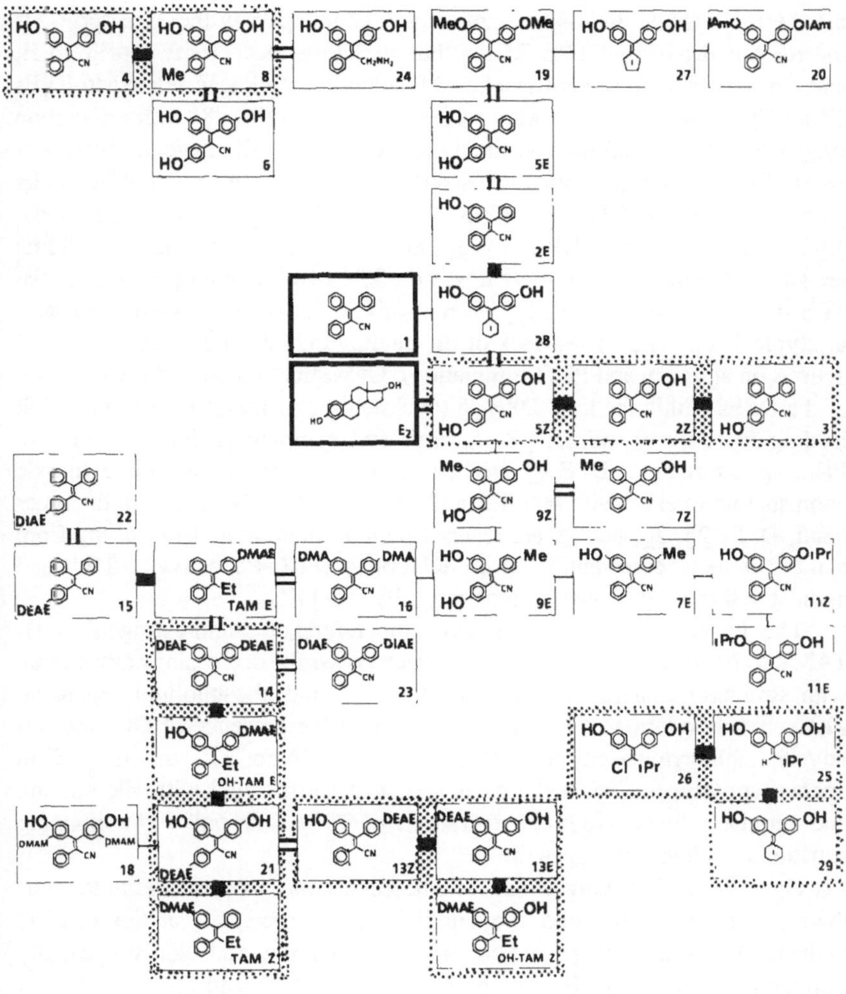

FIGURE 10. Minimum spanning tree of TPEs constructed on the basis of their activities[190] on the following parameters: RBA at 25°C, stimulation of MCF_7 cell proliferation, cytotoxicity, inhibition of Ca^{++}-dependent PKC activity (in the presence or absence of diolein), stimulation of this activity in the presence of diolein. The relative proximity of two TPEs is indicated by the nature of the line linking them as in Figure 6. Clusters of molecules with highly similar activity profiles are surrounded by stippling (ME = OCH_3, Et = C_2H_5, OiPr = $OCH(CH_3)_2$, DEAE = $O(CH_2)_2N(C_2H_5)_2$, DMA = $N(CH_3)_2$, DMAM = $CH_2N(CH_3)_2$, OiAm = $O(CH_2)_2CH(CH_3)_2$, DIAE = $O(CH_2)_2N[CH(CH_3)_2]_2$, DMAE = $O(CH_2)_2N(CH_3)_2$. (Adapted from Doré, J. C., Gilbert, J., Bignon, E., Crastes de Paulet, A., Ojasoo, T., Pons, M., Raynaud, J. P., and Miquel, J. F., *J. Med. Chem.*, 35, 573, 1992.)

The data were also submitted to an MST analysis as described above which compared the selectivity profiles of the individual molecules (Figure 10).[190] This MST has a main trunk and four branches of unequal length. The reference molecule estradiol (E_2) is at the center of the tree, in an offshoot off the main trunk. Its only neighbor, and counterpart in the TPE series, is the αβ-dihydroxylated TPE 5Z that has the closest selectivity profile to E_2 and forms part of the main trunk. TPE 5Z is most closely related to TPEs 2Z and 3, somewhat less closely related to the duo DPE 28 (with a 6-carbon ring) and TPE 2E and more distantly related to TPE 9Z. Several clusters of two to four TPEs with comparable selectivity profiles can be identified: TPEs 4 and 8; TAM E and TPE 15; TPEs 14, OH TAM E, 21, and TAM Z; TPEs 13Z/E and OH TAM Z; DPEs 25, 26, and 29. On the other hand, two TPEs are situated at the end of remote tips, i.e., TPEs 18 (the only metasubstituted TPE in the series) and TPE 20 (an αα'-diisoaminoethyl derivative) because of atypical activity profiles (lack of ER binding in both cases, but differential actions on agonism and PKC stimulation). Of particular note is the dispersion of the DPEs within the tree. DPE 28 (cyclofenil) constitutes the essential link leading from the E_2-mime, 5Z, towards test compounds that are selective PKC agents such as DPE 27 (right-hand branch) or that are noncytotoxic agonists that interact with both ER and PKC (left-hand branch). On the other hand, DPEs 25, 26, and 29 are reached by a bifurcation at 9E resulting from a more marked contribution of stimulation of PKC-PS activity, RBA and agonist activity, and some cytotoxicity (7E and 11Z).

The lower branches leading from the reference "anti-estrogen" OH-TAMZ to 29 are depicted in Figure 11 which shows the decreasing contribution with structural change of inhibition of Ca^{++}- and phospholipid-dependent PKC phosphorylation of histone, especially in the presence of DO, the relatively stable cytotoxicity up to 9E, the dip in RBA for ER, and the peak in agonist activity. α-Hydroxylation is seen to be associated with ER binding and agonist activity. High concentrations of TPEs with bulky N-containing substituents inhibit cell growth.

Overall, the following trends were noted: absence of cytotoxicity towards the top of the tree compared to cytotoxicity of the roots, stimulation of PKC in the top right-hand branch and bottom right-hand root, and agonism virtually throughout with the notable exceptions of OH TAM Z and 20.

V. CONCLUSIONS: BRIDGING THE GAP BETWEEN LIGANDS AND MOLECULAR TARGETS

The two approaches that we have briefly described above, (1) 3D modeling of the hormone-binding domain of the steroid hormone receptor together with localization of domain functions and of amino acids involved in steroid binding and (2) analysis of binding and activity data for large numbers of diversely substituted steroids, are still far from converging into a single unified vision of the mechanisms of steroid action. It is not possible to fit a steroid into a binding site and show how it can, but why its analog cannot, induce the cascade of events necessary for correct binding and efficient transcription via dissociation of the SHR complex, phosphorylation, changes in confor-

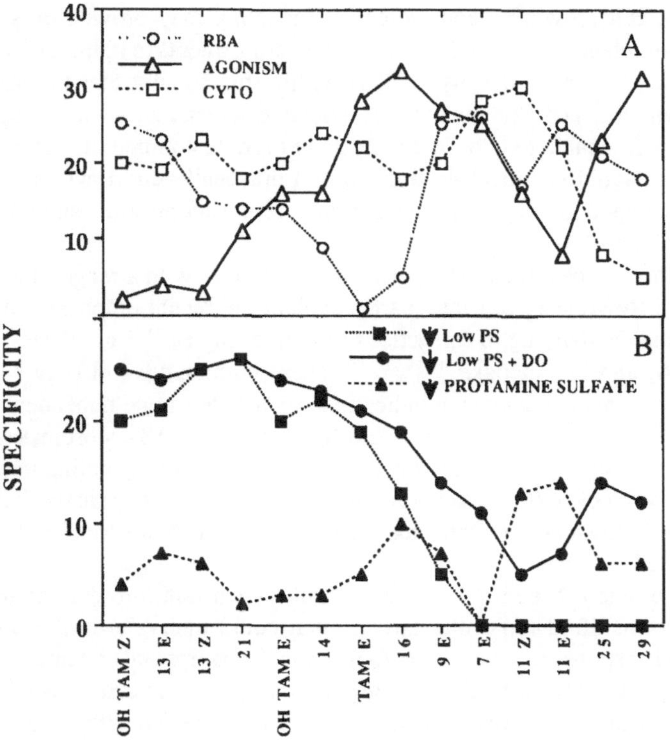

FIGURE 11. Specificity profiles of the TPEs along two branches of the minimum spanning tree of Figure 10. The top panel represents RBA for ER, ability to stimulate MCF_7 cell proliferation (AGONISM), and cytotoxicity in MCF_7 cells (CYTO). The bottom panel represents PKC activity under three different experimental conditions [activity in the presence of phosphatidylserine (PS) with or without diolein (DO) using H_1 histone as substrate, activity when using protamine sulfate as substrate].

mation, etc. Moreover, although homology modeling furthers our understanding of the formation and workings of the steroid-receptor complex, it is unlikely to provide a sufficiently stable basis for rational drug design.

Among the first useful tools for deciphering steroid receptor interactions were potent synthetic steroids (for reviews, see References 15 and 220) that were chosen partly because of their high potency in the traditional biological assays used to evaluate natural hormone activity. They are still in widespread use in modern molecular biology tests (e.g., promegestone), but the chemical series from which these lead compounds were derived are, for the most part, ignored. It is likely that the crystal structures of receptor complexes with *many* compounds will have to be resolved before the relevance of the conformational changes that occur on binding can be assessed. The binding affinities of the activators of the receptor do not directly parallel all the intricate responses of the receptor complex (e.g., DNA binding, specific gene transcription, cell growth[221]) and molecular modeling may provide only part of the answer when it comes to features such as adjustments to optimize hydrogen bonding and electron density.[222] It is within this context that the structural knowledge obtained from analysis of available screening data by multivariate methods will enable to choose the most appropriate compounds for modeling studies, to avoid chemical dead ends, and drastically cut down the number of new compounds to be synthesized thus representing a considerable gain in time and cost.

The difficulties in analyzing series of hormones with a range of activities (e.g., see Reference 223) may partly explain the recent emphasis on developing an understanding of the actions of antihormones,[21] e.g., ICI164384,[147] RU38486, and its analogs.[171,178,224-227] These compounds not only have potential as therapeutic agents in indications such as hormone-dependent cancer but, like very high-affinity reversible ligands (e.g., 11β-chloromethylestradiol)[228] or covalent ligands, are invaluable new tools in pinpointing the step(s) that may be blocked (e.g., receptor dimerization).[143,229] They suffer, however, from the disadvantage of freezing the system rather than revealing how it is coordinated.

The problem becomes even more acute when nonsteroidal analogs that may have manifold molecular targets, that could intervene in the observed biological responses, are used as tools. Such a compound is tamoxifen that binds not only to the estrogen receptor but modulates the activities of several enzymes including kinases. Nonsteroids such as tamoxifen with obvious structural similarities with the natural hormones are indeed liable to interact with the same target, but also with other targets that could limit their specificity. A knowledge of the active steroid-binding site might permit the design of entities that are substantially different from the natural ligands, substantially different from each other, and also more specific with regard to the molecular target. Then the problem of designing the ligand will be superseded by that of identifying the target. Why design, for instance, anti-estrogens for native receptors in cases where variant receptors have the dominant role in malignant cell proliferation? Refined structures for receptors are probably still a few years away, but the chances are that, when available, this research field will witness progress similar to that presently being made in the design and optimization of ligands for enzymes.[230]

REFERENCES

1. **Wittliff, J. L., Wenz, L. L., Dong, J., Nawaz, Z., and Butt, T. R.**, Expression and characterization of an active human estrogen receptor as a ubiquitin fusion protein from *Escherichia coli*, *J. Biol. Chem.*, 265, 22016, 1990.
2. **Young, C. Y. F., Qiu, S., Prescott, J. L., and Tindall, D. J.**, Overexpression of a partial human androgen receptor in *E. coli*: characterization of steroid binding, DNA binding, and immunological properties, *Mol. Endocrinol.*, 4, 1841, 1990.
3. **Power, R. F., Conneely, O. M., McDonnell, D. P., Clark, J. H., Butt, T. R., Schrader, W. T., and O'Malley, B. W.**, High-level expression of a truncated chicken progesterone receptor in *Escherichia coli*, *J. Biol. Chem.*, 265, 1419, 1990.
4. **Mak, P., McDonald, D. P., Weigel, N. L., Schrader, W. T., and O'Malley, B. W.**, Expression of functional chicken oviduct progesterone receptors in yeast *(Saccharomyces cerevisiae)*, *J. Biol. Chem.*, 264, 21613, 1989.
5. **Srinivasan, G. and Thompson, E. B.**, Overexpression of full-length human glucocorticoid receptor in *Spodoptera frugiperda* cells using the baculovirus expression vector system, *Mol. Endocrinol.*, 4, 209, 1990.
6. **Alnemri, E. S., Maksymowych, A. B., Robertson, N. M., and Litwack, G.**, Characterization and purification of a functional rat glucocorticoid receptor overexpressed in a baculovirus system, *J. Biol. Chem.*, 266, 3925, 1991.
7. **Härd, T., Kellenbach, E., Boelens, R., Maler, B. A., Dahlman, K., Freedman, L. P., Carlstedt-Duke, J., Yamamoto, K. R., Gustafsson, J. Å., and Kaptein, R.**, Solution structure of the glucocorticoid receptor DNA-binding domain, *Science*, 249, 157, 1990.
8. **Schwabe, J. W. R., Neuhaus, D., and Rhodes, D.**, Solution structure of the DNA-binding domain of the oestrogen receptor, *Nature*, 348, 458, 1990.
9. **Luisi, B. F., Xu, W., Otwinowski, Z., Freedman, L. P., Yamamoto, K. R., and Sigler, P. B.**, Crystallographic analysis of the interaction of the glucocorticoid receptor with DNA, *Nature*, 352, 497, 1991.
10. **Raynaud, J. P., Bissery, V., Gaboriaud, C., Ojasoo, T., Teutsch, G., and Mornon, J. P.**, An analysis of the steroid binding domains of receptors and of ligand structure and binding affinity, in *Steroid/Thyroid Hormone Receptor Family and Gene Regulation*, Carlstedt-Duke, J., Eriksson, H., and Gustafsson, J. Å., Eds., Birkhäuser Verlag, Basel, 1989, 337.
11. **Mornon, J. P., Bissery, V., Gaboriaud, C., Thomas, A., Ojasoo, T., and Raynaud, J. P.**, Hydrophobic cluster analysis (HCA) of the hormone-binding domain of receptor proteins, *J. Steroid Biochem.*, 34, 355, 1989.
12. **Ojasoo, T., Mornon, J. P., and Raynaud, J. P.**, Steroid hormone receptors, in *Comprehensive Medicinal Chemistry*, Vol. 3, Emmett, J. C., Ed., Pergamon Press, Oxford, 1990, 1175.
13. **Raynaud, J. P., Ojasoo, T., and Mornon, J. P.**, Modelling the hormone-binding domain of steroid receptors including vitamin D_3 receptor, in *Vitamin D — Gene Regulation, Structure-Function Analysis, and Clinical Application*, Norman, A. W., Bouillon, R., and Thomasset, M., Eds., Walter de Gruyter, Berlin, 1991, 64.
14. **Lemesle-Varloot, L., Ojasoo, T., Mornon, J. P., and Raynaud, J. P.**, Spatial distribution of the functions of the steroid binding domain of receptors, *J. Steroid Biochem. Mol. Biol.*, 41, 369, 1992.
15. **Ojasoo, T. and Raynaud, J. P.**, Unique steroid congeners for receptor studies, *Cancer Res.*, 38, 4186, 1978.
16. **Raynaud, J. P., Ojasoo, T., Bouton, M. M., and Philibert, D.**, Receptor binding as a tool in the development of new bioactive steroids, in *Drug Design*, Vol. 8, Ariëns, E. J., Ed., Academic Press, New York, 1979, 169.

17. **Raynaud, J. P. and Ojasoo, T.,** The relevance of structure-affinity relationships in the study of steroid hormone action, in *Steroid Hormone Receptors: Structure and Function,* Eriksson, H. and Gustafsson, J. Å., Eds., Elsevier, Amsterdam, 1983, 141.
18. **Raynaud, J. P., Ojasoo, T., Bouton, M. M., Bignon, E., Pons, M., and Crastes de Paulet, A.,** Structure-activity relationships of steroid estrogens, in *Estrogens in the Environment 2,* McLachlan, J. A., Ed., Elsevier, New York, 1985, 24.
19. **Raynaud, J. P., Fortin, M., Hunt, P., Ojasoo, T., Doré, J. C., Surcouf, E., and Mornon, J. P.,** Approaches to drug development using receptors, in *The Role of Receptors in Biology and Medicine,* Gotto, A. M., Jr. and O'Malley, B., Eds., Raven Press, New York, 1986, 63.
20. **Raynaud, J. P. and Ojasoo, T.,** with the collaboration of Bignon, E., Doré, J. C., and Surcouf, E., Receptor binding as a tool in the development of selective new bioactive steroids and nonsteroids, in *Innovative Approaches in Drug Research,* Harms, A. F., Ed., Elsevier, Amsterdam, 1986, 47.
21. **Raynaud, J. P. and Ojasoo, T.,** The design and use of sex-steroid antagonists, *J. Steroid Biochem.,* 25, 811, 1986.
22. **Ojasoo, T., Delettré, J., Mornon, J. P., Turpin-Vandycke, C., and Raynaud, J. P.,** Towards the mapping of the progesterone and androgen receptors, *J. Steroid Biochem.,* 27, 255, 1987.
23. **Bignon, E., Pons, M., Crastes de Paulet, A., Doré, J. C., Gilbert, J., Abecassis, J., Miquel, J. F., Ojasoo, T., and Raynaud, J. P.,** Effect of triphenylacrylonitrile derivatives on estradiol receptor binding and on human breast cancer cell growth, *J. Med. Chem.,* 32, 2092, 1989.
24. **Baker, A. R., McDonnell, D. P., Hughes, M., Crisp, T. M., Mangelsdorf, D. J., Haussler, M. R., Wesley Pike, J., Shine, J., and O'Malley, B.,** Cloning and expression of full-length c-DNA encoding human vitamin D receptor, *Proc. Natl. Acad. Sci. U.S.A.,* 85, 3294, 1988.
25. **Green, S. and Chambon, P.,** A superfamily of potentially oncogenic hormone receptors, *Nature,* 324, 615, 1986.
26. **Evans, R. M.,** The steroid and thyroid hormone receptor superfamily, *Science,* 240, 889, 1988.
27. **Carson-Jurica, M. A., Schrader, W. T., and O'Malley, B. W.,** Steroid receptor family: structure and functions, *Endocr. Rev.,* 11, 201, 1990.
28. **O'Malley, B.,** The steroid receptor superfamily: more excitement predicted for the future, *Mol. Endocrinol.,* 4, 363, 1990.
29. **Weinberger, C., Thompson, C. C., Ong, E. S., Lebo, R., Gruol, D. J., and Evans, R. M.,** The c-erb-A gene encodes a thyroid hormone receptor, *Nature,* 324, 641, 1986.
30. **Muñoz, A., Zenke, M., Gehring, U., Sap, J., Beug, H., and Vennstrom, B.,** Characterization of the hormone-binding domain of the chicken c-erbA/thyroid hormone receptor protein, *EMBO J.,* 7, 155, 1988.
31. **Green, S., Walter, P., Kumar, V., Krust, A., Bornert, J. M., Argos, P., and Chambon, P.,** Human oestrogen receptor cDNA: sequence, expression and homology to V-erb-A, *Nature,* 320, 134, 1986.
32. **Petkovich, M., Brand, N. J., Krust, A., and Chambon, P.,** A human retinoic acid receptor which belongs to the family of nuclear receptors, *Nature,* 330, 444, 1987.
33. **Giguère, V., Ong, E. S., Segui, P., and Evans, R. M.,** Identification of a receptor for the morphogen retinoic acid, *Nature,* 330, 624, 1987.
34. **Gustafsson, J. Å., Carlstedt-Duke, J., Poellinger, L., Okret, S., Wikström, A. C., Brönnegard, M., Gillner, M., Dong, Y., Fuxe, K., Cintra, A., Härfstrand, A., and Agnati, L.,** Biochemistry, molecular biology, and physiology of the glucocorticoid receptor, *Endocr. Rev.,* 8, 185, 1987.
35. **Isseman, I. and Green, S.,** Activation of a member of the steroid hormone receptor superfamily by peroxisome proliferators, *Nature,* 347, 645, 1990.

36. **Wang, L. H., Tsai, S. Y., Cook, R. G., Beattie, W. G., Tsai, M. J., and O'Malley, B. W.**, COUP transcription factor is a member of the steroid receptor superfamily, *Nature*, 340, 163, 1989.
37. **Ladias, J. A. A. and Karathanasis, S. K.**, Regulation of the apolipoprotein AI gene by ARP-1, a novel member of the steroid receptor superfamily, *Science*, 251, 561, 1991.
38. **Mlodzik, M., Hiromi, Y., Weber, U., Goodman, C. S., and Rubin, G. M.**, The *Drosophila* seven-up gene, a member of the steroid receptor gene superfamily, controls photoreceptor cell fates, *Cell*, 60, 211, 1990.
39. **Oro, A. E., Ong, E. S., Margolis, J. S., Posakony, J. W., McKeown, M., and Evans, R. M.**, The *Drosophila* gene *knirps-related* is a member of the steroid-receptor gene superfamily, *Nature*, 336, 493, 1988.
40. **Nauber, U., Pankratz, M. J., Kienlin, A., Seifert, E., Klemm, U., and Jäckle, H.**, Abdominal segmentation of the *Drosophila* embryo requires a hormone receptor-like protein encoded by the gap gene *knirps*, *Nature*, 336, 489, 1988.
41. **Oro, A. E., McKeown, M., and Evans, R. M.**, Relationship between the product of the *Drosophila ultraspiracle* locus and the vertebrate retinoid X receptor, *Nature*, 347, 298, 1990.
42. **Lavorgna, G., Ueda, H., Clos, J., and Wu, C.**, FTZ-F1, a steroid hormone receptor-like protein implicated in the activation of *fushi tarazu*, *Science*, 252, 848, 1991.
43. **Power, R. F., Lydon, J. P., Conneely, O. M., and O'Malley, B. W.**, Dopamine activation of an orphan of the steroid receptor superfamily, *Science*, 252, 1546, 1991.
44. **Green, S. and Chambon, P.**, Oestradiol induction of a glucocorticoid-responsive gene by a chimearic receptor, *Nature*, 325, 75, 1987.
45. **Carlstedt-Duke, J., Strömstedt, P. E., Wrange, O., Bergman, T., Gustafsson, J. Å., and Jörnvall, H.**, Domain structure of the glucocorticoid receptor protein, *Proc. Natl. Acad. Sci. U.S.A.*, 84, 4437, 1987.
46. **Thompson, E. B.**, The structure of the human glucocorticoid receptor and its gene, *J. Steroid Biochem.*, 27, 105, 1987.
47. **Kumar, V., Green, S., Stack, G., Berry, M., Jin, J. R., and Chambon, P.**, Functional domains of the human estrogen receptor, *Cell*, 51, 941, 1987.
48. **Loebermann, H., Tokuoaka, R., Deisenhofer, J., and Huber, R.**, Human α_1-proteinase inhibitor: crystal structure analysis of two crystal modifications, molecular model and preliminary analysis of the implications for function, *J. Mol. Biol.*, 177, 531, 1984.
49. **Lim, V. I.**, Polypeptide chain folding through a highly helical intermediate as a general principle of globular protein structure formation, *FEBS Lett.*, 89, 10, 1978.
50. **Gaboriaud, C., Bissery, V., Benchetrit, T., and Mornon, J. P.**, Hydrophobic cluster analysis: an efficient new way to compare and analyse amino acid sequences, *FEBS Lett.*, 224, 149, 1987.
51. **Lemesle-Varloot, L., Henrissat, B., Gaboriaud, C., Bissery, V., Morgat, A., and Mornon, J. P.**, Hydrophobic cluster analysis: procedures to derive structural and functional information from 2-D-representation of protein sequences, *Biochimie*, 72, 555, 1990.
52. **Lemesle-Varloot, L.**, (1) Contribution au Développement de la Méthode HCA: Logiciel Interactif MANSEK. (2) Application à la Modélisation Partielle de Récepteurs Hormonaux, Ph.D. thesis, Paris University VI, Paris, 1991.
53. **Henrissat, B., Claessens, M., Tomme, P., Lemesle, L., and Mornon, J. P.**, Cellulase families revealed by hydrophobic cluster analysis, *Gene*, 81, 83, 1989.
54. **Thoreau, E., Petridou, B., Kelly, P. A., Djiane, J., and Mornon, J. P.**, Structural symmetry of the extracellular domain of the cytokine/growth hormone/prolactin receptor family and interferon receptors revealed by hydrophobic cluster analysis, *FEBS Lett.*, 282, 26, 1991.
55. **Rougé, P. and Varloot, L.**, Structural homologies between *leguminosae* lectins as revealed by the hydrophobic cluster analysis (HCA) method, *Biochem. Syst. Ecol.*, 18, 419, 1990.

56. **Walter, P., Green, S., Greene, G., Krust, A., Bornert, J. M., Jeltsch, J. M., Staub, A., Jensen, E., Scrace, G., Waterfield, M., and Chambon, P.**, Cloning of the human estrogen receptor cDNA, *Proc. Natl. Acad. Sci. U.S.A.*, 82, 7889, 1985.
57. **Greene, G. L., Gilna, P., Waterfield, M., Baker, A., Hort, Y., and Shine, J.**, Sequence and expression of human estrogen receptor complementary DNA, *Science*, 231, 1150, 1986.
58. **Misrahi, M., Atger, M., D'Auriol, L., Loosfelt, H., Meriel, C., Fridlansky, F., Guiochon-Mantel, A., Galibert, F., and Milgrom, E.**, Complete amino acid sequence of the human progesterone receptor deduced from cloned cDNA, *Biochem. Biophys. Res. Commun.*, 143, 740, 1987.
59. **Trapman, J., Klassen, P., Kuiper, G. G. J. M., van der Korput, J. A. G. M., Faber, P. W., van Rooij, H. C. J., Geurts van Kessel, A., Voorhorst, M. M., Mulder, E., and Brinkmann, A. O.**, Cloning, structure and expression of a cDNA encoding the human androgen receptor, *Biochem. Biophys. Res. Commun.*, 153, 241, 1988.
60. **Chang, C., Kokontis, J., and Liao, S.**, Structural analysis of complementary DNA and amino acid sequences of human and rat androgen receptors, *Proc. Natl. Acad. Sci. U.S.A.*, 85, 7211, 1988.
61. **Lubahn, D. B., Joseph, D. R., Sullivan, P. M., Willard, H. F., French, F. S., and Wilson, E. M.**, Cloning of the human androgen receptor complementary DNA and localisation on the X chromosome, *Science*, 240, 327, 1988.
62. **Hollenberg, S. M., Weinberger, C., Ong, E. S., Cerelli, G., Oro, A., Lebo, R., Thompson, E. B., Rosenfeld, M. G., and Evans, R. M.**, Primary structure and expression of a functional glucocorticoid receptor cDNA, *Nature*, 318, 635, 1985.
63. **Arriza, J. L., Weinberger, C., Cerelli, G., Glaser, T. M., Handelin, B. L., Housman, D. E., and Evans, R. M.**, Cloning of human mineralocorticoid receptor complementary DNA: structural and functional kinship with the glucocorticoid receptor, *Science*, 237, 268, 1987.
64. **Hammond, G. L., Smith, C. L., Goping, J. S., Underhill, D. A., Harley, M. J., Reventos, J., Musto, N. A., Gunsalus, G. L., and Bardin, C. W.**, Primary structure of human corticosteroid binding globulin, deduced from hepatic and pulmonary cDNAs, exhibits homology with serine protease inhibitors, *Proc. Natl. Acad. Sci. U.S.A.*, 84, 5153, 1987.
65. **Kurachi, K., Chandra, T., Friezner Degen, S. J., White, T. T., Marchioro, T. L., Woo, S. L. C., and Davie, E. W.**, Cloning and sequence of cDNA coding for α_1-antitrypsin, *Proc. Natl. Acad. Sci. U.S.A.*, 78, 6826, 1981.
66. **Mourey, L., Samama, J. P., Delarue, M., Choay, J., Lormeau, J. C., Petitou, M., and Moras, D.**, Antithrombin III: structural and functional aspects, *Biochimie*, 72, 599, 1990.
67. **Stein, P. E., Leslie, A. G. W., Finch, J. T., Turnell, W. G., McLaughlin, P. J., and Carrell, R. W.**, Crystal structure of ovalbumin as a model for the reactive centre of serpins, *Nature*, 347, 99, 1990.
68. **Creighton, T. E. and Chothia, C.**, Selecting buried residues, *Nature*, 339, 14, 1989.
69. **Distelhorst, C. W., Janiga, K. E., Howard, K. J., Strandjord, S. E., and Campbell, E. J.**, Neutrophil elastase produces 52-kD and 30-kD glucocorticoid receptor fragments in the cytosol of human leukaemia cells, *Blood*, 70, 860, 1987.
70. **Hubbard, J. R. and Kalimi, M.**, Influence of proteinase inhibitors on glucocorticoid receptor properties: recent progress and future perspectives, *Mol. Cell. Biochem.*, 66, 101, 1985.
71. **Ganesan, S., Bashirelahi, N., Young, J. D., and Cohen, S. P.**, Phenylmethylsulfonyl fluoride (PMSF) inhibits beta-estradiol binding to estrogen receptor from human prostate, *Life Sci.*, 41, 2767, 1987.

72. **Nigro, V., Medici, N., Abbondanza, C., Minucci, S., Moncharmont, B., Molinari, A. M., and Puca, G. A.**, An aprotinin binding site localized in the hormone binding domain of the estrogen receptor from calf uterus, *Biochem. Biophys. Res. Commun.*, 170, 930, 1990.
73. **Kasayama, S., Noma, K., Sato, B., Nakao, M., Nishizawa, Y., Matsumoto, K., and Kishimoto, S.**, Leupeptin inhibits the transformation of glucocorticoid receptor, *J. Steroid Biochem.*, 28, 273, 1987.
74. **Baker, M. E. and Kimlinger, W. R.**, Protease substrates inhibit binding of ^{3}H-R 5020 to the G-fragment in chicken oviduct cytosol, *Biochem. Biophys. Res. Commun.*, 108, 1067, 1982.
75. **Skern, T., Zorn, M., Blaas, D., and Kuechler, E.**, Protease or protease inhibitor?, *Nature*, 344, 26, 1990.
76. **Flink, I. L., Bailey, T. J., Gustafson, T. A., Markham, B. E., and Morkin, E.**, Complete amino-acid sequence of human thyroxine-binding globulin deduced from cloned cDNA: close homology to the serine antiproteases, *Proc. Natl. Acad. Sci. U.S.A.*, 83, 7708, 1986.
77. **Benvenga, S.**, The 27-kilodalton thyroxine (T_4)-binding protein is human apolipoprotein A-I: identification of a 68-kilodalton high density lipoprotein that binds T_4, *Endocrinology*, 124, 1265, 1989.
78. **Picado-Leonard, J. and Miller, L.**, Homologous sequences in steroidogenic enzymes, steroid receptors and a steroid binding protein suggest a consensus steroid binding sequence, *Mol. Endocrinol.*, 2, 1145, 1988.
79. **Fawell, S. F., Lees, J. A., and Parker, M. G.**, A proposed consensus steroid-binding sequence — a reply, *Mol. Endocrinol.*, 3, 1002, 1989.
80. **Westphal, U.**, in *Steroid-Protein Interactions II*, Monogr. Endocrinol. 27, Springer-Verlag, Berlin, 1986.
81. **Le Gaillard, F., Racadot, A., Aubet, J. P., and Dautrevaux, M.**, Modification of human transcortin by tetranitromethane, evidence for the implication of a tyrosine residue in cortisol binding, *Biochimie*, 64, 153, 1982.
82. **Peter, W., Brüller, H. J., Vriend, G., Beato, M., and Suske, G.**, Identification of residues essential for progesterone binding to uteroglobin by site-directed mutagenesis, *J. Steroid Biochem. Mol. Biol.*, 38, 27, 1991.
83. **Kuhn, R. W. and Raymoure, W. J.**, The role of tyrosine in the binding of steroid by progesterone binding globulin from the guinea-pig, *J. Steroid Biochem.*, 17, 459, 1982.
84. **Koffman, B., Modaress, K. J., Beckerman, T., and Bashirelahi, N.**, Evidence for involvement of tyrosine in estradiol binding by rat uterus estrogen receptor, *J. Steroid Biochem. Mol. Biol.*, 38, 135, 1991.
85. **Inano, H. and Tanaoki, B. I.**, Chemical modification of tyrosine residues in active site of human placental 17β-dehydrogenase by tetranitromethane, *J. Steroid Biochem.*, 20, 887, 1984.
86. **Danielson, M., Northrop, J. P., and Ringold, G. M.**, The mouse glucocorticoid receptor: mapping of functional domains by cloning, sequencing and expression of wild-type and mutant receptor proteins, *EMBO J.*, 5, 2513, 1986.
87. **Holmes, S. D., Van, N. T., Stevens, S., and Smith, R. G.**, Affinity labelling of the human uterine progesterone receptor with 21-, 16α- and 11α-bromoacetoxy progesterones, *Endocrinology*, 109, 670, 1981.
88. **MacDonald, R. G. and Leavitt, W. W.**, Reduced sulphydryl groups are required for activation of uterine progesterone receptor. Possible involvement of an inhibitor of activation, *J. Biol. Chem.*, 257, 311, 1982.
89. **Miller, N. R. and Simons, S. S., Jr.**, Steroid binding of hepatoma tissue culture cell glucocorticoid receptors involves at least two sulfhydryl groups, *J. Biol. Chem.*, 263, 15217, 1988.
90. **Silva, C. M. and Cidlowski, J. A.**, Direct evidence for intra- and intermolecular disulfide bond formation in the human glucocorticoid receptor, *J. Biol. Chem.*, 254, 6638, 1989.

91. **Sheridan, P. J., Blum, K., and Martin, P. M.,** Steroid receptor dynamics: a function of cellular equilibrium, in *Steroid Receptors and Disease Cancer, Autoimmune, Bone, and Circulatory Disorders,* Sheridan, P. J., Blum, K., and Trachtenberg, M. C., Eds., Marcel Dekker, New York, 1988, 49.
92. **King, W. J. and Greene, G. L.,** Monoclonal antibodies localize oestrogen receptor in the nuclei of target cells, *Nature,* 307, 745, 1984.
93. **Welshons, W. V., Lieberman, W. E., and Gorski, J.,** Nuclear localization of unoccupied oestrogen receptors, *Nature,* 307, 747, 1984.
94. **Guiochon-Mantel, A., Loosfelt, H., Lescop, P., Sar, S., Atger, M., Perrot-Applanat, M., and Milgrom, E.,** Mechanisms of nuclear localization of the "progesterone" receptor: evidence for interaction between monomers, *Cell,* 57, 1147, 1989.
95. **Ostermann, J., Horwich, A. L., Neupert, W., and Hartl, F. U.,** Protein folding in mitochondria requires complex formation with hsp60 and ATP hydrolysis, *Nature,* 341, 125, 1989.
96. **Pratt, W. B.,** Interaction of hsp90 with steroid receptors: organizing some diverse observations and presenting the newest concepts, *Mol. Cell. Endocrinol.,* 74, C69, 1990.
97. **Housley, P. R., Sanchez, E. R., Danielson, M., Ringold, G. M., and Pratt, W. B.,** Evidence that the conserved region in the steroid binding domain of the glucocorticoid receptor is required for both optimal binding of hsp90 and protection from proteolytic cleavage, *J. Biol. Chem.,* 265, 12778, 1990.
98. **Pratt, W. B., Scherrer, L. C., Hutchison, K. A., and Dalman, F. C.,** A model of glucocorticoid receptor unfolding and stabilization by a heat shock protein complex, *J. Steroid Biochem. Mol. Biol.,* 41, 223, 1992.
99. **Howard, K. J., Holley, S. J., Yamamoto, K. R., and Distelhorst, C. W.,** Mapping the HSP90 binding region of the glucocorticoid receptor, *J. Biol. Chem.,* 265, 11928, 1990.
100. **Schlatter, L. K., Parker, M. G., and Distelhorst, C. W.,** Mapping the hsp90 binding region of the estrogen receptor, in Abstracts, 73rd Annu. Meet. Endocr. Soc., Washington, DC, no. 650, 1991, 193.
101. **Schowalter, D. B., Sullivan, W. P., and Toft, D. O.,** Characterization of progesterone receptor expressed in cultured cells and receptor binding to the 90 and 70 kDa heat shock proteins. in Abstracts, 73rd Annu. Meet. Endocr. Soc., Washington, DC, no. 373, 1991, 124.
102. **Baker, M. E.,** Similarity between tyrosyl-tRNA synthetase and the estrogen receptor, *FASEB J.,* 3, 2086, 1989.
103. **Baker, M. E.,** Similarity between human retinoic acid receptor and *Escherichia coli* homoserine kinase, *Biochem. J.,* 255, 748, 1988.
104. **Moudgil, V. K.,** Phosphorylation of steroid hormone receptors, *Biochim. Biophys. Acta,* 1055, 243, 1990.
105. **Denner, L. A., Schrader, W. T., O'Malley, B. W., and Weigel, N. L.,** Hormonal regulation and identification of chicken progesterone receptor phosphorylation sites, *J. Biol. Chem.,* 263, 16348, 1990.
106. **Smith, L. I., Mendel, D. B., Bodwell, J. E., and Munck, A.,** Phosphorylated sites within the functional domains of the ~100-kDa steroid-binding subunit of glucocorticoid receptors, *Biochemistry,* 28, 4490, 1989.
107. **Godowski, P. J., Rusconi, S., Miesfeld, R., and Yamamoto, K. R.,** Glucocorticoid receptor mutants that are constitutive activators of transcriptional enhancement, *Nature,* 325, 365, 1987.
108. **Hollenberg, S. M. and Evans, R. M.,** Multiple and cooperative transactivation domains of the human glucocorticoid receptor, *Cell,* 55, 899, 1988.
109. **Fawell, S. E., Lees, J. A., White, R., and Parker, M. G.,** Characterization and colocalization of steroid binding and dimerization activities in the mouse estrogen receptor, *Cell,* 60, 953, 1990.

110. **Picard, D. and Yamamoto, K. R.,** Two signals mediate hormone-dependent nuclear localization of the glucocorticoid receptor, *EMBO J*, 6, 3333, 1987.

111. **Wolff, B., Dickson, R. B., and Hanover, J. A.,** A nuclear localization signal in steroid hormone receptors?, *Trends Pharm. Sci.*, 8, 119, 1987.

112. **Willmann, T. and Beato, M.,** Steroid-free glucocorticoid receptor binds specifically to mouse mammary tumour virus DNA, *Nature,* 324, 688, 1986.

113. **Becker, P. B., Gloss, B., Schmid, W., Strahle, U., and Schutz, G.,** *In vivo* protein-DNA interactions in a glucocorticoid response element require the presence of the hormone, *Nature,* 324, 686, 1986.

114. **Denis, M., Poellinger, L., Wikström, A. C., and Gustafsson, J. Å.,** Requirement of hormone for thermal conversion of the glucocorticoid receptor to a DNA-binding state, *Nature,* 333, 686, 1988.

115. **Lepicard, G., Mornon, J. P., Delettré, J., Ojasoo, T., and Raynaud, J. P.,** Steroid hormone binding to cytoplasmic receptors: additivity of the relative binding affinity (RBA) increments calculated for individual substituents, *J. Steroid Biochem.*, 9, 830, 1978.

116. **Wolff, M. E., Baxter, J. D., Kollman, P. A., Lee, D. L., Kuntz, I. D., Bloom, E., Matulich, D. T., and Morris, J.,** Nature of steroid-glucocorticoid receptor interactions: thermodynamic analysis of the binding reaction, *Biochemistry,* 17, 3201, 1978.

117. **Andrews, P. R., Craik, D. J., and Martin, J. L.,** Functional group contributions to drug-receptor interactions, *J. Med. Chem.*, 27, 1648, 1984.

118. **Pons, M., Bignon, E., Crastes de Paulet, A., Gilbert, J., Ojasoo, T., and Raynaud, J. P.,** Hydroxylated triphenylacrylonitriles adopt a unique orientation within the binding site of the estrogen receptor, *J. Steroid Biochem.*, 36, 391, 1990.

119. **Delettré, J., Mornon, J. P., Ojasoo, T., and Raynaud, J. P.,** An approach to the mapping of the steroid hormone receptor, in *Perspectives in Steroid Receptor Research,* Bresciani, F., Ed., Raven Press, New York, 1980, 1.

120. **Raynaud, J. P., Delettré, J., Ojasoo, T., Lepicard, G., and Mornon, J. P.,** Steps towards mapping of steroid hormone receptors, in *Physiopathology of Endocrine Diseases and Mechanisms of Hormone Action,* Soto, R. J., de Nicola, A., and Blaquier, J., Eds., Alan R. Liss, New York, 1981, 461.

121. **Delettré, J., Mornon, J. P., Lepicard, G., Ojasoo, T., and Raynaud, J. P.,** Steroid flexibility and receptor specificity, *J. Steroid Biochem.*, 13, 45, 1980.

122. **Griffin, J. F., Duax, W. L., and Weeks, C. M.,** in *Atlas of Steroid Structure,* Vol. 2, Plenum Press, New York, 1984.

123. **Mornon, J. P., Delettré, J., Lepicard, G., Bally, R., Surcouf, E., and Bondot, P.,** Interactions of hormonal steroids: progestogens, *J. Steroid Biochem.*, 8, 51, 1977.

124. **Baker, M. E. and Terry, L. S.,** Diethylpyrocarbonate, a histidine selective reagent, inhibits progestin binding in chick oviduct cytosol, *Steroids,* 42, 593, 1983.

125. **Baker, M. E., Sklar, D. H., Terry, L. S., and Hedges, M. R.,** Diethylpyrocarbonate, a histidine selective reagent, inhibits estrogen binding to receptor proteins in rat uterus cytosol, *Biochem. Int.*, 11, 233, 1985.

126. **Borgna, J. L. and Scali, J.,** Differential inhibition of estrogen and antiestrogen binding to the estrogen receptor by diethylpyrocarbonate, *J. Steroid Biochem.*, 31, 427, 1988.

127. **Sweet, F. and Murdock, G. L.,** Affinity labeling of hormone-specific proteins, *Endocr. Rev.*, 8, 154, 1987.

128. **Simons, S. S., Jr., Pumphrey, J. G., Rudikoff, S., and Eisen, H. J.,** Identification of cysteine 656 as the amino acid of hepatoma tissue culture cell glucocorticoid receptors that is covalently labeled by dexamethasone-21-mesylate, *J. Biol. Chem.*, 262, 9676, 1987.

129. **Carlstedt-Duke, J., Strömstedt, P. E., Persson, B., Cederlund, E., Gustafsson, J. Å., and Jörnvall, H.,** Identification of hormone-interacting amino acid residues within the steroid binding domain of the glucocorticoid receptor in relation to other steroid hormone receptors, *J. Biol. Chem.*, 263, 6842, 1988.

130. **Ohara-Nemoto, Y., Strömstedt, P. E., Dahlman-Wright, K., Nemoto, T., Gustafsson, J. Å., and Carlstedt-Duke, J.,** The steroid-binding properties of recombinant glucocorticoid receptor: a putative role for heat shock protein hsp90, *J. Steroid Biochem. Mol. Biol.*, 37, 481, 1990.
131 **Smith, L. I., Bodwell, J. E., Mendel, D. B., Ciardelli, T., North, W. G., and Munck, A.,** Identification of cysteine-644 as the covalent site of attachment of dexamethasone 21-mesylate to murine glucocorticoid receptors in WEHI-7 cells, *Biochemistry*, 27, 3747, 1988.
132. **Vessières, A., Vaillant, C., Salmain, M., and Jaouen, G.,** Organometallic derivatives of estradiol as bioligands: targetted binding of the estradiol receptor, *J. Steroid Biochem.*, 34, 301, 1989.
133. **Harlow, K. W., Smith, D. N., Katzenellenbogen, J. A., Greene, G. L., and Katzenellenbogen, B. S.,** Identification of cysteine 530 as the covalent attachment site of an affinity-labeling estrogen (ketononestrol aziridine) and antiestrogen (tamoxifen aziridine) in the human estrogen receptor, *J. Biol. Chem.*, 264, 17476, 1989.
134. **Hurley, D. M., Accili, D., Stratakis, C. A., Karl, M., Vamvakopoulos, N., Rorer, E., Constantine, K., Taylor, S. I., and Chrousos, G.,** Point mutation causing a single amino acid substitution in the hormone binding domain of the glucocorticoid receptor in familial glucocorticoid resistance, *J. Clin. Invest.*, 87, 680, 1991.
135. **Malloy, P. J., Hughes, M. R., Pike, J. W., and Feldman, D.,** Vitamin D receptor mutations and hereditary 1,25-dihydroxy-vitamin D resistant rickets, in *Vitamin D — Gene Regulation, Structure-Function Analysis, and Clinical Application,* Norman, A. W., Bouillon, R., and Thomasset, M., Eds., Walter de Gruyter, Berlin, 1991, 116.
136. **Chakraborti, P. K., Hoeck, W., Groner, B., and Simons, S. S.,** Localisation of the vicinal dithiols involved in steroid binding to the rat glucocorticoid receptor, *Endocrinology,* 127, 2530, 1990.
137. **Simons, S. S., Jr.,** Specific amino acids involved in steroid binding to rat glucocorticoid receptors, in Abstracts, 73rd Annu. Meet. Endocr. Soc., Washington, DC, 1991, 14.
138. **Conneely, O. M., Sullivan, W. P., Toft, D. O., Birnbaumer, M., Cook, R. G., Maxwell, B. L., Zarucki-Schulz, T., Greene, G. L., Schrader, W. T., and O'Malley, B. W.,** Molecular cloning of the chicken progesterone receptor, *Science,* 233, 767, 1986.
139. **Jeltsch, J. M., Krozowski, Z., Quirin-Stricker, C., Gronemeyer, H., Simpson, R. J., Garnier, J. M., Krust, A., Jacob, A., and Chambon, P.,** Cloning of the chicken progesterone receptor, *Proc. Natl. Acad. Sci. U.S.A.*, 83, 5424, 1986.
140. **Groyer, A., Le Bouc, Y., Joab, I., Radanyi, C., Renoir, J. M., Robel, P., and Baulieu, E. E.,** Chick oviduct glucocorticosteroid receptor. Specific binding of the synthetic steroid RU 486 and immunological studies with antibodies to chick oviduct progesterone receptor, *Eur. J. Biochem.*, 149, 445, 1985.
141. **Gronemeyer, H., Benhamou, B., Berry, M., Bocquel, M. T., Gofflo, D., Garcia, T., Lerouge, T., Metzger, D., Meyer, M. E., Tora, L., Vergezac, A., and Chambon, P.,** Mechanisms of anti-hormone action, *J. Steroid Biochem. Mol. Biol.*, 41, 217, 1992.
142. **Mendel, D. B., Bodwell, J. E., and Munck, A.,** Glucocorticoid receptors lacking hormone-binding activity are bound in nuclei of ATP-depleted cells, *Nature,* 324, 478, 1986.
143. **Meyer, M. E., Pornon, A., Ji, J., Bocquel, M. T., Chambon, P., and Gronemeyer, H.,** Agonistic and antagonistic activities of RU486 on the functions of the human progesterone receptor, *EMBO J.*, 9, 3923, 1990.
144. **Gray, G. O. and Leavitt, W. W.,** RU 486 is not an antiprogestin in the hamster, *J. Steroid Biochem.*, 28, 493, 1987.
145. **Tora, L., Mullick, A., Metzger, D., Ponglikitmongkol, M., Park, I., and Chambon, P.,** The cloned human oestrogen receptor contains a mutation which alters its hormone binding properties, *EMBO J.*, 8, 1981, 1989.

146. **Pakdel, F. and Katzenellenbogen, B. S.,** Mutant human estrogen receptors with altered sensitivities towards estrogens and antiestrogens, in Abstracts, 73rd Annu. Meet. Endocr. Soc., Washington, DC, no. 359, 1991, 120.
147. **Wakeling, A. E. and Bowler, J.,** Novel antiestrogens without partial agonist activity, *J. Steroid Biochem.*, 31, 645, 1988.
148. **Reese, J. C. and Katzenellenbogen, B. S.,** Mutagenesis of cysteines in the hormone binding domain of the human estrogen receptor: effects of hormone binding and transcriptional activation efficiency, in Abstracts, 72nd Annu. Meeting Endocr. Soc., Atlanta, no. 287, 1990, 96.
149. **McGuire, W. L.,** Estrogen receptor variants in clinical breast cancer, in Abstracts, 73rd Annu. Meet. Endocr. Soc., Washington, DC, 1991, 26.
150. **Murphy, L. C.,** Estrogen receptor variants in human breast cancer, *Mol. Cell. Endocrinol.*, 74, C83, 1990.
151. **Miller, L. A., Leslie, K. K., Graham, M. L., II, Francis, M. D., and Horwitz, K. B.,** Hormone resistant breast cancer is linked to genetic instability and mutant estrogen receptors, in Abstracts, 72nd Annu. Meeting Endocr. Soc., Atlanta, no. 288, 1990, 96.
152. **Garcia, T., Lehrer, S., Bloomer, W. D., and Schachter, B.,** A variant estrogen receptor messenger ribonucleic acid is associated with reduced levels of estrogen binding in human mammary tumors, *Mol. Endocrinol.*, 2, 785, 1988.
153. **Fuqua, S. A. W., Fitzgerald, S. D., Chamness, G. C., Tandon, A. K., McDonnell, D. P., Nawaz, Z., O'Malley, B. W., and McGuire, W. L.,** Variant human breast tumor estrogen receptor with constitutive transcriptional activity, *Cancer Res.*, 51, 105, 1991.
154. **Murphy, L. C. and Dotzlaw, H.,** Variant estrogen receptor mRNA species detected in human breast cancer biopsy samples, *Mol. Endocrinol.*, 3, 687, 1989.
155. **Brown, T. R., Lubahn, D. B., Wilson, E. M., French, F. S., Migeon, C. J., and Corden, J. L.,** Functional characterization of naturally occurring mutant androgen receptors from subjects with complete androgen insensitivity, *Mol. Endocrinol.*, 4, 1759, 1990.
156. **Ris-Stalpers, C., Kuiper, G. G. J. M., Faber, P. W., Schweikert, H. U., van Rooij, H. C. J., Zegers, N. D., Hodgins, M. B., Degenhart, H. J., Trapman, J., and Brinkmann, A. O.,** Aberrant splicing of androgen receptor mRNA results in synthesis of a nonfunctional receptor protein in a patient with androgen insensitivity, *Proc. Natl. Acad. Sci. U.S.A.*, 87, 7866, 1990.
157. **Marcelli, M., Tilley, W. D., Wilson, C. M., Griffin, J. E., Wilson, J. D., and McPhaul, M. J.,** Definition of the human androgen receptor gene structure permits the identification of mutations that cause androgen resistance: premature termination of the receptor protein at amino acid residue 588 causes complete androgen resistance, *Mol. Endocrinol.*, 4, 1105, 1990.
158. **De Bellis, A., Quigley, C. A., Cariello, N. F., Ho, H. C., and French, F. S.,** DNA and steroid binding domain mutations of the androgen receptor cause partial or complete androgen insensitivity, in Abstracts, 73rd Annu. Meet. Endocr. Soc., Washington, DC, no. 1141, 1991, 316.
159. **Yarbrough, W. G., Quarmby, V. E., Simental, J. A., Joseph, D. R., Sar, M., Lubahn, D. B., Olsen, K. L., French, F. S., and Wilson, E. M.,** A single base mutation in the androgen receptor gene causes androgen insensitivity in the testicular feminized rat, *J. Biol. Chem.*, 265, 8893, 1990.
160. **Gaspar, M. L., Meo, T., and Tosi, M.,** Structure and size distribution of the androgen receptor mRNA in wild-type and *Tfm/y* mutant mice, *Mol. Endocrinol.*, 4, 1600, 1990.
161. **Asselin, J., Melançon, R., Gourdeau, Y., Labrie, F., Bonne, C., and Raynaud, J. P.,** Specific binding of (^{3}H)-methyltrienolone to both progestin and androgen binding components in human benign prostatic hypertrophy (BPH), *J. Steroid Biochem.*, 10, 483, 1979.

162. **Gyorki, S., James, R., Warne, G. L., and Funder, J. W.**, Photoaffinity labelling of androgen receptors from human foreskin and cultured genital fibroblasts with tritiated methyltrienolone [^{3}H]-R 1881, *J. Steroid Biochem.*, 25, 355, 1986.
163. **Brinkmann, A. O., Kuiper, G. G. J. M., de Boer, W., Mulder, E., Bolt, J., van Steenbrugge, G. J., and van der Molen, H. J.**, Characterization of androgen receptors after photoaffinity labeling with [^{3}H]methyltrienolone (R 1881), *J. Steroid Biochem.*, 24, 245, 1986.
164. **Bannister, P., Sheridan, P., and Losowsky, M. S.**, Use of a new radioactive ligand, 7α,17α-dimethyl[17α-methyl^{3}H] 19-nortestosterone for the estimation of androgen receptors in rat liver cytosol, *J. Steroid Biochem.*, 23, 121, 1985.
165. **Kaufman, M., Pinsky, L., Gottlieb, B., and Trifiro, M.**, The state transitions of normal and mutant androgen-receptor complexes in human genital skin fibroblasts, *J. Steroid Biochem.*, 36, 651, 1990.
166. **Veldscholte, J., Ris-Stalpers, C., Kuiper, G. G. J. M., Jenster, G., Berrevoets, C., Claassen, E., van Rooij, H. C. J., Trapman, J., Brinkmann, A. O., and Mulder, E.**, A mutation in the ligand binding domain of the androgen receptor of human LNCaP cells affects steroid binding characteristics and response to anti-androgens, *Biochem. Biophys. Res. Commun.*, 173, 534, 1990.
167. **Veldscholte, J., Voorhorst-Ogink, M. M., Bolt-de-Vries, J., van Rooij, H. C. J., Trapman, J., and Mulder, E.**, Unusual specificity of the androgen receptor in the human prostate tumor cell line LNCaP: high affinity for progestagenic and estrogenic steroids, *Biochim. Biophys. Acta*, 1052, 187, 1990.
168. **Veldscholte, J., Berrevoets, C. A., Ris-Stalpers, C., Kuiper, G. G. J. M., Jenster, G., Trapman, J., Brinkman, A. O., and Mulder, E.**, The androgen receptor in LNCAP cells contains a mutation in the ligand binding domain which affects steroid binding characteristics and response to antiandrogens, *J. Steroid Biochem. Mol. Biol.*, 41, 665, 1992.
169. **Newcomer, M. E., Jones, T. A., Aqvist, J., Sundelin, J., Ericksson, U., Rask, L., and Peterson, P. A.**, The three-dimensional structure of retinol-binding protein, *EMBO J.*, 3, 1451, 1984.
170. **Westbrook, E. M., Piro, O. E., and Sigler, P. B.**, The 6-Å crystal structure of delta 5-3-ketosteroid isomerase. Architecture and location of the active center, *J. Biol. Chem.*, 259, 9096, 1984.
171. **Vrielink, A., Lloyd, L., and Blow, D.**, Crystal structure of cholesterol oxidase from *Brevi bacterium sterolicum* refined at 1.8Å resolution, *J. Mol. Biol.*, 219, 533, 1991.
172. **Suthers, M. B., Pressley, L. A., and Funder, J. W.**, Glucocorticoid receptors: evidence for a second, non-glucocorticoid binding site, *Endocrinology*, 99, 260, 1976.
173. **Jones, T. R. and Bell, P. A.**, Glucocorticoid-receptor interactions. Studies of the negative co-operativity induced by steroid interactions with a secondary, hydrophobic, binding site, *Biochem. J.*, 188, 237, 1980.
174. **Moguilewsky, M. and Deraedt, R.**, Interrelations between glucocorticoid and progestin receptors, *J. Steroid Biochem.*, 15, 329, 1981.
175. **Svec, F. and Rudis, M.**, Progestin-induced enhancement of dexamethasone dissociation from glucocorticoid hormone receptors, *Arch. Biochem. Biophys.*, 16, 135, 1981.
176. **Svec, F., Teubner, V., and Tate, D.**, Location of the second steroid-binding site on the glucocorticoid receptor, *Endocrinology*, 125, 3103, 1989.
177. **Philibert, D., Deraedt, R., Teutsch, G., Tournemine, C., and Sakiz, E.**, RU38486 — a new lead for steroidal anti-hormones, in Abstracts, 64th Annu. Meet. Endocr. Soc., San Francisco, no. 668, 1982.
178. **Teutsch, G., Ojasoo, T., and Raynaud, J. P.**, 11β-substituted steroids, an original pathway to antihormones, *J. Steroid Biochem.*, 31, 549, 1988.
179. **Anstead, G. M., Peterson, C. S., and Katzenellenbogen, J. A.**, Hydroxylated 2,3-diarylindenes: synthesis, estrogen receptor binding affinity, and binding orientation considerations, *J. Steroid Biochem.*, 33, 877, 1989.

180. **Prim, R. C.**, Shortest connection networks and some generalisations, *Bell Syst. Technol. J.*, 36, 1389, 1957.
181. **Devillers, J. and Doré, J. C.**, Heuristic potency of the minimum spanning tree (MST) method in toxicology, *Ecotox Environ. Safety*, 17, 227, 1989.
182. **Doré, J. C., Viel, C., Lacroix, R., and Lacroix, J.**, Les analyses multivariées comme outil heuristique d'accès à une spécificité pharmacochimique optimale et à des archétypes moléculaires inédits, *J. Pharm. Belg.*, 45, 101, 1990.
183. **Doré, J. C., Lacroix, J., Lacroix, R., and Viel, C.**, De la manipulation informatique des formules développées, à l'étude des relations structure-activité. Application aux médicaments antiparasitaires, *J. Pharm. Belg.*, 45, 375, 1990.
184. **Benzécri, J. P.**, L'analyse des données. II. *L'analyse des Correspondances*, Vol. 2, Dunod, Paris, 1983.
185. **Greenacre, M. J.**, *Theory and Application of Correspondence Analysis*, Academic Press, New York, 1983.
186. **Doré, J. C., Gilbert, J., Ojasoo, T., and Raynaud, J. P.**, Correspondence analysis applied to steroid receptor binding, *J. Med. Chem.*, 29, 54, 1986.
187. **Ojasoo, T., Doré, J. C., Gilbert, J., and Raynaud, J. P.**, Binding of steroids to the progestin and glucocorticoid receptors analyzed by correspondence analysis, *J. Med. Chem*, 31, 1160, 1988.
188. **Doré, J. C. and Miquel, J. F.**, Approche de l'étude des relations structure/activité par analyse des correspondances, *C. R. Acad. Sci. Paris*, 293, 1061, 1981.
189. **Bignon, E., Pons, M., Doré, J. C., Gilbert, J., Ojasoo, T., Miquel, J. F., Raynaud, J. P., and Crastes de Paulet, A.**, Influence of di- and tri-phenylethylene estrogen/antiestrogen structure on the mechanisms of protein kinase C inhibition and activation as revealed by a multivariate analysis, *Biochem. Pharmacol.*, 42, 1373, 1991.
190. **Doré, J. C., Gilbert, J., Bignon, E., Crastes de Paulet, A., Ojasoo, T., Pons, M., Raynaud, J. P., and Miquel, J. F.**, Multivariate analysis by the minimum spanning tree method of the structural determinants of diphenylethylenes and triphenylacrylonitriles implicated in estrogen receptor binding, protein kinase C activity and MCF_7 cell proliferation, *J. Med. Chem.*, 35, 573, 1992.
191. **Kloosterboer, H. J., Vonk-Noordegraaf, C. A., and Turpijn, E. W.**, Selectivity in progesterone and androgen receptor binding of progestagens used in oral contraceptives, *Contraception*, 38, 325, 1988.
192. **Raynaud, J. P., Ojasoo, T., and Labrie, F.**, Steroid hormones — agonists and antagonists, in *Mechanisms of Steroid Action*, Lewis, G. P. and Ginsburg, M., Eds., Macmillan, London, 1981, 145.
193. **Ojasoo, T. and Raynaud, J. P.**, Receptor binding profiles of progestins, in *Steroids and Endometrial Cancer*, Jasonni, V. M., Nenci, I., and Flamigni, C., Eds., Raven Press, New York, 1983, 11.
194. **Raynaud, J. P., Bouton, M. M., Moguilewsky, M., Ojasoo, T., Philibert, D., Beck, G., Labrie, F., and Mornon, J. P.**, Steroid hormone receptors and pharmacology, *J. Steroid Biochem.*, 12, 143, 1980.
195. **Raynaud, J. P. and Ojasoo, T.**, La promégestone, un nouveau progestatif, *J Gyn. Obst. Biol. Reprod.*, 12, 697, 1983.
196. **Paris, J., Thévenot, R., Bonnet, P., and Granero, M.**, The pharmacological profile of TX 066 (17α-acetoxy-6-methyl-19-nor-4,6-pregna-diene-3,20-dione), a new oral progestative, *Arzneim. Forsch.*, 33, 710, 1983.
197. **Naylor, P. H., Gilani, S. S., Milholland, R. J., Ip, M., and Rosen, F.**, *In vivo* antiglucocorticoids: comparison between *in vivo* activity and *in vitro* competition of progestins for the glucocorticoid receptor, *J. Steroid Biochem.*, 14, 1303, 1981.
198. **Teutsch, G., Bélanger, A., Philibert, D., and Tournemine, C.**, Synthesis of 11β-vinyl-19-norsteroids as potent progestins, *Steroids*, 39, 607, 1982.
199. **Azadian-Boulanger, G., Bucourt, R., Nédélec, L., and Nominé, G.**, Stéroïdes triéniques androgènes et anabolisants, *Eur. J. Med. Chem.*, 10, 353, 1975.

200. **Arányi, P.**, Kinetics of the hormone-receptor interaction. Competition experiments with slowly equilibrating ligands, *Biochim. Biophys. Acta*, 628, 220, 1980.
201. **Bouton, M. M. and Raynaud, J. P.**, The relevance of interaction kinetics in the determination of specific binding to the estrogen receptor, *J. Steroid Biochem.*, 9, 9, 1978.
202. **Cullen, K. J. and Lippman, M. E.**, Estrogen regulation of protein synthesis and cell growth in human breast cancer, *Vit. Horm.*, 45, 127, 1989.
203. **Jordan, V. C. and Murphy, C. S.**, Endocrine pharmacology of antiestrogens as antitumor agents, *Endocr. Rev.*, 11, 578, 1990.
204. **Miquel, J. F. and Gilbert, J.**, A chemical classification of nonsteroidal antagonists of sex-steroid hormone action, *J. Steroid Biochem.*, 31, 525, 1988.
205. **Poulin, R., Mérand, Y., Poirier, D., Levesque, C., Dufour, J. M., and Labrie, F.**, Antiestrogenic properties of keoxifene, *trans*-4-hydroxytamoxifen, and ICI 164384, a new steroidal antiestrogen in ZR-75-1 human breast cancer cells, *Breast Cancer Res. Treat.*, 14, 65, 1989.
206. **Kangas, L. and Baum, M., Eds.**, Proc. Toremifene Satellite Symposium. UICC World Cancer Congress, Budapest, 1986, *J. Steroid Biochem.*, 36, 191, 1990.
207. **Pons, M., Michel, F., Crastes de Paulet, A., Gilbert, J., Miquel, J. F., Précigoux, G., Hospital, M., Ojasoo, T., and Raynaud, J. P.**, Influence of new hydroxylated triphenylethylene (TPE) derivatives on estradiol binding to uterine cytosol, *J. Steroid Biochem.*, 20, 137, 1984.
208. **Nishizuka, Y.**, The role of protein kinase C in cell surface signal transduction and tumor production, *Nature*, 308, 693, 1984.
209. **Nishizuka, Y.**, Studies and prospectives of the protein kinase C family for cellular regulation, *Cancer*, 63, 1892, 1989.
210. **O'Brian, C. A. and Ward, N. E.**, Biology of the protein kinase C family, *Cancer Metastasis Rev.*, 8, 199, 1989.
211. **Gescher, A. and Dale, I. L.**, Protein kinase C — a novel target for rational anti-cancer drug design?, *Anti-Cancer Drug Design*, 4, 93, 1989.
212. **O'Brian, C. A., Vogel, V. G., Singletary, S. E., and Ward, N. E.**, Elevated protein kinase C expression in human breast tumor biopsies relative to normal breast tissue, *Cancer Res.*, 49, 3215, 1985.
213. **Bignon, E., Ogita, K., Kishimoto, A., Gilbert, J., Abecassis, J., Miquel, J. F., and Nishizuka, Y.**, Modes of inhibition of protein kinase C by triphenylacrylonitrile antiestrogens, *Biochem. Biophys. Res. Commun.*, 163, 1377, 1989.
214. **Bignon, E., Pons, M., Gilbert, J., and Nishizuka, Y.**, Multiple mechanisms of protein kinase C inhibition by triphenylacrylonitrile antiestrogens, *FEBS Lett.*, 271, 54, 1990.
215. **Ojasoo, T., Bignon, E., Crastes de Paulet, A., Doré, J. C., Gilbert, J., Miquel, J. F., Pons, M., and Raynaud, J. P.**, Relative involvement of the estrogen receptor and protein kinase C in the action of a population of triphenylethylenes on MCF_7 cell proliferation as revealed by correspondence factorial analysis (CFA), *J. Steroid Biochem. Mol. Biol.*, submitted.
216. **Epand, R. M. and Lester, D. S.**, The role of membrane biophysical properties in the regulation of protein kinase C activity, *Trends Pharm. Sci.*, 11, 317, 1990.
217. **Horgan, K., Cooke, E., Hallett, M. B., and Mansel, R. E.**, Inhibition of protein kinase C mediated signal transduction by tamoxifen. Importance for antitumour activity, *Biochem. Pharmacol.*, 35, 4463, 1986.
218. **Clarke, R., van den Berg, H. W., Nelson, J., and Murphy, R. F.**, Pharmacological and suprapharmacological concentrations of both 17β-oestradiol and tamoxifen reduce the membrane fluidity of MCF-7 and MDA-MB-436 human breast cancer cells, *Biochem. Soc. Trans.*, 15, 243, 1987.
219. **Kishimoto, A., Takai, Y., Mori, T., Kikkawa, U., and Nishizuka, Y.**, Activation of calcium- and phospholipid-dependent protein kinase by diacylglycerol, its possible relation to phosphatidylinositol turnover, *J. Biol Chem.*, 255, 2273, 1980.

220. **Ojasoo, T., Moguilewsky, M., Philibert, D., and Raynaud, J. P.,** Study and assay of steroid hormone receptors, in *Advances in Steroid Analysis '87,* Görög, S., Ed., Akadémiai Kiadó, Budapest, 1988, 17.
221. **Bignon, E., Pons, M., Gilbert, J., and Crastes de Paulet, A.,** Analogies and differences in the modulation of progesterone receptor induction and cell proliferation by estrogens and antiestrogens in MCF-7 human breast cancer cells: study with 24 triphenylacrylonitrile derivatives, *J. Steroid Biochem.*, 31, 877, 1988.
222. **Palomino, E.,** A-ring substituted steroids, *Drugs of the Future,* 15, 909, 1990.
223. **Raynaud, J. P. and Bouton, M. M.,** The design of estrogens and/or anti-estrogens on the basis of receptor binding, in *Cytotoxic Estrogens in Hormone Receptive Tumors,* Raus, J., Martens, H., and Leclercq, G., Eds., Academic Press, New York, 1980, 49.
224. **Teutsch, G.,** Analogues of RU 486 for the mapping of the progestin receptor: synthetic and structural aspects, in *The Antiprogestin Steroid RU 486 and Human Fertility Control,* Baulieu, E. E. and Segal, S. J., Eds., Plenum Press, New York, 1985, 27.
225. **Neef, G., Sauer, G., Seeger, A., and Wiechert, R.,** Synthetic variations of the progesterone antagonist RU 38486, *Tetrahedron Lett.*, 25, 3425, 1984.
226. **Henderson, D.,** Antiprogestational and antiglucocorticoid activities of some novel 11β-aryl substituted steroids, in *Pharmacology and Clinical Uses of Inhibitors of Hormone Secretion and Action,* Furr, B. J. A. and Wakeling, A. E., Eds., Baillière Tindall, London, 1987, 106.
227. **Bohl, M., Schubert, G., Ponsold, K., Reck, G., Höhne, E., and Simon, K.,** Molecular mechanics and X-ray crystal structure investigations on conformations of 11β substituted 4,9-dien-3-one steroids, *J. Mol. Graphics,* 7, 122, 1989.
228. **Bindal, R. D., Carlson, K. E., Reiner, G. C. A., and Katzenellenbogen, J. A.,** 11β-chloromethyl-[^{3}H]estradiol-17β: a very high affinity, reversible ligand for the estrogen receptor, *J. Steroid Biochem.*, 28, 361, 1987.
229. **Fawell, S. E., White, R., Hoare, S., Sydenham, M., Page, M., and Parker, M. G.,** Inhibition of estrogen receptor-DNA binding by the "pure" antiestrogen ICI 164,384 appears to be mediated by impaired receptor dimerization, *Proc. Natl. Acad. Sci. U.S.A.*, 87, 6883, 1990.
230. **Hodgson, J.,** Data-directed drug design, *Biotechnology,* 9, 19, 1991.

Chapter 5

STILBESTROL ESTROGENS: MOLECULAR/ STRUCTURAL PROBES FOR UNDERSTANDING ESTROGEN ACTION

Kenneth S. Korach

TABLE OF CONTENTS

I. INTRODUCTION

Compounds with a stilbene chemical structure have been known to possess estrogen hormonal activity since the mid 1930s.[1,2] The most potent of these were diethylstilbestrol (DES) and its saturated isomer hexestrol,[3] which were originally synthesized and characterized by Dodds and associates in search of an inexpensive orally active estrogen. Interestingly, other analogs with slightly different structures such as dimethylstilbestrol (DMS) were also described as having some antiestrogenic activity.[4-6] Since those early reports, the stilbene, chemical structure has been the basis for the synthesis of a number of compounds of varying estrogenic[7-10] and antiestrogenic[11-14] activity. In fact, the well-described triphenylethylene antiestrogens are based on the stilbene structure with substitutions of a derivatized phenyl ring in place of one of the alkyl side chains as reviewed by Jordan.[15]

The stilbene chemical structure bears little resemblance to the estrane steroid nucleus. Structural and molecular similarities have been investigated and postulated for a number of years. Initial studies using molecular modeling and X-ray crystallography have suggested the planar molecular structure and the close approximation of the hydroxyl functions were an explanation for the similar hormonal activity between estradiol and DES.[16] However, DMS, a comparatively weaker estrogen,[17] has a similar molecular structure and hydroxyl group distances as DES except for differences in the alkyl side chains. More recent analysis has indicated that the terminal hydroxyl group distances for estradiol and DES were not identical and that DES was not as planar a structure as previously thought.[18,19] In fact, the two aromatic rings of DES are 30° out of plane to each other.[20] This ring orientation is influenced by the position of the ethyl group side chains on DES.

DES is oxidatively metabolized to a variety of compounds shown in Figure 1, in a number of species,[21-24] including man.[25] Because of its extensive metabolism, we had investigated the estrogenic and receptor binding activity of the metabolites as a possible explanation for the significant biological potency of DES.[26] Previously, we had reported the receptor binding and uterotropic activity of a number of DES metabolites.[27] This report describes some of our latest data obtained while investigating the influence of ligand structure on receptor binding and hormonal responses. In our earlier studies with stilbestrol analogs we reported that the oxidative metabolism of DES resulted in metabolic products which retained different amounts of biochemical and receptor-binding activities.[28] Several of these stilbestrol compounds were found to have equivalent or stronger receptor binding affinity than the most potent endogenous steroidal estrogen, estradiol. It is this group of compounds which we have utilized as structural ligand probes to investigate uterine hormonal activity.

Two sets of compounds, pseudo-DES and indanyl-DES shown in Figure 1, were of particular interest because of the strong receptor binding affinities,

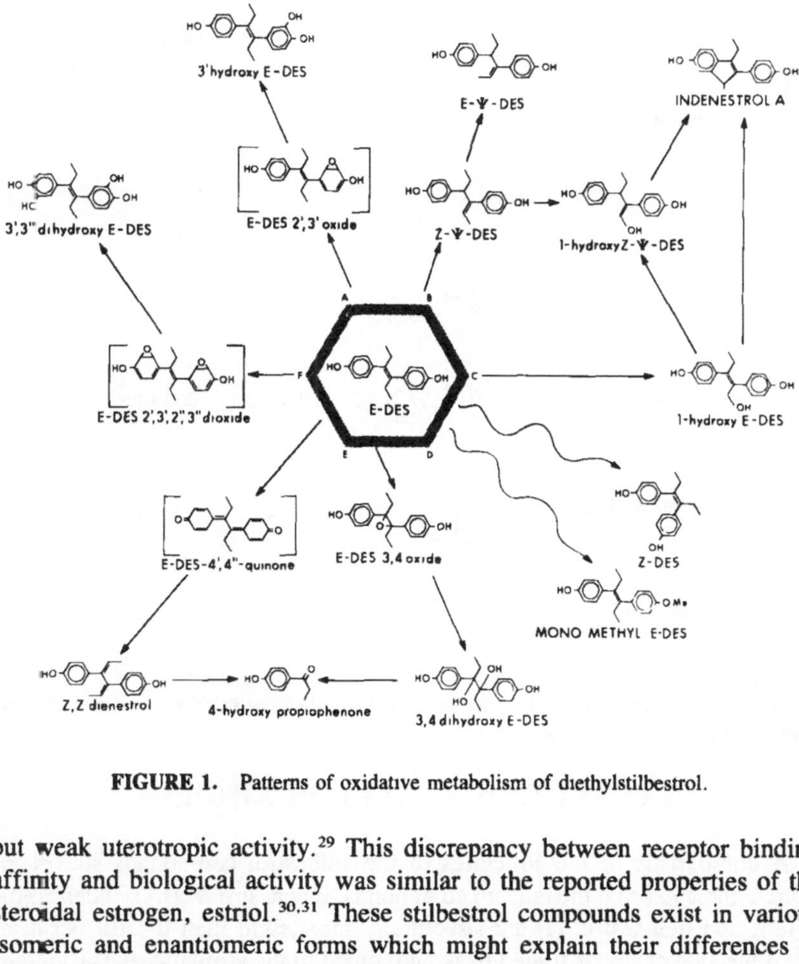

FIGURE 1. Patterns of oxidative metabolism of diethylstilbestrol.

but weak uterotropic activity.[29] This discrepancy between receptor binding affinity and biological activity was similar to the reported properties of the steroidal estrogen, estriol.[30,31] These stilbestrol compounds exist in various isomeric and enantiomeric forms which might explain their differences in biological activity.[32] However, the structural variations of these compounds are subtle, compared to differences in the biological potency of steroidal estrogens such as estradiol and estriol. Therefore, it was suspected that these structural subtleties would not be significant enough to explain the differences in their potency.

An additional structural property of the stilbestrol compounds is their similarity to the class of triphenylethylene antiestrogens such as tamoxifen, clomiphene, and CI-628.[15] In order to examine this discordance of strong receptor binding and poor biological activity in more detail, we have attempted to investigate the receptor-mediated mechanism of estrogen action in the uterus, and evaluate whether certain steps or responses are compromised or altered due to treatment with these various compounds. It is proposed that such an approach should aid our understanding of the interrelationships of events which occur during estrogen stimulation of uterine target tissues.

Compound	Structure	Kd	Doubling Dose
Indenestrol A	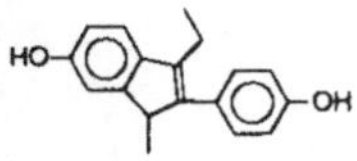	0.7 + 0.1	107 + 14
Indenestrol B	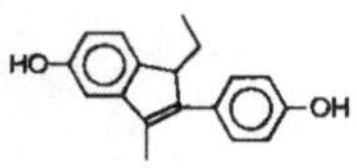	0.8 + 0.1	111 + 8
E - DES	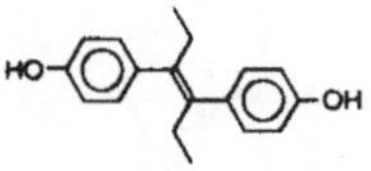	0.3 ± 0.1	7 ± 2
Z - Pseudo - DES	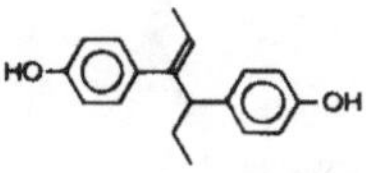	1.0 ± 0.1	600 ± 20
Estradiol - 17 β	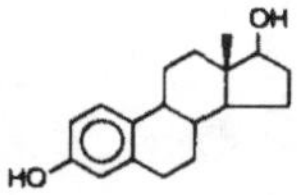	1.0 ± 0.1	5 ± 1
Indanestrol	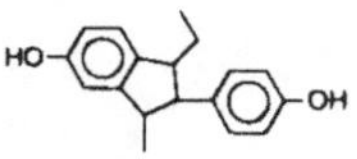	60.0 ± 5.0	1,120 ± 5

FIGURE 2. Binding affinities (K_d), and biological activities (μg/kg dose) of some DES metabolites and analogs in the mouse uterus.

II. ESTROGEN RECEPTOR BINDING

A comparison of the structures of the stilbestrol compounds shown in Figure 2 indicates that the indanyl compounds, indenestrol A (IA), indenestrol B (IB), and indanestrol (I) are structurally more rigid than E-DES because one of the side chains is folded back and joined to the aromatic ring. Binding affinities of mouse uterine estrogen receptor with these different compounds is listed in the figure, as well as the uterotropic activity. Saturation of the indene double bond produces the indanestrol structure, and decreases the receptor binding affinity and hormonal activity of the compound. X-ray crystallographic analysis of indanestrol indicates that saturation of the indene double bond results in a molecular conformation in which the phenolic rings are no longer planar, but have become perpendicular to each other at the indanyl ring C2 carbon atom.[33] The unsaturated indenestrol IA and IB differ only in the position of the double bond, but have equivalent binding affinities which are themselves better than estradiol, and in the same range as DES. IA and IB maintain a planar ring orientation.[34] Pseudo-DES (PD) is a partially restricted structure since the double bond is now shifted to one of the side chains which limits the alkyl side chain rotation of that group. Moreover, PD can exist in E or Z isomeric forms. Studies with these isomers indicated that

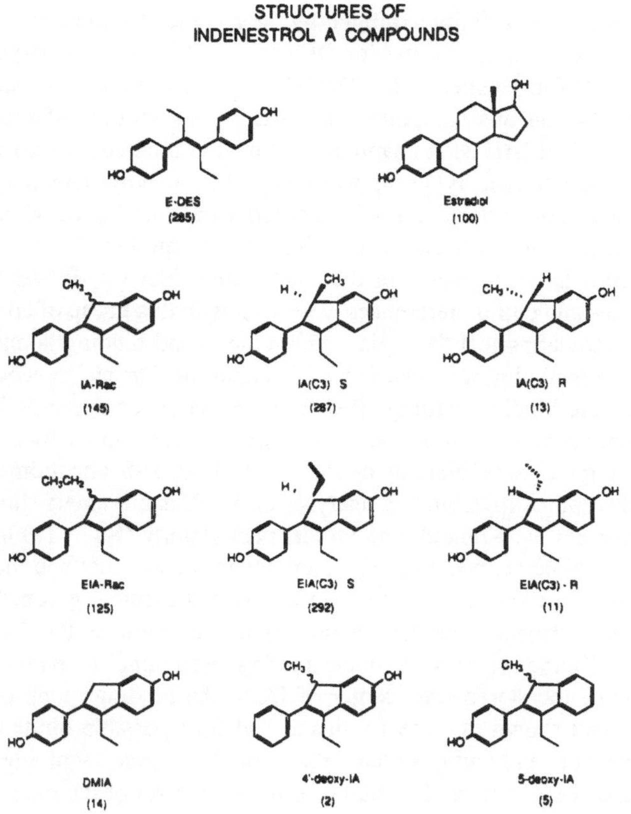

FIGURE 3. Structures and relative binding activities (RBA) of diethylstilbestrol and indenestrol A compounds. RBA values are expressed in parenthesis.

the uterine estrogen receptor binds the two PD isomers with a 15-fold difference in affinity with the Z-isomer being bound more strongly (Figure 2). Surprisingly, Z-PD has the same right angular phenolic ring structure as indanestrol.[33] The receptor binds the Z-PD isomer as well as estradiol, indicating that a planar ligand structure is not necessary for a compound to be a high-affinity ligand, although, as will be mentioned later, there is an influence on biological activity. The NMR spectrum indicates that the position of the terminal methyl group of Z-PD is similar to the side-chain orientation of E-DES.

IA and IB have unique stereogenic centers; the enantiomers of each isomer were separated and purified by chiral column HPLC.[35] Estrogen receptor binding of the individual enantiomers has been analyzed using a competitive cytosol binding assay.[36] Structures of this series of IA compounds are shown in Figure 3. The receptor demonstrated a 10- to 20-fold stereochemical preferance for the IA(C3)-S ($K_d \sim 0.37$ n*M*) enantiomer compared to IA(C3)-R

($K_d \sim 8.1$ n*M*). The affinity exhibited by the receptor for the active IA(C3)-S enantiomer was similar to that for DES ($K_d \sim 0.35$ n*M*). Removal of the methyl group to form demethyl IA (DMIA) eliminated the stereogenic center. The receptor bound this compound ($K_d \sim 3.5$ n*M*) with an affinity similar to that of the weak IA(C3)-R enantiomer. An additional compound was synthesized in which the methyl group was replaced by an ethyl substituent (EIA) in order to determine if a size limitation existed in the binding site. Competitive binding studies of the EIA compounds have been reported.[37]

EIA(C3)-S had a K_d value of 0.37 n*M*, while EIA(C3)-R was 9.0 n*M*. Similar values and chiral preferences were seen with this group of compounds as with IA, indicating that the region within the ligand-binding domain is not restricted to a methyl group. The size of this additional group, however, does not increase the binding affinity. Results with the IA compounds indicated that the stereochemical group on these stilbene compounds functioned to confer the high-affinity binding of the active IA(C3)-S enantiomer. More interestingly, competitive binding analyses of the IB enantiomers showed that both enantiomers were bound with an identical affinity ($K_d \sim 1.0$ n*M*). The receptor showed no stereochemical preference for this isomer and the change in the position of the stereogenic center and stereochemical group.[36]

Since the estrogen receptor bound the enantiomers of the IA and IB compounds differently, we were interested in determining the relationship of the position of the stereogenic center of IA to the binding mode of the IA ligand. We had shown that the IA ligand had four possible binding modes with the receptor, and had postulated that a binding mode involving the free phenyl ring of IA would be the point of primary contact of the ligand.[34] This idea was based on a structural comparison of IA and estradiol from data obtained from X-ray crystallography and is shown in Figure 4. These analyses positioned the stereogenic center of IA in close proximity to carbon 11 of the estrane steroid nucleus. To test the binding mode possibilities and the ligand structure, we synthesized IA with a single phenolic hydroxyl group on each of the different aromatic rings and evaluated the receptor binding affinity. The structures and relative binding affinities (RBA) are shown in Figure 3 and are from previous work.[37] The values presented for the two deoxy IA compounds are based on competitive binding analysis of the racemic mixtures. Recently, we have been successful in separating and purifying the individual enantiomers of these deoxy compounds in order to correlate their binding with their configurations. These data should help to further evaluate the structural orientation and stereochemical binding of this group of compounds. It is apparent from the data that the presence of both phenolic hydroxyls increased the receptor-binding affinity at least two- to six-fold. Comparison of the 4′- or 5-deoxy IA compounds with the IA racemic (RBA ~ 0.8 n*M*) demonstrated a difference of ~30- to 70-fold. Thus, the binding mode of these stilbestrol ligands indicates that a single phenolic hydroxyl will not substitute for the parent IA molecule as it relates to estrogen receptor-ligand

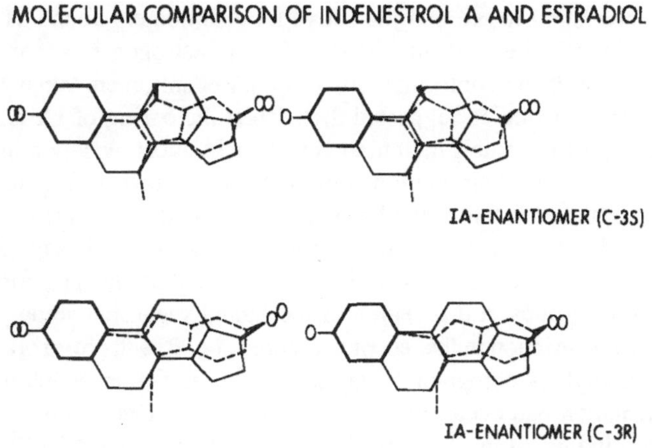

FIGURE 4. Comparison of molecular structures of estradiol and indenestrol A taken from X-ray crystallographic data. (From Duax, W. L., Swenson, D. C., Strong, P. D., Korach, K. S., McLachlan, J. A., and Metzler, M., *Mol. Pharmacol.*, 26, 520, 1984. With permission.)

binding. Furthermore, the lower affinity of 4′-deoxy IA compared to 5-deoxy IA indicates that there is a preference in the binding mode for an orientation involving the 4′-hydroxy phenyl group.

X-ray crystallographic analysis of the indanyl compounds (IA and IB) illustrated that both isomers exhibit a planar structure. Our earlier work had analyzed the molecular structure of these indanyl compounds and compared them to DES and estradiol.[34] It was obvious from those analyses that the lower receptor-binding affinity and subsequent biological activity of indanestrol was due to its nonplanar orientation. Even though PD also had a nonplanar structure, it was bound by the receptor with 60 times higher affinity. The analyses also indicated the planar nature of the indenestrol compounds (IA and IB) was due to the double bond in the indenyl ring and the possibility that the receptor could have different binding modes for these ligands. Our recent findings with the deoxy IA compounds would indicate that our earlier prediction of orientation and the ligand binding mode was correct.[34] Given this, the phenyl ring hydroxyl of IA is the initial contact of the ligand and the indanyl ring hydroxyl is the secondary contact. If this were not the case, either hydroxyl compound would be bound with equivalent affinity, contrary to what was observed. Recent work by Katzenellenbogen and associates studying hydroxylated 2,3 diarylindenes has shown consistency with our findings for the ligand structure of the indenestrol compounds.[38,39] However, other compounds having triphenyl indanyl ring structures may have a different binding orientation. Past studies with steroidal estrogens indicated that the phenolic hydroxyl was more important for receptor binding than the 17β-hydroxyl group,[40,41] even though high affinity, as judged by comparison of IA with the deoxy compounds, indicates a need for both hydroxyl groups.

It was interesting to see that the active IA enantiomers bound as well as DES, one of the highest-affinity ligands for the estrogen receptor-binding site. Comparison of the conformations of the IA enantiomers taken from X-ray crystallography studies suggested that, when the α-ring of IA is aligned with the A-ring of estradiol, the chiral center of IA most closely aligns with carbon 11 of the steroid structure as reported[34] and shown in Figure 4. The IA(C3)-S enantiomer is the active form and agrees with studies from Bouton and Raynaud,[42] who used estradiol derivatives, which indicated that the 11β-epimer of moxestrol binds with higher affinity than the 11α-form. The α-orientation of groups in that region of the ligand suggested some possible stereochemical restriction in the receptor-binding site. Results from our studies with the demethyl IA compound suggest that no restriction or inhibition of binding orientation had occurred, since removal of the methyl group did not increase the binding to the level of the high-affinity ligand.[36,37] Therefore, the methyl group orientation of IA(C3)-S enantiomer must be conferring the resulting high affinity. The space for interaction within the binding site is not critically limiting, since extending the stereochemical alkyl group from a methyl to an ethyl group resulted in the same affinity. A conformational model for a steroid receptor ligand-binding site envisioned from hydrophobic cluster analysis (HCA) and X-ray crystallographic data of known peptide sequences has been reported.[43,44] The steroid receptor HCA analysis was modeled to include a hydrophobic channel which could be the ligand-binding region. Whether this model and binding site could accomodate our proposed binding orientation and the stereochemical group on the IA will require further analyses.

Even though the indene isomers have a similar molecular structure, the estrogen receptor demonstrated no preference in the binding activity or nuclear occupancy of the IB enantiomers.[36] This indicated that orientation of the ethyl group on the stereogenic carbon was not restricted in the binding site based on our predicted indenestrol binding mode which corresponds to the carbon 7 or carbon 8 region of the steroid. Again, this suggests that the binding mode of these stilbene compounds is similar to the steroids and that the binding site may have a border or cleft which fits the upper region of the steroid nucleus in juxtaposition to carbons 11, 12, 17, and 19. It appears that steroid derivatives having groups which occupy the lower region of the steroid, such as at carbons 4 and 6, do not decrease or change activity to a great extent. This may be due to the fact that the binding site is less involved with that portion of the ligand molecule.

From our earlier analysis, it is proposed that the initial ligand contact involves the α-ring hydroxyl illustrated in the model in Figure 5A, resulting in a conformational change around the alkyl group at the chiral center of IA(C3)-S (Figure 5B and C). Structural comparisons of DES with the IA(C3)-S modeled shown in Figure 5 indicate that the DES ethyl side chain would position and function the same way,[33,34] emphasizing the fact that the receptor

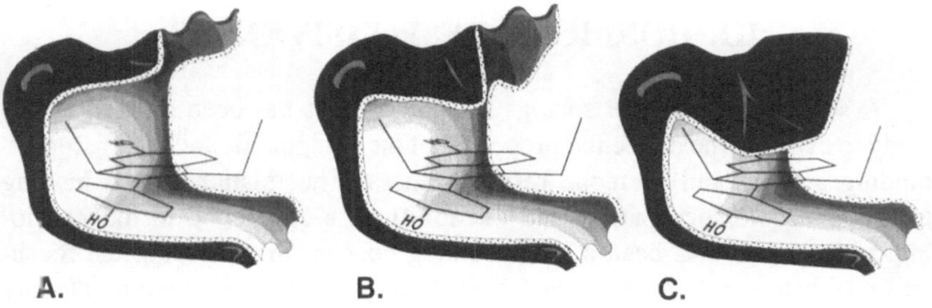

FIGURE 5. Model of ligand binding and binding site conformational change of the active IA enantiomer IA(C3)-S to the estrogen receptor. (A) Binding site with ligand positioned in open conformation. (B) Binding site and protein conformational change. (C) Ligand bound with conformational change around stereochemical group.

exhibits very high binding affinity for both of these ligands and would explain the consistency involved with this conformational change. Triphenylethylene antiestrogens could use the same binding orientation with their phenyl rings positioned like the methyl group of IA.[35] However, it is not certain whether that region of the binding site can accommodate the larger phenyl group. In fact, a recent study has suggested that compounds with a triarylethylene structure probably do not bind in the orientation shown in Figure 5, but are proposed to be turned in position compared to IA in such a way that the stereochemical group would be out of the cleft or pocket.[39] That orientation is consistent with the structure and binding of the IB isomer where no stereochemical preference exists and the IB ethyl group is positioned out of the site that contacts the IA ethyl group. IB is bound with a strong affinity (K_d ~0.8 n*M*) by the receptor. In the IB structure the methyl group, which is important for IA binding, is also in a similar conformation.[34,36]

One can only speculate on what endogenous compounds and chemical groups participate in the stereochemical ligand binding. Studies of steroid ligand structure would suggest that this region of the ER is probably recognizing the C-19 methyl and C-17 hydroxyl of estradiol which influence receptor binding, conformational changes, and hormonal activity. Estrogen receptor binding of 17β-estradiol is at least ten times greater than the affinity for 17α-estradiol or 16α-estradiol.[45,46] Most of the lower affinity is due to increased dissocation rates of the ligand which is consistent with it not possessing the proper functional group or orientation to allow the conformational change proposed in Figure 5. Changes in receptor conformations are believed to be required for the increased transcriptional and biological activity of the protein. We are presently evaluating the stereochemistry of the deoxy IA compounds and synthesizing compounds with larger alkyl groups in order to identify and probe the size and position of this binding-site region.

III. HORMONE RESPONSIVENESS

A selection of stilbene analogs and metabolites has been utilized in this study to evaluate the molecular properties of the compounds and their receptor-binding activity and hormone responsiveness. The relationship of binding affinity, receptor occupancy, and the ability of a compound to induce biological responses has been a long-standing concept in describing the mechanism of estrogen hormone action.[47] Compounds with weak binding affinity produced minimal or transient nuclear receptor complexes resulting in poor biological response. This observation is consistent with past reports that studied the activity of DMS[17] and its steroidol counterparts, estriol[48] or 16α-estradiol.[49] However, the poor uterotropic activity and biological responsiveness with pseudo-DES and the indenestrol compounds are inconsistent with this relationship. Retention or occupancy is not insufficient, since at certain doses PD, IA, and IB maintained nuclear estrogen receptor levels comparable to DES even though biological activity was not stimulated. Cell transfection response assays with estrogen receptor and estrogen-responsive DNA sequences have confirmed the requirement of the hormone to stimulate estrogen receptor-mediated gene transcription.[50,51] Thus, the presence of the receptor protein within the nucleus is insufficient by itself for hormonal stimulation, but probably requires a ligand with proper molecular structure to produce a receptor protein conformation which results in responsiveness.

In earlier studies, it was shown that the binding activity of a ligand for the receptor correlated with nuclear receptor occupancy and resulted in the expression of various tissue responses. Nuclear estrogen receptor levels and their degree of retention in the nucleus have been suggested as a correlative to hormone responsiveness.[52] It might therefore be possible that some of these compounds would not produce similar nuclear receptor levels or retention as the potent synthetic estrogen DES. Such an experimental approach investigating these receptor levels was used with steroidal compounds such as estriol or 16α-estradiol, illustrating poor nuclear receptor retention resulting in weak and incomplete uterine stimulation.[31,48,49] Results of these types of analyses from earlier studies,[32] are shown in Figure 6 and illustrate the receptor distribution pattern after a single 20-μg/kg dose of DES and the three principal stilbestrol compounds with the highest binding affinity. Comparison of the total nuclear receptor levels shows that the pattern of racemic IA, racemic IB, and Z-PD is essentially the same as that of DES. These compounds retain as much total receptor in the nucleus as DES, but are not active biologically.[32] Nuclear estrogen receptor has been characterized biochemically to include two forms, 0.4 *M* KCl salt extractable and resistant receptor forms. Levels of salt-resistant receptor sites were proposed as the best correlative for stimulation of hormonal responses.[52] Recent findings with the stilbestrol compounds shows that racemic mixtures of IA and IB produce similar levels of nuclear salt-resistant estrogen receptor sites as does DES; while Z-PD produces

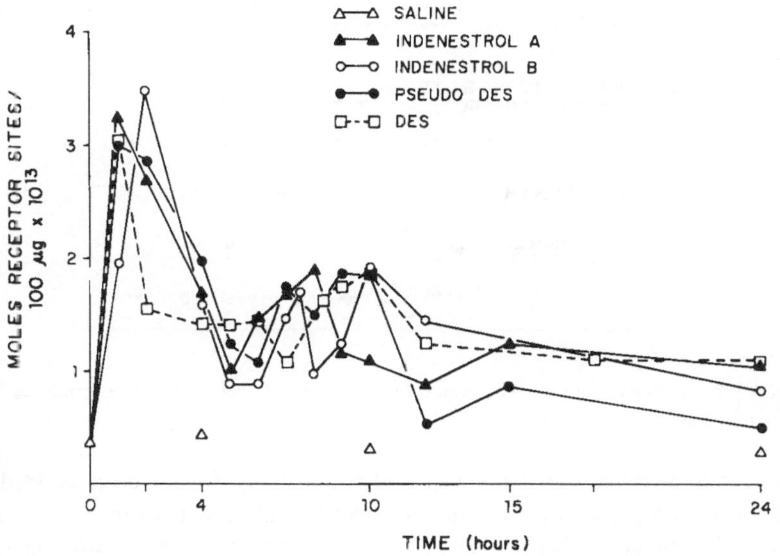

FIGURE 6. Nuclear occupancy pattern after a single 20-μg/kg dose i.p. injection at zero time of stilbestrol compounds. Nuclear estrogen receptor levels were measured as noted in experimental section and as previously reported.[3]

very low levels.[53] One interpretation of these data is that the nonplanar structure of Z-PD, mentioned earlier, results in a ligand receptor conformation that is incapable of producing the tight nuclear binding form.

Nuclear receptor levels were also sensitive to the stereochemistry of the indenestrol enantiomers. We showed that the differences in ligand binding of the IA and IB enantiomers were matched further by differences in nuclear receptor levels.[36,54] The activity of the IA racemic was essentially due to the active enantiomer IA(C3)-S while the weak IA enantiomer IA(C3)-R did not contribute to nuclear receptor levels or occupancy at low doses. IB enantiomers, which were not bound stereospecifically, produced comparable levels of nuclear estrogen receptor. At the present time, analyses of the salt-resistant nuclear sites with the individual enantiomers have not been performed. There are several uterine responses linked to estrogen stimulation involving nuclear eceptor genomic interactions including stimulation of progesterone receptor (PR), glucose 6-phosphate dehydrogenase (G-6-PD), DNA synthesis, and ornithine decarboxylase (ODC). As mentioned earlier, we have investigated the stimulation of several of these responses by the stilbestrol compounds,[54] summarized in Figure 7 as relative levels of activity.

It should be pointed out that the responsiveness studies that were investigated utilized racemic mixtures of IA and IB and were based on doses of compound, which as previously shown,[28,29] resulted in equivalent levels of total nuclear estrogen receptor. Measurement of the individual uterine re-

HORMONAL ACTIVITY OF DES COMPOUNDS

	E2	DES	IA	IB	Z-PD
GLUCOSE-6-PHOSPHATE DEHYDROGENASE	++	+	+	−	−
PROGESTERONE RECEPTOR	++	+	+	+	+
DNA SYNTHESIS	+	+	−	+	−
ORNITHINE DECARBOXYLASE ACTIVITY	+	++	±	±	−
ORNITHINE DECARBOXYLASE mRNA	+	++	−	+	−

FIGURE 7. Survery of the level of stimulation of different hormonal responses produced by the various stilbestrol compounds.

sponses indicated that stimulation occurred differentially.[32] It has been reported[47] that weak estrogens stimulate uterine biochemical responses to a minimal degree. Presumably, this is due to the lower binding affinity resulting in dissociation of the ligand to an unbound receptor form which would have minimal nuclear occupancy and would be biologically inactive. There, it was expected that these DES compounds would demonstrate the same pattern of activity consistent with their poor uterotropic activity.

ODC has been shown to correlate with estrogen stimulation in target tissues. This enzyme controls the rate-limiting step in polyamine biosynthesis.[55] After a single treatment activity is stimulated maximally by 16 h in the rat uterus[56] and mouse uterus. Results of such experiments using the stilbestrol compounds are summarized in Figure 7 where maximal activity with DES is increased to 600 times that of control. Enzymatic activity is not stimulated significantly by PD at this dose, yet it is slightly above control at 20 μg/kg with racemic IA. Higher IA and IB doses resulted in an effect that was only twice control and never reached the level found with 10 to 20 μg/kg DES doses. When the ODC activity was checked with the IA enantiomers there were some similarities to the DNA synthesis activity, but also some differences as reported earlier.[37] With this ODC biological response, racemic IA produced a significantly lower activity, like that seen for DNA synthesis. However, the partial activity of the racemic IA was completely mimicked by the active IA(C3)-S enantiomer, which was not the pattern seen with the DNA synthesis response. The weak IA(C3)-R enantiomer produced no stimulation. ODC activity was increased only slightly by any of the DES compounds. Studies in mouse kidney have indicated an increase in ODC messenger RNA levels by androgen hormones.[57] It is uncertain whether the results seen in our study are due to an effect at a transcriptional or translational level. We are presently investigating, with the use of an ODC cDNA probe, whether the DES compounds are effecting transcriptional regulation of mRNA levels.

Progesterone receptor, DNA synthesis, and growth of estrogen target tissues and cells are examples of responses reported to occur in a coordinated manner.[58] Our findings have indicated that these estrogen responses are perhaps regulated independently. Induction of soluble progesterone receptor has been shown to be relatively estrogen specific in target tissues such as the uterus or mammary gland.[47] Results summarized in Figure 7 were based on saturation levels from Scatchard plots and binding affinity ranged from 2 to 4 nm with no significant differences observed between groups. PD produced stimulation of progesterone receptor levels similar to those of DES, but produced weak DNA synthesis activity. IA also showed this same property of dissociating stimulation of progesterone receptor from DNA synthesis. A similar observation has been made with estrogen responsive cells in culture.[59] Cheng et al.[60] found discordance of RNA, protein, and DNA synthesis in uterine luminal epithelial cells following estriol and estradiol treatment. They concluded that estrogen stimulation of uterine tissue may occur by regulation of two independent pathways. Our results could be interpreted similarly, particularly in the case where protein synthesis is stimulated without DNA synthesis. Recent reports of progesterone receptor stimulation by growth factors in addition to estrogen receptor[61] have suggested that some hormone responses may have multifactorial regulation. The stilbestrol compounds, viewed as estrogen receptor agonists, may indicate additional sites of action of the compounds and for the estrogen receptor previously not considered.

Stimulation of G-6-PD has been used by many laboratories as a marker for estrogen action.[62,63] Results summarized in Figure 7 show estradiol was the most active compound tested[32] while IA was similar to DES but slightly less active than estradiol (180%). None of the other compounds produced activity above the control group. Stimulation by PD was as low as IB, but both of these gave significant nuclear receptor occupancy. Expanded dose response studies have confirmed this difference between IA and IB isomers.[32] There was approximately a doubling of levels of PR after a single injection which was significantly different than the G-6-PD activity where IB and PD were not active in stimulating that response.

A number of studies with a mammary tumor cell line (MCF-7) have correlated progesterone receptor stimulation with estrogen-responsive cell growth. DNA synthesis and mitotic activity of the DES compounds were evaluated to see if they correlated with the progesterone receptor levels. The stimulation of uterine DNA synthesis is summarized in Figure 7 for the 20-μg/kg dose. Experiments investigating the time course of DNA synthesis illustrated that the poor activity of the DES compounds was not due to a shift in the response time.[28,32] DES was the most active, while IB showed some increase in activity and produced comparable stimulation at a five-times higher dose. Interestingly, IA showed a poorer stimulation and slight decrease at higher doses. Most importantly, the two unsaturated isomers (IA and IB) showed very different results which were accentuated at higher doses of 50 to 100 μg/kg.[32] The saturated isomer indanestrol showed very weak uterine activity in all responses investigated which is partly explained by the lower receptor binding affinity. PD produced activity only slightly above control values; although it is bound by the receptor with a high affinity comparable to estradiol.

Considering the receptor-binding differences seen with the IA enantiomers, uterine DNA synthesis was measured after treatment with different IA enantiomers to investigate whether this also resulted in different stimulation in biological responses.[37] Activity with a comparable 20-μg/kg dose of IA racemic produced approximately 65% of the response of DES. The inactive IA(C3)-R enantiomer produced a weaker response. Surprisingly, the active IA(C3)-S enantiomer had only marginally better stimulation which did not equal that of the racemic IA. This was in contrast to the activity of the enantiomers as it related to nuclear receptor levels where the IA(C3)-S enantiomer was responsible for all of the activity of the IA racemic. Demethyl indenestrol A (DMIA) was similar to the IA(C3)-R enantiomer with weaker receptor-binding activity as well as poor stimulation of DNA synthesis.

Recent studies using cell transfection procedures of cDNA for human estrogen receptor[50] and estrogen-responsive reporter gene constructs indicated that the transcriptional activity of the receptor required binding of its cognate hormone. Some additional steps in this mechanism involve receptor dimerization[51] and binding to DNA responsive elements (ERE).[64] Thus, the differential hormonal responsiveness of the stilbestrol compounds may be due to an effect of ligand structure on the conformation of the estrogen receptor protein which influences its ability to bind to DNA. A report from Parker and associates[65] has suggested receptor dimerization as the step in the estrogen mechanism where the antiestrogen ICI 164,384 was effective. In an attempt to test this hypothesis, we measured the binding of various stilbestrol receptor ligand complexes to a vitellogenin A_2 estrogen responsive element (ERE) known to regulate estrogen-responsive gene expression. The analysis used the oligonucleotide band shift technique[64] with stilbestrol estrogen nuclear receptor preparations. Three specific bands have been shown previously to contain the estrogen receptor protein by estrogen receptor antibody binding.[64] All three specific bands and one nonspecific band (D) are formed after incubating nuclear estrogen receptor stilbestrol estrogen extracts with a ^{32}P-labeled 56-bp vitellogenin A_2 ERE (Figure 8). This technique showed no detectable difference in the activity of different ligand bound receptor proteins to bind this DNA sequence.[64] Use of imperfect palidromic ERE sequences from the PS2 regulatory region[66,67] have resulted in similar findings indicating no differences.[68]

Our studies are consistent with the idea that the estrogen receptor binding site demonstrates stereochemical recognition. Moreover, the biological response data described earlier are apparently due to the activity of the one IA enantiomer. Notably, the structure of this enantiomer allows it to bind and

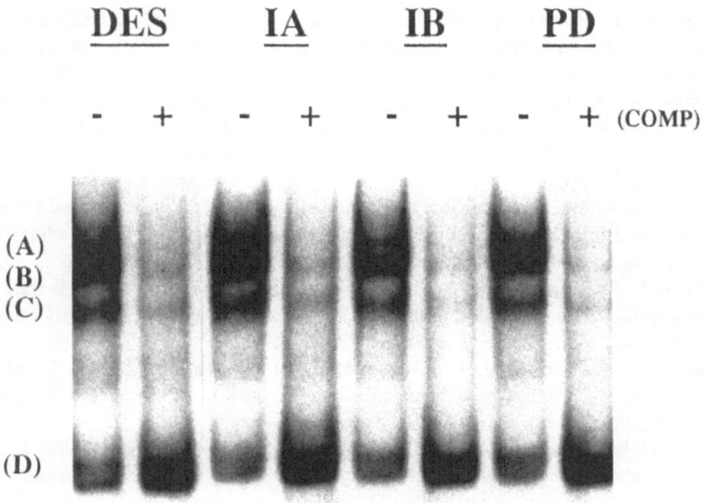

FIGURE 8. Measurement of ^{32}P-labeled oligonucleotide nuclear estrogen receptor complexes. Nuclear receptor preparations were produced 1 h after treatment with 20-μg/kg dose of the stilbestrol compounds. Procedures are described and outlined in Reference 64.

interact with the estrogen receptor as well as DES, but comparable biological activity does not result. This suggests that the ligand structure may play a more significant role in hormone stimulation than simply to occupy or activate the estrogen receptor to bind DNA specific sequences. Further work with additional ligands is under way in an attempt to model the estrogen receptor-binding site and evaluate additional steps in the cellular mechanism of estrogen receptor action.

ACKNOWLEDGMENTS

The author would like to thank the following colleagues and collaborators who have helped with these studies for the past several years: Kun Chae, Pamela Sarver, Sylvia Curtis, Louis Levy, Manfred Metzler, and Bill Duax.

REFERENCES

1. **Dodds, E. C., Goldberg, L., Lawson, W., and Robinson, R.,** Synthetic oestrogenic compounds related to stilbene and diphenylethane, *Proc. R. Soc. London Ser. B*, 127, 140, 1939.
2. **Dodds, E. C., Golberg, L., Lawson, W., and Robinson, R.,** Oestrogenic activity of certain synthetic compounds, *Nature*, 141, 247, 1938.
3. **Campbell, N. R., Dodds, E. C., and Lawson W.,** Nature of the estrogenic substances produced during the demethylation of anethol, *Proc. R. Soc. London Ser. B*, 128, 253, 1940.

4. **Emmens, C. W. and Carr, W. I.,** Further studies of compounds exhibiting prolonged antiestrogenic and antifertility activity in the mouse, *J. Reprod. Fert*, 34, 29, 1973.
5. **Emmens, C. W., Collins, D. J., Hobbs, J. J., Miller, B. G., and Owen, W. J.,** Antiestrogenic and antifertility properties of some basic ether derivatives of dimethylstilbestrol and related compounds, *J. Pharmacol. Exp. Ther.*, 165, 52, 1969.
6. **Martin, L.,** Dimethylstilbestrol and 16-oxo-estradiol: antiestrogens or estrogens?, *Steroids*, 13, 1, 1969.
7. **Hartman, R. W., Schwarz, W., and Schonenberger, H.,** Ring substituted 1,2-dialkylated 1,2-bis(hydroxyphenyl) ethanes. Synthesis and estrogen receptor binding affinity of 2,2'- and 3,3'-disubstituted hexestrols, *J. Med. Chem.*, 26, 1137, 1983.
8. **Pento, J. T., Magarian, R. A., Wright, R. J., King, M. M., and Benjamin, E. J.,** Nonsteroidal estrogens and antiestrogens: biological activity of cyclopropyl analogs of stilbene and stilbenediol, *J. Pharm. Sci.*, 70, 399, 1981.
9. **Dodds, E. C., Goldberg, L., Grunfeld, E. I., Lawson, W., Saffer, C. M., and Robinson, R.,** Synthetic estrogenic compounds related to stilbene and diphenylethane II., *Proc. R. Soc. London*, 131, 83, 1943.
10. **Clark, J. H., Guthrie, S. C., McCormack, S. A., and Markaverich, B. M.,** Agonistic and antagonistic properties of clomiphene, in *Hormone Antagonists*, Agarwal, M. K., Ed., Walter deGruyter, New York, 1982, 89.
11. **von Angerer, E., Prekajac, J., and Strohmeier, J.,** 2-Phenylindoles. Relationship between structure, estrogen receptor affinity and mammary tumor inhibiting activity in the rat, *J. Med. Chem.*, 27, 1439, 1984.
12. **Kranzfelder, G., Hartmann, R. W., von Angerer, E., Schonenberger, H., and Bogden, A. E.,** 3,4-bis (3'-Hydroxyphenyl) hexane. A new mammary tumor inhibiting compound, *J. Cancer Res. Clin. Oncol.*, 103, 165, 1982.
13. **Black, L. J.,** Biological actions and binding properties of a new estrogen antagonist, LY117018, in *Hormone Antagonists*, Agarwal, M. K., Ed., Walter deGruyter, New York, 1982, 109.
14. **Jordan, V. C., Lieberman, M. E., Cormier, E., Koch, R., Bagley, J. R., and Ruentiz, P. C.,** Structural requirements for the pharmacological activity of nonsteroidal antiestrogens *in vitro*, *Mol. Pharmacol.*, 26, 272, 1984.
15. **Jordan, V. C.,** Biochemical pharmacology of antiestrogen action, *Pharmacol. Rev.*, 36, 245, 1984.
16. **Keasling, H. H. and Schueler, F. W.,** The relationship between estrogenic action and chemical constitution in a group of azomethine derivatives, *J. Am. Pharm. Assoc.*, 39, 87, 1950.
17. **Katzenellenbogen, B. S., Iwamoto, H. S., Heiman, D. F., Lan, N. C., and Katzenellenbogen, J. A.,** Stilbestrols and stilbestrol derivatives: estrogenic potency and temporal relationships between estrogen receptor binding and uterine growth, *Mol. Cell. Endocrinol.*, 10, 103, 1978.
18. **Busetta, B., Courseille, C., and Hospital, M.,** Structure cristalline et moleculaire de lœstradiol hemihydrate, *Acta Cryst.*, B28, 560, 1971.
19. **Busetta, B., Courseille, C., and Hospital, M.,** Solvatees du diethylstilbestrol, *Acta Cryst.*, B29, 2456, 1973.
20. **Darden, T., McKinney, J. D., Gottschalk, K., Maynard, A. T., and Pedersen, L.,** A theoretical study of the minimum energy structures of diethylstilbestrol and its analogs by molecular mechanics (MM2P), MNDO and ab initio calculations, *J. Am. Chem. Soc.*, 108, 560, 1986.
21. **Gawienowski, A. M., Knoche, H. W., and Moser, H. C.,** The metabolism of ^{14}C and ^{3}H-labeled diethylstilbestrol in the rat, *Biochem. Biophys. Acta*, 65, 150, 1962.
22. **Gottschlich, R. and Metzler, M.,** Metabolic fate of diethylstilbestrol in the syrian hamster, a susceptible species for diethylstilbestrol carcinogenicity, *Xenobiotica*, 10, 317, 1979.

23. **Metzler, M.,** Metabolic activation of diethylstilbestrol: indirect evidence for the formation of a stilbene oxide intermediate in hamster and rat, *Biochem. Pharmacol.*, 24, 1449, 1975.
24. **Metzler, M.,** The metabolism of diethylstilbestrol, *CRC Critical Rev. Biochem.*, 10, 171, 1981.
25. **Metzler, M.,** Metabolic activation of carcinogenic diethylstilbestrol in rodents and humans, *J. Toxicol. Environ. Health*, Suppl. 1, 21, 1976.
26. **Dodds, E. C., Lawson, W., and Noble, R. L.,** Biological effects of the synthetic oestrogenic substance 4,4′ dihydroxy-α,β diethyl stilbene, *Lancet*, 1, 1389, 1938.
27. **Korach, K. S., Metzler, M., and McLachlan, J. A.,** Estrogenic activity *in vivo* and *in vitro* of some diethylstilbestrol metabolites and analogs, *Proc. Natl. Acad. Sci U.S.A.*, 75, 468, 1978.
28. **Korach, K. S.,** Biochemical and estrogenic activity of some diethylstilbestrol metabolites and analogs in the mouse uterus, in *Hormones and Cancer*, Leavitt, E., Ed., Plenum Press, New York, 1982, 39.
29. **Korach, K. S., Metzler, M., and McLachlan, J. A.,** Diethylstilbestrol metabolites and analogs: new probes for the study of hormone action, *J. Biol. Chem.*, 254, 8963, 1979.
30. **Korach, K. S., Fox-Davies, C., and Baker, V.,** Different response to estriol and estradiol in the mouse uterus: correlation to an additional nuclear event, *Endocrinology*, 106, 1900, 1980.
31 **Lan, N. C. and Katzellenbogen, B. S.,** Temporal relationships between hormone receptor binding and biological responses in the uterus: studies with short and long acting derivatives of estriol, *Endocrinology*, 98, 220, 1976.
32. **Korach, K. S., Fox-Davis, C., Quarmby, V. E., and Swaisgood, M. H.,** Diethylstilbestrol metabolites and analogs: biochemical probes for differential stimulation of uterine estrogen responses, *J. Biol. Chem.*, 261, 15420, 1985.
33. **Duax, W. L., Griffin, J. L., Weeks, C. M., and Korach, K. S.,** Molecular conformation, receptor binding and hormone action of natural and synthetic estrogens and antiestrogens, *Environ. Health Perspect.*, 61:111, 1985.
34. **Duax, W. L., Swenson, D. C., Strong, P. D., Korach, K. S., McLachlan, J. A., and Metzler, M.,** Molecular structures of metabolites and analogues of diethylstilbestrol and their relationship to receptor binding and biological activity, *Mol. Pharmacol.*, 26, 520, 1984.
35. **Chae, K., Levy, L. A., and Korach, K. S.,** Chromatographic separation and isolation of enantiomeric form of diethylstilbestrol metabolites, *J. Chromatogr.*, 439, 484, 1988.
36. **Korach, K. S., Chae, K., Levy, L. A., Duax, W. L., and Sarver, P. J.,** Diethylstilbestrol metabolites and analogs: stereochemical probes for the estrogen receptor binding site, *J. Biol. Chem.*, 264, 5642, 1989.
37. **Korach, K. S., Chae, K., Gibson, M., and Curtis, S.,** Estrogen receptor stereochemistry: ligand binding and hormonal responsiveness, *Steroids*, 56, 263, 1991.
38. **Anstead, G. M., Peterson, C. S., and Katzenellenbogen, J. A.,** Hydroxylated 2,3 diarylindenes: synthesis estrogen receptor binding activity and binding orientation considerations, *J. Steroid Biochem.*, 33, 877, 1989.
39. **Anstead, G. M., Wilson, S. R., and Katzenellenbogen, J. A.,** 2-Aryindenes and 2-arylindenones: molecular structures and considerations in the binding orientations of unsymetrical nonsteroidal ligands to the estrogen receptor, *J. Med. Chem.*, 32, 2163, 1989.
40. **Franchenko, N. D., Stutchak, S. V., Shchedrina, R. N., Pivitisky, K. K., Novikou, E. A., and Ishkov, V. L.,** The specificity of the human uterine receptor, *Acta Endocrinol.*, 90, 167, 1979.
41. **Chernyaev, G. A., Barkova, T. I., Egorova, V., Sorokina, I. B., Anachenko, S. N., Matarodze, G. D., Sokolova, N. A., and Rozen, V. B.,** A series of optical structural and isometic analogues of estradiol: a comparative study of the biological activity and affinity to cytosol receptor of rabbit uterus, *J. Steroid Biochem.*, 6, 1483, 1975.

42. **Bouton, M. M. and Raynaud, J. P.,** The relevance of interaction kinetics in determining biological response to estrogens, *Endocrinology,* 105, 509, 1979.
43. **Ojasoo, T., Raynaud, J.-P., and Mornon, J.-P.,** Steroid hormone receptors, in *Comprehensive Medicinal Chemistry,* Vol. 3, Emmett, J. C., Ed., Pergamon Press, Elmsford, NY, 1990, 1175.
44. **Mornon, J. P., Bissey, V., Gaboriaud, C., Thomas, A., Ojasoo, T., and Raynaud, J.-P.,** Hydrophobic cluster analysis (HCA) of the hormone-binding domain of receptor proteins, *J. Steroid Biochem.,* 34, 355, 1989.
45. **Stack, G. and Gorski, J.,** Relationship of estrogen receptors and protein synthesis to the mitogenic effect of estrogens, *Endocrinology,* 117, 2024, 1985.
46. **Axelson, M., Clark, J. H., Eriksson, H. A., and Sjovall, J.,** Estrogen binding in target tissues; a GC/MS method for assessing uptake, retention and processing of estrogens in target cell nuclei under *in vivo* conditions, *J. Steroid Biochem.,* 14, 1253, 1981.
47. **Clark, J. H. and Peck, E. J., Jr.,** *Female Sex Steroid Receptors and Function,* Springer-Verlag, New York, 1979.
48. **Clark, J. H., Paszko, Z., and Peck, E. J., Jr.,** Nuclear binding and retention of the receptor estrogen complex: relation to agonist and antagonist properties of estriol, *Endocrinology,* 100, 91, 1977.
49. **Stack, G. and Gorski, J.,** Estrogen-stimulated deoxyribonucleic acid synthesis: a ratchet model for the prereplicative period, *Endocrinology,* 117, 2017, 1985.
50. **Green, S. and Chambon, P.,** Nuclear receptors enhance our understanding of transcription regulation, *Trends Genet.,* 4, 309, 1988.
51. **Kumar, V. and Chambon, P.,** The estrogen receptor binds tightly to its responsive element as a ligand-induced homodimer, *Cell,* 55, 145, 1988.
52. **Clark, J. H. and Peck, E. J.,** Nuclear retention of receptor-oestrogen complex and nuclear acceptor sites, *Nature,* 260, 635, 1976.
53. **Metzger, D. A., Curtis, S., and Korach, K. S.,** Diethylstilbestrol metabolites and analogs: differential ligand effects on estrogen receptor interactions with nuclear matrix sites, *Endocrinology,* 128, 1785, 1991.
54. **Korach, K. S., Levy, L. A., and Sarver, P. J.,** Stereochemical analysis of stilbene estrogens: receptor binding and hormone responsiveness, in *Estrogen in the Environment II. Influences on Development,* McLachlan, I. A., Ed., Elsevier, New York, 1985, 43.
55. **Tabor, C. W. and Tabor, H.,** 1,4-Diaminobutane (putrescine), spermidine and spermine, *Annu. Rev. Biochem.,* 45, 285, 1976.
56. **Rorke, E. A., Kendra, K. L., and Katzenellenbogen, B. S.,** Relationship among uterine growth, ornithine decarboxylase activity and polyamine levels: studies with estradiol and antiestrogens, *Mol. Cell. Endocrinol.,* 38, 31, 1984.
57. **Janne, O. A., Kontula, K. K., Isomaa, V. V., Torkkeil, T. K., and Bardin, C. W.,** Androgen receptor-dependent regulation of ornithine decarboxylase gene expression in mouse kidney, in *Steroid Hormone Receptors: Structure and Function,* Eriksson, H. and Gustafsson, J.-A., Eds., Elsevier Science, New York, 1983, 461.
58. **Horwitz, K. B. and McGuire, W. L.,** Estrogen control of progesterone receptor in human breast cancer, *J. Biol. Chem.,* 253, 2223, 1978.
59. **Eckert, R. L. and Katzenellenbogen, B. S.,** Effects of estrogens and antiestrogens on estrogen receptor dynamics and the induction of progesterone receptor in MCF-7 human breast cancer cells, *Cancer Res.,* 42, 139, 1982.
60. **Cheng, S. V. Y., MacDonald, B. S., Clark, B. F., and Pollard, J. W.,** Cell growth and cell proliferation may be dissociated in the mouse uterine luminal epithelium treated with female sex steroids, *Exp. Cell Res.,* 160, 459, 1985.
61. **Katzenellenbogen, B. S. and Norman, M. J.,** Multihormonal regulation of the progesterone receptor in MCF-7 human breast cancer cells: interrelationships among insulin/insulin-like growth factor-I, serum and estrogen, *Endocrinology,* 126, 891, 1990.

62. **Moulton, B. C. and Barker, K. L.,** Synthesis and degradation of glucose-6 phosphate dehydrogenase in the rat uterus, *Endocrinology,* 89, 1131, 1971.
63. **Barker, K. L., Adams, D. J., and Donohue, T. M.,** Regulation of the levels of mRNA of glucose-6 phosphate dehydrogenase and its rate of translation in the uterus by estradiol, in *Cellular and Molecular Aspects of Implanatation,* Glasser, S. R. and Bullock, D. W., Eds., Plenum Press, New York, 1981, 269.
64. **Curtis, S. W. and Korach, K. S.,** Uterine estrogen receptor interaction with estrogen responsive DNA sequences *in vitro:* effects of ligand binding on receptor-DNA complexes, *Mol. Endocrinol.,* 4, 276, 1990.
65. **Fawell, S. E., White, R., Hoare, S., Sydenham, M., Page, M., and Parker, M. G.,** Inhibition of estrogen receptor-DNA binding by the "pure" antiestrogen ICI 164,384 appears to be mediated by impaired receptor dimerization, *Proc. Natl. Acad. Sci. U.S.A.,* 87, 6883, 1990.
66. **Nunez, A.-M., Berry, M., Imler, J.-L., and Chambon, P.,** The 5′ flanking region of the pS2 gene contains a complex enhancer region responsive to oestrogens, epidermal growth factor, a tumor promoter (TPA), the c-Ha-ras oncoprotein and the c-jun protein, *EMBO J.,* 8, 823, 1989.
67. **Berry, M., Nunez, A.-M., and Chambon, P.,** Estrogen-responsive element of the human pS2 gene is an imperfectly palidromic sequence, *Proc. Natl. Acad. Sci. U.S.A.,* 86, 1218, 1989.
68. **Curtis, S. and Korach, K. S.,** Uterine estrogen receptor-DNA complexes: effect of different ERE sequences, ligands and receptor forms, *Mol. Endocrinol.,* 5, 959, 1991.

Chapter 6

AROMATASE: MECHANISM AND INHIBITION

Philip A. Cole and Cecil H. Robinson

TABLE OF CONTENTS

I. INTRODUCTION

Estrogens play an important role in the growth and differentiation of a variety of mammalian tissues. In recent years, great strides have been made in the understanding of their mechanism of action. The estrogen receptor has been isolated and purified. It is now generally acknowledged that the receptor lies in the cell nucleus, and that upon binding estrogen steroid hormone it can interact productively with the DNA transcriptional apparatus to regulate the synthesis of messenger RNA of select genes.[1]

Of particular current medical concern is the concept that estrogen can serve as a positive regulator of certain neoplastic tissue. The historic studies of Beatson[2] in 1896 first demonstrated that oophorectomy could lead to breast cancer remission in women. Breast cancer is currently one of the two leading causes of cancer deaths in women in the U.S. Of the approximately 140,000 cases of breast cancer diagnosed each year in this country, about one third are responsive to anti-estrogen therapy.

The application of estrogen receptor studies to neoplastic tissues has helped predict the hormone therapy responsiveness among breast cancer patients. About two thirds of all primary breast cancer patients' tumors are estrogen receptor positive (i.e., they contain significant quantities of receptor based on *in vitro* assays). Of those patients who are positive, about 50 to 70% will show an objective response to hormone therapy. In contrast, estrogen-receptor negative cancer will show an objective response in less than 10% of cases.[3a]

One can envision that there are many levels at which therapeutic intervention could lead to the inhibition of estrogen action. For practical reasons, most of the research effort has been concentrated on two approaches. The first is the development of estrogen receptor antagonists. The stilbene derivative, tamoxifen, is currently the most widely used drug of this type. The second approach has been directed at lowering the serum levels of estrogen. Originally, this was accomplished by surgical ablation therapy, oophorectomy as discussed above, as well as adrenalectomy and hypophysectomy. Due to the rather harsh nature of these surgical therapies, alternative medical approaches have been sought for the reduction of estrogen levels. In particular, inhibition of estrogen biosynthesis has been of continued interest. Aromatase is the enzyme complex which catalyzes the final step in estrogen production in humans (the conversion of steroidal androgens 1 to estrogens 2, see Scheme 1), and therefore it has been the principal target. The remainder of this chapter will be devoted to a discussion of aromatase mechanism and inhibition.[3] Special attention will be paid to the interrelationship between steroidal structure and chemical/biological reactivity, an interrelationship which has been and will continue to be crucial to inhibitor design and mechanistic understanding.

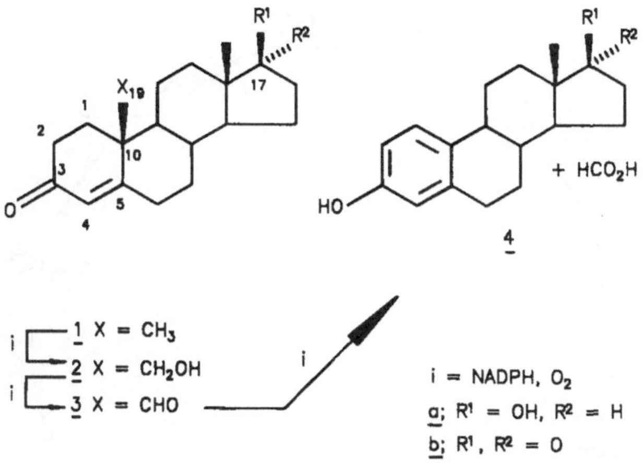

SCHEME 1.

II. AROMATASE MECHANISM

Before discussing aromatase inhibition, it seems important to outline current knowledge of aromatase mechanism. Besides the benefits to rational drug design, interest in the elucidation of the mechanism of aromatase has come from the chemical novelty of the androgen to estrogen conversion. Human placental aromatase, the most extensively studied form, is a cytochrome P-450 enzyme that functions in association with an NADPH-dependent reductase. Like all microsomal cytochrome P-450 enzyme systems, aromatase consists of a specific heme protein (P-450) and a generic flavoprotein NADPH-dependent reductase. The name cytochrome P-450 comes from the fact that, when these heme proteins are in their reduced (Fe^{+2}) form, their complex with carbon monoxide leads to a UV-absorbing Soret band maximum at about 450 nm. Much has been learned about the chemistry of cytochrome P-450s, primarily from mechanistic studies of bacterial camphor hydroxylase,[4] liver microsomal detoxifying enzymes,[5] and porphyrin model systems.[6] From these studies, and because the NADPH-dependent reductases interact non-specifically with different microsomal P-450s (liver-detoxifying enzymes, sterol, and steroid biosynthetic enzymes), while members of the P-450 family show amino acid sequence and presumed structural homology,[7] a general scheme for the mechanism of cytochrome P-450 is thought to hold. A widely accepted version of the hydroxylation reaction is shown (Scheme 2).[4a,8] Other types of reactions (e.g., epoxidations) presumably follow the initial stages of this pathway but involve modifications in the later steps.

The "resting state" of the enzyme (in the absence of substrate) has a hexacoordinate heme iron (III) in the low spin state (one unpaired electron).

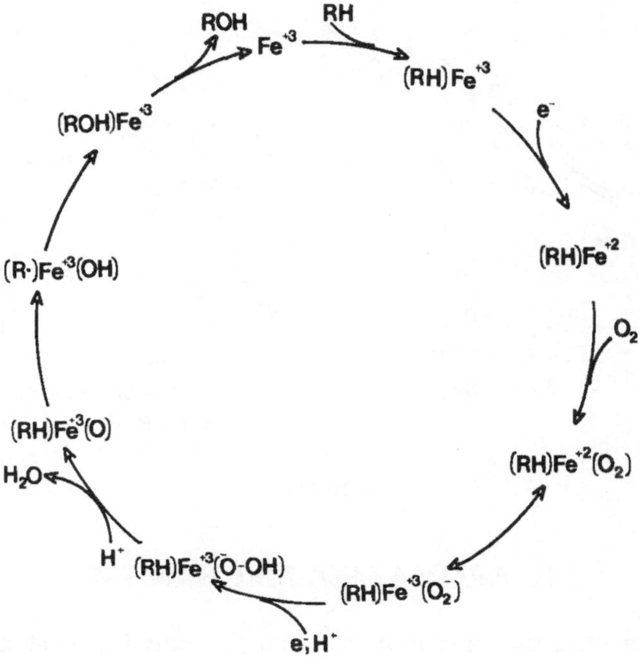

SCHEME 2.

Ligands of the iron include four nitrogens from the porphyrin, one thiolate from a well-conserved cysteine residue, and one water molecule. Upon substrate binding, the iron becomes pentacoordinate — expelling the sixth ligand water molecule (presumably for steric reasons). The heme iron also becomes high spin (five unpaired electrons) and this change is accompanied by a characteristic UV-absorption spectrum change of the Soret band maximum from about 420 nm to about 390 nm. In the presence of NADPH and reductase, the iron is then reduced by one electron and formally is Fe^{+2}. A molecule of oxygen is then rapidly accepted into the sixth coordination site. This O_2-Fe interaction has been formulated as involving either unreduced oxygen (with Fe^{+2}) or superoxide (with Fe^{+3}), as the available evidence can support either picture. Next, believed to be the rate-limiting step for the cycle, a second electron is passed to the iron (making one full equivalent of NADPH necessary per cycle) and the ferric peroxide species is generated. Heterolytic fragmentation of this intermediate occurs (possibly with acyl activation) to generate the iron-oxo species. It is the iron-oxo species which is believed to perform the hydroxylation by a radical recombination event. The hydroxylated product is released and the cycle may be repeated. Overall, one equivalent each of O_2 and NADPH is utilized per turn, with one atom of oxygen incorporated into the substrate while the other oxygen atom is washed into the aqueous medium, along with a proton from NADPH and from the substrate.

The overall process of androgen to estrogen conversion is comprised of three steps, each apparently requiring one molar equivalent of NADPH and O_2.[9] The first two steps are relatively well understood. In the first step, the androgen (androstenedione 1b, testosterone 1a) is hydroxylated at C-19 to afford the 19-hydroxy intermediate 2.[10] The reaction occurs with retention of configuration, and is accompanied by a significant kinetic isotope effect.[11] It is considered to be a "classical" cytochrome P-450 hydroxylation. In the second step the 19-hydroxy intermediate 2 is oxidized to the 19-oxo compound 3. This reaction involves stereospecific removal of the C-19 pro-R hydrogen,[12] with no isotope effect.[13] It has also been shown that the aldehyde oxygen atom is derived solely from the first mole of oxygen consumed.[14] These results imply that there is either a second "classical" hydroxylation leading to the gem-diol, followed by stereospecific dehydration, or a direct dehydrogenation of the 19-alcohol group. The former mechanism is probably correct, however, based on the elegant work of Beusen et al.[15] with artificial substrates.

In the mysterious third step, the bond between C-10 and C-19 is oxidatively cleaved to afford estrogen (estradiol 4a, estrone 4b) and formic acid. Oxygen atoms from each of the first and third moles of oxygen consumed,[14] as well as one of the original C-19 hydrogens, are incorporated into the formic acid. Based on the early work of Fishman and of Brodie, it has been thought for many years that the androgen 1β- and 2β-hydrogens are lost stereospecifically to the aqueous medium.[16] There is now thought to be a substrate-dependent variation in stereospecificity at C-2.[17] In the case of testosterone (1a) in contrast to androstenedione (1b), there is significant 2α-H loss. Since it has been shown that placental microsomes readily interconvert testosterone (1a) and androstenedione (1b) in the presence of NADPH, the classical tritium release aromatase activity assay of Thompson and Siiteri[9] (measurement of the rate of $1\beta,2\beta\text{-}^3H$ incorporation from substrate into water) may not be completely accurate. The recent shift to the use of $1\beta\text{-}^3H$-androstenedione in assays is therefore a necessary change.

As far as is known, all of the aromatase transformations occur in the same active site of the same protein molecule. The recent cloning of the human placental aromatase cDNA and its transfection into COS cells has led to expression of the complete enzymic activity.[18] This provides the strongest evidence that one protein molecule is responsible for all three steps. Furthermore, there is presently no evidence that more than one aromatase gene exists in human tissues.[18]

Since aromatase bears less than 36% overall sequence similarity to any member of the P-450 superfamily, it belongs to a unique family, designated P-450 XIX. Nevertheless, there are substantial portions of the aromatase amino acid sequence that show much higher similarity to the conserved portions of the P-450 gene such as the Ozols region, the aromatic region, the heme-binding region, and the I-helix.[18] Furthermore, many of the secondary structural domains of aromatase would appear to resemble those of camphor hydroxylase,[7a] the lone P-450 enzyme whose three-dimensional structure has been solved. The hypothesis that one active site is responsible for all three oxidative transformations for androgen to estrogen conversion therefore appears quite likely, but may not be rigorously established until a crystallographic structure of aromatase itself is obtained.

Several mechanistic theories for the third step have been shown experimentally to be unlikely. These include 2β-hydroxylation,[19] 4,5-epoxidation,[20] Baeyer-Villiger oxidation at C-10,[14] 10β-hydroxyestr-4-ene-3,17-dione formation,[21] and any process that involves water attack on the 19-aldehyde group. One theory which has remained consistent with all known aromatase data and cytochrome P-450 theory was proposed by Akhtar et al.[14] It features attack of an enzyme-bound ferric peroxide on the 19-aldehyde group, to produce the geminal hydroxyferric peroxide 5 (Scheme 3). This unstable intermediate could collapse to estrogen and formic acid by hydride shift,[14] proton transfer,[22] or free radical[23] pathways. A key point is that the high electrophilicity of the aldehyde leads to circumvention of normal ferric peroxide breakdown and alters the normal hydroxylation cycle. Evidence for the existence and nucleophilic reactivity of the proposed heme ferric peroxide intermediate comes from model studies. While present technology prevents a rigorous evaluation of the Akhtar mechanism, analogues of the proposed Akhtar intermediate 6a and 6b[24] have been synthesized and studied.

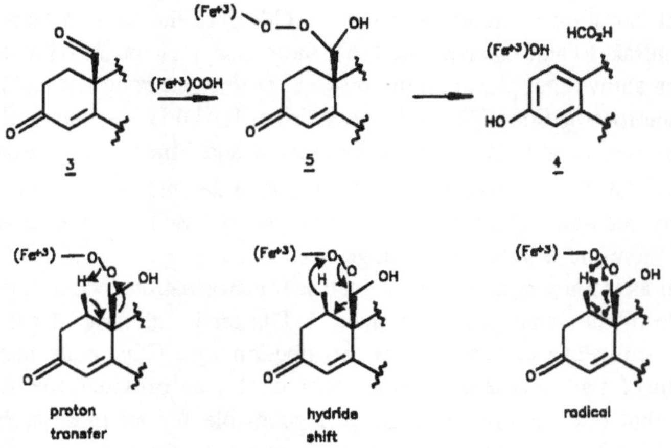

SCHEME 3.

The 19,19-methoxyhydroperoxide 6a was generated by ozonolysis of 10β-[2-methoxy]vinyl-estr-4-ene-3,17-dione (7) in the presence of MeOH (see Scheme 4). This peroxide, stable at −60°C, was found to undergo an unusual internal epoxidation forming 19-oxoandrost-4β,5β-oxido-3,17-dione (8), after warming to room temperature (see Scheme 4). Similarly, reaction of 19-oxoandrostenedione (3b) with HOOH was found to generate the same 4β,5β-

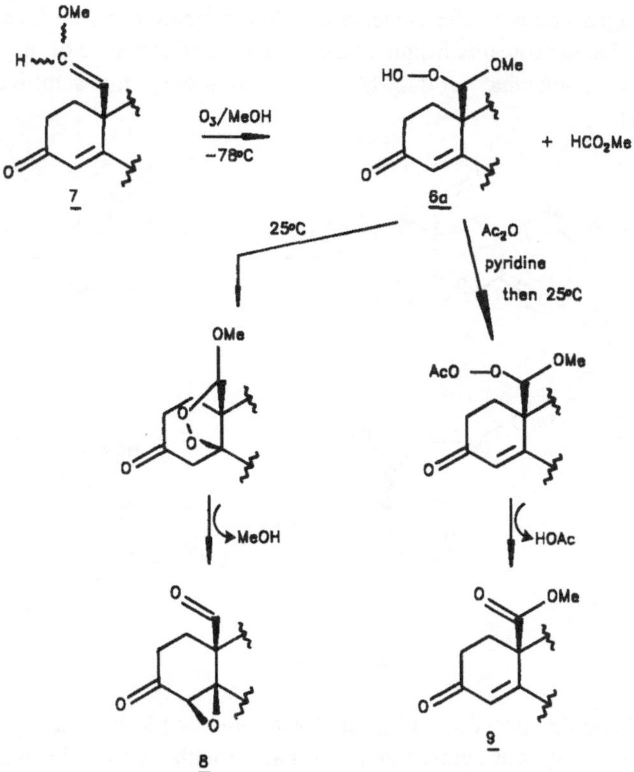

SCHEME 4.

epoxide 8 via 6b (see Scheme 5). Reaction of the methoxyhydroperoxide 6a with acetic anhydride/pyridine afforded 10β-carboxymethyl-estr-4-ene-3,17-dione (9), presumably after forming the peroxy acetate (see Scheme 4). Furthermore, treatment of 6a with Fe^{+2} led to formation of the same 19-methyl ester 9, which was rationalized as occurring by a radical mechanism.[24]

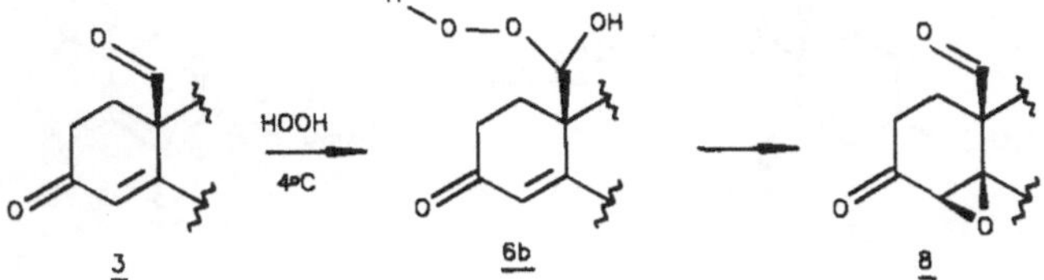

SCHEME 5.

In none of the above cases was aromatization detected, suggesting that the driving force for C-10/C-19 bond cleavage and 1-hydrogen removal was absent. Consequently, an alternative to the Akhtar theory was proposed, wherein enolization of the 3-ketone occurred prior to, or simultaneous with, peroxide fragmentation[24] (see Scheme 6). Enol formation would be expected to lower the barrier towards fragmentation if the transition state of the reaction were at least somewhat product-like (i.e., possessing the stabilization of aromaticity).

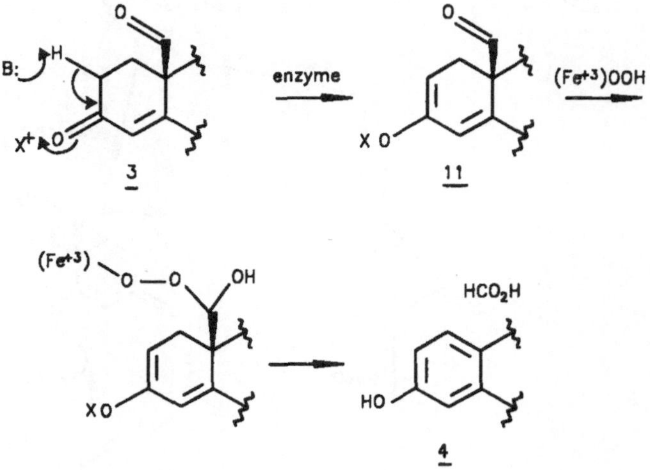

SCHEME 6.

To evaluate this proposal, a 2,3-enolized 19-oxoandrogen analogue of 11 as the enol ether 10 was synthesized and reacted with HOOH.[25] It was found to undergo aromatization to the corresponding estrogen derivative 13 with stoichiometric formic acid production. Furthermore, a substantial body of data supports the intermediacy of the 19,19-hydroxyhydroperoxide 12 in this reaction (see Scheme 7). This has therefore been termed an aromatase model reaction, and it has been found to be faithful to the actual aromatase reaction with regard to stereochemical removal of 1β-hydrogen and ^{18}O-incorporation patterns.[25b]

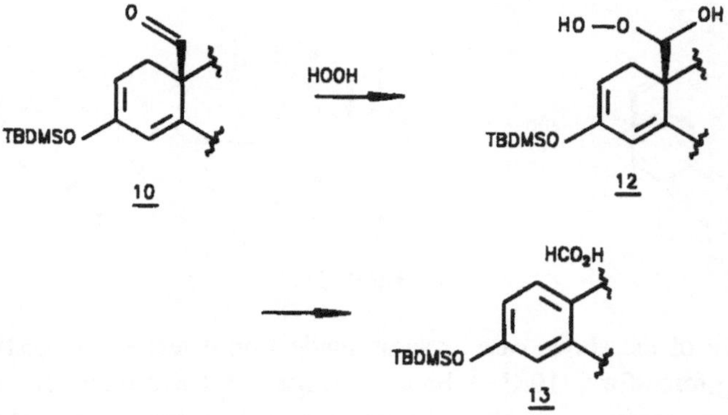

SCHEME 7.

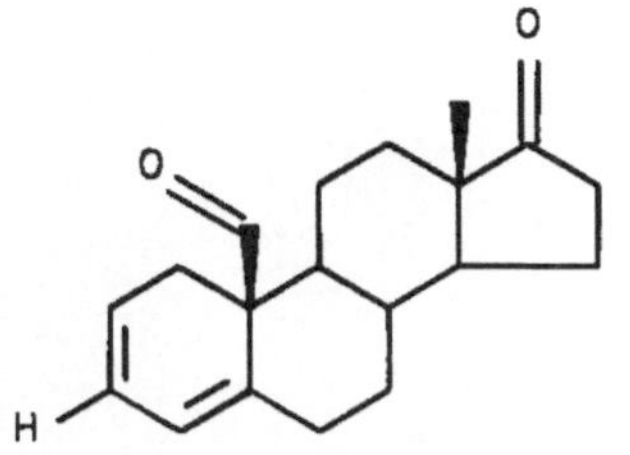

14

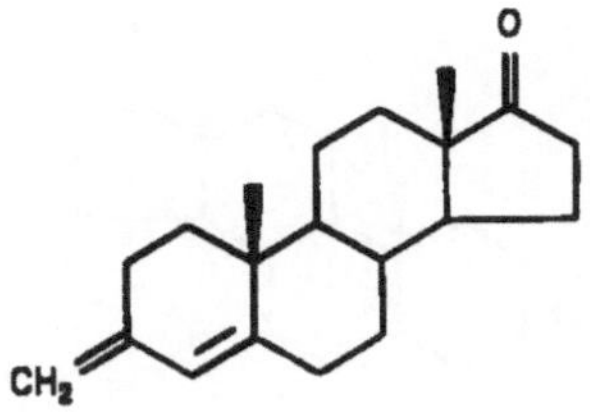

18

Ki=4.7 nM

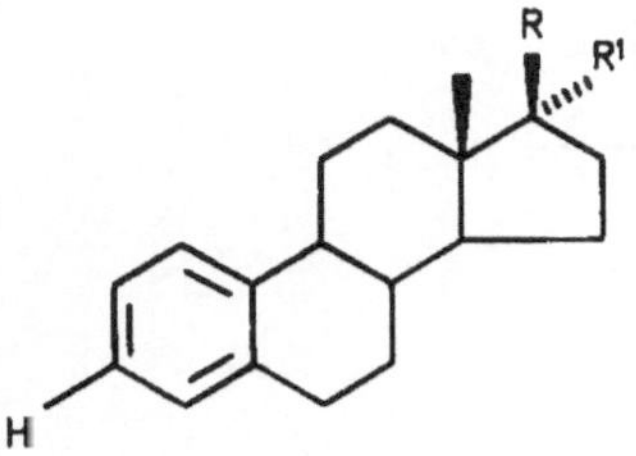

15

a; R = OH, R[1] = H

b; R, R[1] = O

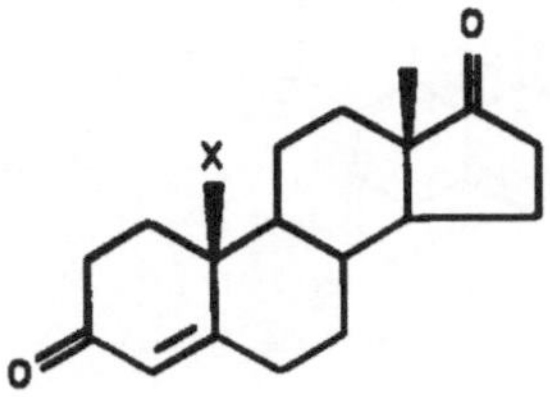

19 X = vinyl

Ki=3 nM (from IC_{50})

20 X = ethyl

Ki=9 nM (from IC_{50})

21 X = CH_3CHOH

22 X = propyl

Ki=3000 nM (from IC_{50})

23 X = CH_2=CH_2CH_2

Ki=56 nM (from IC_{50})

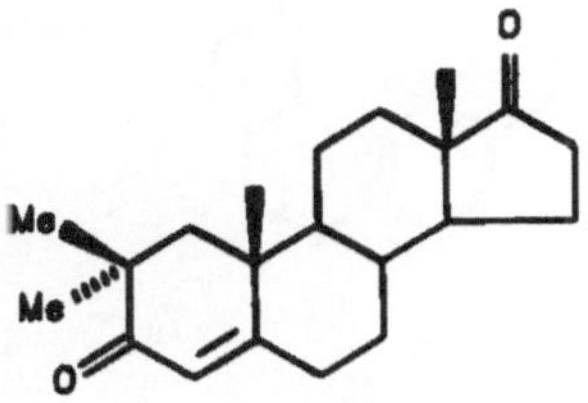

16

Ki=2.3 nM

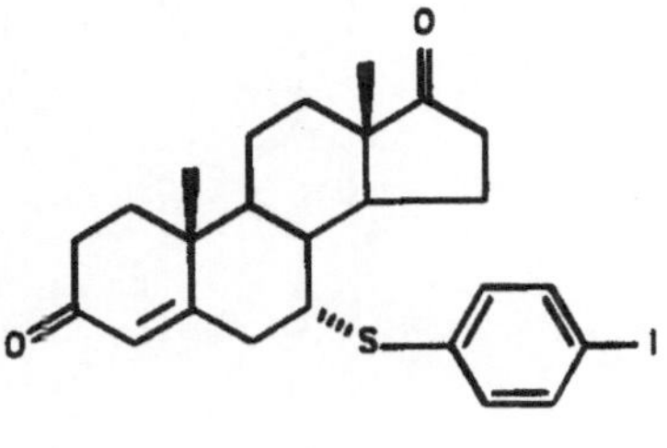

17

Ki=12 nM

SCHEME 7. (continued)

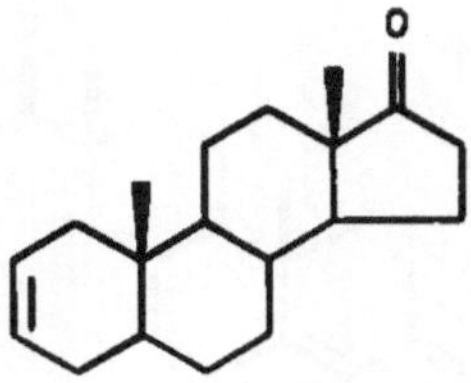

24

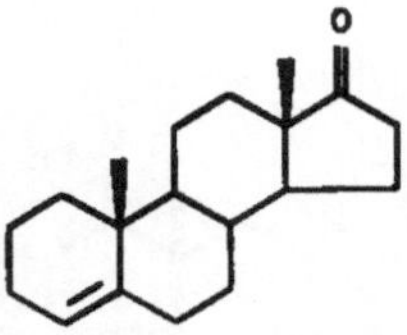

25
Ki=37 nM

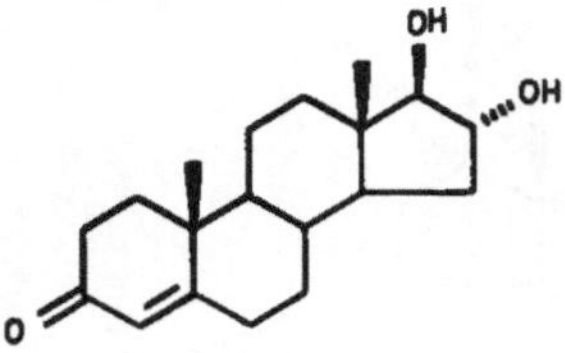

27

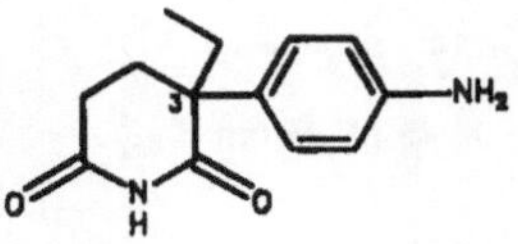

28
Ki = 600 nM

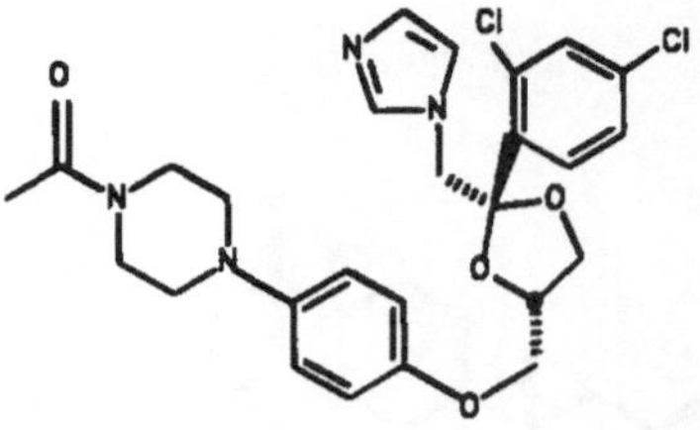

29

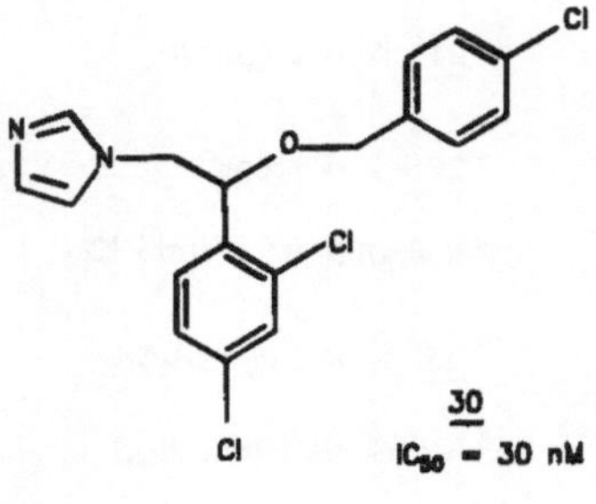

30
IC_{50} = 30 nM

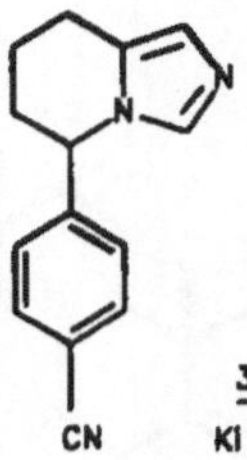

31
Ki = 1.5 nM

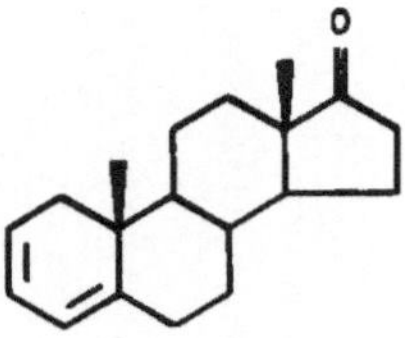

26
Ki = 28 nM

SCHEME 7. (continued)

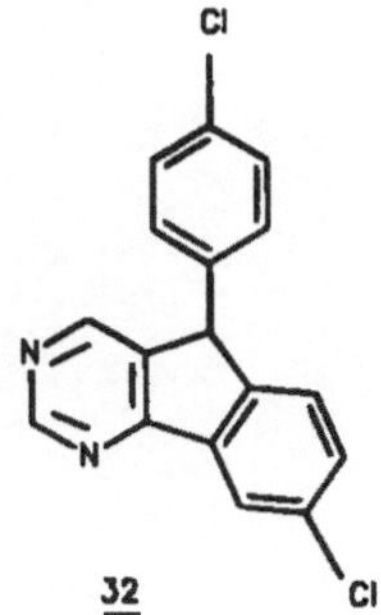

32

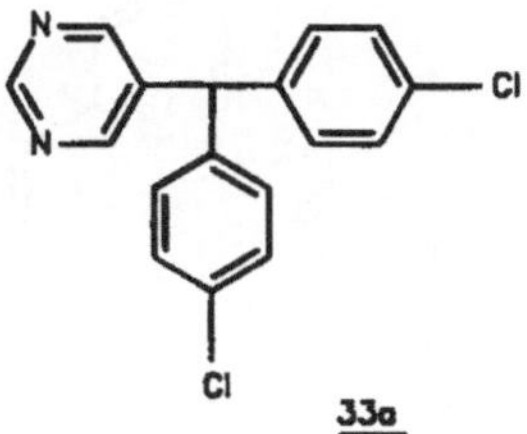

33a

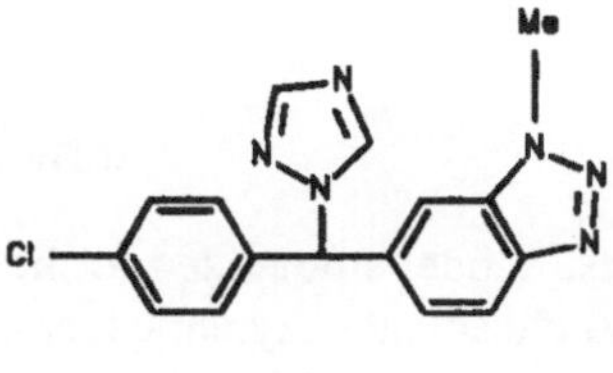

33b

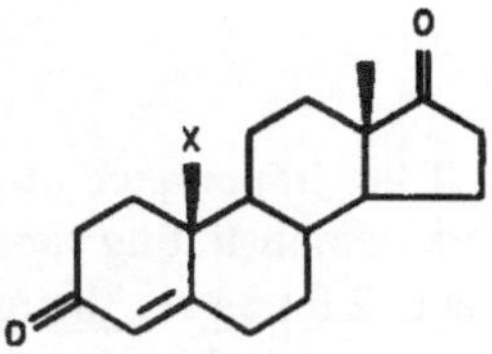

35 X = $MeSCH_2$

Ki=1 nM

36 X = 

a: 19R, Ki=1 nM

b: 19S, Ki=75 nM

37 X = 

a: 19R, Ki=7 nM

b: 19S, Ki=75 nM

38 X = $HSCH_2$

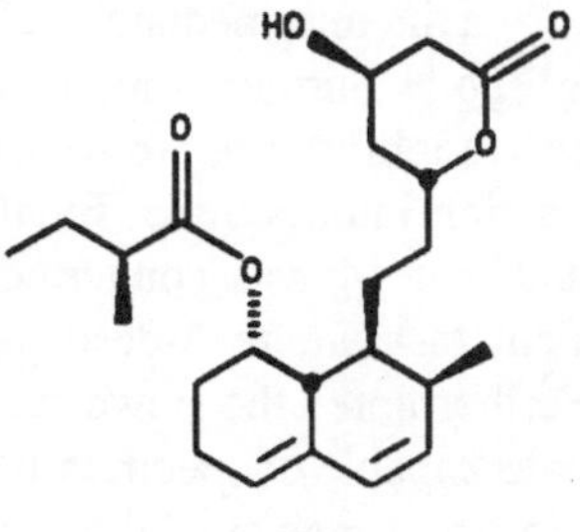

34

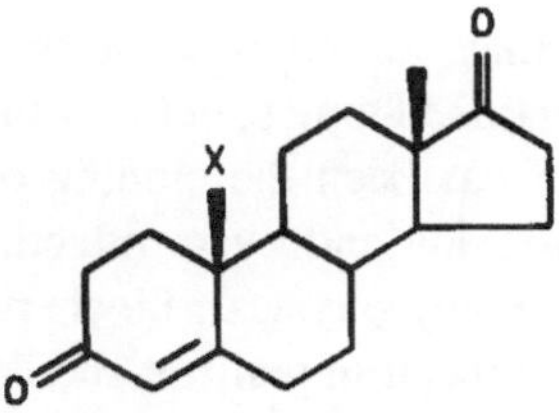

39 X = 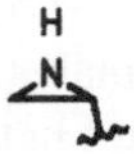

40 X = H

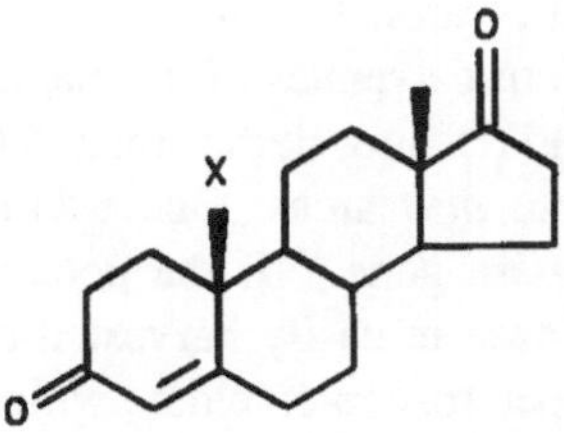

41 X = $MeOCH_2$

42 X = $EtSSCH_2$

SCHEME 7. (continued)

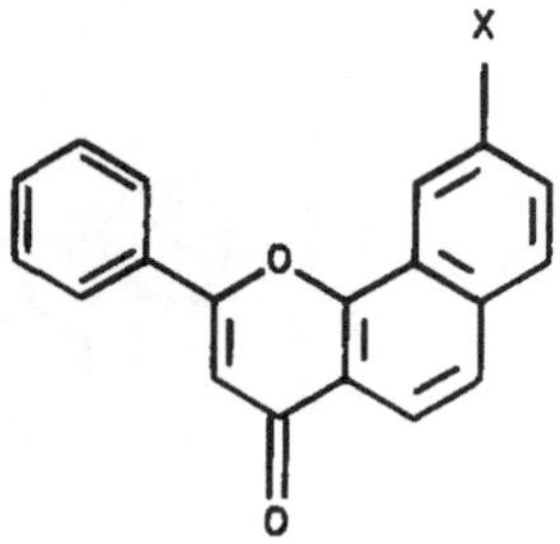

43 X = H

44 X = OH

SCHEME 7. (continued)

These model studies led to the investigation of the importance of enolization to the actual enzymatic reaction. Recent work demonstrating substrate-dependent stereospecificity for hydrogen removal at C-2 for 3a vs. 3b provides evidence that aromatase does mediate hydrogen removal at this position.[17b]

Another set of experiments which supports this proposal involved the study of 3-desoxy-2,4-dien-19-oxo androgen analogue 14.[26] Compound 14 was viewed as an analogue of the 2,4-dien-3-ol tautomer of 3b. It was found that 14 was a nonenzymatic "model reaction" substrate which reacted with HOOH (via the intermediate 19,19-hydroxyhydroperoxide) to form 3-desoxyestrone 15b.[26] Furthermore, it was shown that compound 14 was a potent competitive inhibitor of aromatase (K_i = 73 n*M*), as expected for an analogue of a reaction intermediate. Finally, in an unprecedented event, it was demonstrated that 14 was converted by placental microsomes, in an NADPH-dependent fashion, to 3-desoxyestrogen 10 at an enzymatically significant rate. Furthermore, the powerful aromatase inhibitor 19R-thiirane 36a (*vide infra*) was capable of specifically inhibiting microsomal-catalyzed conversion of 14 to 15, suggesting that aromatase was, in fact, the catalyst for this process. These results implicate enzyme-induced enolization as an important aspect of aromatization.

Future avenues of mechanistic exploration may not only involve continuing and refining studies related to the above mentioned topics, but may branch out into new areas. One exciting development has been the cloning of the aromatase gene.[18] If the potential for overexpression and overproduction of aromatase in easily harvested cells is realized, many physical, biochemical, and spectroscopic studies which have heretofore been unrealistic should become feasible. In addition, site-directed mutagenesis, coupled with new improvements in molecular modeling and structural analysis, may provide insights into the key active site amino acids which result in the unique chemical reactions catalyzed by aromatase. In particular, it will be of great interest to understand this carbon-carbon bond cleavage reaction in the context of the class of cytochrome P-450 lyases operative in steroid biosynthesis. As recently reviewed by Akhtar, there are several lyases which may utilize a similar mechanism to aromatase.[23] These include 14α-demethylase, 17,20-lyase, and the 16,17-ene-producing enzyme. All can be formulated as involving nucleophilic attacks of ferric peroxide species on carbonyl groups, followed by fragmentation of the resultant hydroxy-ferric peroxide species. A more detailed study of this group of lyases may lead to unifying principles of enzyme mechanism and inhibition. To date, no metalloporphyrin models have been shown to perform this class of reactions and this question, too, could be worthy of investigation.

Another potentially intriguing area concerns comparison of aromatases in different species. There has been a recent surge of interest in New World Monkey aromatase, based on the discovery that levels of estrogen in the males are very high.[27a] Furthermore, there has been preliminary investigation of aromatase from goldfish brain[27b] and equine testis.[28] The aromatase gene from chicken has recently been cloned and shown to be highly homologous to the human enzyme, based on primary sequence.[29] Rigorous evaluation of aromatases from these different sources will not only help in animal model studies of inhibitor design, but can be of value in structure/function analysis of the proteins themselves.

III. AROMATASE INHIBITION

As a cytochrome P-450 protein and a steroid-processing enzyme, aromatase is an exciting and multifaceted target for inhibition. Many steroids and nonsteroids have been designed, synthesized, and tested as aromatase inhibitors. A complete discussion of all aromatase inhibitors would not be wise because of space limitations, nor would it be particularly helpful since it would dilute the important developments. For these reasons we have chosen to describe trends in inhibitor design, with particular emphasis on rational design. Attention will be focused on those compounds where mechanism of action has at least been partly explored, those that show exceptional inhibitory properties, and those that may show clinical promise. There is now a very large number of steroidal inhibitors of aromatase, and it should be pointed out that a pioneering paper by the Brodie laboratory in 1973 led the way in this field.[30] It should be stated that the aromatase inhibition assays are performed differently in various labs and that a range of K_m values for androstenedione has been obtained. Unless otherwise mentioned, the inhibitory data are based on assays with human placental aromatase.

IV. TYPE I-COMPETITIVE INHIBITORS

Compounds that bind reversibly to the active site of aromatase as steroid substrate analogues, and fail to turn over very rapidly, may be useful inhibitors. If these compounds bind as normal androgen substrates, they may induce analogous type I changes in the UV-absorption spectrum Soret band of the enzyme-bound heme. This phenomenon is a general characteristic for cytochrome P-450 enzymes. Type I binders induce a shift in the Soret band maximum from about 420 nm to about 390 nm.

There are many compounds which have either been demonstrated or are presumed to act as type I inhibitors. Several of the most impressive in this category include 2,2-dimethyl-androstenedione (16),[31] 7α-S-[*p*-iodophenyl]-androstenedione (17),[32] 3-exomethylene-androst-4-en-17-one (18),[33] 10β-vinyl-estr-4-ene-3,17-dione (19),[34] and the 10β-ethyl-estr-4-ene-3,17-dione (20).[34] The K_i values for these compounds represent nearly the best that have been achieved for aromatase inhibitors (K_m for androstenedione (1b) about 30 n*M*). Furthermore, it is interesting that the best inhibitors all have similar K_i values despite their different structures. It is possible that these values may be near the lower limit. There are few examples in studies of small molecule-large molecule interactions where affinity constants below 0.1 n*M* are found. While transition-state analogues in some enzymes can have dissociation constants on the order of 10^5 times lower than the K_m, it is not clear that cytochrome P-450 reactions achieve accelerations by binding the transition-state more tightly. Rate constants for P-450-type reactions catalyzed by porphyrin model compounds are not much lower than the fastest rates in the actual P-450 enzyme reactions. For example, an iron tetramesitylporphyrin analogue epoxidizes styrene with a turnover number of 8.6 s^{-1} at 20°C.[35] Reconstituted camphor hydroxylase, one of the most catalytically active P-450 enzymes, has a turnover number of 17 s^{-1} at 25°C.[4a] For comparison, reconstituted aromatase has a turnover number of about 5 min^{-1} at 37°C.[36] Furthermore, the 19-hydroxy 2 and 19-oxo 3 compounds, which may be considered "high-energy intermediates" for aromatase, actually have higher K_m values than does the 10β-methyl starting material 1.[36]

Another important point is that a K_i of 1 n*M* shown by a seemingly powerful aromatase inhibitor is only a factor of 30 (about 2 kcal/mol) lower than the Michaelis constant (K_m) for androstenedione 1b. Aromatase has evolved to have a rather sluggish k_{cat} (5 min^{-1}) but a very low K_m. Thus, it would seem that mechanism-based inactivators (k_{cat} inhibitors) might be better candidates for *in vivo* inhibition (*vide infra*). Still, if adequate drug levels can be achieved in patients without producing toxic side effects, competitive inhibitors may still be a therapeutic option.

Little metabolic work has been reported on the above compounds. The 10β-ethyl substituent of 20 was shown not to be converted to the corresponding 10β-hydroxyethyl 21 in placental microsomes.[15] The 3-exomethylene compound 18 does not apparently get oxidized at C-19.[33] Thus, the mode of binding of these competitive inhibitors may be slightly altered compared to normal substrates, and therefore the former are not processed rapidly.

The varied structures of the above inhibitors make it difficult to deduce key active site features. One clear trend appears to be that hydrophobic functionalities lead to increased affinities. The optimum placement of these functional groups is difficult to predict, however. For example, simply increasing the 10β-ethyl chain length by a CH_2 group (as seen in 22)[34] leads to an increase in the dissociation constant by a factor of 300 (3.4 kcal/mol). Inserting a CH_2-group into the C-10:C-19 bond of 10β-vinyl compound 19 (as in 23)[34] increases the K_i by a factor of 20 (1.8 kcal/mol). Furthermore, A-ring conformation seems to play a critical role, as 5α-androst-2-en-17-one (24) is a mediocre binder,[30] whereas androst-4-en-17-one (25) is a potent inhibitor (K_i = 37 n*M*).[37] Androsta-2,4-dien-17-one (26) is also a good competitive inhibitor (K_i = 28 n*M*) suggesting that C4-5 unsaturation is particularly important for binding.[26]

Bohl et al.[38] have performed an extensive theoretical analysis of aromatase inhibitors and noted that the 4,5-double bond can significantly affect the conformation of C-19 substitutents. They found that the preferred calculated conformation of 19-hydroxy-5α-androstane-3,17-dione had the 19-hydroxy group in the out-of-ring position, in agreement with X-ray crystal structure data. In contrast, the analogous 4-en-3-one compound 2b (the aromatase intermediate) is known crystallographically to have the 19-hydroxy group lying over the steroid A-ring and was predicted as such by Bohl's calculations. Bohl's calculations would appear to be congruous with the above inhibitory data emphasizing the importance of the 4,5-double bond.

Another important aspect of competitive inhibitor design which has received attention recently concerns the conformation of and substitution in the D-ring. It is known that the three substrates androstenedione (1b), testosterone (1a), and 16α-hydroxytestosterone (27) have similar k_{cat} values for aromatase but widely different K_m values. According to Kellis and Vickery[36] the K_m values for 1b, 1a, and 27 are 60, 210, and 58,000 n*M*, respectively with purified aromatase. This suggests that the best steroidal competitive inhibitors might be expected to be 17-ketones. As will be discussed in the time-dependent inactivator section, D-ring modifications have also been found to be important in inactivator binding and inactivation rate.

V. BATHOCHROMIC SHIFT (TYPE II) COMPETITIVE INHIBITORS

There are some organic compounds containing suitably positioned heteroatoms which are capable of binding to cytochrome P-450 enzymes such that their heteroatoms coordinate to the heme iron. This special type of binding is reflected in Soret band changes (usually bathochromic with respect to type I binders). Furthermore, the precise Soret band displacement is often diagnostic of the heteroatom type (N, S, O, S^-). When these inhibitors coordinate to the heme iron *via* a nitrogen atom, they are often referred to as type II inhibitors. For convenience, we will use the term type II to refer to all the Fe-coordinating inhibitors, regardless of the nature of the heteroatom.

There are several examples of nonsteroidal type II binding aromatase inhibitors. One of the earliest nonsteroidal aromatase inhibitors is aminoglutethimide (28), which has been used clinically to treat breast cancer.[39] It is a moderately potent aromatase inhibitor (K_i = 600 n*M*) previously known for its role in adrenocortical steroidogenesis blockade and treatment of Cushing's syndrome. The aminophenyl nitrogen appears to be key for binding, although it is not certain which atom is actually coordinated to the iron.[40] Compounds with increased size of the 3-alkyl side chain also bind more tightly.[41] Interestingly, the d-enantiomer binds about 40 times more efficiently than the l-enantiomer.[42]

Several of the imidazole antimycotics have been tested for aromatase inhibition. These compounds have been shown to block ergosterol biosynthesis in yeast by inhibiting 14α-methyl demethylase, a P-450 lyase mentioned above. Ketoconazole (29), for example, is a mildly effective aromatase inhibitor (IC_{50} = 65 μ*M*, using 1.5 μ*M* androstenedione substrate).[43] Another member of this class, econazole (30), has a very potent IC_{50} (0.03 μ*M*, using 1.5 μ*M* androstenedione substrate).[43]

Other promising compounds include CGS 16949a 31,[44] LY 113174 32 (IC_{50} = 24 nM, using 0.1 μ*M* androstenedione substrate and rat ovarian microsomal aromatase),[45a] and the related pyrimidine LY56110 33a (IC_{50} = 29 n*M* using 0.1 μ*M* androstenedione substrate and rat ovarian microsomal aromatase).[45a] The recently reported triazole derivative 33b also shows very potent aromatase inhibition (IC_{50} = 3 n*M* using 1 μ*M* androstenedione substrate and a rat granulosa cell aromatase assay).[45b] The Ciba compound 31 has appeared very impressive in rat tumor model studies and phase I clinical trials.[46] CGS 16949a 31 appears to be 500 times more effective than aminoglutethimide on a per milligram basis at lowering estrogen levels in breast cancer patients.[46]

Good inhibitors appear to contain two domains: an Fe-coordinating domain and a hydrophobic domain. The structural requirements of the hydrophobic domain appear to be critical, but the chemical details of the specificity are unknown. This sort of dual interaction is comparable to the interaction with and inhibition of HMG CoA reductase by compactin 34. Abeles[47] has proposed that the decalin moiety of compactin serves as a hydrophobic anchor and the lactonic portion as a mevalonate analogue. The entropic gain for linking these two portions in compactin was estimated to correspond to about 50,000-fold decrease in K_d.

It is tempting to speculate that the hydrophobic domain of the nonsteroidal type II binders is binding in the pocket normally occupied by the steroid. Banting et al.[48] describe an interesting model for the nonsteroidal inhibitors proposing that the hydrophobic domain utilizes a common binding site with ring A of steroidal substrates. This suggests another approach to aromatase inhibitor design, that is, steroids containing heteroatoms appropriately placed near the iron might be expected to behave as type II competitive inhibitors.

Such an approach with inhibitors for the cholesterol side-chain cleavage cytochrome P-450 has been successfully accomplished by Vickery and coworkers.[49] Recently, several tight-binding compounds of this type have also been reported for aromatase.

This class of compounds now includes the 10β-$MeSCH_2$ 35,[50] 10β-thiirane 36,[51] 10β-oxirane 37,[51] and the 10β-$HSCH_2$ 38[52] derivatives of estr-4-ene-3,17-dione. Most recently, the 10β-aziridine diastereomers 39[51e] have also been synthesized and preliminarily evaluated. All of these compounds are potent competitive inhibitors of aromatase, and show UV-absorption spectra which suggest Fe coordination. Furthermore, bathochromic Soret band shifts are observed which are consistent with the type of heteroatom involved. Binding of the thioether compounds 35 and 36 to aromatase caused the Soret maxima to shift to 425 nm in each case, the epoxide 37 to 411 nm, and the thiol 38 to 474 nm (indicative of a thiolate anion). The placement of the heteroatoms on these steroids suggests that the steroid is binding in a manner similar to the natural androgen substate, and that the heme iron is located near the androgen 19-carbon as expected for a direct attack of the iron-oxo species on the methyl group.

Careful analysis of the effects of androstenedione (1b) and 19-norandrostenedione (40) on cyanide binding to aromatase has also led to similar conclusions.[53] In these experiments, it was found that 19-norandrostenedione (40) promoted cyanide ion binding to the aromatase heme iron whereas androstenedione (1b) competed with cyanide for binding to aromatase. It was inferred that, in the 19-nor case, cyanide was situated in the space normally occupied by the 19-methyl group.

Another rather intriguing report demonstrated that isotope-induced metabolic switching of aromatase processing of [19-$^3H_3$]-androstenedione (and [19-$^3H_3$]-testosterone) can occur.[54] Thus, a mixture of 1β-OH, 2β-OH, and 19-OH labeled androgens was produced from the labeled substrates by aromatase. Based on these findings, Osawa and colleagues reasoned that the heme iron was positioned over the A-ring near C-19, C-1, and C-2. Interestingly, the ratio of 1β- to 2β-hydroxyandrogen products was different for the labeled androstenedione substrate vs. the labeled testosterone substrate. This substrate-dependent variability is analogous to that seen for the aromatase-catalyzed 2-hydrogen removal discussed earlier.

It is also worth noting that neither the 19-OH 2b nor the 19-OMe 41 compounds coordinate to heme iron, as evidenced by the type I spectrum observed when added to substrate-free enzyme.[55] Furthermore, the 19R-epoxide 37a and 19R-thiirane 36a coordinate to iron, and are better inhibitors than the corresponding 19S-epoxide 37b and 19S-thiirane 36b which coordinate marginally under the experimental conditions.[51c] Thus, there are still subtleties about the nature of the enzyme active site-inhibitor interactions which may only be understood when larger quantities of pure enzyme are available for physical study. As mentioned in the previous section, the crystal structure of

19-hydroxyandrostenedione (2b) shows that the 19-oxygen atom lies directly over the A-ring. In the 19R-epoxide 37a the 19-oxygen is also positioned over the A-ring whereas in the 19S-epoxide 37b the 19-oxygen sits over the B-ring.[51a] The suggestion can be made that the 19R-epoxide 37a is a better aromatase inhibitor than the 19S-epoxide 37b because its preferred conformation resembles the natural aromatase intermediate 2b. The 19R-epoxide 37a and 19R-thiirane 36a have been shown to be potent *in vivo* inhibitors in a nude mouse tumor model.[56]

An interesting pro-drug 42, which has the 19-thiol grouping derivatized as a sidulfide, has entered phase I clinical trials.[57] Presumably this derivatization allows for improved stability extracellularly, while intracellular reduction of the disulfide would produce the potent 19-thiol 38.

Flavones, a class of plant natural products, and some of their *in vivo* metabolites have also been tested as aromatase inhibitors.[58] α-Napthoflavone (43) (7,8-benzoflavone) is a potent competitive inhibitor of aromatase (K_i = 20 n*M*).[58a] Furthermore, a structure-activity study of flavone derivatives revealed that 9-hydroxy-7,8-benzoflavone (44) was an even more potent competitive inhibitor (K_i = 5 n*M*).[58b] These compounds also appeared to act as type II binders, suggesting that a heteroatom coordinates to the heme iron as a sixth ligand. Whether this heteroatom is an oxygen from the inhibitor molecule or comes from water or an amino acid residue is still unknown. This question may not be answerable until further structural elucidation of the aromatase active site is achieved.

VI. SUICIDE INACTIVATORS

Because aromatase has a relatively low K_m, high dosage levels may be needed even for competitive inhibitors with apparently low K_i. In the search for increased efficacy, as well as specificity, a number of groups has developed suicide or mechanism-based inhibitors. Such compounds which contain a latent electrophilic group, intended to be activated by the target enzyme, can provide high specificity. The irreversible nature of the enzyme inactivation process due to covalent modification at the active site provides several advantages. One lies in the expectation that lower amounts of inhibitor will be needed. Furthermore, the destruction of an enzyme which has a low turnover rate means that the inhibitor's effect can persist after its clearance from the system. Finally, such compounds can serve as valuable mechanistic probes and as active site mapping agents. Unfortunately, there have been no reports as yet on the successful use of such inhibitors to label purified aromatase. In the absence of such labeling, the details of their mechanism cannot be established.

The 10β-propargyl compound 45, synthesized independently in three laboratories, was shown to be a potent suicide inactivator.[59] Inactivation is dependent upon NADPH. The K_i of competitive inhibition (68 n*M*) was found to be in rough agreement with the K_i extrapolated from a Kitz-Wilson analysis (23 n*M*), suggesting that the inactivation process occurs at the active site.[59a] Two major classes of mechanisms have been postulated to account for the inactivation. The first mechanistic proposal (Scheme 8) postulates that the 19-methylene group is oxidized to a 19-carbonyl, and the resulting αβ-acetylenic ketone can then react as an electrophilic 1,4 acceptor of enzyme active-site nucleophiles.[59a] While the acetylenic ketone 46 actually proved to be a time-dependent inactivator of aromatase, its rate of inactivation (k_{inact} = 5.35 × 10^{-4} s^{-1}) was somewhat smaller than that shown by the parent propargyl compound (k_{inact} = 1.11 × 10^{-3} s^{-1}).[59a] Furthermore, Covey has shown that 10β-ethyl compound 20 is not hydroxylated at C-19 by aromatase.[15] This casts doubt on the likelihood of aromatase-catalyzed hydroxylation of the propargyl methylene group. Johnston and co-workers have found that there is no detectable isotope effect associated with the inactivation process for 19-deuterium-labeled propargyl compound.[59b] These results taken together argue against the acetylenic ketone mechanism.

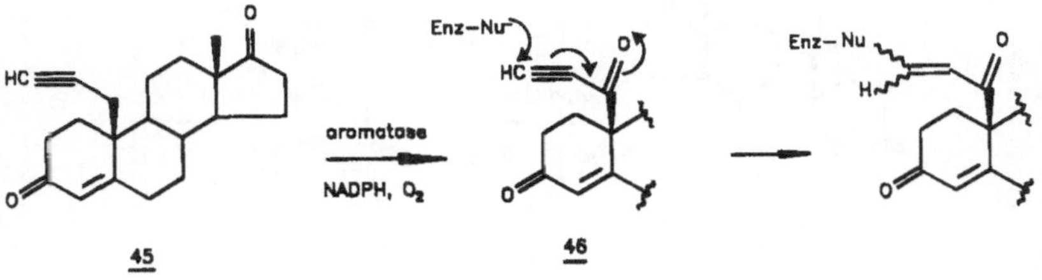

SCHEME 8.

The second mechanistic proposal of inactivation involves iron-oxo attack on the acetylenic grouping to afford an oxirene species (Scheme 9).[59b] The unstable oxirene could then directly, or after rearrangement, react with an active-site grouping. Such a mechanism is analogous to those proposed by Ortiz de Montellano for inactivation of liver microsomal cytochrome P-450 enzymes by terminal acetylenic compounds.[60] While the oxirene mechanism is interesting, more supporting evidence is needed to assess its validity, in the case of aromatase. Arguing against the oxirene mechanism, no oxirane formation was detected upon treatment of allyl compound 23 with aromatase[25c] as might have been expected based on the work of Thornburg and Stubbe.[85]

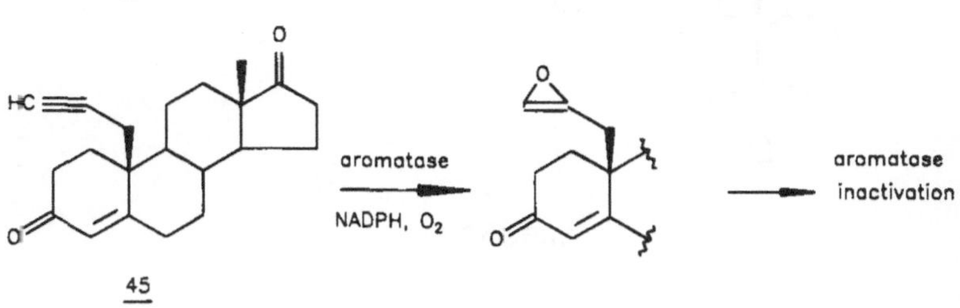

SCHEME 9.

The 10β-propargyl compound has shown excellent success in many *in vivo* settings including oral activity in a nude mice study in which the mice harbored an aromatase-producing human choriocarcinoma tumor. Compound 45 showed IC_{50} = 3 mg/kg for the inhibition of aromatase activity, 6 h after oral administration of drug.[61] This should be compared to the clinically used aromatase inhibitor aminoglutethimide (28) which has IC_{50} = 250 mg/kg under the same conditions.[61] The 10β-propargyl compound 45 is presently undergoing phase I clinical trials in the U.S.

Another and even closer structural analogue of androstenedione (1b) is 19,19-difluoroandrost-4-ene-3,17-dione (47), which was designed as a mechanism-based inactivator of aromatase.[34] Enzymatic hydroxylation at the 19-carbon would give a geminal fluorohydrin, which should spontaneously lose HF yielding an acyl fluoride 48 (Scheme 10). The latter might react covalently with an active site nucleophile and thus inactivate the enzyme.

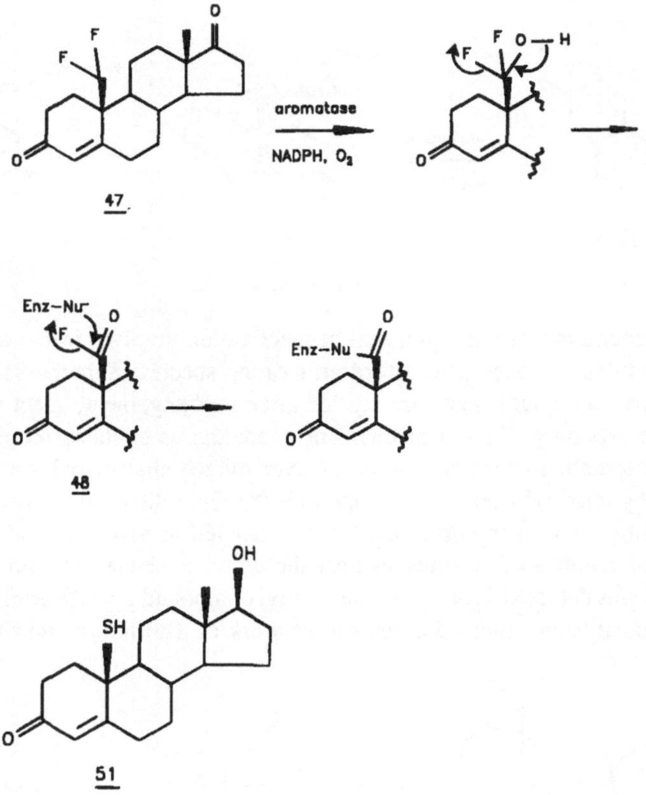

SCHEME 10.

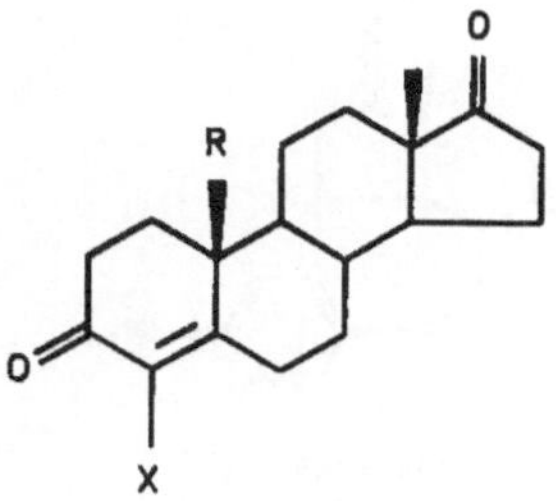

49 X = H, R = CH_2F

50 X = OH, R = CHF_2

52 X = OH, R = CH_3

53 X = OH, R = H

54 X = OH, R = CHO

55 X = OH, R = CH_2OH

56 X = OAc, R = CH_3

57 X = OMe, R = CH_3

58 X = Cl, R = CH_3

59 X = SH, R = CH_3

60 X = NH_2, R = CH_3

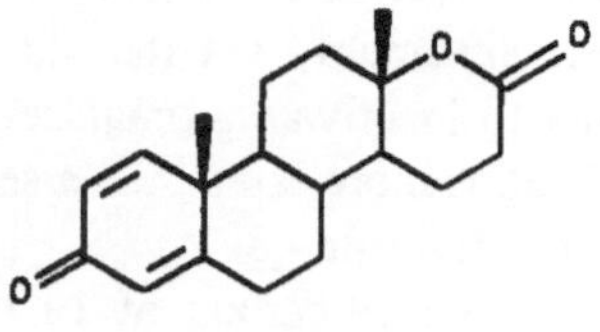

61

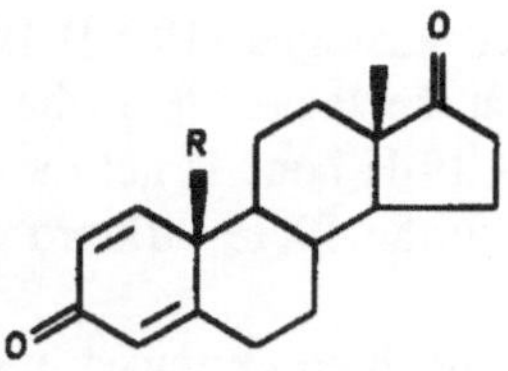

62 R = CH_3

63 R = CH_2OH

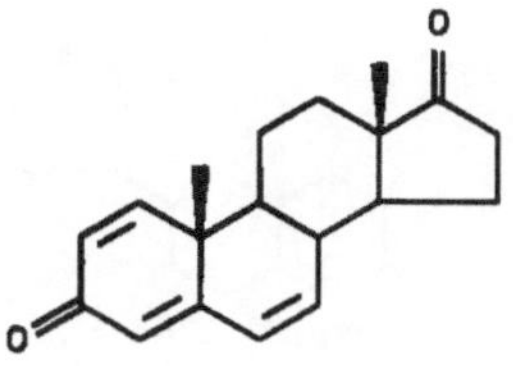

64

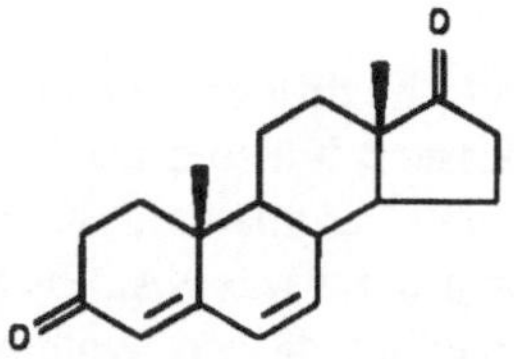

65

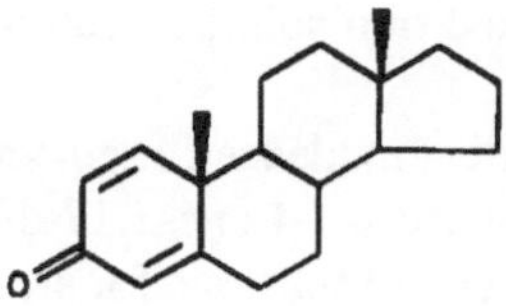

66

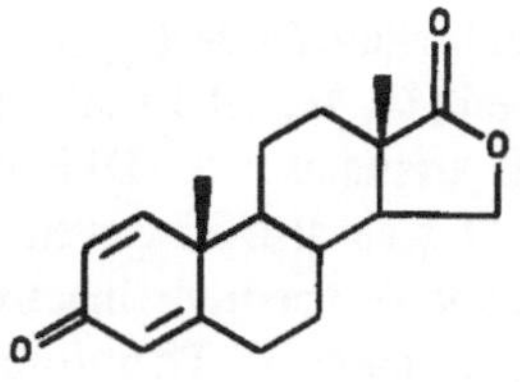

67

SCHEME 10. (continued)

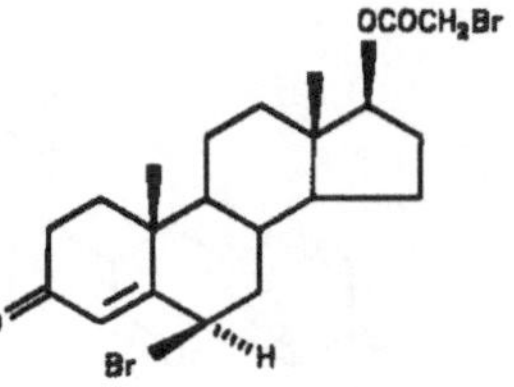

68

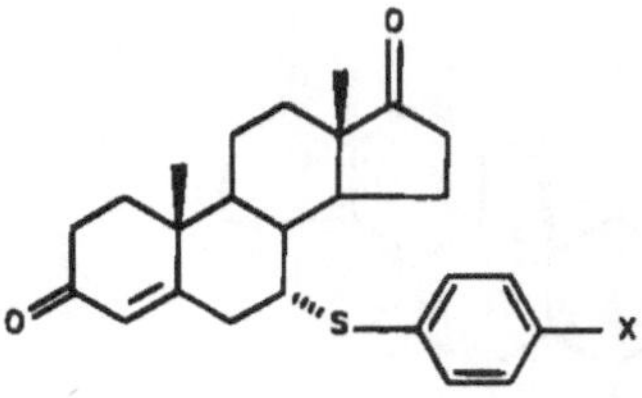

69 X = $HNCOCH_2Br$

70 X = N_3

SCHEME 10. (continued)

In fact, the difluoro compound 47 was found to be a competitive inhibitor of aromatase.[34] It also caused time-dependent irreversible inactivation of human placental aromatase. (K_i = 1.0 μM; k_{inact} = 0.023 min^{-1}).[62] The inactivation process was NADPH dependent. Although the postulated acyl fluoride intermediate 48 was synthesized and was shown to cause inactivation of the enzyme, the instability of the compound under the experimental conditions did not provide conclusive evidence about its involvement.[62] Addition of scavenging nucleophiles such as dithiothreitol or ethanolamine to incubations of the difluoro compound with aromatase had no effect on the enzyme inactivation rates. This excludes the possibility that an electrophilic intermediate is released into solution, followed by rebinding to the enzyme and inactivation.[62]

Indirect evidence for enzymatic attack on C-19 came from studies[62] with 19-fluoroandrost-4-ene-3,17-dione (49). This *mono*fluoro analogue proved to be an alternate substrate for aromatase, and was converted by human placental microsomes in high yield (52%) to estrone, presumably via the aldehyde intermediate 3b. Compound 49 was found not to inactivate aromatase. Presumably the rate of production of 19-aldehyde 3b is more rapid than a second hydroxylation at C-19 which would lead to acyl fluoride 48.

Direct evidence for attack by aromatase at the 19-carbon of 19,19-difluoroandrostenedione (47) came from studies[63] using tritium labeling at C-19. When [19-^{3}H]-19,19-difluoroandrostenedione was incubated with human pacental aromatase, NADPH-dependent release of tritium into solution occurred. It was also observed that the deuteriated analogue [19-^{2}H]-19,19-difluoroandrostenedione inactivated the enzyme at the same rate as the non-deuteriated parent.[63] Therefore, cleavage of the C19-H bond is not the rate-limiting step in the inactivation process caused by the 19,19-difluoro compound.

The effects of 19,19-difluoroandrostenedione have been examined in other aromatase systems, such as baboon and rhesus monkey placental aromatase, and chicken ovarian aromatase, as well as in a human trophoblast cell culture system.[56] The K_i and $t_{1/2}$ data are tabulated in Table 1. The 4-hydroxy analogue

TABLE 1
Inhibitory Data for 19,19-Difluoroandrost-4-Ene-3,17-Dione

	K_I (μM)	$t_{1/2}$ (min)
Human trophoblast (cell culture)	0.10	56
Human placental	1.0	30
Baboon placental	0.244	31
Rhesus monkey (placental)	0.445	v. slow

50 has been synthesized.[64] It is reported to be an inhibitor of placental aromatase (IC_{50} = 3.3 μM) although no experimental details are provided.

Another class of time-dependent inactivators is the thio-substituted compounds represented by 10β-mercapto 51 (K_I = 106 nM, k_{inact} = 0.0032 s^{-1}) and 19-mercapto compounds 38 (K_I = 34 nM, k_{inact} = 0.0012 s^{-1}).[65] Bednarski and Nelson found that these compounds obeyed many of the classical criteria of suicide inactivators. It is also very interesting that the 19-thiol 38 is capable of coordinating to the heme iron via sulfur (*vide supra*). Such a complex would presumably be unrelated to the time-dependent inactivation, since its formation does not require NADPH.

The compound 4-hydroxyandrostenedione 52 (4-OHA), was originally described as a competitive inhibitor of human placental aromatase by Brodie et al.[66] in 1977. It was later shown that 4-OHA 52 causes time-dependent inactivation ($t_{1/2}$ = ~2 min, K_I = ~50 nM).[67] The time-dependent inactivation required NADPH. The inactivation was irreversible as demonstrated by examining activity after prolonged dialysis to remove the inhibitor. The inactivation rate was slowed when substrate (androstenedione) was included in the preincubation step, evidence for an active site process. Covey and Hood[67b] showed that 19-nor-4-hydroxyandrostenedione (53), while capable of binding to the enzyme, did not bring about aromatase inactivation. This finding led them to suggest that 19-oxygenated intermediates could be the inactivating species. On the other hand, 4-hydroxy-19-oxoandrostenedione (54) has been found to be a poor inactivator relative to 4-OHA 52 which argues against their idea.[68] No studies have been reported with the 19-hydroxy analogue 55.

The 4-O-acetyl derivative 56 of 4-OHA 52 was also demonstrated to be a time-dependent inactivator of aromatase.[67] It should be emphasized that it is likely that this compound is hydrolyzed to 4-OHA 52 before inactivation occurs. It has been shown that human placental microsomes possess ample esterase activity that can convert 4-O-acetyl derivative 56 to the 4-OH compound 52.[67b] Furthermore, the 4-OMe 57 and 4-Cl 58 compounds are not time-dependent inactivators, underlining the importance of the free O-H.[69]

The 4-substituted thio compounds (e.g., 59) also have been synthesized but details of time-dependent inactivation experiments have not yet been reported.[70] Although 4-NH_2-androstenedione (60) was shown to be a time-dependent inactivator ($t_{1/2}$ = 31 min, K_i = 37 n*M*),[71] its mechanism of inactivation and that of 4-OHA 52 have not been satisfactorily accounted for. Interestingly, 2,2-dimethyl-4-hydroxy-androst-4-ene-3,17-dione, a competitive inhibitor of aromatase, is not a time-dependent inactivator.[31] This suggests that either A-ring conformation or C-2 reactivity may be critical to the inactivation mechanism of 52.

In clinical trials in Britain, 4-OHA 52 has shown promise in promoting tumor regression in patients with advanced breast cancer.[72] A clinical response was seen in 34% of patients receiving therapy. Side effects were less frequent than with aminoglutethimide. A problem which exists is rapid glucuronidation which may need to be addressed for optimal dosing.

The class of inhibitors containing extra units of unsaturation in the A and B rings has expanded rapidly since a report on the aromatase inhibitory properties of testolactone (61) appeared in 1979.[73] It was subsequently shown that androsta-1,4-diene-3,17-dione (62) ($t_{1/2}$ = 12.7 min, K_i = 320 n*M*) as well as testolactone (61) ($t_{1/2}$ = 32 min, K_i = 35 μ*M*) behave as time-dependent aromatase inactivators (inactivation dependent upon NADPH) due to the presence of the 1,2-double bond.[74] It is known that androsta-1,4-diene-3,17-dione (62) is converted to estrone (4b) by placental microsomes, but this could presumably occur without enzymic assistance by a retro-aldol type reaction after formation of the 19-hydroxy-dienone (63).[75] A recently proposed mechanism of inactivation suggests that the 1,2-double bond is oxidized to a radical cation, which is then intercepted by an active site grouping.[76] There are little published data to support this theory, however. The more highly conjugated androsta-1,4,6-triene-3,17-dione (64) is also a time-dependent inactivator ($t_{1/2}$ = 10.5 min, K_i = ~180 n*M*),[77] whereas the androsta-4,6-diene-3,17-dione (65) (K_i = 42 n*M*) is not.[74]

More highly substituted 1,4-dienone compounds have been synthesized and evaluated as aromatase inhibitors. Substitutions include 4-hydroxy,[69] 7α-S-[*p*-aminophenyl],[78] 1-methyl,[79] 6-exo-methylene,[80] and 4-amino.[71] There have also been similar types of substitutions attempted on the 1,4,6-triene systems. While some of these compounds display potent aromatase inhibition, their properties do not shed much light on the basic mechanism of inactivation. Results of aromatase inactivation by the 7α-substituted compounds with purified enzyme have recently been reported,[78b] suggesting that active-site labeling of aromatase may be on the horizon.

Recently, Sherwin and co-workers[81] have synthesized modified D-ring analogues of androsta-1,4-dienones. They found that the reduced compound 17-methylene 66 (K_i = 1.1 μ*M*, $t_{1/2}$ = 15.2 min) showed less than a fourfold decrease in inactivation rate or binding affinity relative to 62. Similarly, the corresponding D-ring butyrolactone 67 (K_i = 390 n*M*, $t_{1/2}$ = 6.2 min) was still a potent suicide inactivator. In contrast, analogous inhibitors with open D-rings were poor inhibitors and in general were very poor inactivators. Covey inferred that the open D-ring chain may sterically interfere with binding of the inhibitor to the enzyme, but that the precise substitution of the intact D-ring does not appear to be critical for binding or enzyme processing of these inhibitors.[81]

VII. AFFINITY LABELS

Such compounds are expected to play an important role in mapping out the active site of aromatase. Ideally these labels should show low K_i and high k_{inact} values (although a high k_{inact} is the more important) and require the use of purified enzyme. Denatured enzyme, even though pure, would clearly be unsuitable.

A number of affinity labels has been synthesized and evaluated as time-dependent inactivators of aromatase. Osawa and co-workers[82] have reported that the 2α-, 6α-, and 19-bromacetoxy derivatives of androstenedione induce time-dependent inactivation of aromatase at micromolar levels, but K_i values were not recorded. In addition, testosterone 17β-bromoacetate and its 6β-methoxy-, 6β-bromo-, and 6-keto derivatives were found[82] to cause time-dependent loss of activity. One of the best of the above derivatives was 6β-bromotestosterone 17β-bromoacetate (68), which had a poor K_i (25 μ*M*) but a good k_{inact} (0.0274 min^{-1}). There are no reports of the use of these reagents for labeling of aromatase preparations.

Finally, Snider and Brueggemeier[83] have synthesized a bromacetamido affinity label 69, which has a poor K_i (33 μ*M*) but a good k_{inact} of 0.148 min^{-1}. The ^{14}C-labeled compound was shown to bind irreversibly to microsomal placental aromatase, although specific binding was not obtained.

VIII. PHOTOAFFINITY LABEL

Brueggemeier and co-workers[84] have described a 7α-(4′azido) phenylthio-derivative 70 of androstenedione, and have shown that, in the dark, the compound is a powerful competitive inhibitor of aromatase (K_i = 1.3 n*M*). When incubated with placental microsomes under UV irradiation, time-dependent inactivation of aromatase is observed (K_i = 0.89 μ*M*; k_{inact} = 0.0993 min^{-1}). This inactivation process, presumably involving a nitrene intermediate, is slowed in the presence of androstenedione.

IX. FUTURE DIRECTIONS

The rational design of a new generation of highly specific mechanism-based inhibitors will require detailed understanding of the mechanism of action both of aromatase and of the present group of inhibitors.

In more general terms, given the relatively large number of available inhibitors with K_i values in the nanomolar range, potency is not a major issue. Instead, major questions are oral effectiveness, and enzyme and tissue specificity *in vivo*, as well as stability towards metabolic inactivation. Potential problems with steroidal inhibitors of aromatase are illustrated by the possibility of interaction with androgen-transforming enzymes (e.g., 5α-reductase) and with androgen receptors or on the other hand with cytochrome P-450 enzymes of the adrenal.

In addition, steroidal aromatase inhibitors with 17-keto groups may also prove to be substrates for 17β-hydroxysteroid dehydrogenases. The resulting 17β-ols would be expected to be less effective inhibitors of aromatase than the parent 17-ketones, and they also could be conjugated and excreted as water-soluble 17-conjugates. Covey et al.[81] have already drawn attention to possible ways of circumventing this problem (see the use of D-secosteroids).

ACKNOWLEDGMENTS

We thank the National Institute of Child Health and Human Development (Grant HD-11840 to Cecil H. Robinson) and the Medical Scientist Training Program of the National Institute of General Medical Sciences (Grant GM-07309 for Philip A. Cole) for financial support.

REFERENCES

1. (a) **Webster, N. J. G., Green, S., Tasset, D., Ponglikitmongkol, M., and Chambon, P.,** *EMBO J.*, 8, 1441, 1989. (b) **Green, S. and Chambon, P.,** *Nature*, 325, 75, 1987. (c) **Walter, P., Green, S., Greene, G., Krust, A., Bornert, J. M., Jeltsch., J. M., Staub, A., Jensen, E., Scrace, G., Waterfield, M., and Chambon, P.,** *Proc. Natl. Acad. Sci. U.S.A.*, 82, 7889, 1985.
2. **Beatson, G. T.,** *Lancet*, ii, 104, 1896.
3. (a) **Cowan, K. and Lippman, M.,** *Arch. Intern. Med.*, 142, 363, 1982. (b) **Brodie, A. M. H.,** *Biochem. Pharmacol.*, 34, 3213, 1985. (c) **Santen, R. J. and Misbin, R. I.,** *Pharmacotherapy*, 1, 95, 1981. (d) Van Wauwe, J. P. and Janssen, P. A. J., *J. Med. Chem.*, 32, 2231, 1989. (e) **Johnston, J. O. and Metcalf, B. W.,** in *Approaches to Cancer Chemotherapy*, Prasad, S. S., Ed., Academic Press, New York, 1984, 307.
4. (a) **Sligar, S. G. and Murray, R. I.** In *Cytochrome P-450: Structure, Mechanism, and Biochemistry*, Ortiz de Montellano, P. R., Ed., Plenum Press, New York, 1986, 429. (b) **Poulos, T. L.,** In *Cytochrome P-450: Structure, Mechanism, and Biochemistry*, Ortiz de Montellano, P. R., Ed., Plenum Press, New York, 1986, 505.
5. **Waxman, D. J.,** In *Cytochrome P-450: Structure, Mechanism, and Biochemistry*, Ortiz de Montellano, P. R., Ed., Plenum Press, New York, 1986, 525.
6. **McMurry, T. J. and Groves, J. T.,** In *Cytochrome P-450: Structure, Mechanism, and Biochemsitry*, Ortiz de Montellano, P. R., Ed., Plenum Press, New York, 1986, 1.
7. (a) **Nelson, D. R. and Stobel, H. W.,** *Biochemistry*, 28, 656, 1989. (b) **Nebert, D. W. and Gonzales, F. J.,** *Annu. Rev. Biochem.*, 56, 945, 1987.

8. White, R. E. and Coon, M. J., *Annu. Rev. Biochem.*, 49, 315, 1980.
9. (a) Thompson, E. A. and Siiteri, P. K., *J. Biol. Chem.*, 249, 5364, 1974. (b) Siiteri, P. K. and Thompson E. A., *J. Steroid Biochem.*, 6, 317, 1975.
10. (a) Meyer, A. S., *Biochim. Biophys. Acta*, 17, 441, 1955. (b) Meyer, A. S., *Experientia*, 11, 99, 1955.
11. Caspi, E., Arunachalam, T., and Nelson, P. A., *J. Am. Chem. Soc.*, 108, 1847, 1986.
12. (a) Osawa, Y., Shibata, K., Rohrer, D., Weeks, C., and Duax, W. L., *J. Am. Chem. Soc.*, 97, 4400, 1975. (b) Arigoni, D., Battaglia, R., Akhtar, M., and Smith, T., *J. Chem. Soc., Chem. Commun.*, 185, 1975.
13. Miyairi, S. and Fishman, J., *J. Biol. Chem.*, 260, 320, 1985.
14. Akhtar, M., Calder, M. R., Corina, D. L., and Wright, J. N., *Biochem. J.*, 201, 569, 1982.
15. Beusen, D. D., Carrell, H. L., and Covey, D. F., *Biochemistry*, 26, 7833, 1987.
16. (a) Brodie, H. J., Kripalani, K. J., and Possanza, G., *J. Am. Chem. Soc.*, 91, 1241, 1969. (b) Fishman, J. and Guzik, H., *J. Am. Chem. Soc.*, 91, 2805, 1969. (c) Fishman, J., Guzik, H., and Dixon, D., *Biochemistry*, 8, 4304, 1969.
17. (a) Fronckowiak, M. D. and Osawa, Y., *The 70th Endocrine Society Meeting*, New Orleans, June 8 to 11, 1988 (Abstract 430). (b) Cole, P. A. and Robinson, C. H., *Biochem. J.*, 268, 553, 1990.
18. (a) Corbin, C. J., Graham-Lorence, S., McPhaul, M., Mason, J. I., Mendelson, C. R., and Simpson, E. R., *Proc. Natl. Acad. Sci. U.S.A.*, 85, 8948, 1988. (b) Means, G. D., Mahendroo, M. S., Corbin, C. J., Mathis, J. M., Powell, F. E., Mendelson, C. R., and Simpson, E. R., *J. Biol. Chem.*, 264, 19385, 1989.
19. (a) Hosoda, H. and Fishman, J., *J. Am. Chem. Soc.*, 96, 7325, 1974. (b) Goto, J. and Fishman, J., *Science*, 195, 80, 1977. (c) Fishman, J. and Raju, M. S., *J. Biol. Chem.*, 256, 4472, 1981. (d) Hahn, E. F. and Fishman, J., *J. Biol. Chem.*, 259, 1689, 1984. (e) Caspi, E., Wicha, J., Arunachalam, T., Nelson, P., and Spiteller, G., *J. Am. Chem. Soc.*, 106, 7282, 1984.
20. (a) Morand, P., Williamson, D. G., Layne, D. S., Lompa-Krzymien, L., and Salvador, J., *Biochemistry*, 14, 635, 1975. (b) Mastalerz, H. and Morand, P., *J. Chem. Soc. Perkin Trans. 1*, 2611, 1982.
21. (a) Covey, D. F., Hood, W. F., Beusen, D. D., and Carrell, H. L., *Biochemistry*, 23, 5398, 1984. (b) Townsley, J. D., Possanza, G. J., and Brodie, H. J., *Fed. Proc. Fed. Am. Soc. Exp. Biol.*, 25 (abstr. 520), 282, 1966. (c) Caspi, E., Dharmaratne, H. R. W., and Shackleton, C., *J. Chem. Soc. Chem. Commun.*, 1699, 1989.
22. Stevenson, D. E., Wright, J. N., and Akhtar, M., *J. Chem. Soc. Chem. Commun.*, 1078, 1985.
23. Stevenson, D. E., Wright, J. N., and Akhtar, M., *J. Chem. Soc. Perkin Trans. 1*, 2043, 1988.
24. (a) Cole, P. A. and Robinson, C. H., *J. Chem. Soc. Chem. Commun.*, 1651, 1986. (b) Cole, P. A. and Robinson, C. H., *J. Chem. Soc. Perkin Trans. 1*, 2119, 1990.
25. (a) Cole, P. A. and Robinson, C. H., *J. Am. Chem. Soc.*, 110, 1284, 1988. (b) Cole, P. A. and Robinson, C. H., *J. Am. Chem. Soc* , 113, 8130, 1991. (c) Cole, P. A. and Robinson, C. H., unpublished observations.
26. Cole, P. A., Bean, J. M., and Robinson, C. H., *Proc. Natl. Acad. Sci. U.S.A.*, 87, 2999, 1990.
27. (a)McCamant, S. K., Klosterman, L. L., Goldman, E. S., Murai, J. T., and Siiteri, P. K., *Steroids*, 50, 549, 1987. (b) Callard, G. V. and Pasmanik, M., *Steroids*, 50, 475, 1987.
28. Gaillard, J.-L. and Silberzahn, P., *J. Biol. Chem.*, 262, 5717, 1987.
29. McPhaul, M. J., Noble, J. F., Simpson, E. R., Mendelson, C. R., and Wilson, J. D., *J. Biol. Chem.*, 263, 16358, 1988.
30. Schwarzel, W. C., Krugel, W. G., and Brodie, H. J., *Endocrinology*, 92, 866, 1973.

31. **Furth, P. S., Rosenberger, J. G., Marcotte, P. A., and Robinson, C. H.,** *J. Enz. Inhibition,* 4, 131, 1990.
32. **Brueggemeier, R. W., Li, P.-K., Snider, C. E., Darby, M. V., and Katlic, N. E.,** *Steroids,* 50, 163, 1987.
33. **Miyairi, S. and Fishman, J.,** *J. Biol. Chem.,* 261, 6772, 1986
34. **Marcotte, P. A. and Robinson, C. H.,** *Steroids,* 39, 325, 1982.
35. **Naruta, Y. and Maruyama, K.,** *Tetrahedron Lett.,* 28, 4553, 1987.
36. **Kellis, J. T., Jr. and Vickery, L. E.,** *J. Biol. Chem.,* 262, 4413, 1987.
37. **Numazawa, M., Mutsumi, A., Hoshi, K., and Koike, R.,** *Biochem. Biophys. Res. Commun.,* 160, 1009, 1989.
38. **Bohl, M., Simon, Z., and Lochmann, J. R.,** *Z. Naturforsch.,* 44c, 217, 1989.
39. **(a) Santen, R. J., Worgul, W. J., Samojlik, E., Interrante, A., Boucher, A. E., Lipton, A., Harvey, H. A., White, D. S., Smart, E., Cox, C., and Wells, S.,** *N. Engl. J. Med.,* 305, 545, 1981. **(b) Shaw, M. A., Nicholls, P. J., and Smith, H. J.,** *J. Steroid Biochem.,* 31, 137, 1988.
40. **Foster, A. B., Jarman, M., Leung, C. S., Rowlands, M. G., Taylor, G. N., Plevey, R. G., and Sampson, P.,** *J. Med. Chem.,* 28, 200, 1985.
41. **Hartmann, R. W. and Batzl, C.,** *J. Med. Chem* , 29, 1362, 1986.
42. **Graves, P. E. and Salhanick, H. A.,** *Endocrinology,* 105, 52, 1979.
43. **Mason, J. I., Carr, B. R., and Murry, B. A.,** *Steroids,* 50, 179, 1987.
44. **Steele, R. E., Mellor, L. B., Sawyer, W. K., Wasvary, J. M., and Browne, L. J.,** *Steroids,* 50, 147, 1987.
45. **(a) Hirsch, K. S., Jones, C. D., Lindstrom, T. D., Stamm, N. B., Sutton, G. P., Taylor, H. M., and Weaver, D. E.,** *Steroids,* 50, 201, 1987. **(b) Wouters, W., De Coster, R., Krekels, M., van Dun, J., Beerens, D., Haelterman, C., Raeymaekers, A., Freyne, E., Van Gelder, J., Venet, M., and Janssen, P. A. J.,** *J. Steroid Biochem.,* 32, 781, 1989.
46. **(a)Lipton, A., Harvey, H. A., Demers, L. M., Hanagan, J., Mulagha, M., Sanders, S., Fitzsimmons, S., and Santen, R. J.,** 13th Congress of the European Society for Medical Oncology, Lugano, Switzerland, October 30, 1988. **(b) Lipton, A., Harvey, H. A., Demers, L. M., Hanagan, J. R., Mulagha, M. T., Kochak, G. M., Fitzsimmons, S., Sanders, S. I., and Santen, R. J.,** *Cancer,* 65, 1279, 1990. **(c) Stein, R. C., Dowsett, M., Davenport, J., Hedley, A., Ford, H. T., Gazet, J. C., and Coombes, R. C.,** *Cancer Res.,* 50, 1381, 1990.
47. **Abeles, R. H.,** *Drug Dev. Res.,* 10, 221, 1987.
48. **Banting, L., Smith, H. J., James, M., Jones, G., Nazareth, W., Nicholls, P. J., Hewlins, M. J. E., and Rowlands, M. G.,** *J. Enzyme Inhibition,* 2, 215, 1988.
49. **(a) Sheets, J. J. and Vickery, L. E.,** *Proc. Natl. Acad. Sci. U.S.A.,* 79, 5773, 1982. **(b) Sheets, J. J. and Vickery, L. E.,** *J. Biol. Chem.,* 258, 1720, 1983. **(c) Sheets, J. J. and Vickery, L. E.,** *J. Biol. Chem.,* 258, 11446, 1983.
50. **(a) Wright, J. N., van Leersum, P. T., Chamberlin, S. G., and Akhtar, M.,** *J. Chem. Soc. Perkin Trans. 1,* 1647, 1989. **(b) Wright, J. N., Calder, M. R., and Akhtar, M.,** *J. Chem. Soc. Chem. Commun.,* 1733, 1985.
51. **(a) Shih, M.-J., Carrell, M. H., Carrell, H. L., Wright, C. L., Johnston, J. O., and Robinson, C. H.,** *J. Chem. Soc. Chem. Commun.,* 213, 1987. **(b) Childers, W. E. and Robinson, C. H.,** *J. Chem. Soc. Chem. Commun.,* 320, 1987. **(c) Kellis, J. T., Childers, W. E., Robinson, C. H., and Vickery, L. E.,** *J. Biol. Chem.,* 262, 4421, 1987. **(d) Childers, W. E., Shih, M.-J., Furth, P. S., and Robinson, C. H.,** *Steroids,* 50, 121, 1987. **(e) Njar, V., Safi, E., and Robinson, C. H.,** 199th National Meeting of the American Chemical Society, Division of Organic Chemistry, April, 1990, Boston Abstract #44.
52. **Bednarski, P. J. and Nelson, S. D.,** *J. Med. Chem.,* 32, 203, 1989.
53. **Kellis, J. T. and Vickery, L. E.,** *J. Biol. Chem.,* 262, 8840, 1987.

54. **Osawa, Y., Yoshida, N., Fronckowiak, M., and Kitawaki, J.,** *Steroids,* 50, 11, 1987.
55. **Vickery, L. E.,** personal communication.
56. **Johnston, J. O.,** unpublished work.
57. Organon company communication, March 1988.
58. (a) **Kellis, J. T., Jr. and Vickery, L. E.,** *Science,* 225, 1032, 1984. (b) **Kellis, J. T., Jr. and Vickery, L. E.,** *Biochem. Pharmacol.,* 35, 2887, 1986.
59. (a) **Covey, D. F., Hood, W. F., and Parikh, V. D.,** *J. Biol. Chem.,* 256, 1076, 1981. (b) **Metcalf, B. W., Wright, C. L., Burkhart, J. P., and Johnston, J. O.,** *J. Am. Chem. Soc.,* 103, 3221, 1981. (c) **Marcotte, P. A. and Robinson, C. H.,** *Fed. Proc. Fed. Am. Soc. Exp. Biol.,* 40, 1867, 1981.
60. (a) **Ortiz de Montellano, P. R., Kunze, K. L., Beilan, H. S., and Wheeler, C.,** *Biochemistry,* 21, 1331, 1982. (b) **Kunze, K. L., Mangold, B. L. K., Wheeler, C., Beilan, H. S. and Ortiz de Montellano, P. R.,** *J. Biol. Chem.,* 258, 4202, 1983. (c) **Ortiz de Montellano, P. R. and Komives, E. A.,** *J. Biol. Chem.,* 260, 3330, 1985.
61. **Johnston, J. O.,** *Steroids,* 50, 105, 1987.
62. **Marcotte, P. A. and Robinson, C. H.,** *Biochemistry,* 21, 2773, 1982.
63. **Furth, P. S. and Robinson, C. H.,** *Biochemistry,* 28, 1254, 1989.
64. **Mann, J. and Pietrzak, B.,** *J. Chem. Soc. Perkin Trans. 1,* 385, 1987.
65. **Bednarski, P. J., Porubek, D. J., and Nelson, S. D.,** *J. Med. Chem.,* 28, 775, 1985.
66. **Brodie, A. M. H., Schwarzel, W. C., Shaikh, A. A., and Brodie, H. J.,** *Endocrinology,* 100, 1684, 1977.
67. (a) **Brodie, A. M. H., Garrett, W. M., Hendrickson, J. R., Tsai-Morris, C., Marcotte, P. A., and Robinson, C. H.,** *Steroids,* 38, 693, 1981. (b) **Covey, D. F. and Hood, W. F.,** *Mol. Pharmacol.,* 21, 173, 1982.
68. **Newitt, J. A. and Robinson, C. H.,** *Fed. Proc. Fed. Am. Soc. Exp. Biol.,* 45 (abstr. 2362), 1885, 1986.
69. **Marsh, D. A., Brodie, H. J., Garrett, W., Tsai-Morris, C.-H., and Brodie, A. M. H.,** *J. Med. Chem.,* 28, 788, 1985.
70. (a)**Abul-Hajj, Y. J.,** *J. Med. Chem.,* 29, 582, 1986. (b) **Abul-Hajj, Y.,** *J. Steroid Biochem.,* 35, 139, 1990.
71. **di Salle, E., Giudici, D., Ornati, G., Briatico, G., and Lombardi, P.,** Conference on Aromatase: Future Perspectives, Miami, March 4, 1987, Abstract 18.
72. (a) **Coombes, R. C., Goss, P. E., Dowsett, M., Hutchinson, G., Cunningham, D., Jarman, M., and Brodie, A. M. H.,** *Steroids,* 50, 245, 1987. (b) **Reed, M. J., Lai, L. C., Owen, A. M., Singh, A., Coldham, N. G., Purohit, A., Ghilchik, M. W., Shaikh, N. A., and James, V. H. T.,** *Cancer Res.,* 50, 193, 1990.
73. **Barone, R. M., Shamonki, I. M., Siiteri, P. K., and Judd, H. L.,** *J. Clin. Endocrinol. Metab.,* 49, 672, 1979.
74. **Covey, D. F. and Hood, W. F.,** *Cancer Res.,* 42 (suppl.), 3327, 1982.
75. (a) **Ryan, K. J.,** *J. Biol. Chem.,* 221, 268, 1959. (b) **Gual, C., Morato, T., Hayano, M., Gut, M., and Dorfman, R. I.,** *Endocrinology,* 71, 920, 1962.
76. **Covey, D. F.,** In *Sterol Biosynthesis Inhibitors: Pharmaceutical and Agrochemical Aspects,* Berg, D. and Plempel, M., Eds., Ellis Horwood, Chichester, England, 1988, 534.
77. **Covey, D. F. and Hood, W. F.,** *Endocrinology,* 108, 1597, 1981.
78. (a) **Snider, C. E. and Brueggemeier, R. W.,** *J. Biol. Chem.,* 262, 8685, 1987. (b) **Moh, P. P., Li, P.-K., and Brueggemeier, R. W.,** The 72nd Endocrine Society Meeting, Atlanta, June, 1990, Abstract 972.
79. **Henderson, D., Norbisrath, G., and Kerb, U.,** *J. Steroid Biochem.,* 24, 303, 1986
80. **Giudici, D., Ornati, G., Briatico, G., Buzzetti, F., Lombardi, P., and di Salle, E.,** 8th International Symposium of the Journal of Steroid Biochemistry, Paris, 1987, Abstract 53P.
81. **Sherwin, P. F., McMullan, P. C., and Covey, D. F.,** *J. Med. Chem.,* 32, 651, 1989.
82. **Numazawa, M., Tsuji, M., and Osawa, Y.,** *Steroids,* 48, 347, 1986.

83. **Snider, C. E. and Brueggemeier, R. W.,** *J. Steroid Biochem.*, 22, 325, 1985.
84. **Brueggemeier, R. W., Snider, C. E., and Kimball, J. G.,** *Steroids*, 40, 679, 1982.
85. **Thornburg, L. D. and Stubbe, J.,** *J Am. Chem. Soc.*, 111, 7632, 1989.

Chapter 7

STRUCTURE-ACTIVITY RELATIONSHIPS OF GLUCOCORTICOIDS: IMPORTANCE OF THE REGULATED GENE AND *TRANS*-ACTING FACTORS IN DETERMINING GLUCOCORTICOID AND ANTIGLUCOCORTICOID ACTIVITY

S. Stoney Simons, Jr.

TABLE OF CONTENTS

I. INTRODUCTION

Glucocorticoids evoke many biological responses in the human body. Consequently, pharmacologists have long dreamed of constructing from modular features a steroid with precisely defined and highly specific biological activity. Initially it seemed logical that there should be a straightforward relationship between steroid structure and biological activity. This view was strengthened by the initial successes in discerning common structural features among steroids which were grouped by similar biological activities (e.g., glucocorticoid, progestational, estrogenic). With the discovery of specific, intracellular receptor proteins that mediated the actions of steroid hormones, it became possible to compare the biological activity of various steroids with their affinity for the cognate receptor. When this was done with glucocorticoids, it was generally found that those steroids with the higher binding affinity for the receptor were also the more potent agonists.[1] The problem could then be reduced to determining which structural features afforded a higher affinity.

These same early structure-activity studies also defined a class of steroids called antiglucocorticoids, or antagonists, which blocked the actions of glucocorticoids, or agonists. However, progress was much slower in defining

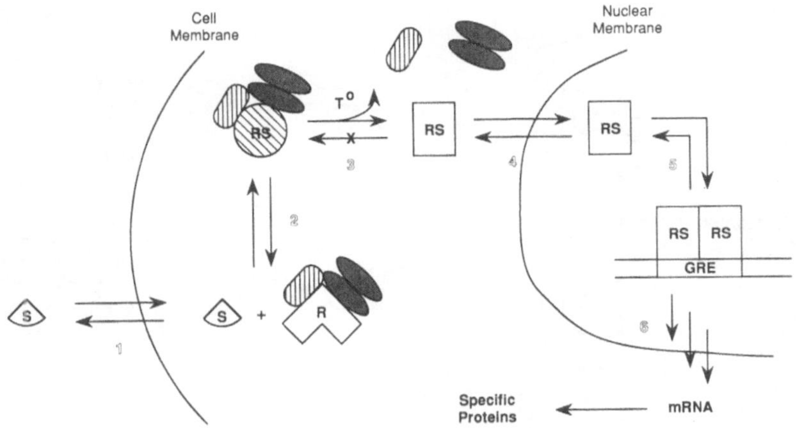

FIGURE 1. Model of mechanism of glucocorticoid hormone action. The receptor protein (R) is located predominantly in the cytoplasm and is associated with several nonreceptor proteins (hatched and shaded species). Steroid (S) enters the cell by passive diffusion (step 1) and binds to the receptor (step 2) to give the receptor-steroid complex (RS). This binding is thought to be accompanied by a conformational change in the receptor, as indicated by the change in shape of the receptor symbol. The temperature-dependent step (step 3) of activation converts the RS complex to a species with increased affinity for DNA and nuclei. Activation is accompanied by a dissociation of the nonreceptor proteins and possibly a conformational change in the RS complex. After translocation into the nucleus (step 4), monomers of the activated complex bind to both the biologically active glucocorticoid regulatory elements (GREs) and to nonspecific DNA (not shown) (step 5). The initially bound monomer facilitates the formation of the final dimeric complex (step 5). A series of ensuing steps (step 6) give rise to alterations in the levels of mRNA encoded by those genes which are regulated by the GRE.

structure-activity relationships among antiglucocorticoids. There were few obvious common structural features and the correlation between the affinity of a given antagonist for the glucocorticoid receptor and its ability to block glucocorticoid induction of a responsive gene was dubious.[1,2] An examination of the models for the mechanism of steroid hormone action (e.g., Figure 1) reveals one obvious explanation for this poor correlation. While all agonists must follow a single series of steps from binding to the receptor through receptor-steroid complex activation of gene transcription, an antagonist can function by blocking any of these steps. In fact, clinically useful "antiglucocorticoids" (such as aminoglutethamide, ketoconazole, metyrapone, mitotane, and trilostane) act at steps prior to those of Figure 1 by preventing steroidogenesis.[3]

Most studies of antisteroid action have concentrated on those compounds that bind to the receptor protein to give a complex that is defective in some required activity. Almost all of the steps in Figure 1 have been proposed as possible sites of antiglucocorticoid action. It was initially suggested that agonists and antagonists bound to two separate forms of the receptor.[4] This hypothesis is no longer accepted but it does appear that steroid binding (step

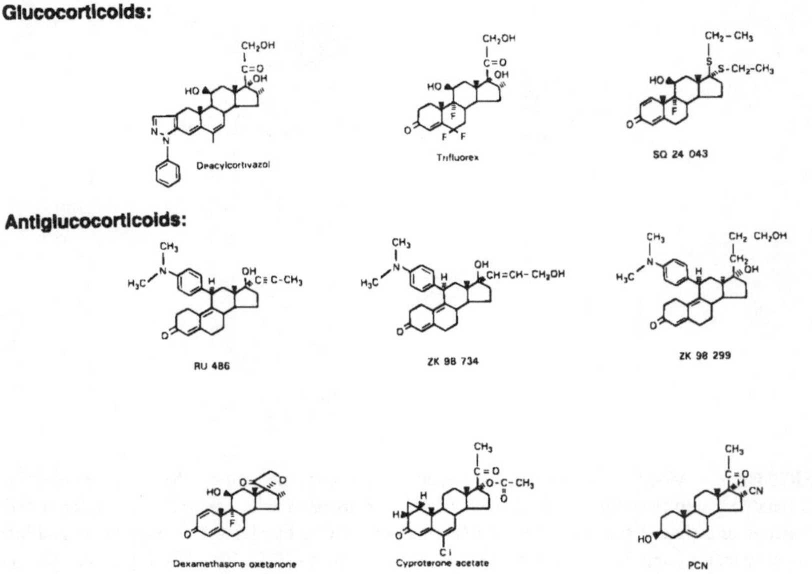

FIGURE 2. Various synthetic glucocorticoids and antiglucocorticoids with uncommon structures.

2) causes a conformational change in the receptor.[5,6] A recent report that receptors lacking hsp90 are not constitutively active[7] suggests that this steroid-induced conformational change in receptors may be necessary for the expression of biological activity. Antagonists have also been proposed to act by rapidly dissociating from the receptor[8,9] (step 2), by causing increased dissociation of agonists[10-12] (step 2), by preventing dissociation of hsp90[13] (step 3), by having defective nuclear translocation[14] (step 4), by inhibiting the DNA binding and/or the subsequent dimerization of monomeric receptor-steroid complexes[15] (step 5), and finally by what may be the most popular explanation of altering pretranscriptional events[16] and/or being unable to regulate transcription[17-19] (step 6). A variant defect in step 5 would be a steroid that altered the specificity of receptor binding to nonspecific DNA vs. the biologically active glucocorticoid regulatory, or responsive, elements (GREs). Such altered specificity appears to occur in mutant receptors[20-21] but has not yet been seen for wild-type receptors associated with antagonists.[22-24] One interesting and possibly useful mechanism for expressing antisteroid activity that has not been actively pursued is to control the entry/retention of specific steroids in cells (step 1).[25]

Throughout all of the above studies, there has remained one perplexing and irrepressible, albeit largely ignored, dilemma: Why is it that nearly all antagonists can display partial, and sometimes full, agonist activity? The antiglucocorticoids progesterone, dexamethasone 21-mesylate (Dex-Mes), dexamethasone oxetanone (Dex-ox) (Figure 2), and cortexolone all show

>50% agonist activity with regard to the induction of tyrosine aminotransferase (TAT) in Fu5-5 rat hepatoma tissue culture cells.[26-27] Such behavior is not limited to antiglucocorticoids. Thus, tamoxifen is a full agonist in mouse, a partial agonist in rat, and a full antagonist in chick (see Reference 28 and references therein). Hydroxy-tamoxifen has ~50% agonist activity with both transfected human estrogen receptors and a transfected reporter gene in chicken embryo fibroblasts[29] and displays 100% agonist activity with a chimeric receptor formed by substitution of the estrogen receptor steroid-binding domain into the chick progesterone receptor.[30] Some antiestrogens display both agonist and antagonist effects in rat uterus.[31] RU 486, usually considered a pure antagonist, can act both as a progestin and as an anti-progestin in the same cells.[30,32] Cyproterone acetate, conventionally regarded as an antiandrogen, acts as an agonist for repression of androgen receptor mRNA.[33] Thus, the classification of a given steroid as an antagonist is not a straightforward matter.

The purpose of this chapter is twofold. First, I will briefly review the progress in determining which structural features dictate the type of activity, and the potency, of various synthetic glucocorticoids. Only steroidal structures will be considered since very few nonsteroidal compounds have been described which will bind to the glucocorticoid receptor.[34-37] Second, I will summarize the evidence that several quantitative aspects of steroid activity depend not as much on the structure of the ligand as on the responsive gene that is examined and/or the cellular environment of that gene. In particular, I will discuss the evidence and possible causes for variations both in the amount of agonist activity displayed by antiglucocorticoids and in the potency of glucocorticoids (i.e., the concentration of agonist that is required for half-maximal activity, or EC_{50}). These variations cannot be explained by the current model of glucocorticoid action (Figure 1). Thus, a better understanding of these phenomena is essential for a more precise description of glucocorticoid action, is necessary for developing an accurate and predictive model, and may be helpful in limiting the undesirable side effects of glucocorticoids in various therapeutic regimes. Only the data from *in vitro* studies (e.g., tissue culture cells) will be considered. This is because, for some yet-undetermined reason, most of those steroids which are antiglucocorticoids *in vitro* are agonists *in vivo*.[38-40] This indicates a further level of complexity *in vivo* which is beyond the scope of this chapter.

II. CLASSICAL ATTEMPTS TO ESTABLISH STRUCTURE-ACTIVITY RELATIONSHIPS

A. GLUCOCORTICOIDS

The rate-determining step in glucocorticoid hormone action has been found to be activation,[9,41,42] in which the receptor-steroid complex acquires an increased affinity for nuclei and DNA. However, the amount of activated complex (and its subsequent nuclear binding and biological activity) is a function of the rate of steroid dissociation from the receptor, which is also the major determinant of steroid affinity.[9,43,44] Thus, the affinity of steroid for receptor has been a predictor of potency as a glucocorticoid. This view is supported by the two observations that those steroids having higher affinity for receptors yield more activated complexes and more biological activity[9,43,45] and that the amount of receptor is the limiting factor for glucocorticoid induction of transcription, so that more available receptor affords a greater biological response to the same steroid.[46,47]

In theory, determining the affinity of a given steroid for the glucocorticoid receptor should be a straightforward matter. Typically, affinity measurements have been performed using receptors from rat liver, thymus, or hepatoma tissue culture (HTC) cells,[2,48,49] with the assumption that the same values, within experimental error, will apply to all other systems. However, the glucocorticoid receptors from different tissues[45,48] and species[50] do not show the same absolute or even relative affinities for glucocorticoid steroids. This is presumably due to tissue-specific changes in metabolism permeability but could also result from minor differences in amino acid sequences in the steroid-binding domain[51,52,52a] which is known to vary in different species.[51,53,54] Another common problem in accurately determining steroid affinity is that many reactions are not carried out to equilibrium, which usually requires incubations of at least 24 h at 0°C. For some steroids with very high affinity, even longer incubations would be required.[55,56] That shorter (e.g., 2.5 h) incubations work at all probably reflects the fact that, to a first approximation, all rates of association are similar and the K_d of a given steroid is therefore proportional to its rate of dissociation.[44,57]

Since the potency of glucocorticoid agonists is generally proportional to the affinity of their binding to the receptor, structure-activity relationships can be simplified to structure-affinity relationships. Thus, the question becomes "What structural feature(s) determines higher affinity?" Unfortunately this question still has not been answered. It has been proposed that the A ring of steroids controls affinity[58] but how this might be effected is not clear since the introduction of a simple methyl group at C-1 of prednisolone causes a 660-fold reduction in affinity[2] while fusion of the extremely bulky phenyl pyrazole group to the C-2/C-3 positions of a dexamethasone (Dex)-like structure gives deacylcortivazol (Figure 2), which is the most potent glucocorticoid known with an affinity that is ≥9 times that of Dex.[55] Furthermore, an X-ray structure of cortivazol[58a] shows that the conformation of the A ring is less bent from the plane of the steroid than expected for high-affinity binding.[59,60] Certain functional groups have been found to increase the binding affinity of a given steroid[2,45,49,61-64] but it is not yet established that a group which increases the affinity of one steroid will always cause the same increase when introduced into another steroid. Given that the number of structural modifications of the "basic glucocorticoid structure" is almost infinite, a rigorous study of the effect of single modifications, one at a time, on the activity of several different steroids would be unimaginably laborious just at the synthetic level.

The Roussel-Uclaf group has collected a vast amount of data on numerous steroids, most of which contain unusual substituents.[45,49,61,63] In many cases the structure of the steroid has been determined, either directly by X-ray diffraction or indirectly by calculations, and a qualitative, theoretical picture of the various steroid-binding cavities has emerged.[49,60,63] I will not try to summarize these data, which will obviously be of great use in the future. However, their current benefit is still largely correlative and their predictive value has yet to be established.

B. ANTIGLUCOCORTICOIDS

Briefly, it has not yet proved possible to find any functional group that will always convert a glucocorticoid into an antiglucocorticoid. It has been proposed that the substituents on the D ring determine the activity of the steroid[58] but this was not confirmed when the same D-ring substituents were added to cortisol, Dex, and deacylcortivazol.[56] Furthermore, 11β-substituted steroids are usually,[60] but not always (see below), antiglucocorticoids.

As discussed above (Section II.A), a direct relationship has been established between the affinity of glucocorticoids binding to the receptor and the potency of each steroid as an agonist. Unfortunately no such correlation has evolved for antiglucocorticoids.[1,2,65] Nor has it been possible to establish a single defect in the steps of hormone action (Figure 1) that is common to all antiglucocorticoids (see also Section I). The proposal that rapidly dissociating steroids are antagonists[8,9] is clearly incomplete since the affinity labeling steroid Dex-Mes binds covalently to the receptor but is an irreversible antiglucocorticoid.[14,66-68] Similarly, studies on the activation,[69,70] nuclear binding,[14,67] and DNA binding[22,69] of covalent, Dex-Mes-labeled receptors argue that alterations in any of these steps are not required for antagonist activity. Such structure-activity problems are not unique for antiglucocorticoids but have been encountered for other antisteroids (e.g., see Reference 71). Recent results suggest that this lack of a common determinant of antisteroid activity arises not from the ability of antagonists to act at different steps in hormone action (see Figure 1) but rather from the capacity of receptor-steroid complexes to *interact differently with individual regulated genes*.

III. BIOLOGICAL ACTIVITY OF A GIVEN GLUCOCORTICOID DEPENDS ON THE GENE EXAMINED

A. CLASSICAL QUANTITATION AND ASSIGNMENT OF BIOLOGICAL ACTIVITY

A tenet of the current models of steroid hormone action has been that 50% receptor occupancy should always produce no more, and no less, than

50% biological activity. Thus, a left shift in the dose-response curve relative to the dose-binding curve, such that 50% of the maximal biological response is produced by less than 50% receptor occupancy, should never be observed for steroid hormones. In fact, a close correlation in intact cells between the dose-dependent curves for steroid binding to its receptor and induction of a biological response has classically been regarded as evidence for the direct involvement of receptor-steroid complexes in the production of the response.[72] This doctrine fits nicely with the conclusion that activation of steroid-bound receptors is the rate-limiting step of glucocorticoid action[9,41,42] since the amount of activated complexes for a given steroid is directly proportional to the affinity of that steroid for receptors[9,43] and continuous steroid occupancy of receptors appears required for biological activity.[43a,b] Thus, half saturation of the whole cell receptors should afford half of the maximal amount of activated complexes, of nuclear bound complexes, and of biological response. It should be noted, however, that these models are acknowledged to ignore the contributions to gene regulation of any events distal to the nuclear binding of complexes[73] (see also Section III.B).

Typically, a given steroid has been classified as a glucocorticoid, partial or full antiglucocorticoid, or inactive steroid based on its activity in a single bioassay. The same assay is often used to determine the potency, or EC_{50}, of the steroid. The implicit assumption in such determinations is that the same activity would be observed for the same steroid interacting with glucocorticoid receptors in different cells and tissues. The single, most commonly used bioassay is the induction of TAT in HTC cells.[1,2,27,45,74,75] Usually there has been a good correlation between whole cell occupancy of receptors and induction of TAT in HTC cells.[76,77] However, as more genes have been examined and as more quantitative measurements have been made, discrepancies were noted.

B. NONCLASSICAL BEHAVIOR OF AGONISTS AND ANTAGONISTS

1. Left Shift in Dose-Response Curves for Gene Induction by Glucocorticoids

The maximal response to saturating levels of glucocorticoids, and probably all steroids, is determined by the amount of receptor available.[46,47] However, there is mounting evidence that the amount of induction seen with *subsaturating concentrations* of glucocorticoids is influenced by other factors. This latter phenomenon should therefore be distinguished from the observations that the *maximal level of induction* of a selected gene can vary among cell lines,[78-80] that the expression of a specific glucocorticoid-inducible gene can be extinguished without affecting other inducible genes,[80-83] and that other factors can *switch the response* to agonist complexes from gene induction to gene repression.[84] In contrast to these phenomena, we have found that the EC_{50} for Dex induction of TAT is not the same in two rat hepatoma cell lines

(HTC and Fu5-5). Thus, while approximately 10^{-8} M Dex is required for half-maximal occupation of the whole cell receptors in HTC and Fu5-5 cells, half-maximal induction of TAT enzyme activity occurs at about 2×10^{-8} M Dex in HTC cells (which would occupy ~60% of the receptors) and at about 0.2×10^{-8} M Dex in Fu5-5 cells (which would occupy ~30% of the receptors).[77] However, given that the affinity of Dex for a specific glucocorticoid receptor is thought to be constant,[2] it was even more surprising to find that the EC_{50} for Dex induction of TAT was not constant in each cell line[26,27,77,85-87a] but varied over several weeks from 0.4 to 3×10^{-8} M in HTC cells and from 0.08 to 0.4×10^{-8} M in Fu5-5 cells.[26,85,87a] These differences in the dose-response curves in Fu5-5 vs. HTC cells were also seen at the level of TAT mRNA.[85,87] Further studies revealed that the induction of TAT mRNA at low Dex concentrations in Fu5-5 cells resulted from an increased rate of transcription[87] and utilized the same transcriptional start site as in rat liver cells.[87b]

The unequal behavior of Fu5-5 and HTC cells could be due to sequence differences in the TAT gene of the two rat cell lines; however, such an explanation cannot explain the variable responses *within* each cell line. Inheritable, covalent modifications of the TAT gene are unlikely since the variations in EC_{50} are reversible.[85] The dissimilar dose-response curves in Fu5-5 and HTC cells do not arise from preferential metabolism of Dex in one cell line[26] and are not unique to Dex but were seen with four other glucocorticoids of very different structures, i.e., cortisol, trifluorex, deacylcortivazole,[77] and SQ 24,043[27] (Figure 2). Other explanations which can be eliminated include (1) differential steroid permeability[25] in Fu5-5 vs. HTC cells since Dex binding to the whole cell receptors is equal in both cell lines,[77] (2) disparate amounts of glucocorticoid receptor,[77] (3) modified physicochemical properties of the receptors in the two cell lines,[68] (4) changed amounts of a TAT-inactivating factor[88] in the two cell lines,[77] (5) unequal TAT basal levels,[85] (6) altered kinetics of TAT enzyme induction,[77] and (7) differences in the media, since the same behavior was seen in Swim's S77 and Richter's IMEM media.[85]

A left shift in the dose-response curve relative to the receptor-occupancy curve is not, however, seen for all glucocorticoid-inducible genes in Fu5-5 cells. The dose-response curves for Dex induction of TAT and glutamine synthetase (GS) are not superimposable in Fu5-5 cells but are almost the same in HTC cells (Figure 3). Similarly, the dose-response curve for Dex induction of TAT mRNA is left shifted from that for mouse mammary tumor virus (MMTV) RNA in Fu5-5 cells which had been stably transfected with MMTV (Figure 4). It thus appears that the current model of hormone action (Figure 1) must be amended to include factors which could somehow modulate the parameters for glucocorticoid induction of at least the TAT gene, but not all glucocorticoid inducible genes, in hepatoma cells. Some of these factors may interact with portions of the transcriptional machinery that do not involve the

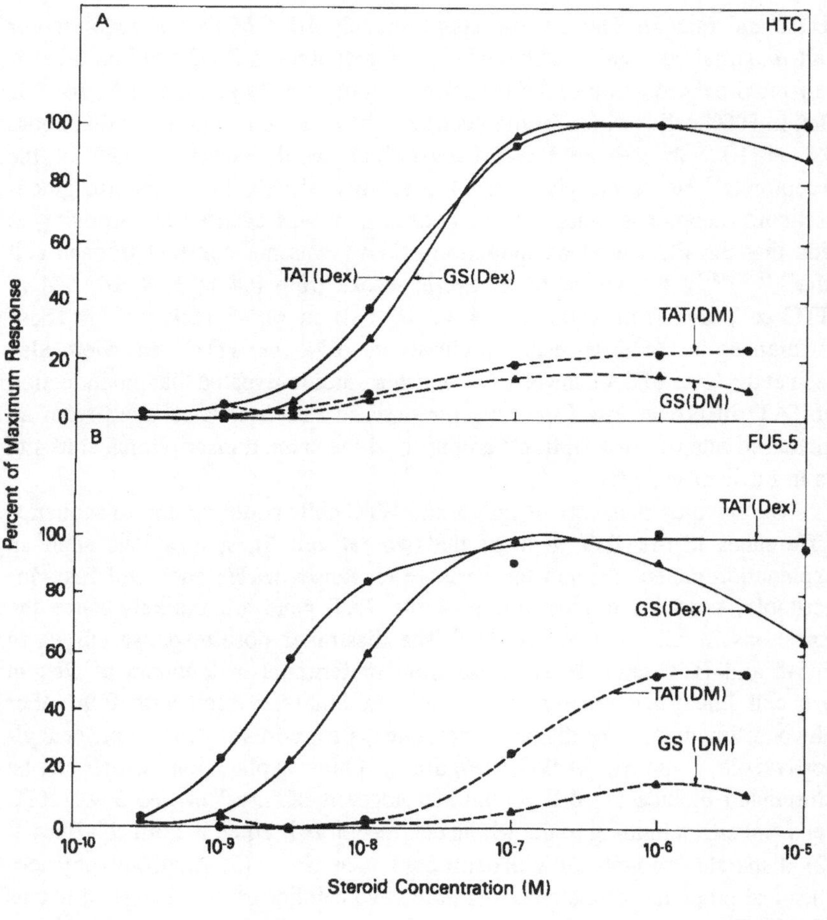

FIGURE 3. Dose-response curves for induction of TAT and GS in HTC and Fu5-5 cells by Dex and Dex-Mes. The specific activities of TAT (●) and GS (▲) in the lysates of duplicate dishes of cells that were incubated with various concentrations of Dex (——) or Dex-Mes (---) for 16 to 20 h at 37°C were expressed as percent of maximal induction and plotted as a function of steroid concentration. (From Mercier, L., Miller, P. A., and Simons, S. S., Jr., *J. Steroid Biochem.*, 25, 11, 1986. With permission.)

glucocorticoid receptor since the dose-response curve for cAMP induction of TAT in Fu5-5 cells is also left shifted from that in HTC cells.[90]

Other examples of left shifts in dose-response curves for glucocorticoid induction do exist but are not numerous, partially because of the relatively few studies at subsaturating concentrations of steroid. Left shifts relative to the curves for steroid occupancy of receptors have been reported for glucocorticoid induction of α-lactalbumin in rat mammary gland explants,[91] of alanine aminogransferase (AAT) and TAT vs. γ-glutamyltransferase (GGT) in Fao cells,[92] of pulmonary surfactant protein A mRNA in fetal human lung

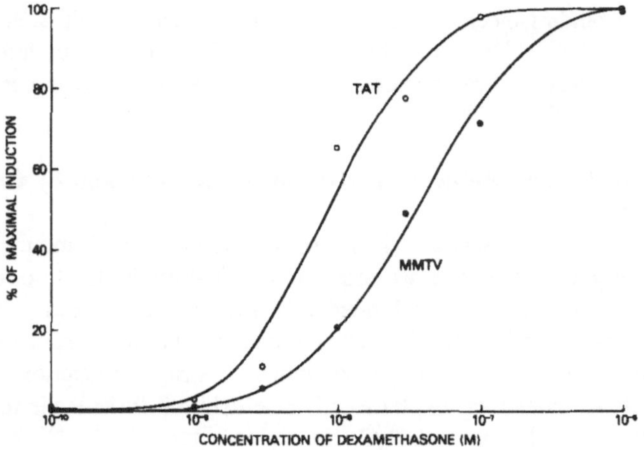

FIGURE 4. Dose-response curves for induction of TAT and MMTV RNA in clone 24 of MMTV infected Fu5-5 cells. The amount of TAT (○) and MMTV (●) RNA was determined by autoradiography (and optical scanning) of Northern blots containing ^{32}P-labeled TAT-cDNA or MMTV DNA that had been hybridized with the total cytoplasmic RNA from cells treated for 6 h at 37°C with various concentrations of Dex. The average value from duplicate cultures was expressed as percent of maximal induction and plotted as a function of steroid concentration. (From Wasner, G., Oshima, H., Thompson, E. B., and Simons, S. S., Jr., *Mol. Endocrinol.*, 2, 1009, 1988. With permission.)

cultures,[93] and of chloramphenicol acetyl transferase (CAT) from a transiently transfected MMTV/CAT reporter gene in T47D human breast cancer cells containing a stably transfected glucocorticoid receptor expression vector.[94] While most of these examples have not been rigorously demonstrated to be primary GRE-mediated responses, they do suggest that left-shifted dose-response curves may be much more common than previously suspected. Whether or not the EC_{50} of these left-shifted responses changes over time, as seen above for TAT induction, has not been reported.

Many more left-shifted dose-response curves can be found among genes that are repressed by glucocorticoids. Examples include Dex inhibition of prostaglandin secretion in rat renomedullary interstitial tissue culture cells,[95] of vasoactive intestinal peptide (VIP)-induced effects in rat pituitary cells,[96] of cellular proliferation of rat hepatoma[97] and mammary tumor cells,[98] of plasmin receptor activity in human sarcoma cells,[99] of phorbol ester-induced expression of CAT from a transiently transfected collegenase/CAT hybrid gene in HeLa-tk$^-$ cells,[100] and possibly of glucocorticoid receptor mRNA levels in COS cells that were transiently transfected with a glucocorticoid expression vector.[101] It is not known whether the EC_{50} for repression of these genes can change with time, as has been documented for TAT.

Very little is known about glucocorticoid repression of genes. Many of the effects are mediated by negative (or composite) GREs (nGREs) and/or by non-DNA binding mechanisms that are quite different from those employed for induction.[84,100,102-104] It may be significant that no cases of left-shifted dose-response curves have appeared so far for genes where the receptor acts by binding to nGREs.[84,105]

2. Left Shift in Dose-Response Curves for Gene Induction by Other Steroids

The above "nonclassical" behavior of glucocorticoid induction and repression of gene expression, as reflected by left shifts in the dose-response curves, does not appear to be limited to a few, glucocorticoid responsive genes. Reports of similar left shifts can be found for all of the steroid hormones (except mineralocorticoids) and for vitamin D. Estrogen induction of vitellogenin A2/CAT constructs in HeLa cells is left shifted,[106] as is the induction of two inducible mRNAs in GH_4C_1 rat pituitary tumor cells,[107] of several mRNAs in human breast cancer cells,[108,109] of five yolk protein genes in avian liver,[110] of estrogen receptor mRNA in *Xenopus laevis* liver cultures,[111] and of a transiently transfected hybrid chick vitellogenin II/CAT construct in T47D cells when examined in the presence of progesterone.[112] For progestins, the dose-response curve for induction of transiently transfected MMTV/CAT constructs by R5020[113] and by progesterone[114] appears to be left shifted. Repression of α-lactalbumin synthesis in mouse mammary gland explants[91,115] and of progesterone receptor mRNA by Org 2058 in human breast cancer cells occurs at lower concentrations than required for receptor occupancy.[116] Finally, the curves for both androgen induction of kidney androgen protein (KAP) in the kidneys of intact mice[117] and vitamin D induction of osteocalcin synthesis in rat osteoblast-like cells[118] are left shifted. Thus, it appears that induction of selected genes at lower concentrations than are required for comparable levels of receptor occupancy may be a common feature of all the steroid hormones. Whether or not the EC_{50} for the above genes varies over time, as seen for glucocorticoid induction of TAT (see Section III.B.1), remains to be determined.

3. Variation in Amount of Agonist Activity for Antiglucocorticoids

A well-known phenomenon among antisteroids is that they can display partial agonist activity, the amount of which is dramatically different in various systems (see Section I). In our studies of TAT induction, we have yet to find a pure antiglucocorticoid. Every steroid examined so far, which binds to the glucocorticoid receptor with any reasonable affinity, can display some agonist activity[86,89] (Table 1). Thus, even the purported pure antiglucocorticoid RU 486 (=RU 38,486) affords some agonist activity in selected systems[17,24,101] (Table 1). Additional examples have also been reported in which progesterone[119,120] and several other antiglucocorticoids[92] exhibit considerable to complete agonist activity.

What was remarkable in the studies of TAT induction, however, was that (1) each antiglucocorticoid except PCN displayed more agonist activity in

TABLE 1
Relative Agonist Activity of Antiglucocorticoids for Induction of TAT in HTC and Fu5-5 Cells

Steroid	HTC cells	Fu5-5 cells
RU 486	2.6	8.0
ZK 98,299	2.5 (n = 4)	7.5 (n = 4)
ZK 98,734	0.0 (n = 4)	8.9 (n = 4)
Dex	100	100
Dex-Mes	56 (n = 4)	94 (n = 4)
Dexamethasone oxetanone	36	67
Progesterone	10 (n = 3)	58 (n = 3)
Cortexolone	35	73
Cortisol 21-mesylate	4.3 (n = 4)	40 (n = 5)
Cyproterone acetate	2.5	2.8 (n = 4)
Pregenolone 16α-carbonitrile (PCN)	2.5 (n = 6)	0.0

Note: The induction of TAT enzyme activity by Dex and different antiglucocorticoids in at least two experiments (number of experiments is given in parentheses when n ≠ 2) was determined as in Figure 3. The percentage of agonist activity was calculated by dividing the value of maximal induction above basal level for each steroid (usually observed at 10^{-5} *M*) by the full induction value obtained with 2 to 3 × 10^{-6} *M* Dex.[26,27,77]

Fu5-5 cells than in HTC cells (Table 1), (2) the increased agonist activity of antiglucocorticoids was not seen for all glucocorticoid-inducible genes in the same cells, both at the level of protein[26,85,92] (Figure 3) and mRNA transcripts[89] (Figure 4), (3) the amount of agonist activity of a given antiglucocorticoid in either Fu5-5 or HTC cells was not constant but changed over time in a reversible manner,[26,85,86] and (4) an inverse correlation existed between the amount of agonist activity with Dex-Mes and the EC_{50} for Dex in *both* cell lines (Figure 5). These properties are very similar to those discussed in Section III.B.1 for glucocorticoid induction of TAT and suggest that a common mechanism is modulating both the EC_{50} for Dex induction and the amount of agonist activity of antiglucocorticoids. However, the current model of steroid hormone action (Figure 1) appears incapable of suggesting a viable mechanism because other glucocorticoid-inducible genes *in the same cell* are not similarly affected and because the magnitude of agonist activity for the same antagonist complex is not constant in a given cell line but varies over time in a reversible manner.

IV. INABILITY OF EXISTING MODELS FOR ANTISTEROID ACTION TO EXPLAIN MODULATION OF TAT GENE EXPRESSION

Despite the apparent linkage of changes in Dex EC_{50} and amount of Dex-Mes agonist activity (Figure 5), the two phenomena may be unrelated. Thus,

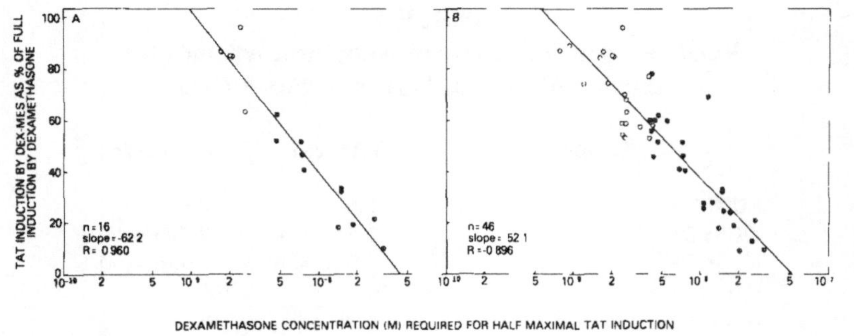

FIGURE 5. Correlation between the level of Dex-Mes agonist activity and the $\log_{10}$(EC_{50} for Dex) for the induction of TAT. Experiments were performed as in Figure 3 to simultaneously determine the Dex EC_{50} value from the Dex dose-response curve and the level of Dex-Mes agonist activity (Dex-Mes concentration = 10^{-6} *M*) in duplicate plates of Fu5-5 (○) and two lines of HTC (●,◑) cells. A. Data were obtained from 16 separate experiments by one investigator over a 15-month period. B. Data were obtained from 46 separate experiments (including data in A) by 4 different investigators over a period of 8 years (R = correlation coefficient determined by linear least squares program). (From Simons, S. S., Jr., Miller, P. A., Wasner, G., Miller, N. R., and Mercier, L., *J. Steroid Biochem.*, 31, 1, 1988. With permission.)

while the current model of hormone action is unable to account for the responses with agonists, it is conceivable that one of the several models for antisteroid action might explain the variations in agonist activity for antiglucocorticoids.

A. RATE OF DISSOCIATION OF RECEPTOR-STEROID COMPLEX

Many antagonists have a more rapid rate of dissociation from the receptor than Dex or cortisol, which led to the hypothesis that a rapid rate of dissociation from receptors was diagnostic of antisteroids.[8,44] Conversely, other studies indicated that increased agonist activity of antiglucocorticoids is associated with increased affinity for the receptor.[56,64] Thus, a cellular factor(s) that could influence the affinity of antagonists for glucocorticoid receptors might explain the modulation of agonist activity. However, subsequent studies have failed to establish the generality of this correlation.[121,122] The most blatant exception is Dex-Mes, which forms a nondissociable, covalent receptor-steroid complex[70,123] but is an antagonist.[14,67]

B. NUCLEAR CONCENTRATION OF RECEPTOR-STEROID COMPLEXES

The most important determinant of transcriptional activity for those transcription factors that act by binding to chromatin is their nuclear concentration.[124] Similarly, the concentration of nuclear-bound receptor-steroid complexes is the determining factor concerning the magnitude of glucocorticoid-induced gene transcription.[46,47] While activation and binding to GREs proceeded normally for Dex-Mes-labeled complexes,[22,69,70] the amount of whole cell nuclear binding was reduced.[67,68] This decrease in the quantity of nuclear-bound covalent complexes was initially thought to reflect the antagonist activity of Dex-Mes.[67,68] However, further studies in a line of bovine papilloma virus (BPV) transformed murine fibroblasts, where ≥90% of the receptors in intact cells were covalently labeled, showed that this could not be the entire explanation. Under conditions where Dex-Mes caused nuclear translocation of 30% of the receptors, Dex-Mes displayed only 3 and 7% agonist activity in the induction of two different mRNAs.[14] A similar dissociation of antagonist activity and amount of nuclear-bound complexes has been reported for RU-486 in human leukemic cells.[125]

C. OTHER MODELS

One problem with all of the other conventional models for antisteroid hormone action is that they ignore the possible role of events distal to the nuclear binding of activated complexes. Part of the reason for this stems from the fact that many of the early studies were conducted in the same system (i.e., TAT induction in HTC cells) so that there were few occasions to test this model rigorously until recently. Furthermore, even in the early experiments, quantitative discrepancies between repeated experiments and between dose-response vs. receptor-occupancy curves[72] appear to have been overlooked. Regardless of the reasons, though, such models will never be able to account for different activities of the same steroid with different genes in the same cell (for glucocorticoids, see References 26,89, 120, and 126 to 128, and for estrogens and progestins, see References 19, 29 to 32 and 109), not to mention variations in the activity of the same steroid with the same gene in the same cell (see Sections III.B.1 and 3), when there is only one receptor gene[53,129] encoding one receptor protein.[68] Thus, a consideration of new models that depend on the individuality of the induced genes appears unavoidable.

V. POSSIBLE GENE-SPECIFIC MODELS TO EXPLAIN THE MODULATION OF GLUCOCORTICOID INDUCTION OF SELECTED GENES

A. DIFFERENT GREs

The nucleotide sequence of all GREs is not identical.[130] Furthermore, changing just two nucleotides can convert an nGRE into a GRE.[84] Thus, minor changes in the GREs of responsive genes could conceivably alter the affinity for receptor-steroid complexes and thus the dose-response curves of agonists. In fact, such a model of high-affinity, biologically active nuclear sites was popular when the nuclear binding of receptor-steroid complexes was first observed to be saturable but was discarded when these results were found to be artifactual.[131,132] Recently, such differences in affinities for hormone regulatory, or responsive, elements (HREs) have been observed for receptor-estrogen complex binding to uteroglobin vs. vitellogenin EREs.[106] It is not yet known if such differences could give rise to different dose-response curves in intact cells. Nevertheless, such explanations would be unable to account for reversible changes in dose-response curves or for variations in the amount of agonist activity for antagonists.

B. MULTIPLE *TRANS*-ACTIVATION DOMAINS IN THE RECEPTOR

Molecular dissection of the receptor proteins has determined that sequences at opposite ends of the protein are involved in the regulation of gene transcription.[17,133,134] Recent results[15,19,29,135] indicate that these two domains can be differentially activated by bound antagonists, probably by steroid-induced conformational changes in the receptor. The number of activated domains (0, 1, or 2) in combination with yet-unidentified factors specific to the responsive gene and cellular environment would then dictate the amount of agonist activity seen for a given steroid. Such a model is capable of explaining many of the above observations and, it should be noted, implies receptor-protein interactions. It thus represents a major departure from previous models (see Figure 1) and should facilitate numerous new and testable hypotheses regarding hormonal control of gene transcription. However, even this model does not appear to be sufficiently flexible to explain the variations either in the agonist activity of a given antagonist, or in the EC_{50} of an agonist, for the same gene in a given cell.

C. PROTEIN-PROTEIN INTERACTIONS OF RECEPTOR WITH *TRANS*-ACTIVATING FACTORS

Several studies have implicated receptor-protein interactions in the control of gene transcription.[136-140] Recently, direct evidence has been obtained that other molecules can contact steroid receptors to modify their transcriptional activity. Jun and fos interact with glucocorticoid receptors to either inactivate the receptor complex[100,102] or switch the activity between induction and repression.[84] These interactions of glucocorticoid receptor-agonist complexes have been observed both with "composite" GREs, which have previously been classified as nGREs,[84] and with conventional GREs, such as the MMTV GRE.[102] It is not yet known whether fos and/or jun can alter the dose-response curves of glucocorticoid-regulated responses.

Another class of *trans*-acting factors, which comes closer to reproducing the modulated responses discussed above for TAT, is the steroid receptors themselves. In particular, observed left shifts in the dose-response curves for steroid induction of transiently transfected hybrid CAT constructs were proposed to result from cooperative interactions both of glucocorticoid receptors at two closely spaced TAT GREs[141] and of estrogen and glucocorticoid or progesterone complexes at adjacent ERE and GRE or PREs.[112,142]

While the above results indicate that the mechanism of steroid hormone action should be modified, at least in some situations, to include protein-protein interactions with the receptor, the incorporation of such interactions in the model is still unable to account for the modulation of TAT induction parameters that has been described above for glucocorticoids and antiglucocorticoids. Nevertheless, these results do provide a precedent for a model which can explain the modulation of TAT induction parameters. This new model involves varying levels or activities of a different *trans*-acting factor(s) which both binds to a *cis*-acting sequence(s) and interacts with GRE-bound complexes.

D. VARYING LEVELS OR ACTIVITIES OF A *TRANS*-ACTING FACTOR(S) ACTING AT A NON-GRE, *CIS*-ACTING SEQUENCE

1. Evidence for a *trans*-Acting Factor(s) Acting at a Non-GRE, *cis*-Acting Sequence

Two lines of evidence support the involvement of a *trans*-acting factor in the modulation of TAT induction by glucocorticoids. First, the properties of Dex-Mes induction of a stably transfected hybrid TAT/CAT construct in Fu5-5 and HTC cells are the same as for the endogenous TAT gene. Thus, Dex-Mes induction of both TAT and CAT was greater in transfected Fu5-5 cells than in transfected HTC cells.[87b] Since the same sequence was less active in HTC cells, it is unlikely that the effects in Fu5-5 cells are due solely to the presence of a *cis*-acting sequence. Instead, we favor the action of a *trans*-acting factor which binds to a *cis*-acting sequence. Second, a candidate *cis*-acting sequence has been identified in transient transfections of Fu5-5 cells. Thus, a 62-bp sequence from the 5′ region of the TAT gene can convey the same behavior to a heterologous promotor and gene as is seen for the endogenous TAT gene.[142a]

2. Evidence for Varying Levels or Activities of a *trans*-Acting Factor(s)

The linear semilog plot of Dex-Mes agonist activity vs. $\log_{10}$ Dex EC_{50} (Figure 5) shows that the variations of these two parameters in the two cell lines are not restricted to discrete ranges and are not even quantized but rather form a continuum of values. These values changed reversibly, and in both directions, over several weeks in culture for the same cells. Furthermore, the amount of Dex-Mes agonist activity in cloned HTC cells was randomly distributed over the range defined by HTC cells in Figure 5. Such continuous and reversible small changes in these induction parameters are difficult to account for by modifications of the TAT gene sequence. A simpler explanation involves changing levels or activities of a *trans*-acting factor that is proposed to interact with the above identified *cis*-acting sequence. Further support for this mechanism came from experiments in which the amount of Dex-Mes agonist activity could be controlled simply by varying the cell density

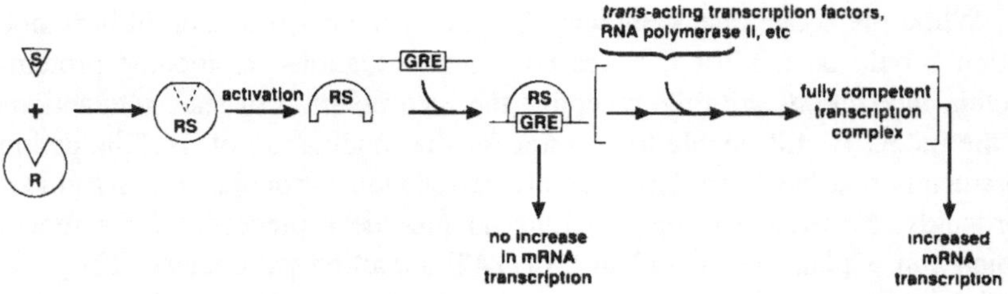

FIGURE 6. Simplified model of glucocorticoid receptor-steroid complex induction of gene transcription, which requires additional events distal both to steroid (S) binding to receptor (R) and to activated complex (RS) binding to the biologically active DNA sequences of the glucocorticoid regulatory element (GRE). (From Simons, S. S., Jr., Mercier, L., Miller, N. R., Miller, P. A., Oshima, H., Sistare, F. D., Thompson, E. B., Wasner, G., and Yen, P. M., *Cancer Res.*, 49(suppl.), 2244, 1989. With permission.)

and/or the volume of medium in the culture dish. Cell density may be the dominant determinant since regions of cells that were plated at different densities *in the same culture dish* exhibited different amounts of agonist activity.[87a] Since these experiments are conducted over a period of ~40 h (less than the time for two doublings of the cells), mechanisms that require changes in the level or activity of our proposed *trans*-acting factor are much more attractive than those involving modifications of the TAT gene sequence.

3. Possible Mechanism of Action

The *cis*-acting element that we have identified is nowhere near either the GRE or the promoter of the TAT gene. Thus, direct interactions, similar to those occurring between receptor-steroid complexes at closely spaced HREs to cause a left shift in the dose-response curves,[112,141,142] could not occur. However, our *trans*-acting factor could interact with the GRE and/or the TAT promoter region by a looping of the DNA[143] and/or through the intermediacy of an "adapter"[144] or "co-activator"[145] molecule. The precise mechanism is likely to be different, though, from that of an "adapter", which was originally invoked to explain synergism, and from that of a "co-activator", which is thought to be required for *trans*-activation.

Loeb and Stickland have described mathematically how hormone binding to receptors to give a directly proportional number of biologically active second messengers (e.g., cAMP, IP_3, Ca^{++}) will give rise to left-shifted dose-response curves.[146,147] While steroids do not generate these second messengers, it has been pointed out that steroid hormone action does involve at least one second messenger in mathematical terms.[86,147] This is because steroid binding to receptors, and even receptor-glucocorticoid complex binding to GREs, does not, of itself, cause an increase in the rate of specific gene transcription (Figure 6). Other intervening events, such as the activation of receptor-steroid complexes, the induction of DNase I hypersensitive sites,[148-150] and the formation of the transcription initiation complex,[151,152]

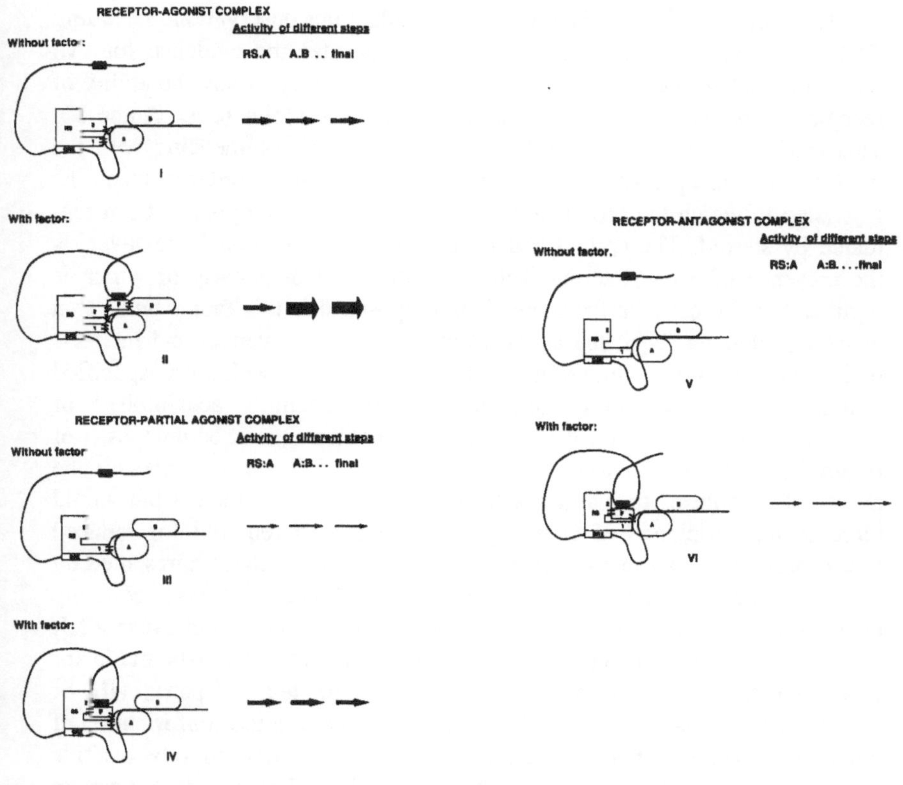

FIGURE 7. Proposed model for modulation by factor of TAT induction by glucocorticoids and antiglucocorticoids. After the activated receptor-steroid complex (RS) binds to the glucocorticoid regulatory element (GRE), at least three undefined but ordered interactions, or steps, are required to alter the transcription of the regulated gene. The sequential steps shown consist of interactions between RS and A, and between A and B, plus a final unspecified step. The amount of activity generated by a *subsaturating* concentration of steroid at each step is depicted by the arrows. The factor (stippled box labeled F) binds to a specific DNA sequence (hatched box) and can interact both with domain 2 of the receptor and with a common region of molecules A or B (but not both at once). Domains 1 and 2 of receptor-agonist complexes can both interact with A. All protein-protein interactions of receptor and factor F are indicated by the double bars. Due to conformational changes in domain 2 upon binding either a partial or a full antagonist, domain 2 no longer can contact A but the binding site for F remains intact; bound factor F interacts with A. The binding of a full antagonist also prevents the interaction of domain 1 with A. See text for further discussion.

occur before transcription rates are altered. One example of the consequences of these intervening events is that glucocorticoid induction of MMTV involves a synergistic interaction of GREs while the prior receptor-steroid complex binding to these GREs is not cooperative.[153] These intervening steps are mathematically equivalent to those where a conventional second messenger is generated and thus can be expected to cause a left shift in the dose-response curve relative to the dose-binding curve.

The model in Figure 7 is advanced to explain our observations regarding TAT induction. This speculative model incorporates the evidence for two transcriptional activating domains (1 and 2) in the receptor and the ability of receptor to interact with general and gene-specific proteins (e.g., A and B). This model also requires an ordered series of steps, commencing with the interaction of receptor-steroid complex with protein A, in order to reach the final step in which the production of specific gene transcripts has been regulated by steroid. The crucial and novel feature of our model, however, is the presence of a *trans*-acting factor (F), the level or activity of which is regulated in the cell. Furthermore, it is proposed that this factor binds to a *cis*-acting element (e.g., the above identified 62-bp sequence) and interacts both with a *trans*-activating domain of the receptor and with the unspecified transcription factors A or B via general protein-protein,[136] acid blob,[143] or leucine zipper[154] interactions. Interaction of receptor-agonist complexes (I of Figure 7) with this DNA-bound factor F to give II would not increase the ability of complexes to interact with other transcriptional factors but would increase the efficiency of coupling in the next step required for induced transcription (i.e., an interaction between A and B, which is represented by the second arrow in ''activity of different steps''). It is this increased coupling efficiency *at subsaturating concentrations of steroid* that would cause a left shift in the dose-response curve.[147] Identical interactions of DNA-bound factor F with receptor would not be possible with the complexes of partial (III) or full (V) antagonists due to steroid-induced differences in the conformation of transcription activating domain 2. Domain 2 has been proposed to be inactive with all partial and full antiglucocorticoids.[15,19,29,135] However, if that region of domain 2 which interacts with factor F in agonist complexes remains intact in the other complexes, the factor could still interact with the receptor and additionally contact the ''first'' transcription factor (i.e., factor A) to give a structure (IV) which resembles that seen with a full agonist (i.e., I in Figure 7). In this manner, the activity of the complex with partial antagonists would approximate that of an agonist complex. Similarly, complexes with full antagonists (V), in which domains 1 and 2 are both inactive, could interact with F to give VI, which would exhibit some agonist activity. In both cases, there would be no change in the dose-response curves. Alterations in the intracellular levels or activities of factor F would affect the responses both to agonists (to shift the dose-response curves) and to partial and full antagonists (to alter the amount of agonist activity). This model can be tested with the use of truncated receptors.[19,29]

This factor could be unique for glucocorticoid induction of the TAT gene. On the other hand, this factor may be a common component for glucocorticoid induction of many responsive genes if, as detailed by Loeb and Strickland,[147] the factor itself ''induces its physiological effects by reacting with more distal intracellular effector units characterized by differing intrinsic affinities for the mediator'' (i.e., the factor). It should be noted that all of the GREs affected by this factor could now be classified as ''composite'' GREs.[84]

Alternatively, this *trans*-acting factor could be a type of adaptor molecule which increases the affinity of agonist complexes for GREs. Even though there are not normally enough receptor molecules inside the cell to bind all of the GREs,[46,47] factors causing a sufficient increase in agonist complex binding to GREs would give a left shift in dose-response curve. Most studies have concentrated on the binding of receptor-steroid complexes to naked DNA. However, it is clear that other proteins can increase the affinity of complexes for nonspecific[132] and specific[155,156] DNA sequences. Since we do not see any evidence of a left shift in the dose-response curves of antagonists (see Figure 3), the interaction of our factor could not affect the DNA binding affinity of antagonist complexes, possibly due to an improper receptor conformation. Instead, the factor may, as in the model of Figure 7, increase the activity of the normally inactive transcriptional activating domain 2 to cause more agonist activity. Regardless of whether these, or other models, are proved to be correct, it is clear that modifications in the current models of glucocorticoid hormone action are required. The data for other steroid hormones (see Section III.B.2) suggest that similar adjustments are required for all of the steroid receptors.

E. CANDIDATE *TRANS*-ACTING FACTOR(S)

Numerous *trans*-acting factors which influence the activity of receptor-steroid complexes have recently been described. Because detailed investigations of dose-response curves and of changes in amount of agonist activity for antagonists are not common, it is possible that one of these already identified species could be our proposed factor. However, the properties of some of these species are inconsistent with our observations. Thus, a new class of transcriptional intermediary factors that are involved in synergism and squelching of transcription,[128] but have no reported effect on the dose-response curve, can be eliminated. In fact, these factors have been proposed to be necessary for the function of enhancers[128] and thus are similar, if not identical, to "co-activators".[145]

A factor from enteric bacteria causes a left shift in the dose-response curve for Dex induction of TAT in rat liver[157] and HTC[158] cells. Similarly, metyrapone causes a left shift in TAT induction by Dex in primary hepatocytes[159] and cyclosporin A causes a left shift in the dose-response curve for the killing of T-cells by Dex.[160] None of these substances have been added in our experiments on TAT induction and can thus be excluded. However, these substances may modify the level of our factor.

The expression of p37v-mos interferes with glucocorticoid induction of responsive genes.[161] However, a v-mos-encoded factor appears an unlikely candidate for modulation of TAT induction since only glucocorticoids are affected. Thus, complexes bound with Dex, but not with RU 486, undergo a redistribution to the cytoplasm at permissive temperatures in v-mos-transformed NRK cells, as determined by whole cell immunofluorescence.[162] Furthermore, it is difficult to envisage how a redistribution of complexes to the cytoplasm could cause a left shift in the dose-response curve for gene induction.

Some of the more promising candidates for our factor are the many species that have recently been reported to increase the chromatin[162a] or DNA[155,163-168] binding of receptor-steroid complexes. The molecular weight of these components vary from high (~70 kDa[155,163]) to medium (~48 kDa[162a]), low (>5 kDa[164,164] and 10 to 12 kDa[166]), and very low (0.7 to 3 kDa[167,168]). The most extensively studied has been a very low M_r species of 0.7 to 3 kDa which, as proposed for the model of Figure 7, does not affect the affinity of complexes for nonspecific DNA[168] and acts by conferring DNA-binding capacity to a subpopulation of the activated noncovalent receptor-Dex[167,168] and covalent receptor-Dex-Mes complexes (Cavanaugh and Simons, unpublished results). This factor is most abundant in extracts of crude HTC cell nuclei. We do not know whether the amount of this low M_r factor changes in HTC cells or is different from that found in Fu5-5 cells. This low M_r factor may be the same as, or very similar to, a nuclear factor that increases the binding of progesterone complexes[169] (Cavanaugh and Simons, unpublished results). Similarly, factors have been described that participate in the DNA binding of estrogen[170,171] and thyroid[172-175] receptors. It will be very interesting to examine the activity of such factors in the cell-free transcription assays that have recently been developed.[152]

VI. CONCLUSIONS

A large amount of data has been gathered over the last 2 decades in an effort to relate glucocorticoid steroid structure with biological activity. These studies have been prefaced by the assumption that all of the information for the potency and type (i.e., agonist, antagonist, or inactive) of activity of a given steroid is contained entirely within the steroid itself. However, recent studies, especially with the TAT gene and with hybrid receptors, have called this assumption into question. It now appears clear that, at least for some genes, the structure of the steroid is a necessary but not sufficient determinant. Other factors must be involved since the same receptor-glucocorticoid complex can exhibit quantitatively different amounts of activity with selected genes in the same cell. These activity differences are most easily documented as unequal values for the E_{50} of an agonist or for the amount of agonist activity expressed by an antagonist. This behavior is not unique for glucocorticoid steroids but also has been reported for almost all steroid hormones. Furthermore, at least for glucocorticoids, but probably for all steroid hormones, the specific value both for the EC_{50} of agonist and for the amount of agonist activity expressed by an antagonist *is not constant for the same gene in the same cell.* These observations cannot be accommodated by the current models of steroid hormone action. Based mostly on our studies of the control of TAT gene expression, a provisional model has evolved (Figure 7) which incorporates the recent evidence for differentially activated transcription domains of the receptor. The most unusual feature of this model, however, is the proposal that variations in the intracellular levels or activities of a *trans*-acting factor(s), which interacts with some component(s) of the transcriptional regulatory mechanism, can modulate the induction parameters for both glucocorticoids and antiglucocorticoids. Mathematical models exist which predict that such interactions would alter the observed EC_{50} of an agonist and probably the amount of agonist activity expressed by an antagonist. This model also offers exciting new prospects for the selective transcriptional control of specific genes during development, in different tissues, and in response to external factors in fully differentiated tissues.

In order to validate this model, we must further define the function of the 62-bp *cis*-acting element that is believed to bind our *trans*-acting factor, establish the crucial nucleotide sequence, and compare this sequence to that of other known *trans*-acting factors. We also must identify and clone the *trans*-acting factor, document the proposed changes in intracellular concentration or activity of this *trans*-acting factor, and try to determine the mechanisms that control these changes of the *trans*-acting factor. Similar investigations are required in the other models that are emerging in order to identify the gene-specific factors which modulate the activity of receptor-agonist and -antagonist complexes. This will entail major, and long-term, studies in an area with little precedent. However, these efforts will not be in vain since resolution of these questions promises to uncover new mechanisms for the control of gene expression that may prove to be general for all steroid hormones and perhaps even other ligand-regulated *trans*-acting factors. Once these new mechanisms are better understood, it should be much easier to determine which structural features of glucocorticoid steroids contribute to the expression of glucocorticoid and antiglucocorticoid activity.

ACKNOWLEDGMENTS

I would like to thank Drs. E. W. Czerwinski, G. L. Firestone, D. K. Granner, H. Gronemeyer, P. Herrlich, T. Ojasoo, M. G. Parker, E. B. Thompson, B. R. Westley, and K. R. Yamamoto for generously making numerous reprints and/or preprints of their work available, Drs. J. Drouin, and H. Oshima for helpful discussions, and Drs. A. H. Cavanaugh, J. M. Harmon, J. N. Loeb, and E. B. Thompson for critical review of the manuscript.

REFERENCES

1. **Samuels, H. H. and Tomkins, G. M.,** Relation of steroid structure to enzyme induction in hepatoma tissue culture cells, *J. Mol. Biol.*, 52, 57, 1970.
2. **Rousseau, G. G. and Schmit, J.-P.,** Structure-activity relationships for glucocorticoids. Determination of receptor binding and biological activity, *J. Steroid Biochem.*, 8, 911, 1977.
3. **Haynes, R. C., Jr.,** Adrenocorticotropic hormone: adrenocortical steroid and their synthetic analogs: inhibitors of the synthesis and actions of adrenocortical hormones, in *The Pharmacological Basis of Therapeutics*, 8th ed., Gilman, A. G., Rall, T. W., Nies, A. S., and Taylor, P., Ed., Pergamon Press, Elmsford, NY, 1990, 1458.
4. **Higgins, S. J., Rousseau, G. G., Baxter, J. D., and Tomkins, G. M.,** Early events in glucocorticoid action, *J. Biol. Chem.*, 248, 5866, 1973.
5. **Wrange, O. and Gustafsson, J.-A.,** Separation of the hormone- and DNA-binding sites of the hepatic glucocorticoid receptor by means of proteolysis, *J. Biol. Chem.*, 253, 856, 1978.
6. **Simons, S. S., Jr., Sistare, F. D., and Chakraborti, P. K.,** Steroid binding activity is retained in a 16-kDa fragment of the steroid binding domain of rat glucocorticoid receptors, *J. Biol. Chem.*, 264, 14493, 1989.
7. **Picard, D., Khursheed, B., Garabedian, M. J., Fortin, M. B., Kindquist, S., and Yamamoto, K. R.,** Reduced levels of hsp90 compromise steroid receptor action *in vivo*, *Nature*, 348, 166, 1990.
8. **Raynaud, J. P., Bouton, M. M., and Ojasoo, T.,** The use of interaction kinetics to distinguish potential antagonists from agonists, *TIPS (Trends in Pharm. Sci.)*, 324, 1980.
9. **Munck, A. and Holbrook, N. J.,** Glucocorticoid-receptor complexes in rat thymus cells. Rapid kinetic behavior and a cyclic model, *J. Biol. Chem.*, 259, 820, 1984.
10. **Jones, T. R. and Bell, P. A.,** Glucocorticoid-receptor interactions. Studies of the negative co-operativity induced by steroid interactions with a secondary, hydrophobic, binding site, *Biochem. J.*, 188, 237, 1980.
11. **Bell, P. A., Weatherill, P. J., and Jones, T. R.,** Influence of hydrophobic interactions on the structure and steroid-binding properties of glucocorticoid receptors, *J. Steroid Biochem.*, 24, 57, 1986.
12. **Svec, F., Teubner, V., and Tate, D.,** Location of the second steroid-binding site on the glucocorticoid receptor, *Endocrinology*, 125, 3103, 1989.
13. **Groyer, A., Schweizer-Groyer, G., Cadepond, F., Mariller, M., and Baulieu, E.-E.,** Antiglucocorticosteroid effects suggest why steroid hormone is required for receptors to bind DNA *in vivo* but not *in vitro*, *Nature*, 328, 624, 1987.
14. **Sistare, F. D., Hager, G. L., and Simons, S. S., Jr.,** Mechanisms of dexamethasone 21-mesylate antiglucocorticoid action. I. Receptor-antiglucocorticoid complexes do not competitively inhibit receptor-glucocorticoid complex activation of gene transcription *in vivo*, *Mol. Endocrinol.*, 1, 648, 1987.
15. **Fawell, S. E., White, R., Hoare, S., Sydenham, M., Page, M., and Parker, M. G.,** Inhibition of estrogen receptor-DNA binding by the "pure" antiestrogen ICI 164,384 appears to be mediated by impaired receptor dimerization, *Proc. Natl. Acad. Sci. U.S.A.*, 87, 6883, 1990.
16. **Richard-Foy, H., Sistare, F. D., Riegel, A. T., Simons, S. S., Jr., and Hager, G. L.,** Mechanism of dexamethasone 21-mesylate antiglucocorticoid action. II. Receptor-antiglucocorticoid complexes do not interact productively with mouse mammary tumor virus long terminal repeat chromatin, *Mol. Endocrinol.*, 1, 659, 1987.
17. **Webster, N. J. G., Green, S., Jin, J. R., and Chambon, P.,** The hormone-binding domains of the estrogen and glucocorticoid receptors contain an inducible transcription activation function, *Cell*, 54, 199, 1988.

18. **Seyfred, M. A. and Gorski, J.,** An interaction between the 5′ flanking distal and proximal regulatory domains of the rat prolactin gene is required for transcriptional activation by estrogens, *Mol. Endocrinol.*, 4, 1226, 1990.
19. **Meyer, M.-E., Pornon, A., Ji, J., Bocquel, M.-T., Chambon, P., and Gronemeyer, H.,** Agonistic and antagonistic activities of RU 486 on the functions of the human progesterone receptor, *EMBO*, 9, 3923, 1990.
20. **Yamamoto, K. R., Gehring, U., Stampfer, M. R., and Sibley, C. H.,** Genetic approaches to steroid hormone action, *Recent Prog. Horm. Res.*, 32, 3, 1976.
21. **Dieken, E. S., Meese, E. U., and Miesfeld, R. L.,** Glucocorticoid receptor transcripts lack sequences encoding the amino-terminal transcriptional modulatory domain, *Mol. Cell. Biol.*, 10, 4574, 1990.
22. **Miller, P. A., Ostrowski, M. C., Hager, G. L., and Simons, S. S., Jr.,** Covalent and noncovalent receptor-glucocorticoid complexes preferentially bind to the same regions of the long terminal repeat of murine mammary tumor virus proviral DNA, *Biochemistry*, 23, 6883, 1984.
23. **Rajpert, E. J., Lemaigre, F. P., Eliard, P. H., Place, M., Lafontanine, D. A., Economidis, I. V., Belayew, A., Martial, J. A., and Rousseau, G. G.,** Glucocorticoid receptors bound to the antagonist RU486 are not downregulated despite their capacity to interact *in vitro* with defined gene regions, *J. Steroid Biochem.*, 26, 513, 1987.
24. **Schweizer-Groyer, G., Cadepond, F., Groyer, A., Idziorek, T., Mariller, M., and Baulieu, E. E.,** Stimulation of specific transcription and DNA binding studies suggest that *in vitro* transformed RU 486-glucocorticosteroid receptor complexes display agonist activity, *J. Steroid Biochem.*, 30, 291, 1988.
25. **Johnson, D. M., Newby, R. F., and Bourgeois, S.,** Membrane permeability as a determinant of dexamethasone resistance in murine thymoma cells, *Cancer Res.*, 44, 2435, 1984.
26. **Mercier, L., Miller, P. A., and Simons, S. S., Jr.,** Antiglucocorticoid steroids have increased agonist activity in thise hepatoma cell lines that are more sensitive to glucocorticoids, *J. Steroid Biochem.*, 25, 11, 1986.
27. **Simons, S. S., Jr. and Yen, P. M.,** Variations in agonist activity among antiglucocorticoid steroids and its relation to glucocorticoid regulated genes, in *Steroid and Sterol Hormone Action*, Spelsberg, T. C. and Kumar, R., Ed., Martinius Nijhoff, Boston, 1987, 251.
28. **Kupfer, D. and Bulger, W. H.,** Inactivation of the uterine estrogen receptor binding of estradiol during P-450 catalyzed metabolism of chlorotrianisene (TACE). Speculation that TACE antiestrogenic activity involves covalent binding to the estrogen receptor, *FEBS Lett.*, 261, 59, 1990.
29. **Berry, M., Metzger, D., and Chambon, P.,** Role of the two activating domains of the oestrogen receptor in the cell-type and promoter-context dependent agonistic activity of the anti-oestrogen 4-hydroxytamoxifen, *EMBO J.*, 9, 2811, 1990.
30. **Turcotte, B., Meyer, M.-E., Bocquel, M.-T., Belanger, L., and Chambon, P.,** Repression of the α-fetoprotein gene promoter by progesterone and chimeric receptors in the presence of hormones and antihormones, *Mol. Cell. Biol.*, 10, 5002, 1990.
31. **Sundstrom, S. A., Komm, B. S., Xu, Q., Boundy, V., and Lyttle, C. R.,** The stimulation of uterine complement component C3 gene expression by antiestrogens, *Endocrinology*, 126, 1449, 1990.
32. **El-Ashry, D., Onate, S. A., Nordeen, S. K., and Edwards, D. P.,** Human progesterone receptor complexed with the antagonist RU 486 binds to hormone response elements in a structurally altered form, *Mol. Endocrinol.*, 3, 1545, 1989.
33. **Quarmby, V. E., Yarbrough, W. G., Lubahn, D. B., French, F. S., and Wilson, E. M.,** Autologous down-regulation of androgen receptor messenger ribonucleic acid, *Mol. Endocrinol.*, 4, 22, 1990.
34. **Van Bohemen, C. G. and Rousseau, G. G.,** Calmodulin antagonists competitively inhibit dexamethasone binding to the glucocorticoid receptor, *FEBS Lett.*, 143, 21, 1982.

35. **Loose, D. S., Stover, E. P., and Feldman, D.**, Ketoconazole binds to glucocorticoid receptors and exhibits glucocorticoid antagonist activity in cultured cells, *J. Clin. Invest.*, 72, 404, 1983.
36 **Schmidt, T. J., Miller-Diener, A., and Litwack, G.**, Beta-Lapachone, a specific competitive inhibitor of ligand binding to the glucocorticoid receptor, *J. Biol. Chem.*, 259, 9536, 1984.
37. **Goldman, A. S. and Katsumata, M.**, Murine glucocorticoid receptors: new evidence for a discrete receptor influenced by H-2, *Arch. Biochem. Biophys.*, 249, 316, 1986.
38. **Omrani, G. R., Tohmeh, J. F., and Loeb, J. N.**, Paradoxical induction of tyrosine aminotransferase by progesterone *in vivo:* potentiation of effects of low levels of glucocorticoid hormone, *Proc. Soc. Exp. Biol. Med.*, 162, 254, 1979.
39. **Naylor, P. H. and Rosen, F.**, Antiglucocorticoids: differential antagonism of *in vivo* responses to dexamethasone, in *Hormone Antagonists,* Agarwal, M. K., Ed., Walter de Gruyter, Berlin, 1982, 365.
40. **Moguilewsky, M. and Philibert, D.**, RU 38486: potent antiglucocorticoid activity correlated with strong binding to the cytosolic glucocorticoid receptor followed by an impaired activation, *J. Steroid Biochem.*, 20, 271, 1984.
41. **Munck, A. and Foley, R.**, Activation of steroid hormone-receptor complexes in intact target cells in physiological conditions, *Nature,* 278, 752, 1979.
42. **Markovic, R. D. and Litwack, G.**, Activation of liver and kidney glucocorticoid-receptor complexes occurs *in vivo, Arch. Biochem. Biophys.*, 202, 374, 1980.
43. **Munck, A. and Foley, R.**, Activated and non-activated glucocorticoid-receptor complexes in rat thymus cells: kinetics of formation and relation to steroid structure, *J. Steroid Biochem.*, 12, 225, 1980.
43a. **Steinberg, R. A., Levinson, B. B., and Tomkins, G. M.**, Kinetics of steroid induction and deinduction of tyrosine aminotransferase synthesis in cultured hepatoma cells, *Proc. Natl. Acad. Sci.*, U.S.A., 72, 2007, 1975.
43b. **Becker, P. B., Gloss, B., Schmid, W., Strahle, U., and Schutz, G.**, *In vivo* protein-DNA interactions in a glucocorticoid response element require the presence of the hormone, *Nature,* 324, 686, 1986.
44. **Jones, T. R. and Bell, P. A.**, Glucocorticoid-receptor interactions. Discrimination between glucocorticoid agonists and antagonists by means of receptor-binding kinetics, *Biochem. J.*, 204, 721, 1982.
45. **Ojasoo, T., Dore, J.-C., Gilbert, J., and Raynaud, J.-P.**, Binding of steroid to the progestin and glucocorticoid receptors analyzed by correspondence analysis, *J. Med. Chem.*, 31, 1160, 1988.
46. **Vanderbilt, J. N., Miesfeld, R., Maler, B. A., and Yamamoto, K. R.**, Intracellular receptor concentration limits of glucocorticoid-dependent enhancer activity, *Mol. Endocrinol.*, 1, 68, 1987.
47. **Hirst, M. A., Northrop, J. P., Danielsen, M., and Ringold, G. M.**, High level expression of wild type and variant mouse glucocorticoid receptors in chinese hamster ovary cells, *Mol. Endocrinol.*, 4, 162, 1990.
48. **Raynaud, J. P., Ojasoo, T., Bouton, M. M., and Philibert, D.**, Receptor binding as a tool in the development of new biactive steroids, in *Drug Design,* Vol. 8, Ariens, E. J., Ed., Academic Press, New York, 1979, 170.
49. **Raynaud, J.-P. and Ojasoo, T.**, The relevance of structure-affinity relationships in the study of steroid hormone action, in *Steroid Hormone Receptors: Structure and Function,* Eriksson, H. and Gustafsson, J.-A., Eds., 1983, 141.
50. **Reul, J. M. H. M., de Kloet, E. R., van Sluijs, F. J., Rijnberk, A., and Rothuizen, J.**, Binding characteristics of mineralocorticoid and glucocorticoid receptors in dog brain and pituitary, *Endocrinology,* 127, 907, 1990.
51. **Danielsen, M., Northrop, J. P., and Ringold, G. M.**, The mouse glucocorticoid receptor: mapping of functional domains by cloning, sequencing and expression of wild-type and mutant receptor proteins, *EMBO J.*, 5, 2513, 1986.

52. **Hurley, D. M., Accili, D., Stratakis, C. A., Karl, M., Vamvakopoulos, N., Rorer, E., Constantine, K., Taylor, S. I., and Chrousos, G. P.,** Point mutation causing a single amino acid substitution in the hormone binding domain of the glucocorticoid receptor in familial glucocorticoid resistance, *J. Clin. Invest.*, in press.
52a. **Chakraborti, P. K., Garabedian, M. J., Yamamoto, K. R., and Simons, S. S., Jr.,** Creation of "super" glucocorticoid receptors by point mutations in the steroid binding domain, *J. Biol. Chem.*, 266, 22,075, 1991.
53. **Hollenberg, S. M., Weinberger, C., Ong, E. S., Cerelli, G., Oro, A., Lebo, R., Thompson, E. B., Rosenfeld, M. G., and Evans, R. M.,** Primary structure and expression of a functional human glucocorticoid receptor cDNA, *Nature*, 318, 635, 1985.
54. **Miesfeld, R., Godowski, P. J., Maler, B. A., and Yamamoto, K. R.,** Genetic complementation of a glucocorticoid receptor deficiency by expression of cloned receptor cDNA, *Cell*, 46, 389, 1986.
55. **Simons, S. S., Jr., Thompson, E. B., and Johnson, D. F.,** Anti-inflammatory pyrazolosteroids: potent glucocorticods containing bulky A-ring substituents and no C3-carbonyl, *Biochem. Biophys. Res. Commun.*, 86, 793, 1979.
56. **Lamontagne, N., Mercier, L., Pons, M., Thompson, E. B., and Simons, S. S., Jr.,** Glucocorticoid vs. antiglucocorticoid activity: can a single functional group modification of glucocorticoid steroids always convey antiglucocorticoid activity?, *Endocrinology*, 114, 2252, 1984.
57. **Pratt, W. B., Kaine, J. L., and Pratt, D. V.,** The kinetics of glucocorticoid binding to the soluble specific binding protein of mouse fibroblasts, *J. Biol. Chem.*, 250, 4584, 1975.
58. **Duax, W. L., Griffin, J. F., Rohrer, D. C., Weeks, C. M., and Ebright, R. H.,** Steroid hormone action interpreted from x-ray crystallographic studies, *Biochem. Actions Horm.*, 11, 187, 1984.
58a. **Czerwinski, E. W.,** Structure of Cortivazol, 11β,17α,21-trihydroxy-6,16α-dimethyl-2′-phenyl-2′*H*-pregna-2,4,6-trieno[3,2-c]pyrazol-20-one 21-acetate, *Acta Cryst.*, C47, 2598, 1991.
59. **Duax, W. L., Weeks, C. M., and Rohrer, D. C.,** Crystal structure of steroids: molecular conformation and biological function, in *Recent Progress in Hormone Research*, Vol. 32, Academic Press, New York, 1976, 81.
60. **Teutsch, G., Ojasoo, T., and Raynaud, J. P.,** 11β-Substituted steroids, an original pathway to antihormones, *J. Steroid Biochem.*, 31, 549, 1988.
61. **Ojasoo, T. and Raynaud, J. P.,** Receptor binding profiles of progestins, in *Steroids and Endometrical Cancer*, Jasonni, V. M. et al., Eds., Raven Press, New York, 1983, 11.
62. **Dore, J.-C., Gilbert, J., Ojasoo, T., and Raynaud, J.-P.,** Correspondence analysis applied to steroid receptor binding, *J. Med. Chem.*, 29, 54, 1986.
63. **Raynaud, J. P., Fortin, M., Hunt, P., Ojasoo, T., Dore, J. C., Surcouf, E., and Mornon, J. P.,** Approaches to drug development using receptors, in *The Role of Receptors in Biology and Medicine*, Gotto, A. M., Jr. and O'Malley, B. W., Eds., Raven Press, New York, 1986, 63.
64. **Lefebvre, P., Formstecher, P., Rousseau, G. G., Lustenberger, P., and Dautrevaux, M.,** Improvement in glucocorticoid receptor binding affinity concomitant to shift from antagonist to agonist activity in a series of 17β-carboxamide derivatives of dexamethasone, *J. Steroid Biochem.*, 33, 557, 1989.
65. **Giesen, E. M. and Beck, G.,** Hormonal deinduction of tyrosine aminotransferase, *Horm. Metab. Res.*, 14, 252, 1982.
66. **Simons, S. S., Jr. and Thompson, E. B.,** Dexamethasone 21-mesylate: an affinity label of glucocorticoid receptors from rat hepatoma tissue culture cells, *Proc. Natl. Acad. Sci. U.S.A.*, 78, 3541, 1981.
67. **Simons, S. S., Jr. and Miller, P. A.,** Affinity-labeling steroids as biologically active probes of antiglucocorticoid hormone action, *J. Steroid Biochem.*, 24, 25, 1986.

68. **Miller, P. A. and Simons, S. S., Jr.,** Comparison of glucocorticoid receptors in two rat hepatoma cell lines with different sensitivities to glucocorticoids and antiglucocorticoids, *Endocrinology,* 122, 2990, 1988.
69. **Simons, S. S., Jr., Schleenbaker, R. E., and Eisen, H. J.,** Activation of covalent, affinity labeled, glucocorticoid receptor-steroid complexes, *J. Biol. Chem.*, 258, 2229, 1983.
70. **Simons, S. S., Jr. and Miller, P. A.,** Comparison of DNA binding properties of activated, covalent and noncovalent glucocorticoid receptor-steroid complexes from HTC cells, *Biochemistry,* 23, 6876, 1984.
71. **Qian, X. and Abul-Hajj, Y. J.,** Synthesis and biologic activities of 11β-substituted estradiol as potential antiestrogens, *Steroids,* 55, 238, 1990.
72. **Rousseau, G. G. and Baxter, J. D.,** Glucocorticoid receptors, in *Glucocorticoid Hormone Action,* Baxter, J. D. and Rousseau, G. G., Eds., Springer-Verlag, New York, 1979, 49.
73. **Munck, A. and Holbrook, N. J.,** Steroid hormone antagonism and a cyclic model of receptor kinetics, *J. Steroid Biochem.*, 26, 173, 1987.
74. **Neef, G., Beier, S., Elger, W., Henderson, D., and Wiechert, R.,** New steroids with antiprogestational and antiglucocorticoid activities, *Steroids,* 44, 349, 1985.
75. **Danhaive, P. A. and Rousseau, G. G.,** Binding of glucocorticoid antagonists to androgen and glucocorticoid hormone receptors in rat skeletal muscle, *J. Steroid Biochem.*, 24, 481, 1986.
76. **Bloom, E., Matulich, D. T., Lan, N. C., Higgins, S. J., Simons, S. S., Jr., and Baxter, J. D.,** Nuclear binding of glucocorticoid receptors: relations between cytosol binding, activation and the biological response induced effects, *J. Steroid Biochem.*, 12, 175, 1980.
77. **Mercier, L., Thompson, E. B., and Simons, S. S., Jr.,** Dissociation of steroid binding to receptors and steroid induction of biological activity in a glucocorticoid responsive cell, *Endocrinology,* 112, 601, 1983.
78. **Aviv, D. and Thompson, E. B.,** Variation in tyrosine aminotransferase induction in HTC cell clones, *Science,* 177, 1201, 1972.
79. **Ivarie, R. D. and O'Farrell, P. H.,** The glucocorticoid domain: steroid-mediated changes in the rate of synthesis of rat hepatoma proteins, *Cell,* 13, 41, 1978.
80. **Thompson, E. B., Granner, D. K., Gelehrter, T., Erickson, J., and Hager, G. L.,** Unlinked control of multiple glucocorticoid-induced processes in HTC cells, *Mol. Cell. Endocrinol.*, 15, 135, 1979.
81. **Killary, A. M. and Fournier, R. E. K.,** A genetic analysis of extinction: trans-dominant loci regulate expression of liver-specific traits in hepatoma hybrid cells, *Cell,* 38, 523, 1984.
82. **Schmid, W., Muller, G., Schutz, G., and Gluecksohn-Waelsch, S.,** Deletions near the albino locus on chromosome 7 of the mouse affect the level of tyrosine aminotransferase mRNA, *Proc. Natl. Acad. Sci. U.S.A.*, 82, 2866, 1985.
83. **DeFranco, D., Morris, S. M., Jr., Leonard, C. M., and Gluecksohn-Waelsch, S.,** Metallothionein mRNA expression in mice homozygous for chromosomal deletions around the albino locus, *Proc. Natl. Acad. Sci. U.S.A.*, 85, 1161, 1988.
84. **Diamond, M. I., Miner, J. N., Yoshinaga, S. K., and Yamamoto, K. R.,** Transcription factor interactions: selectors of positive or negative regulation from a single DNA element, *Science,* 249, 1266, 1990.
85. **Simons, S. S., Jr., Miller, P. A., Wasner, G., Miller, N. R., and Mercier, L.,** Inverse correlation between dexamethasone 21-mesylate agonist activity and sensitivity to dexamethasone for induction of tyrosine aminotransferase in rat hepatoma cells, *J. Steroid Biochem.*, 31, 1, 1988.

86. **Simons, S. S., Jr., Mercier, L., Miller, N. R., Miller, P. A., Oshima, H., Sistare, F. D., Thompson, E. B., Wasner, G., and Yen, P. M.**, Differential modulation of gene induction by glucocorticoids and antiglucocorticoids in rat hepatoma tissue culture cells, *Cancer Res.*, 49 (suppl.), 2244, 1989.
87. **Thompson, E. B., Gadson, P., Wasner, G., and Simons, S. S., Jr.**, Differential regulation of tyrosine aminotransferase by glucocorticoids: transcriptional and post-transcriptional control, in *Gene Regulation by Steroid Hormones IV*, Roy, A. K. and Clark, J. H., Eds., Springer-Verlag, New York, 1989, 63.
87a. **Oshima, H. and Simons, S. S., Jr.**, Modulation of glucocorticoid induction of tyrosine aminotransferase gene expression by variations in cell density, *Endocrinology*, in press.
87b. **Szapary, D., Oshima, H., and Simons, S. S., Jr.**, Modulation of glucocorticoid induction of stably transfected tyrosine aminotransferase gene constructs involves elements upstream of the glucocorticoid responsive element, *Endocrinology*, in press.
88. **Beneking, M., Schmidt, H., and Weiss, G.**, Subcellular distribution of a factor inactivating tyrosine aminotransferase. Study of its mechanism and relationship to different forms of the enzyme, *Eur. J. Biochem.*, 82, 235, 1978.
89. **Wasner, G., Oshima, H., Thompson, E. B., and Simons, S. S., Jr.**, Unlinked regulation of the sensitivity of primary glucocorticoid-inducible responses in mouse mammary tumor virus infected Fu5-5 rat hepatoma cells, *Mol. Endocrinol.*, 2, 1009, 1988.
90. **Wasner, G. and Simons, S. S., Jr.**, Differential sensitivity of HTC and Fu5-5 cells for induction of tyrosine aminotransferase by 3′,5′-cyclic adenosine monophosphate, *Mol. Endocrinol.*, 1, 109, 1987.
91. **Quirk, S. J. and Funder, J. W.**, Steroid receptors, and the generation of closely coupled/biphasic dose-response curves, *J. Steroid Biochem.*, 30, 1, 1988.
92. **Gagne, D., Labhilli, M., and Pons, M.**, Description and analysis of differential sensitivity to glucocorticoid in Fao cells, *J. Steroid Biochem.*, 31, 917, 1988.
93. **Liley, H. G., White, R. T., Benson, B. J., and Ballard, P. L.**, Glucocorticoids both stimulate and inhibit production of pulmonary surfactant protein A in fetal human lung, *Proc. Natl. Acad. Sci. U.S.A.*, 85, 9096, 1988.
94. **Nordeen, S. K., Kuhnel, B., Lawler-Heavner, J., Barber, D. A., and Edwards, D. P.**, A quantitative comparison of dual control of a hormone response element progestins and glucocorticoids in the same cell line, *Mol. Endocrinol.*, 3, 1270, 1989.
95. **Russo-Marie, F., Paing, M., and Duval, D.**, Involvement of glucocorticoid receptors in steroid-induced inhibition of prostaglandin secretion, *J. Biol. Chem.*, 254, 8498, 1979.
96. **Rotsztejn, W. H., Dussaillant, M., Nobou, F., and Rosselin, G.**, Rapid glucocorticoid inhibition of vasoactive intestinal peptide-induced cyclic AMP accumulation and prolactin release in rat pituitary cells in culture, *Proc. Natl. Acad. Sci. U.S.A.*, 78, 7584, 1981.
97. **Cook, P. W., Swanson, K. T., Edwards, C. P., and Firestone, G. L.**, Glucocorticoid receptor-dependent inhibition of cellular proliferation in dexamethasone-resistant and hypersensitive rat hepatoma cell variants, *Mol. Cell. Biol.*, 8, 1449, 1988.
98. **Webster, M. K., Guthrie, J., and Firestone, G. L.**, Suppression of rat mammary tumor cell growth *in vitro* by glucocorticoids requires serum proteins, *J. Biol. Chem.*, 265, 4831, 1990.
99. **Pollanen, J.**, Down-regulation of plasmin receptors on human sarcoma cells by glucocorticoids, *J. Biol. Chem.*, 264, 5628, 1989.
100. **Jonat, C., Rahmsdorf, H. J., Park, K.-K., Cato, A. C. B., Gebel, S., Ponta, H., and Herrlich, P.**, Antitumor promotion and antiinflammation: down-modulation of AP-1 (Fos/Jun) activity by glucocorticoid hormone, *Cell*, 62, 1189, 1990.
101. **Burnstein, K. L., Jewell, C. M., and Cidlowski, J. A.**, Hormone glucocorticoid receptor cDNA contains sequences sufficient for receptor down-regulation, *J. Biol. Chem.*, 265, 7284, 1990.

102. **Yang-Yen, H.-F., Chambard, J.-C., Sun, Y.-L., Smeal, T., Schmidt, T. J., Drouin, J., and Karin, M.,** Transcriptional interference between c-Jun and the glucocorticoid receptor: mutual inhibition of DNA binding due to direct protein-protein interaction. *Cell,* 62, 1205, 1990.
103. **Lucibello, F. C., Slater, E. P., Jooss, K. U., Beato, M., and Muller, R.,** Mutual transrepression of fos and the glucocorticoid receptor: involvement of a functional domain in fos which is absent in fosB, *EMBO J.*, 9, 2827, 1990
104. **Schule, R., Rangarajan, P., Kliewer, S., Ransone, L. J., Bolado, J., Yang, N., Verma, I. M., and Evans, R. M.,** Functional antagonism between oncoprotein c-Jun and the glucocorticoid receptor, *Cell,* 62, 1217, 1990.
105. **Drouin, J., Trifiro, M. A., Plante, R. K., Nemer, M., Eriksson, P., and Wrange, O.,** Glucocorticoid receptor binding to a specific DNA sequence is required for hormone-dependent repression of pro-opiomelanocortin transcription, *Mol. Cell. Biol.*, 9, 5305, 1989.
106. **Slater, E. P., Redeuihl, G., Theis, K., Suske, G., and Beato, M.,** The uteroglobin promoter contains a noncanonical estrogen responsive element, *Mol. Endocrinol.*, 4, 604, 1990.
107. **Amara, J. F., Itallie, C. V., and Dannies, P. S.,** Regulation of prolactin production and cell growth by estradiol: difference in sensitivity to estradiol occurs at level of messenger ribonucleic acid accumulation, *Endocrinology,* 120, 264, 1987.
108. **May, F. E. B. and Westley, B. R.,** Identification and characterization of estrogen-regulated RNAs in human breast cancer cells, *J. Biol. Chem.*, 263, 12901, 1988.
109. **Westley, B. R., Holzel, F., and May, F. E. B.,** Effects of oestrogen and the antioestrogens, tamoxifen and LY117018, on four oestrogen-regulated RNAs in the EFM-19 breast cancer cell line, *J. Steroid Biochem.*, 32, 365, 1989.
110. **Evans, M. I., O'Malley, P. J., Krust, A., and Burch, J. B. E.,** Developmental regulation of the estrogen receptor and the estrogen responsiveness of five yolk protein genes in the avian liver, *Proc. Natl. Acad. Sci. U.S.A.*, 84, 8493, 1987.
111. **Barton, M. C. and Shapiro, D. J.,** Transient administration of estradiol-17β establishes an autoregulatory loop permanently inducing estrogen receptor mRNA, *Proc. Natl. Acad. Sci. U.S.A.*, 85, 7119, 1988.
112. **Cato, A. C. B., Heitlinger, E., Ponta, H., Klein-Hitpass, L., Ryffel, G. U., Bailly, A., Rauch, C., and Milgrom, E.,** Estrogen and progesterone receptor-binding sites on the chicken vitellogenin II gene: synergism of steroid hormone action, *Mol. Cell. Biol.*, 8, 5323, 1988.
113. **Cato, A. C. B., Miksicek, R., Schutz, G., Arnemann, J., and Beato, M.,** The hormone regulatory element of mouse mammary tumour virus mediates progesterone induction, *EMBO J.*, 5, 2237, 1986.
114. **Conneely, O. M., Dobson, A. D. W., Tsai, M.-J., Beattie, W. G., Toft, D. O., Huckaby, C. S., Zarucki, T., Schrader, W. T., and O'Malley, B. W.,** Sequence and expression of a functional chicken progesterone receptor, *Mol. Endocrinol.*, 1, 517, 1987.
115. **Quirk, S. J., Gannell, J. E., Fullerton, M. J., and Funder, J. W.,** Progestins specifically suppress alpha-lactalbumin synthesis and secretion, *J. Steroid Biochem.*, 23, 901, 1985.
116. **Alexander, I. E., Clarke, C. L., Shine, J., and Sutherland, R. I.,** Progestin inhibition of progesterone receptor gene expression in human breast cancer cells, *Mol. Endocrinol.*, 3, 1377, 1989.
117. **Catterall, J. F., Watson, C. S., Kontula, K. K., Janne, O. A., and Bardin, C. W.,** Differential regulation of specific gene expression in mouse kidney by androgens and antiandrogens, in *Steroid Hormone Resistance. Mechanisms and Clinical Aspects,* Chrousos, G. R., Loriaux, D. C., and Lipsett, M. B., Eds., Plenum Press, New York, 1986, 213.

118. **Chen, T. L., Hauschka, P. V., and Feldman, D.,** Dexamethasone increases 1,25-dihydroxyvitamin D3 receptor levels and augments bioresponses in rat osteoblast-like cells, *Endocrinology,* 118, 1119, 1986.
119. **Hackney, J. F., Holbrook, N. J., and Grasso, R. J.,** Progesterone as a partial glucocorticoid agonist in L929 mouse fibroblasts: effects on cell growth, glutamine synthetase induction and glucocorticoid receptors, *J. Steroid Biochem.,* 14, 971, 1981.
120. **Winguth, S. D. and Firestone, G. L.,** Dual regulation of protein maturation in viral infected rat HTC hepatoma cells by glucocorticoids and progesterone, *Mol. Endocrinol.,* 1, 823, 1987.
121. **Bignon, E., Pons, M., de Paulet, A. C., Dore, J.-C., Gilbert, J., Abecassis, J., Miquel, J.-F., Ojasoo, T., and Raynaud, J.-P.,** Effect of triphenylacrylonitrile derivatives on estradiol-receptor binding and on human breast cancer cell growth, *J. Med. Chem.,* 32, 2092, 1989.
122. **McCague, R., Leclerq, G., Legros, N., Goodman, J., Blackburn, M., Jarman, M., and Foster, A. B.,** Derivatives of tamoxifen. Dependence of antiestrogenicity on the 4-substituent, *J. Med. Chem.,* 32, 2527, 1989.
123. **Simons, S. S., Jr., Pumphrey, J. g., Rudikoff, S., and Eisen, H. J.,** Identification of cysteine-656 as the amino acid of HTC cell glucocorticoid receptors that is covalently labeled by dexamethasone 21-mesylate, *J. Biol. Chem.,* 262, 9676, 1987.
124. **Hunt, T.,** Cytoplasmic anchoring proteins and the control of nuclear localization, *Cell,* 59, 949, 1989.
125. **Schmidt, T. J.,** Comparison of *in vivo* activation of triamcinolone acetonide- and RU 38486-receptor complexes in the CEM-C7 and IM-9 human leukemic cell lines, *Cancer Res.,* 49, 4390, 1989.
126. **Dong, J.-M., Nordloh, P. W., and Chiu, J.-F.,** The mechanism of the bidirectional regulation of the rat alpha-fetoprotein gene by glucocorticoid hormone, *Mol. Cell. Endocrinol.,* 66, 109, 1989.
127. **Harbour, D. V., Chambon, P., and Thompson, E. B.,** Steroid mediated lysis of lymphoblasts requires the DNA binding region of the steroid hormone receptor, *J. Steroid Biochem.,* 35, 1, 1990.
128. **Tasset, D., Tora, L., Fromental, C., Scheer, E., and Chambon, P.,** Distinct classes of transcriptional activating domains function by different mechanisms, *Cell,* 62, 1177, 1990.
129. **Gehring, U.,** Genetics of glucocorticoid receptors, *Mol. Cell. Endocrinol.,* 48, 89, 1986.
130. **Beato, M., Chalepakis, G., Schauer, M., and Slater, E. P.,** DNA regulatory elements for steroid hormones, *J. Steroid Biochem.,* 32, 737, 1989.
131. **Chamness, G. C., Jennings, A. W., and McGuire, W. L.,** Estrogen receptor binding to isolated nuclei. A nonsaturable process, *Biochemistry,* 13, 327, 1974.
132. **Simons, S. S., Jr., Martinez, H. M., Garcea, R. L., Baxter, J. D., and Tomkins, G. M.,** Interactions of glucocorticoid receptor-steroid complexes with acceptor sites, *J. Biol. Chem.,* 251, 334, 1976.
133. **Picard, D., Salser, S. J., and Yamamoto, K. R.,** A movable and regulable inactivation function within the steroid binding domain of the glucocorticoid receptor, *Cell,* 54, 1073, 1988.
134. **Hollenberg, S. M. and Evans, R. M.,** Multiple and cooperative trans-activation domains of the human glucocorticoid receptor, *Cell,* 55, 899, 1988.
135. **Lees, J. A., Fawell, S. E., and Parker, M. G.,** Identification of two transactivation domains in the mouse oestrogen receptor, *Nucleic Acids Res.,* 17, 5477, 1989.
136. **Schule, R., Muller, M., Kaltschmidt, C., and Renkawitz, R.,** Many transcription factors interact synergistically with steroid receptors, *Science,* 242, 1418, 1988.
137. **Schule, R., Muller, M,, Otsuka-Murakami, H., and Renkawitz, R.,** Cooperativity of the glucocorticoid receptor and the CACCC-box binding factor, *Nature,* 332, 87, 1988.
138. **Strahle, U., Schmid, W., and Schutz, G.,** Synergistic action of the glucocorticoid receptor with transcription factors, *EMBO J.,* 7, 3389, 1988.

139. **Schatt, M. D., Rusconi, S., and Schaffner, W.,** A single DNA-binding transcription factor is sufficient for activation from a distant enhancer and/or from a promoter position, *EMBO J.*, 9, 481, 1990.
140. **Imai, E., Stronstedt, P.-E., Quinn, P. G., Carlstedt-Duke, J., Gustafsson, J.-A., and Granner, D. K.,** Characterization of a complex glucocorticoid response unit in the phosphoenolpyruvate carboxykinase gene, *Mol. Cell. Biol.*, 10, 4712, 1990.
141. **Schmid, W., Strahle, U., Schutz, G., Schmitt, J., and Stunnenberg, H.,** Glucocorticoid receptor binds cooperatively to adjacent recognition sites, *EMBO J.*, 8, 2257, 1989.
142. **Ankenbauer, W., Strahle, U., and Schutz, G.,** Synergistic action of glucocorticoid and estradiol responsive elements, *Proc. Natl. Acad. Sci. U.S.A.*, 85, 7526, 1988.
142a. **Oshima, H. and Simons, S. S., Jr.,** Modulation of transcription factor activity by a distant steroid modulatory element, *Mol. Endocrinol.*, in press.
143. **Ptashne, M.,** How eukaryotic transcriptional activators work, *Nature*, 335, 683, 1988.
144. **Lin, Y.-S., Carey, M., Ptashne, M., and Green, M. R.,** How different eukaryotic transcriptional activators can cooperate promiscuously, *Nature*, 345, 359, 1990.
145. **Pugh, B. F. and Tjian, R.,** Mechanism of transcriptional activation by Sp1: evidence for coactivators, *Cell*, 61, 1187, 1990.
146. **Strickland, S. and Loeb, J. N.,** Obligatory separation of hormone binding and biological response curves in systems dependent upon secondary mediators of hormone action, *Proc. Natl. Acad. Sci. U.S.A.*, 78, 1366, 1981.
147. **Loeb, J. N. and Strickland, S.,** Hormone binding and coupled response relationships in systems dependent on the generation of secondary mediators, *Mol. Endocrinol.*, 1, 75, 1987.
148. **Zaret, K. S. and Yamamoto, K. R.,** Reversible and persistent changes in chromatin structure accompany activation of a glucocorticoid-dependent enhancer element, *Cell*, 38, 29, 1984.
149. **Becker, P., Renkawitz, R., and Schutz, G.,** Tissue-specific DNaseI hypersensitive sites in the 5′-flanking sequences of the tryptophan oxygenase and the tyrosine aminotransferase genes, *EMBO J.*, 3, 2015, 1984.
150. **Richard-Foy, H. and Hager, G. L.,** Sequence-specific positioning of nucleosomes over the steroid-inducible MMTV promoter, *EMBO J.*, 6, 2321, 1987.
151. **Buratowski, S., Hahn, S., Guarente, L., and Sharp, P. A.,** Five intermediate complexes in transcription initiation by RNA polymerase II, *Cell*, 56, 549, 1989.
152. **Tsai, S. Y., Srinivassan, G., Allan, G. F., Thompson, E. B., O'Malley, B. W., and Tsai, M.-J.,** Recombinant human glucocorticoid receptor induces transcription of hormone response genes *in vitro*, *J. Biol. Chem.*, 265, 17055, 1990.
153. **Perlmann, T., Eriksson, P., and Wrange, O.,** Quantitative analysis of the glucocorticoid receptor-DNA interaction at the mouse mammary tumor virus glucocorticoid response element, *J. Biol. Chem.*, 265, 17222, 1990.
154. **Landschulz, W. H., Johnson, P. F., and McKnight, S. L.,** The leucine zipper: a hypothetical structure common to a new class of DNA binding proteins, *Science*, 240, 1759, 1988.
155. **Payvar, F. and Wrange, O.,** Relative selectivities and efficiencies of DNA binding by purified intact and protease-cleaved glucocorticoid receptor, in *Steroid Hormone Receptors: Structure and Function*, Eriksson, H. and Gustafsson, J. A., Eds., Elsevier, Amsterdam, 1983, 267.
156. **Dahlman-Wright, K., Siltala-Roos, H., Carlstedt-Duke, J., and Gustafsson, J.-A.,** Protein-protein interactions facilitate DNA binding by the glucocorticoid receptor DNA-binding domain, *J. Biol. Chem.*, 265, 14030, 1990.
157. **Kido, H., Higashi, T., and Katanuma, N.,** A new factor from enteric bacteria of rats amplifying induction of liver enzyme by glucocorticoid. Mechanism of action, *Eur. J. Biochem.*, 78, 541, 1977.
158. **Katunuma, N. and Kido, H.,** Chemical structure and biological function of amplifying factors of glucocorticoid actions, *Proc. Jpn. Acad.*, 54, 645, 1978.

159. **Tongiani, R., Piazzolla, S., and Poalicchi, A.,** Metyrapone modulation of tyrosine aminotransferase induction by dexamethasone in cultured hepatocytes, *Biochem. Biophys. Res. Commun.*, 166, 801, 1990.
160. **Zacharchuk, C. M., Mercep, M., Chakraborti, P. K., Simons, S. S., Jr., and Ashwell, J. D.,** Programmed T lymphocyte death: cell activation- and steroid-induced pathways are mutually antagonistic, *J. Immunol.*, 145, 4037, 1990.
161. **Jaggi, R., Friis, R., and Groner, B.,** Oncogenes modulate cellular gene expression and repress glucocorticoid regulated gene transcription, *J Steroid Biochem.*, 29, 457, 1988.
162. **Qi, M., Stasenko, L. J., and DeFranco, D. B.,** Recycling and desensitization of glucocortiocoid receptors in v-mos transformed cells depend on the ability of nuclear receptors to modulate gene expression, *Mol. Endocrinol.*, 4, 455, 1990.
162a. **Okamoto, K., Isohashi, F., Ueda, K., Kokufu, I., and Sakamoto, Y.,** Purification and characterization of an adenosine triphosphate-stimulated factor that enhances the nuclear binding of activated glucocorticoid-receptor complex from rat liver, *Endocrinology*, 123, 2752, 1988.
163. **Tymoczko, J. L., Ahern, S., Unger, A. L., Colby, J. L., and Ede, G.,** Identification of a protein from rat liver cytosol that enhances activation of the glucocorticoid receptor, *Biochem. J.*, 256, 47, 1988.
164. **Schmidt, T. J., Miller-Diener, A., Webb, M. L., and Litwack, G.,** Thermal activation of the purified rat hepatic glucocorticoid receptor. Evidence for a two-step mechanism, *J. Biol. Chem.*, 260, 16255, 1985.
165. **Harmon, J. M., Elsasser, M. S., Urda, L. A., and Eisen, L. P.,** Activation of the human glucocorticoid receptor: evidence for a two-step model, *J. Steroid Biochem.*, 31, 275, 1988.
166. **Grippo, J. F., Holmgren, A., and Pratt, W. B.,** Proof that the endogenous heat-stable glucocorticoid receptor-activating factor is thioredoxin, *J. Biol. Chem.*, 260, 93, 1985.
167. **Cavanaugh, A. H. and Simons, S. S., Jr.,** Glucocorticoid receptor binding to calf thymus DNA. I. Identification and characterization of a macromolecular factor involved in receptor-steroid complex binding to DNA, *Biochemistry*, 29, 989, 1990.
168. **Cavanaugh, A. H. and Simons, S. S., Jr.,** Glucocorticoid receptor binding to calf thymus DNA. II. Role of a DNA-binding activity factor in receptor heterogeneity and a multistep mechanism of receptor activation, *Biochemistry*, 29, 996, 1990.
169. **Edwards, D. P., Kuhnel, B., Estes, P. A., and Nordeen, S. K.,** Human progesterone receptor binding to mouse mammary tumor virus deoxyribonucleic acid: dependence on hormone and nonreceptor nuclear factor(s), *Mol. Endocrinol.*, 3, 381, 1989.
170. **Curtis, S. W. and Korach, K. S.,** Uterine estrogen receptor interaction with estrogen-responsive DNA sequences *in vitro*: effects of ligand binding on receptor-DNA complexes, *Mol. Endocrinol.*, 4, 276, 1990.
171. **Mukherjee, R. and Chambon, P.,** A single-stranded DNA-binding protein promotes the binding of the purified oestrogen receptor to its responsive element, *Nucleic Acids Res.*, 18, 5713, 1990.
172. **O'Donnell, A. L. and Koenig, R. J.,** Mutational analysis identifies a new functional domain of the thyroid hormone receptor, *Mol. Endocrinol.*, 4, 715, 1990.
173. **Burnside, J., Darling, D. S., and Chin, W. W.,** A nuclear factor that enhances binding of thyroid hormone receptors to thyroid hormone response elements, *J. Biol. Chem.*, 265, 2500, 1990.
174. **Crone, D. E., Kim, H.-S., and Spindler, S. R.,** Alpha and beta hormone receptors bind immediately adjacent to the rat growth hormone gene TATA box in a negatively hormone-responsive promoter region, *J. Biol. Chem.*, 265, 10851, 1990.
175. **Murray, M. B. and Towle, H. C.,** Identification of nuclear factors that enhance binding of the thyroid hormone receptor to a thyroid hormone response element, *Mol. Endocrinol.*, 3, 1434, 1989.

Chapter 8

MOLECULAR STRUCTURE AND BIOLOGICAL ACTIVITY OF VITAMIN D METABOLITES AND THEIR ANALOGS

Nobuo Ikekawa and Seiichi Ishizuka

TABLE OF CONENTS

I. INTRODUCTION

Basic research on the metabolism of vitamin D_3 (**1**) with the application of modern tools has brought about the development of an important new field in medicinal chemistry. A considerable number of vitamin D metabolites and their analogs have been synthesized and the use thereof has prompted the clarification of the mode of action of vitamin D hormone and the finding of new therapeutically useful compounds as well. It appears that new vitamin D analogs are useful not only for calcium metabolism disorder and bone diseases, but also for differentiation of myelocytic-type leukemias and the treatment of psoriasis. It is indeed a most dramatic example of an investigation on basic science leading to so rapidly a practical application to medicine.

It has been proved that vitamin D must first be 25-hydroxylated in the liver to give 25-hydroxyvitamin D_3 (25-OH-D_3) (**2**) and subsequently 1α-hydroxylated in the kidney prior to exert the classical function. It is now well established that 1,25-dihydroxyvitamin D_3 [1,25-$(OH)_2D_3$] (**3**) is the physiologically active metabolite of vitamin D_3 in stimulating bone mineral mobilization and the intestinal absorption of calcium and phosphate.

The kidney can also produce a second hydroxylated metabolite, 24*R*,25-$(OH_2)D_3$ (**4**), from 25-OH-D_3 with the aid of 1,25-$(OH)_2D_3$ which is known to activate 25-OH-D_3-24-hydroxylase. Under circumstances of relative vitamin D depletion, only 1,25-$(OH)_2D_3$ is produced in the kidney; however, when the supply of vitamin D is adequate, both 1,25- and 24*R*,25-$(OH)_2D_3$ are produced in kidney. Beside the 24*R*-hydroxylation, 23*S*- and 26-hydroxylations also occur in kidney. Thus, 25-OH-D_3, 1,25-, 24*R*,25-, 23S,25- (**5**), 25,26-$(OH)_2D_3$ (**6**), 1,24*R*,25-$(OH)_3D_3$ (**8**), 25-OH-D_3-26,23-lactone (**7**), and 1,25-$(OH)_2D_3$-26,23-lactone (**13**) are metabolites found in human serum, and the metabolic pathways of these compounds are shown in Figure 1, where the heavy solid lines indicate the major pathway.[1-3]

Some 50 metabolites of vitamin D_3 and vitamin D_2 have now been isolated and identified.[1] Most of these have been isolated from *in vitro* incubation systems, kidney perfusions, and cell cultures. To date any unique biological properties have not been shown to most of these metabolites, which are derived from the metabolic transformations on the side chain and are apparently associated with the inactivation or breakdown of the parental metabolites. The metabolites 23*S*,25-, and 25*S*,26-, and 25*R*,26-$(OH)_2D_3$ are biologically inert but the possibility remains that some special function is yet to be discovered. An important pathway involved 23- and 26-hydroxylations of 1,25-$(OH)_2D_3$, followed by additional 26-oxidation and lactonization, which give a unique biologically active substance, 1,25-$(OH)_2D_3$-26,23-lactone (**13**).

Although the biological activities of these metabolites and analogs were already described in terms of calcium metabolism in the previous review,[1] in the last several years, a new physiological function of vitamin D metabolites has been discovered. Thus, most appropriately, this review focuses mainly on the new biological activities and new compounds of interest.

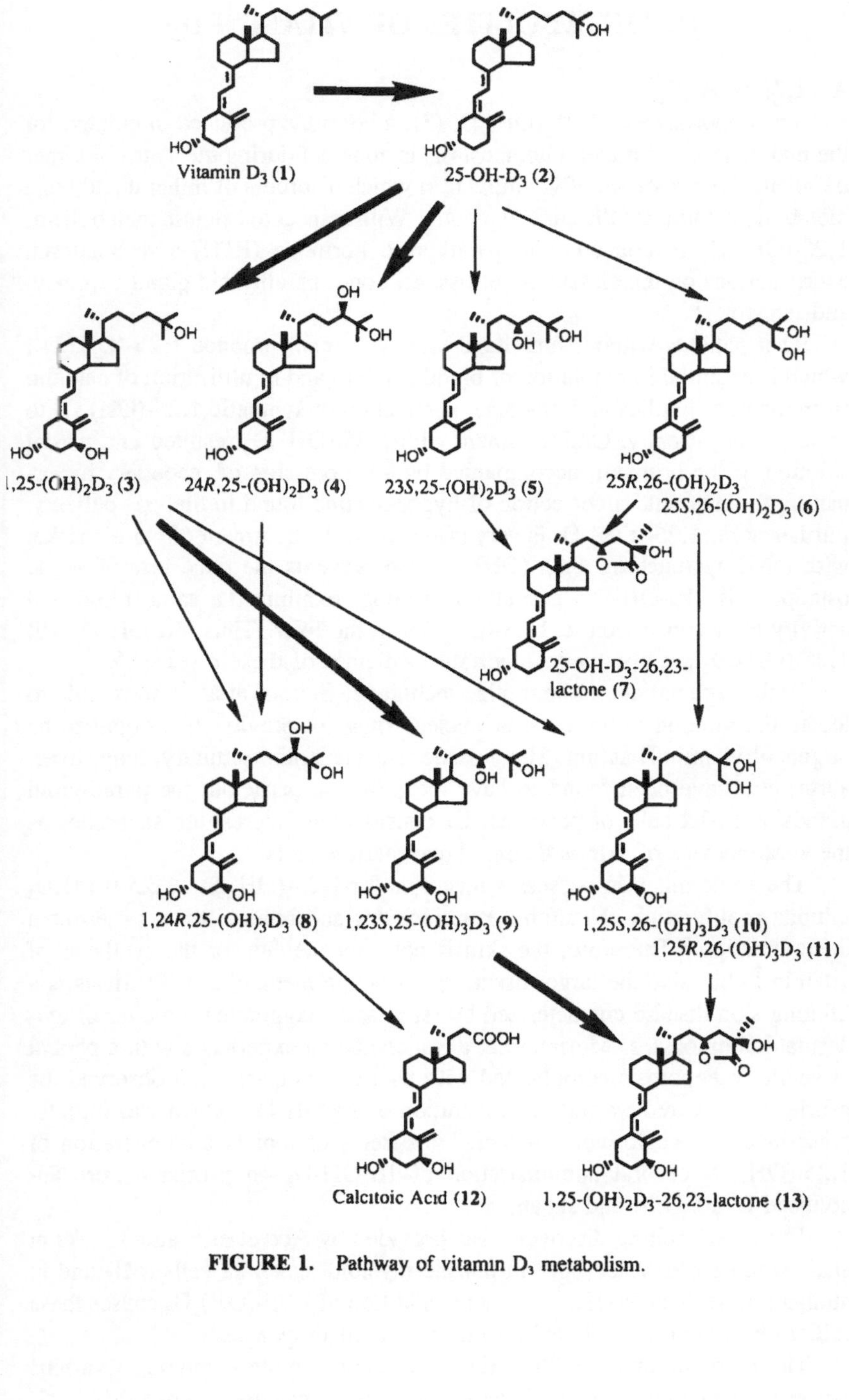

FIGURE 1. Pathway of vitamin D_3 metabolism.

II. METABOLITES OF VITAMIN D_3

A. 1,25-$(OH)_2D_3$

The importance of 1,25-$(OH)_2D_3$ (**3**), a hormone produced in kidney, for the maintenance of mineral homeostasis is apparent during the states of either deficiency or resistance of vitamin D, in which disorders of mineral and bone metabolism dominate the clinical picture. With respect to calcium metabolism, 1,25-$(OH)_2D_3$ in concert with parathyroid hormone (PTH) and calcitonin exerts actions on the classic target tissues: bone, parathyroid glands, kidney, and intestine.[3]

That patients without functional kidneys cannot produce 1,25-$(OH)_2D_3$, which is essential in regulation of blood calcium and in utilization of calcium from the diet, implies an immediate application of synthetic 1,25-$(OH)_2D_3$ to renal osteodystrophy. Oral treatment with 1,25-$(OH)_2D_3$ resulted in a partial solution of the problem, accompanied by a suppression of secondary hyperparathyroidism, and a correction of hypocalcemia found in dialysis patients. Furthermore, 1,25-$(OH)_2D_3$ is very effective in the treatment of child's ricket with renal dysfunction. 1,25-$(OH)_2D_3$ also prevents the bone loss of senile osteoporosis. 1α-OH-D_3, a synthetic analog, exhibits the same biological activity after conversion to 1,25-$(OH)_2D_3$ in the liver. Thus, 1α-OH-D_3 and 1,25-$(OH)_2D_3$ are now very important medicines of these diseases.[4-6]

With a special autoradiographic technique, Stumpf et al.[7,8] were able to locate the vitamin D hormone at nuclear sites not previously thought to be targets of vitamin D action. Many tissues such as brain, pituitary, lung, liver, aorta, etc. have been found to have receptors, in particular the parathyroid glands and islet cells of pancreas. Of considerable interest are such sites as the keratinocytes of skin and some bone marrow cells.

The epidermis also possesses receptors for 1,25-$(OH)_2D_3$. 1,25-$(OH)_2D_3$ inhibits proliferation of cultured keratinocytes and induces them to terminal differentiation.[9] Therefore, the skin is not only the site for the synthesis of vitamin D but also the target tissue for its active metabolite.[10] Psoriasis is a lifelong skin disease characterized by rapid and exaggerated turnover of epidermal keratinocytes. Morimoto et al. once had an experience with a patient of senile osteoporosis complicated with psoriasis vulgaris, and observed the psoriasis was cured by oral administration of 1α-OH-D_3, which was initially prescribed for the former disorder.[11,12] Effects of topical administration of 1,25-$(OH)_2D_3$ or oral administration of 1α-OH-D_3 on psoriasis were announced in the U.S. and Japan.

The most exciting discovery was provided by Abe et al.[13] and Tanaka et al.,[14] who demonstrated both in murine myeloid leukemia cells (MI) and in human promyelocytes (HL-60) that the addition of 1,25-$(OH)_2D_3$ causes these cells to become differentiated into monocyte-macrophages.

The presence of the 1,25-$(OH)_2D_3$ receptor is quite common in various cancer cells including those of epithelial origin. It has been shown that 1,25-

$(OH)_2D_3$ inhibits proliferation of tumor cell growth, induces differentiation of tumor cells, induces differentiation of myeloid leukemia cells into monocyte-macrophages, and prolongs survival time of mice inoculated with myeloid leukemia MI cells.[15,16] These observations have prompted the exploration of possible therapeutic agents based on the vitamin D structure.

The combined effects of sodium butyrate (NaB) and 1,25-$(OH)_2D_3$ on differentiation of cell growth and initiation of enterocytic differentiation were reported in the established HT-29 human colonic adenocarcinoma cells. Thus, the exposure of HT-29 cells to the combined formulation significantly augmented the frequency of differentiated colonocytes, growth inhibition, the extent of goblet cell maturation attained, and the level of alkaline phosphatase (AP) activity over those seen with NaB alone.[17]

B. 24,25-$(OH)_2D_3$

The stereochemical configuration of natural 24,25-$(OH)_2D_3$ was determined as 24*R*,25-$(OH)_2D_3$ (**4**) by comparison with synthetic 24*R*- and 24*S*-epimers by Ikekawa.[1] The biological role of 24,25-$(OH)_2D_3$ is a matter of continuing controversy. Some investigators have postulated that it plays an important role in bone formation, while others consider that it is a catabolic product of vitamin D metabolite without a major biological function.[3] Tam reported the *in vivo* study supporting that 24,25-$(OH)_2D_3$ has a role in the regulation of bone formation and mineralization.[18] Thus, when vitamin D-deficient animals were thyroparathyroidectomized before supplementation with vitamin D metabolites, 24,25-$(OH)_2D_3$ administration was without effect on the linear rate of bone mineral apposition (BMAR). The combined administration of PTH and 24,25-$(OH)_2D_3$ to such animals led, however, to a restoration of BMAR to normal. Simultaneous treatment of vitamin D-sufficient animals with PTH and 24,25-$(OH)_2D_3$ led to a greater increase in BMAR than normal.

Schwartz reported a direct effect of 24,25-$(OH)_2D_3$ and 1,25-$(OH)_2D_3$ on modeling of fetal mice long bones in the *in vitro* study.[19] The results showed that 24,25-$(OH)_2D_3$ at concentrations of 10^{-10} to 10^{-8} M stimulated the growth of the bones as indicated by their increased diaphyseal length, periosteal bone area, and hydroxyproline content. However, bones incubated in media containing 10^{-6} M 24,25-$(OH)_2D_3$ respond in a similar fashion to bones incubated in media containing 10^{-10} to 10^{-8} M 1,25-$(OH)_2D_3$ with inhibition of bone growth.

Recently Nakamura and Orimo reported the increase of bone volume in vitamin D-replete rats by the long-term administration of 24,25-$(OH)_2D_3$.[20] Male Wistar rats were fed a diet containing 24,25-$(OH)_2D_3$ for 2 years. Radiographies revealed a striking increase in the density on the bones of the animals treated with 1.25 ppm or more 24,25-$(OH)_2D_3$. Histological examination revealed a marked increase in the cortical thickness of the femur as well as in the cancellous bone volume of the coccyx. It was also reported that the increase in the bone density in animals administered high doses of 24,25-$(OH)_2D_3$ was accompanied by an increase in mechanical strength of the bone.[21] The mechanism of the increase in the bone mass after administration of high doses 24,25-$(OH)_2D_3$ is not clear.

Under the vitamin D-replete conditions, as in this case, a large dose of 24,25-$(OH)_2D_3$ may show pharmacological actions in the presence of 1,25-$(OH)_2D_3$. Wilhelm recently showed that, although there are no 24,25-$(OH)_2D_3$ receptors in the intestines, 24,25-$(OH)_2D_3$ exerts an allosteric effect on 1,25-$(OH)_2D_3$ to increase intestine calcium absorption.[22] Weintroub also demonstrated the increase of bone γ-carboxyglutamic acid-containing protein in rat bone by the administration of 24,25-$(OH)_2D_3$.[23]

It is conceivable that 24,25-$(OH)_2D_3$ is converted to active metabolites, which cause such a marked increase in the bone mass. 1,24(*R*),25-$(OH)_3D_3$, for example, has calcium mobilization activity and, therefore, may be such a metabolite capable of inducing an increase in the bone mass.[1] Accurate evaluation of changes in the blood and tissue concentration of vitamin D metabolites after massive administration of 24,25-$(OH)_2D_3$ is necessary for clarification of the action mechanism of the agent.

C. 1,25-$(OH)_2D_3$-26,23-LACTONE

1,25-$(OH)_2D_3$-26,23-lactone (**13**) has been isolated and identified as a major metabolite of 1,25-$(OH)_2D_3$ in the serum of rats and dogs given pharmacologic doses of 1,25-$(OH)_2D_3$. Recently Ishizuka et al. demonstrated that 1,25-$(OH)_2D_3$-26,23-lactone is a main metabolite of 1,25-$(OH)_2D_3$ in beagle dogs under the steady-state vitamin D-replete physiological conditions.[24] The stereochemical configuration of the naturally occurring 1,25-$(OH)_2D_3$-26,23-lactone was unequivocally determined to be 23*S* and 25*R* by comparison with all diastereomers of synthetic 1,25-$(OH)_2D_3$-26,23-lactone.[25]

There are two metabolic pathways from 1,25-$(OH)_2D_3$ to 1,25-$(OH)_2D_3$-26,23-lactone. One is from 1,25-$(OH)_2D_3$ to 1,23*S*,25*R*,26-$(OH)_4D_3$ by way of 1,23*S*,25-$(OH)_3D_3$ (**9**) and the other is by way of 1,25*R*,26-$(OH)_3D_3$ (**11**). 1,23*S*,25*R*,26-$(OH)_4D_3$ is further metabolized to 1,25-$(OH)_2D_3$-26,23-lactone via 1,25-$(OH)_2D_3$-26,23-lactol.[26] However, the further metabolism of 1,25-$(OH)_2D_3$-26,23-lactone has not been entirely understood.

The naturally occurring 1,25*R*-$(OH)_2D_3$-26,23*S*-lactone was found to slightly stimulate intestinal calcium absorption and to decrease serum calcium concentration in vitamin D-deficient rats fed a low calcium diet and in normal rats. Furthermore, it inhibits the formation of multinucleated cells with several characteristics of osteoclasts from mouse bone marrow mononuclear cells, mouse hematopoietic blast cells, and human bone marrow mononuclear cells, all of which are induced by 1,25-$(OH)_2D_3$. Bone resorption induced by 1,25-$(OH)_2D_3$ is also inhibited by the lactone *in vivo* and *in vitro*.[27-29]

Recently, Ishizuka et al. reported that the lactone increases AP activity and collagen synthesis in osteoblastic clone MC3T3-E1 cells *in vitro* and it stimulates collagen synthesis at the stage of matrix formation and mineralization at the stage of bone formation *in vivo* using bone morphogenetic protein-derived Dunn osteosarcomas.[30,31] Also, in dynamic histomorphometric study using experimental osteoporosis model rats induced by static neurotomy on the right hind leg combined with ovariectomy, the lactone significantly increased bone formation rate. These results demonstrate that the lactone stimulates bone formation by increasing bone matrix content.[28,29]

Biological activities of all four possible diastereomers of 1,25-$(OH)_2D_3$-26,23-lactone were compared with that of 1,25-$(OH)_2D_3$.[27] The receptor binding affinities of four diastereomers of 1,25-$(OH)_2D_3$-26,23-lactone to chick intestinal 1,25-$(OH)_2D_3$ receptor were as follows: 1,25-$(OH)_2D_3$ (relative competitive index, RCI, 100), 23*S*,25*S*- (7.90), 23*R*,25*R*- (2.27), 23*S*,25*R*- (0.17), and 23*R*,25*S*- (0.22). The RCI is defined as 100 for 1,25-$(OH)_2D_3$. All the four diastereomers of 1,25-$(OH)_2D_3$-26,23-lactone stimulated intestinal calcium transport but less actively than 1,25-$(OH)_2D_3$. The 23*S*,25*S*- and 23*R*,25*R*-lactone were estimated to be 3 and 20 times less active than 1,25-$(OH)_2D_3$ in elevation of serum calcium. However, the 23*S*,25*R*- and 23*R*,25*S*-lactone decreased the serum calcium level at 24 h after administration. The 23*S*,25*R*-lactone reduced serum calcium concentration to a greater extent than the 23*R*,25*S*-lactone.

The four lactone diastereomers had little effect on the protein and DNA content of the osteoblastic clone MC3T3-E1 cells. The 23*S*,25*S*- and 23*R*,25*R*-lactone increased AP activity in a linear dose-dependent fashion. Maximal effects were observed at 100 and 1000 pg/ml, respectively. In contrast, the 23*S*,25*R*- and 23*R*,25*S*-lactone showed biphasic stimulatory effects on AP activity. At 80 pg/ml, both the 23*S*,25*R*- and 23*R*,25*S*-isomers stimulated an increase in collagen synthesis, while at 10000 pg/ml these isomers and 1,25-$(OH)_2D_3$ were without effect.[30] These results indicated that the biological activities of 1,25-$(OH)_2D_3$-26,23-lactone were quite different among four diastereomers.

III. MODIFIED SIDE CHAIN

A. SIDE CHAIN HYDROXY ANALOGS

24*R*-OH-D_3 and 24*R*,25-$(OH)_2D_3$ are almost as active as 25-OH-D_3 in stimulating intestinal calcium transport and calcification of bone in rachitic rats. On the other hand, 24*S*-OH-D_3 and 24*S*,25-$(OH)_2D_3$ have little or no activity in these tests. The high potency of the *R*-isomer was found to be attributed to the specificity of 1α-hydroxylation in kidney which favors the *R*-isomer.[1] This is similar to the case of 25,26-$(OH)_2D_3$ in which only the 25*S*,26-$(OH)_2D_3$ isomer is hydroxylated.[32] However, in the cases of 25,26- and 23,25-$(OH)_2D_3$, both *R*- and *S*-isomers showed nearly same biological activity, which is less than 24*R*,25-$(OH)_2D_3$.[1]

It is extremely interesting to note that synthetic 22-hydroxylated analogs of 25-OH-D_3 and 1,25-$(OH)_2D_3$ exhibited no vitamin D activity. Confor-

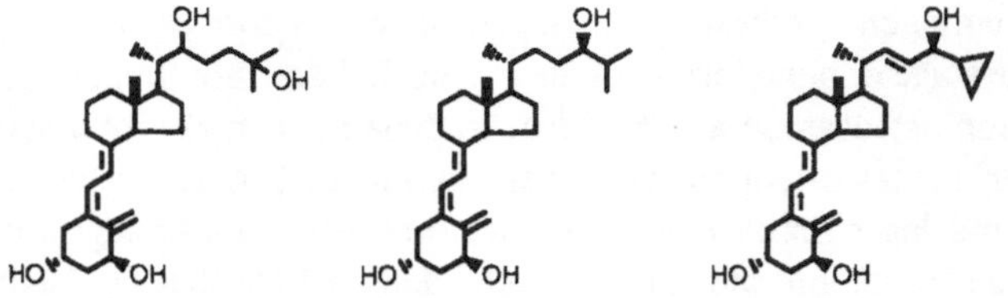

(14) (15) (16)

FIGURE 2. Side chain hydroxy analogs discussed in this chapter.

mational difference between 22*R*- and 22*S*-hydroxy compounds and their biological activities have been discussed by Eguchi et al.[33] Although the unsubstituted sterol side chain bears a zigzag conformation generally, there are several lines of evidence suggesting that derivatives with a C-22 substituent have different side chain conformations depending on the C-22 configuration. Thus, introduction of a substituent onto the C-22 position of 1,25-$(OH)_2D_3$ seems to induce conformational changes, which serve as a clue in examining the possible importance of the side chain conformation for vitamin D activity. The synthetic (22*R* and 22*S*)-22-hydroxy- (**14**) and 22-methoxy analogs of 1,25-$(OH)_2D_3$ induced differentiation of HL-60 cells at a concentration more than 1 μg/ml in the nitroblue tetrazolium (NBT) test. Especially noteworthy was that the 22*S*-isomers were at least 30 times more effective than the 22*R*-isomer. Binding affinity of the (22*S*)-22-methoxy derivative to chick intestinal receptor for 1,25-$(OH)_2D_3$ was about three times as potent as the 22*R*-isomer.

Conformational difference between the C-22 epimers was elucidated by MM2 force field calculation and NMR studies. A zigzag conformation turned out to be sterically most favorable for 22*S*-isomers, whereas such a conformation is energically unfavorable for 22*R*-isomers due to the interaction between the 22-substituent and the 16-methylene group. It is therefore suggested that the zigzag conformation plays an important role for the biological activities (Figure 2).[33]

B. 1,24-$(OH)_2D_3$

1,24*R*-$(OH)_2D_3$ (**15**) and 1,24*S*-$(OH)_2D_3$ were synthesized in 1975 as synthetic mimics of 1,25-$(OH)_2D_3$ by Morisaki et al.[34] The C-24 epimers of 1,24-$(OH)_2D_3$ were compared to 1,25-$(OH)_2D_3$ in terms of their stimulation of intestinal calcium transport and mobilization of calcium from bone in vitamin D-deficient rats, and their relative binding affinity to the chick intestinal cytosol receptor for 1,25-$(OH)_2D_3$. Both epimers remarkably stimulated intestinal calcium transport and bone calcium mobilization. The activities of the 24*R*- and 24*S*-epimers were about one half and one tenth those of 1,25-$(OH)_2D_3$, respectively. However, these activities were two to five times stronger than those of their respective metabolites, 1,24*R*,25-$(OH)_3D_3$ and 1,24*S*,25-$(OH)_3D_3$.[35]

The binding affinity of $1,24R\text{-}(OH)_2D_3$ to $1,25\text{-}(OH)_2D_3$ receptor of chick intestine was 1.3 times stronger than that of $1,25\text{-}(OH)_2D_3$, whereas those of $1,24S\text{-}(OH)_2D_3$, $1,24R,25\text{-}(OH)_3D_3$ and $1,24S,25\text{-}(OH)_2D_3$ were 10, 5, and 39 times weaker than that of $1,25\text{-}(OH)_2D_3$, respectively.[35] However, it was not yet clear whether $1,24\text{-}(OH)_2D_3$ is active by itself or activated by the conversion into $1,24R,25\text{-}(OH)_3D_3$ in the liver. In order to explore this possibility, hepatectomized rats were administered $1,24R\text{-}(OH)_2D_3$, which markedly stimulates the intestinal calcium transport. These results indicated that $1,24R\text{-}(OH)_2D_3$ is active by itself, although it is further metabolized to $1,24R,25\text{-}(OH)_2D_3$ in the liver.[36]

$1,24\text{-}(OH)_2D_3$ also inhibited cell proliferation and induced differentiation of HL-60 cells into monocyte-macrophages. Cell differentiation activity of $1,24R\text{-}(OH)_2D_3$ was 2 times stronger than that of $1,25\text{-}(OH)_2D_3$, whereas the activity of $1,24S\text{-}(OH)_2D_3$ was 5.7 times weaker than that of $1,25\text{-}(OH)_2D_3$.[37]

The effect of $1,24R\text{-}(OH)_2D_3$ on the growth of human keratinocytes cultured in serum free medium was investigated.[38] The growth of cultured normal human keratinocytes was inhibited by 65% at 10^{-8} *M* and 90% at 10^{-7} *M* of $1,24R\text{-}(OH)_2D_3$ and the cell growth was almost completely inhibited at 10^{-6} *M*. $1,24R\text{-}(OH)_2D_3$ also inhibited the growth of keratinocytes derived from patients with psoriasis vulgaris and the inhibitory effect was again more remarkable with $1,24R\text{-}(OH)_2D_3$ than with $1,25\text{-}(OH)_2D_3$. These results can be correlated with the fact that $1,24R\text{-}(OH)_2D_3$ possesses a higher affinity for the $1,25\text{-}(OH)_2D_3$ receptor. Recently, psoriasis was successfully treated with topical application of $1,24R\text{-}(OH)_2D_3$ by Kato et al.[39,40] The double blind test of $1,24R\text{-}(OH)_2D_3$ ointment containing 2 μg/g of petrolatum showed better results than corticosteroids. The $1,24R\text{-}(OH)_2D_3$ ointment will soon appear commercially.

C. CYCLOPROPYL DERIVATIVE OF Δ^{22}-1,24S-$(OH)_2D_3$ (MC903)

By the modification of the structure of $1,24R\text{-}(OH)_2D_3$, a cyclopropyl derivative (MC903) (**16**) was developed by Leo Pharmaceutical Products; this is one of the most successful compounds in separating the biological activity with a relatively weak systematic effect on calcium metabolism while maintaining potent regulatory effects on cell proliferation and differentiation.[41,42] MC903 showed similar effect to $1,25\text{-}(OH)_2D_3$ in a culture system of human histiocytic lymphoma cells (U937) and was 100 to 200 times less potent than $1,25\text{-}(OH)_2D_3$ when tested for the effects on calcium metabolism in rats. Thus, as a promising candidate for topical treatment of psoriasis, many clinical tests and double blind studies were carried out and it turned out to be useful agent for psoriasis.[42,43]

Recently, Thavarajah reported that MC903 is at least as potent as $1,25\text{-}(OH)_2D_3$ in inducing the formation of human osteoclast-like cells *in vitro*.[44] This suggested that the decreased ability of MC903 to induce hypercalcemia *in vivo* is not due to a result of a less marked effect than $1,25\text{-}(OH)_2D_3$ on the regulation of osteoclast formation.

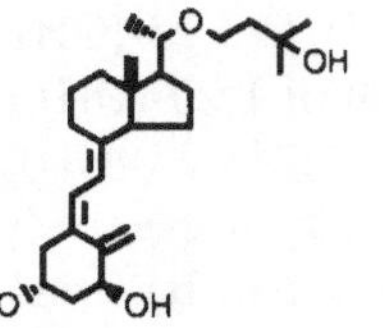

(17)

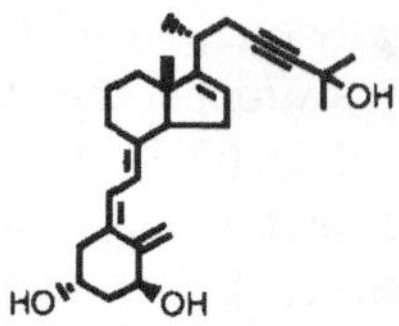

(18)

FIGURE 3. Side chain analogs discussed in this chapter.

The activity of MC903 on bone cells was reported by Marie et al.[45] MC903 was 10 to 100 times less effective than 1,25-$(OH)_2D_3$ in stimulating the synthesis of osteoblast-specific protein, osteocalcin, by human bone cells depending on the basal osteocalcin production. 1,25-$(OH)_2D_3$ and MC903 exert equipotent effects on DNA synthesis and AP activity but distinct effects on the synthesis of osteocalcin.

D. 22-OXA-1,25-$(OH)_2D_3$

Another promising compound which indicated the separation of the differentiation-inducing activity from the bone-resorbing activity at least *in vitro* is 22-oxa-1,25-$(OH)_2D_3$ **(17)**, which was synthesized by Chugai research group.[46] The activities of reducing NBT in HL-60 cells and in suppressing proliferation and inducing differentiation of mouse myelomonocytic leukemia cells (WEHI-3) were about 10 times more potent than 1,25-$(OH)_2D_3$, whereas the *in vitro* bone-resorbing activity of 22-oxa-1,25-$(OH)_2D_3$ was estimated to only 1/50 to 1/100 than of 1,25-$(OH)_2D_3$.[47] It was also reported that oral administration of the 22-oxa compound augmented the primary immune response induced by immunization with a suboptimal number of sheep erythrocytes without inducing hypercalcemia.[48]

1,25-$(OH)_2D_3$ directly suppresses the secretion and synthesis of PTH *in vivo* and in cell culture, and has been used to treat secondary hyperparathyroidism associated with renal failure, but in some patients prolonged treatment with 1,25-$(OH)_2D_3$ results in hypercalcemia. However, the 22-oxa compound is a very effective suppressor of PTH secretion with virtually no calcemic activity. This analog may be a valuable tool for the treatment of secondary hyperparathyroidism (Figure 3).[49]

E. 1,25-$(OH)_2$-16-EN-23-yne-D_3

Among the compounds synthesized by Hoffmann-La Roche, seven new analogs, including 1,25-$(OH)_2O_3$-16-en-23-yne-D_3 (**18**), 1,23,25-$(OH)_3D_3$, and 1,25,26-$(OH)_3D_3$, were found to be either equivalent or more potent than 1,25-$(OH)_2D_3$ as assessed by inhibition of clonal proliferation and induction of differentiation of HL-60 cells. Each analog mediated markedly less active (10- to 200-fold) in intestinal calcium transport and bone calcium mobilization.[50-52]

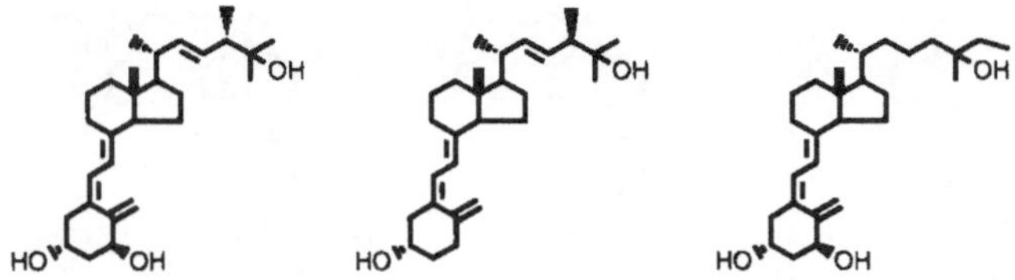

(19) (20) (21)
(22) Δ^{22}

FIGURE 4. Side chain homologs discussed in Section I of this chapter.

1,25-$(OH)_2$-16-en-23-yne-D_3 was found to be sevenfold more potent than the parent 1,25-$(OH)_2D_3$ with respect to inhibition of clonal proliferation of HL-60 cells as well as induction of differentiation of HL-60 cells. A variant which possesses six deuterium atoms on carbon 26 and 27 were slightly less active than the parent compound. However, replacement of 1α-hydroxy group by a fluoro group resulted in analogs that were somewhat less effective than the parent 1,25-$(OH)_2$-16-en-23-yne-D_3.[53]

Further, Zhou reported that the 23-yne compound (**18**) is very potent in inhibiting proliferation and inducing differentiation of myeloid leukemic cells *in vitro,* and is 300 times less active in mediating intestinal calcium absorption and bone calcium moblization as compared to 1,25-$(OH)_2D_3$.[54] The 23-yne was 10 to 25 times less active than 1,25-$(OH)_2D_3$ in causing hypercalcemia *in vivo.* This difference allowed administration of greater amounts of the 23-yne than 1,25-$(OH)_2D_3$ to leukemia mice. Therapeutic potency of this compound was explored using three leukemia models, and significant increase of survival of mice who had myeloid leukemia was observed.

IV. SIDE CHAIN HOMOLOGS

Vitamin D_2 also undergoes metabolic activation in the liver and kidney to 1,25-$(OH)_2D_2$. The biological activity of 1,25-$(OH)_2D_2$ (**19**) is similar to that of 1,25-$(OH)_2D_3$ in mammals, whereas in birds it is about one tenth as potent as the corresponding vitamin D_3 counterpart.[1] In order to study the functional role of the 24-methyl group of vitamin D_2,25-OH-D_2 and its epimer, 24-epi-25-OH-D_2 (**20**), were synthesized. From the incubation of 24-epi-25-OH-D_2 with chick kidney homogenate, a new active metabolite was obtained, in addition to anticipated 24-epi-1,25-$(OH)_2D_2$ and 24-epi-24,25-$(OH)_2D_2$, and was ultimately identified as (22*E*)-22-dehydro-1,25-dihydroxy-26-methylvitamin D_3 (Δ^{22}-26-methyl-1,25-$(OH)_2D_3$) (**22**). This unexpected metabolite turned out to be as active as natural 1,25-$(OH)_2D_3$ and 1,25-$(OH)_2D_2$ in binding to chick intestinal cytosol receptor (Figure 4).[55]

Following this study, 24-homo-1,25-$(OH)_2D_3$ (**26**) and 26-methyl-1,25-$(OH)_2D_3$ (**21**), and their Δ^{22} derivatives were synthesized and the subsequent investigation of their biological activities opened a new area of vitamin D analogs.

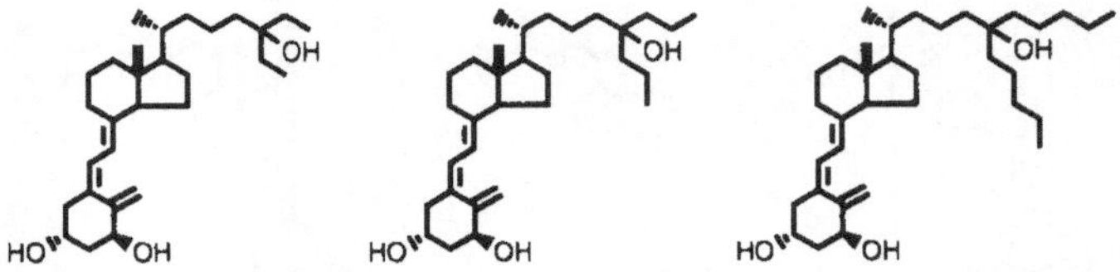

(23) (24) (25)

FIGURE 5. Side chain homologs discussed in Section II of this chapter.

It was found that 24-homo-1,25-$(OH)_2D_3$ and 26-methyl-1,25-$(OH)_2D_3$ are tenfold more potent than 1,25-$(OH)_2D_3$ in inducing differentiation of HL-60 cells *in vitro*.[56-58] These analogs are similar to 1,25-$(OH)_2D_3$ in stimulating intestinal calcium transport in vitamin D-deficient rats. The 24-homo analogs are significantly less active, whereas the 26-methyl derivatives are more active than the natural hormone in mobilizing calcium from bone.[56]

In conjunction with the stereochemistry at the 25-position of Δ^{22}-26-methyl-1,25-$(OH)_2D_3$, both 25-epimers of this unique metabolite were synthesized and their biological activities were investigated.[59] Both epimers are more potent than 1,25-$(OH)_2D_3$ in calcium-replacing activity test. Of interest is that the difference of the activities was observed between the epimers in that the 25*R*-isomer was more potent than the 25*S*-congener. This difference of *in vivo* activities might reflect the binding affinities for 1,25-$(OH)_2D_3$ receptor.

Worthy to note is that both epimers showed long-lasting activity. Δ^{22}-26-Methyl-1,25-$(OH)_2D_3$ may be degraded in rats more slowly than 1,25-$(OH)_2D_3$ due to steric hindrance against a catabolic hydroxylation step of the side chain moiety of the vitamin D_3.[59]

Other activities tested were inhibition of cell growth, induction of phenotypic differentiation, and reduction of *c-myc* mRNA synthesis of HL-60 cells. Both epimers were more potent than 1,25-$(OH)_2D_3$ in these assays, and interstingly the 25*S*-isomer showed higher activity than the 25*R*-isomer.[60]

Based on the anticipated slow deactivation of alkylated derivatives *vide supra*, 26,27-dialkylated analogs were synthesized and their biological activities were tested.[61-65] Affinities of 26,27-dimethyl- (**23**), 26,27-diethyl- (**24**), and 26,27-dipropyl-1,25-$(OH)_2D_3$ (**25**), to the chick 1,25-$(OH)_2D_3$-specific receptor were estimated to be about 1.1-, 0.25-, and 0.05-fold of 1,25-$(OH)_2D_3$, respectively. The 26,27-dialkyl analogs were also tested in a culture system for osteoclast formation from blast cell colonies. The dimethyl and diethyl analogs induced the formation of the multinucleated cells (MNCs) from blast cells in a dose-dependent manner, but the dipropyl compound did not show the ability to form the MNCs at the concentrations tested (Figure 5).[61]

The 1,25-$(OH)_2D_3$ and the dimethyl analog significantly increased the serium calcium concentration in the vitamin D-deficient rats in a dose-related

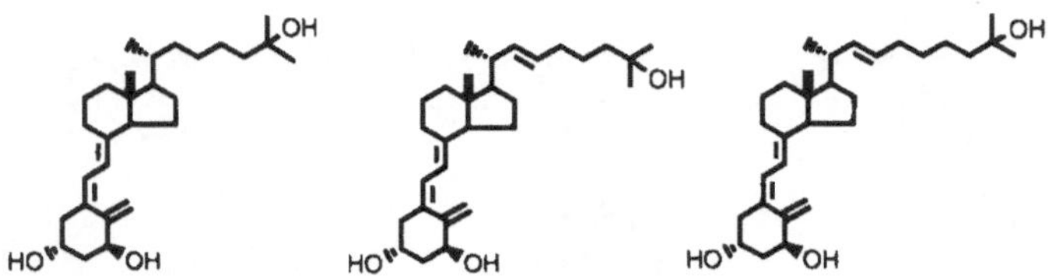

(26)
(27) Δ^{22}

(28)

(29)

FIGURE 6. Side chain homologs discussed in Section III of this chapter.

manner.[61,62] However, the diethyl and dipropyl analogs did not show any increase.[61] Of interest is that the dimethyl analog showed slightly less effectiveness than 1,25-$(OH)_2D_3$ at 12 h after administration, but at 24 h the reverse is the case. This may be due to the longer life of dimethyl analog by prevention of the deactivation step of metabolic hydroxylation at 26 or 27 positions as is the case with multifluoro derivatives of 1,25-$(OH)_2D_3$ *vide post*.

The effects of the dimethyl and diethyl analogs on the differentiation of HL-60 cells were 2.5 and 10 times more active than 1,25-$(OH)_2D_3$, respectively.[62,63] However, in the comparison of vitamin D analogs in *in vivo* and *in vitro* (calcium regulatory activity and cell differentiation activity) tests, care must generally be taken of the effect of serium vitamin D-binding proteins.[63,64] Thus, monocytic differentiation-inducing activity of the chain-lengthened 26,27-dialkyl analogs of 1,25-$(OH)_2D_3$ was examined in HL-60 cells in the serum-supplemented and in the serum-free culture. The order of *in vitro* potency for reducing NBT was the dimethyl $\geq$ 1,25-$(OH)_2D_3$ $>$ the diethyl $>$ the dipropyl derivatives under the serum-free culture conditions. Under serum-supplemented culture conditions, the lack of correlation between biological activity and analog-binding affinity for receptor was observed, which might be caused by differences in binding affinity of individual analogs for serum vitamin D-binding proteins. In any case, these results suggested that serum vitamin D-binding proteins apparently modulate monocytic differentiation of HL-60 cells by these analogs under serum-supplemented culture conditions.[66]

A series of 24-homologated 1,25-$(OH)_2D_3$ analogs were also synthesized and studied with regard to their activity in inducing differentiation of HL-60 cells to monocytes-macrophages and in calcium-mobilizing activity in vitamin D-deficient rats.[67] Homologation of 1,25-$(OH)_2D_3$ and its Δ^{22} analogs by 1 or 2 carbons (**26, 27,** and **28**) increases the activities by tenfold, but 3-carbon homologation (**29**) reduced by one half the activity in causing differentiation of HL-60 cells. On the other hand, homologation causes a substantial decrease in *in vivo* calcium mobilization activity. The addition of 1 or 2 carbons at the 24 position decreases binding to the HL-60 receptor or rat intestinal receptor by 5- to 10-fold and the binding affinity of the trihomo compound for the receptors is 130 times less that of 1,25-$(OH)_2D_3$ (Figure 6).[67]

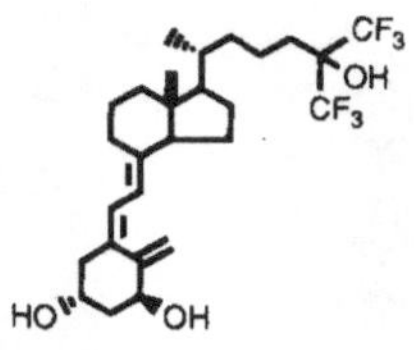

(30)

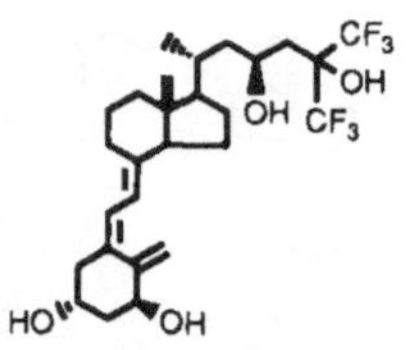

(31)

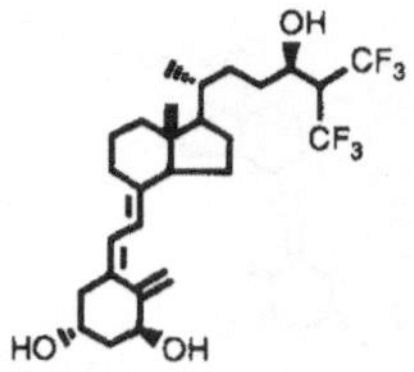

(32)

FIGURE 7. Fluorinated analogs discussed in Section I of this chapter.

Elongation of the steroidal side chain of 1,25-$(OH)_2D_3$ by addition of one carbon at C-24 or C-26 improves the potency by an order of magnitude as described above. Truncation of the steroidal side chain leads to a tenfold reduction in activity for each carbon removed. Elimination of both the C-26 and C-27 methyl groups reduces activity 100-fold.[57]

Δ^{22}-24,24-dihomo (**28**) and Δ^{22}-24,24,24-trihomo-1,25-$(OH)_2D_3$ (**29**) caused much more increase than 1,25-$(OH)_2D_3$ in calbindin-D 9-k mRNA without stimulating intestinal calcium transport.[68]

V. FLUORINATED ANALOGS

A. 26,27-F_6-1,25-$(OH)_2D_3$

Synthesis of fluorinated compounds was undertaken in order to clarify the mode of action of vitamin D, and also to obtain potent analogs which may be therapeutically useful. Thus, two types of fluorinated compounds have been synthesized: (1) compounds fluorinated at the C-25, C-1, or C-3 position in order to understand the biological significance of hydroxylation thereon, since such hydroxy groups are essential for eliciting the biological activity of vitamin D and (2) compounds fluorinated at the C-23, C-24, or C-26 position, hydroxylation of which can be considered as deactivation.

Among others, the well-investigated synthetic fluorinated compound with the strongest activity is 26,27-F_6-1,25-$(OH)_2D_3$ (**30**). This compound shows ten times higher activity than the native hormone, 1,25-$(OH)_2D_3$, in calcium-regulating activity, and much long-lasting activity. The receptor-binding activity is about as same as 1,25-$(OH)_2D_3$. The reason for increased potency has been ascribed to the less rapid metabolism, in which the deactivating hydroxylation at the side chain is blocked by fluorine (Figure 7).[69]

26,27-F_6-1,25-$(OH)_2D_3$ and 26,27-F_6-1,24-$(OH)_2D_3$ (**32**) were reported to be about ten times as potent as 1,25-$(OH)_2D_3$ in suppressing HL-60 cell proliferation and inducing cell differentiation.[70,71,80] These two compounds were also ten times more potent than 1,25-$(OH)_2D_3$ in suppressing interleukin-2 production.[72] The most significant effect in cell differentiation was enhancement observed with the combined use of NaB and 26,27-F_6-1,25-$(OH)_2D_3$. The prominent effect was achieved using cells differentiated with NaB and

10^{-9} *M* 26,27-F_6-1,25-$(OH)_2D_3$ which exhibited a seven-fold prolonged reversal time over colonocytes differentiated by NaB alone.[73]

The effect of 26,27-F_6-1,25-$(OH)_2D_3$ on experimental osteoporosis in rats induced by a combination of immobilization and ovarietomy was evaluated by Okumura et al.[74] Its administration significantly increased the dry weight, ash weight, and ash content of the bone.

The enhanced activity of 26,27-F_6-1,25-$(OH)_2D_3$ was first ascribed to a prolonged half-life *in vivo* as discussed above. However, the serum concentration of 26,27-F_6-1,25-$(OH)_2D_3$ after intrajugular administration to vitamin D-deficient rats was found to be similar to that of 1,25-$(OH)_2D_3$. The result suggested that the serum concentration was not directly related to the higher potency of the F_6 compound. By the distribution studies in rats, much higher radioactivity in the intestine and the kidney was observed with [^{3}H]-26,27-F_6-1,25-$(OH)_2D_3$ administration than with [^{3}H]-1,25-$(OH)_2D_3$.[75]

The major metabolite in the intestine was identified as 26,27-F_6-1,23*S*,25-$(OH)_3D_3$ (**31**) by the Sumitomo group. Binding affinity of 26,27-F_6-1,23*S*,25-$(OH)_3D_3$ to the chick intestinal receptor was similar to that of 26,27-F_6-1,25-$(OH)_2D_3$, being about two to three times less than that of 1,25-$(OH)_2D_3$. Biological activity of 26,27-F_6-1,23S,25-$(OH)_2D_3$, such as stimulation of intestinal calcium transport, was found to be comparable to that of 1,25-$(OH)_2D_3$, despite that metabolites of 1,25-$(OH)_2D_3$ are generally biologically less active than 1,25-$(OH)_2D_3$. These results suggest that accumulation of the 23-hydroxylated active metabolite in the target tissues is an important mechanism for the higher potency of 26,27-F_6-1,25-$(OH)_2D_3$.[77] 26,27-F_6-1,25-$(OH)_2D_3$ had a direct stimulatory effect on AP activity in the human bone-derived cells. This stimulatory effect was far more potent than that of 1,25-$(OH)_2D_3$.[76] Therefore, one of the most likely explanations for the higher activity of 26,27-F_6-1,25-$(OH)_2D_3$ than 1,25-$(OH)_2D_3$ *in vivo* may be that 26,27-F_6-1,25-$(OH)_2D_3$ is metabolized to the fully active 26,27-F_6-1,23*S*,25-$(OH)_2D_3$ in intestine, whereas 1,25-$(OH)_2D_3$ has been degraded to inactive metabolites.

Thus, analysis of the degradation process of 1,25-$(OH)_2D_3$-receptor complex in its target cells, in addition to that in the kidney, may be important in understanding physiological and pathological mechanisms of regulation of cellular responsiveness to 1,25-$(OH)_2D_3$. The fluorinated 1,25-$(OH)_2D_3$ compounds, whose inactivation process seems to be blocked, can be of considerable use as tools in unraveling the mechanisms of action and metabolism of 1,25-$(OH)_2D_3$.

Recently Inaba et al. reported that 26,27-F_6-1,25-$(OH)_2D_3$ and 1,25-$(OH)_2D_3$ recognize the same binding site of the receptor and that the 26,27-F_6,1,25-$(OH)_2D_3$ receptor complex binds DNA more tightly than the 1,25-$(OH)_2D_3$ receptor complex.[77]

Evans has reported that 26,27-F_6-Δ^{22}-1,25-$(OH)_2D_3$ (**33**) and 26,27-F_6-Δ^{23}-1,25-$(OH)_2D_3$ (**34**) exhibited a similar spectrum of activity on human

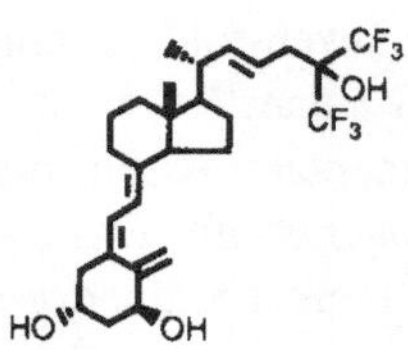

(33)

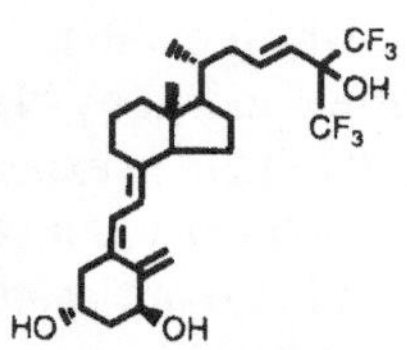

(34)

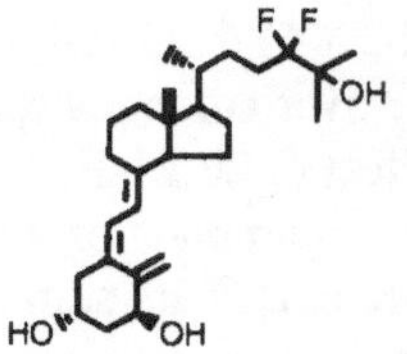

(35)

FIGURE 8. Fluorinated analogs discussed in Section II of this chapter.

osteoblast-like cells to that of 1,25-$(OH)_2D_3$, and were approximately 100-fold more potent in their actions on cell proliferation, but the effect of a double bond in the side chain on their potency is not clear.[78,79]

B. 24-F_2 AND 23-F_2 ANALOGS AND OTHERS

The importance of 24-hydroxylation of the vitamin D_3 molecule has been examined by studying the biological activities of 24,24-F_2-25-OH-D_3 and the corresponding 1-hydroxylated analog. The biological activity of 24,24-F_2-1,25-$(OH)_2D_3$ (**35**) was reported the same as that of 26,27-F_6-1,25-$(OH)_2D_3$ on calcium regulation and cell differentiation.[80] 24,24-F_2-1,25-$(OH)_2D_3$ has the identical binding affinity as 1,25-$(OH)_2D_3$ for the 1,25-$(OH)_2D_3$ receptor from chick intestine and HL-60 cells. 24,24-F_2-1,25-$(OH)_2D_3$ has a four- to nine-fold greater potency than 1,25-$(OH)_2D_3$ in such activities as stimulation of intestinal calcium transport, bone mineral mobilization, epiphyseal plate calcification in vitamin D-deficient rats, promotion of cell differentiation in HL-60 cells, etc.[80] The reason for the enhanced biopotency of 24,24-F_2-1,25-$(OH)_2D_3$ is unclear at present, but solubility in lipid, affinity for the blood vitamin D-binding protein, and rate of catabolism may contribute to this difference (Figure 8).

It was observed by synthesis of 23,23-F_2-1,25-$(OH)_2D_3$ (**36**) that, unlike fluoro substition at C-24, C-26, and C-27, the fluorine atoms at C-23 actually reduce biological activity including the binding to the chick intestinal receptor and the rat plasma-binding protein.[81] This suggests that the electron-withdrawing property of fluroine atoms on C-23 may significantly alter binding, thereby changing target organ response.[81] The reason why fluorine atoms on this carbon should have this marked effect on binding and biological activity as compared to C-26 and C-24 fluorinations is unknown.

24,24-F_2-26,27-dimethyl- (**37**) and 24,24-F_2-24-homo-1,25-$(OH)_2D_3$ (**38**) were synthesized. The former showed more potent activity than 1,25-$(OH)_2D_3$ on the calcium-regulating activity, whereas the latter was much less active.[82,83] However, the latter was at least ten times more active than 1,25-$(OH)_2D_3$ for cell growth inhibitor of HL-60 cells.[63,82]

Introduction of fluorine at the C-22 position to yield 22-fluorovitamin D_3 (**39**) resulted in decreased biological activity when compared to vitamin D_3.[84]

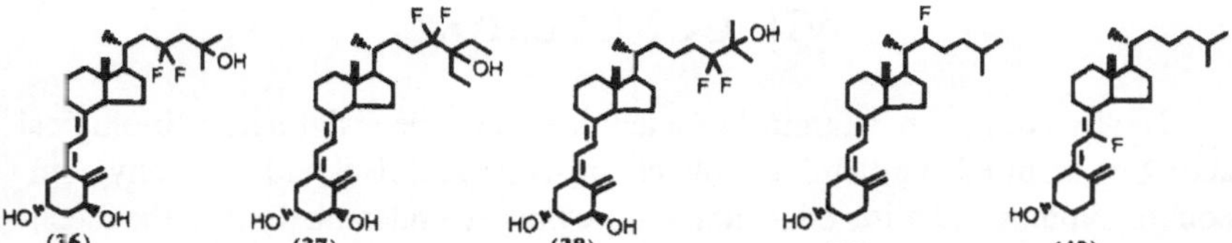

(36) (37) (38) (39) (40)

FIGURE 9. Fluorinated analogs discussed in Section III of this chapter.

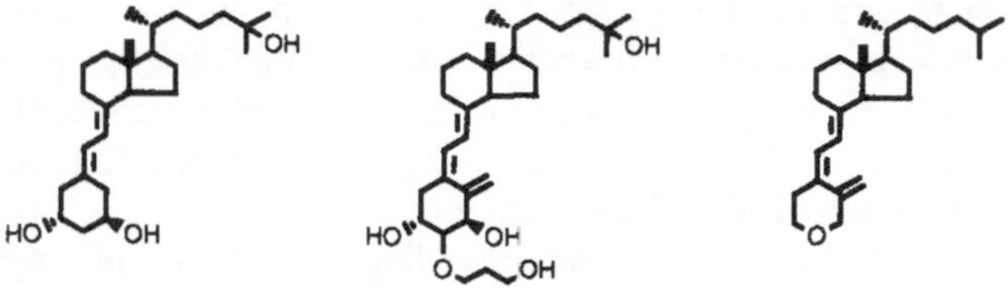

(41) (42) (43)

FIGURE 10. Modified A-ring analogs discussed in this chapter.

Only 6-fluorovitamin D_3 (**40**) was synthesized as an analog with a fluorine group in the nucleus. This compound was by no means an agonist in intestine or bone activity.[85] However, it is of interest that, when this is administered with 1,25-$(OH)_2D_3$, it serves as an antagonist. This is the first analog found to be an *in vivo* antagonist which shows competitive *in vitro* binding to the 1,25-$(OH)_2D_3$ receptor (Figure 9).

VI. MODIFIED RING STRUCTURE

19-Nor-1,25-$(OH)_2D_3$ (**41**), synthesized by DeLuca et al., showed a selective activity profile, i.e., high potency in differentiation of malignant cells, but very low or no bone calcification.[86]

A 2β-(3-hydroxypropoxy)-1,25-$(OH)_2D_3$ (ED-71) (**42**) was synthesized by Chugai research group.[87] The calcium-releasing activity of ED-71 in plasma was not greater than that of 1,25-$(OH)_2D_3$ and a mild round shape time-course curve was shown compared with that of 1,25-$(OH)_2D_3$. The therapeutic effect of ED-71 for the animal models with osteoporosis seemed to be better than that of 1,25-$(OH)_2D_3$.

Biological activity of synthetic 9,11-dehydrovitamin D_3 analogs were approximately half of that for 1,25-$(OH)_2D_3$ in an *in vitro* receptor binding assay.[88] Some of the A-ring oxa compound (**43**) and seven membered A-ring analogs were also synthesized by Okamura,[89,90] but their biological activity was not reported. A new metabolite of vitamin D_3 8α,25-dihydroxy-9,10-seco-4,6,10(19)-cholestatrien-3-one was identified,[91,92] but its biological activity is unknown (Figure 10).

VII. CONCLUSION

New findings on vitamin D metabolites and their significant biological activities inspired synthetic organic chemists a great deal and the many compounds synthesized allowed to attain more precise understanding on the mechanism of function of vitamin D hormone. Recently, new biological activities have been emerging in these synthetic analogs. Thus, interdisciplinary cooperative efforts have successfully opened a new field in medicinal chemistry of steroids.

To be emphasized is that the structure, regulation, and molecular function of the receptor of vitamin D hormone has shed light on the similarity in biochemical as well as signal transduction systems to other steroidal hormones and retinoic acid.[93-95] This development may significantly influence the area of the structure-biological activity relationship.

Based on the more-detailed understanding of such receptor-agonist interaction on the molecular level, vitamin D hormone will be extremely beneficial in solving various medical problems.

REFERENCES

1. **Ikekawa, N.,** Structures and biological activities of vitamin D metabolites and their analogs, *Med. Res. Rev.*, 7, 333, 1987.
2. **Reichel, H., Koeffler, H. P., and Norman, A. W.,** The role of the vitamin D endocrine system in health and disease, *N. Engl. J. Med.*, 320, 980, 1989.
3. **Brommage, M. and DeLuca, H. F.,** Evidence that 1,25-dihydroxyvitamin D_3 is the physiologically active metabolite of vitamin D_3, *Endocr. Rev.*, 6, 491, 1985.
4. **Shiraki, M., Orimo, H., Ito, H., Akiguchi, I., Nakao, J., Takahashi, R., and Ishizuka, S.,** Long-term treatment of postmenopausal osteoporosis with active vitamin D_3, 1-alpha-hydroxycholecalciferol (1α-OHD_3) and 1,24 dihydroxycholecalciferol (1,24-$(OH)_2D_3$), *Endocrinol. Jpn.*, 32, 305, 1985.
5. **Orimo, H., Shiraki, M., Hayashi, T., and Nakamura, T.,** Reduced occurrence of vertebral crush fractures in senile osteoporosis treated with 1α(OH)-vitamin D_3, *Bone Miner.*, 3, 47, 1987.
6. **Orimo, H. and Shiraki, M.,** Long term use of 1α$(OH)D_3$ in involutional osteoporosis, in *Osteoporosis: Physiological Basis, Assessment, and Treatment,* DeLuca, H. F. and Mazess, R., Eds., Elsevier, Amsterdam, 1990, 223.
7. **Narbaitz, R., Stumpf, W. E., and Sar, M.,** The role of autoradiographic and immunocytochemical techniques in the clarification of sites of metabolism and action of vitamin D, *J. Histochem. Cytochem.*, 29, 91, 1981.
8. **Stumpf, W. E.,** Steroid hormones and the cardiovascular system: direct actions of estradiol, progesterone, testosterone, gluco- and mineralcorticoids, and siltriol [vitamin D] on central nervous regulatory and peripheral tissue, *Experientia,* 46, 13, 1990.
9. **Hosomi, J., Hosoi, J., Abe, E., Suda, T., and Kuroki, T.,** Regulation of terminal differentiation of cultured mouse epidermal cells by 1α,25-dihydroxyvitamin D_3, *Endocrinology,* 113, 1950, 1983.

10. **Holick, M. F., Smith, E., and Pincus, S.**, Skin as the site of vitamin D synthesis and target tissue for 1,25-dihydroxyvitamin D_3, *Arch Dermatol.*, 123, 1677, 1987.
11. **Morimoto, S., Onishi, T., Imanaka, S., Yukawa, H., Kozuka, T., Kitano, Y., Yoshikawa, K., and Kumahara, Y.**, Topical administration of 1,25-dihydroxyvitamin D_3 for psoriasis: report of five cases, *Calcif. Tissue Int.*, 38, 119, 1986.
12. **Takamoto, S., Onishi, T., Morimoto, S., Imanaka, S., Yukawa, S., Kozuka, T., Kitano, Y., Seino, Y., and Kumahara, Y.**, Effect of 1α-hydroxycholecalciferol on psoriasis vulgaris: a pilot study, *Calcif. Tissue Int.*, 39, 360, 1986.
13. **Abe, E., Miyaura, C., Sakagami, H., Takeda, M., Konno, K., Yamazaki, T., Yoshiki, S., and Suda, T.**, Differentiation of mouse myeloid leukemia cells induced by 1α,25-dihydroxyvitamin D_3, *Proc. Natl. Acad. Sci. U.S.A.*, 78, 4990, 1981.
14. **Tanaka, H., Abe, E., Miyaura, C., Shiina, Y., and Suda, T.**, 1α,25-Dihydroxyvitamin D_3 induces differentiation of human promyelotic leukemia cells (HL-60) into monocyte macrophages but not into granulocytes, *Biochem. Biophys. Res. Commun.*, 117, 86, 1983.
15. **Honma, Y., Hozumi, M., Abe, E., Konno, K., Fukushima, M., Hata, S., Nishii, Y., DeLuca, H. F., and Suda, T.**, 1α,25-Dihydroxyvitamin D_3 and 1α-hydroxyvitamin D_3 prolong survival time of mice inoculated with myeloid leukemia cells, *Proc. Natl. Acad. Sci. U.S.A.*, 80, 201, 1983.
16. **Suda, T., Miyaura, C., Abe, E., and Kuroki, T.**, Modulation of cell differentiation, immune responses and tumor promotion by vitamin D compounds, in *Bone and Mineral Research*, Peck, W. A., Ed., Elsevier, Amsterdam, 1986, chap. 1.
17. **Tanaka, Y., Bush, K. K., Klauck, T. M., and Higgins, P. J.**, Enhancement of butyrate-induced differentiation of HT-29 human colon carcinoma cells by 1,25-dihydroxyvitamin D_3, *Biochem. Pharmacol.*, 38, 3859, 1989.
18. **Tam, C. S., Heersche, J. N. M., Jones, G., Murray, T. M., and Rasmussen, H.**, The effect of vitamin D on bone *in vivo*, *Endocrinology*, 118, 2217, 1986.
19. **Schwartz, Z., Soskolne, W. A., Atkin, I., Goldstein, M., and Ornoy, A.**, A direct effect of 24,25-$(OH)_2D_3$ and 1,25-$(OH)_2D_3$ on the modeling of fetal mice long bones *in vitro*, *J. Bone Miner. Res.*, 4, 157, 1989.
20. **Nakamura, T., Kurokawa, T., and Orimo, H.**, Increase of bone volume in vitamin D-repleted rats by massive administration of 24R,25$(OH)_2D_3$, *Calcif. Tissue Int.*, 43, 235, 1988.
21. **Nakamura, T., Kurokawa, T., and Orimo, H.**, Increased mechanical strength of the vitamin D-replete rat femur by the treatment with a large dose of 24R,25$(OH)_2D_3$, *Bone*, 10, 117, 1989.
22. **Wilhelm, R., Ross, F. P., and Norman, A. W.**, Specific binding of 24R,25-dihydroxy-vitamin D_3 to chick intestinal mucosa: 24R,25-dihydroxyvitamin D_3 is an allosteric effector of 1,25-dihydroxyvitamin D_3 binding, *Arch. Biochem. Biophys.*, 249, 88, 1986.
23. **Weintroub, S., Price, P. A., and Reddi, A. H.**, The dichotomy in the effects of 1,25-dihydroxyvitamin D_3 and 24,25-dihydroxyvitamin D_3 on bone gamma-carboxyglutamic acid containing protein in serum and bone in vitamin D-deficient rats, *Calcif. Tissue Int.*, 40, 166, 1987.
24. **Ishizuka, S., Ohba, T., and Norman, A. W.**, 1α,25-$(OH)_2D_3$-26,23-lactone is a major metabolite of 1α,25-$(OH)_2D_3$ under physiological conditions, in *Vitamin D: Molecular, Cellular and Clinical Endocrinology*, Norman, A. W., Schaefer, K., Grigoleit, H. G., and Herrath, D. V., Eds., Walter de Gruyter, Berlin, 1988, 143.
25. **Ishizuka, S., Oshida, J., Tsuruta, H., and Norman, A. W.**, The stereochemical configuration of the natural 1α,25-dihydroxyvitamin D_3-26,23-lactone, *Arch. Biochem. Biophys.*, 242, 82, 1985.
26. **Ishizuka, S. and Norman, A. W.**, Metabolic pathways from 1α,25-dihydroxyvitamin D_3 to 1α,25-dihydroxyvitamin D_3-26,23-lactone. Stereo-retained and stereo-selective lactonization, *J. Biol. Chem.*, 262, 7165, 1987.

27. **Ishizuka, S. and Norman, A. W.**, The difference of biological activity among four diastereoisomers of 1α,25-dihydroxycholecalciferol-26,23-lactone, *J. Steroid Biochem.*, 25, 505, 1986.
28. **Ishizuka, S., Kurihara, N., Hakeda, Y., Maeda, N., Ikeda, K., Kumegawa, M., and Norman, A. W.**, 1α,25-Dihydroxyvitamin D_3[1α,25-$(OH)_2D_3$]-26,23-lactone inhibits 1,25-$(OH)_2D_3$-mediated fusion of mouse bone marrow mononuclear cells, *Endocrinology*, 123, 781, 1988.
29. **Kurihara, N., Civin, C., and Roodman, G. D.**, Osteoporotic factor responsiveness of highly purified populations of early and late precursors for human multinucleated cells expressing the osteoclast phenotype, *J. Bone Miner. Res.*, 6, 257, 1991.
30. **Ishizuka, S., Kiyoki, M., Kurihara, N., Hakeda, Y., Ikeda, K., Kumegawa, M., and Norman, A. W.**, Effects of diastereoisomers of 1,25-dihydroxyvitamin D_3-26,23-lactone on alkaline phosphatase and collagen synthesis in osteoblastic cells, *Mol. Cell. Endocrinol.*, 55, 77, 1988.
31. **Shima, M., Tanaka, H., Norman, A. W., Yamaoka, K., Yoshikawa, H., Takaoka, K., Ishizuka, S., and Seino, Y.**, 23(S)25(R)-1,25-Dihydroxyvitamin D_3-26,23-lactone stimulates murine bone formation *in vivo*, *Endocrinology*, 126, 832, 1990.
32. **Ishizuka, S. and Ikekawa, N.**, unpublished data.
33. **Eguchi, T., Yoshida, M., and Ikekawa, N.**, Synthesis and biological activities of 22-hydroxy and 22-methoxy derivatives of 1α,25-dihydroxyvitamin D_3: importance of side chain conformation for biological activities, *Bioorg. Chem.*, 17, 294, 1989.
34. **Morisaki, M., Koizumi, N., Ikekawa, N., Takeshita, T., and Ishimoto, S.**, Synthesis of active forms of vitamin D. IX. Synthesis of 1α,24-dihydroxycholecalciferol, *J. Chem. Soc. Perkin Trans. 1*, 1421, 1975.
35. **Ishizuka, S., Ishimoto, S., and Orimo, H.**, Biological activity, metabolism and mechanism of action of 1α,24-dihydroxyvitamin D_3, in *Vitamin D: Chemical, Biochemical and Clinical Endocrinology of Calcium Metabolism*, Norman, A. W., Schaefer, K., Grigoleit, H. G., and Herrath, D. v., Eds., Walter de Gruyter, Berlin, 1982, 34.
36. **Kawashima, H., Hoshina, K., Saito, N., Hashimoto, Y., Ishimoto, S., Noguchi, T., and Orimo, H.**, Effect and metabolism of 1α,24-hydroxyvitamin D_3 in hepatectomized rats, *FEBS Lett.*, 104, 367, 1979.
37. **Yoshida, M., Ishizuka, S., and Hoshi, A.**, Biological activity of vitamin D_3 derivatives in inducing differentiation of HL-60 human promyelocytic leukemia cells, *J. Pharm. Dyn.*, 7, 962, 1984.
38. **Matsumoto, K., Hashimoto, K., Kiyoki, M., Yamamoto, M., and Yoshikawa, K.**, Effect of 1,24R-dihydroxyvitamin D_3 on the growth of human keratinocytes, *J. Dermatol.*, 17, 97, 1990.
39. **Kato, T., Rokugo, M., Terui, T., and Tagami, H.**, Successful treatment of psoriasis with topical application of active vitamin D_3 analogue, 1α,24-dihydroxycholecalciferol, *Br. J. Dermatol.*, 115, 431, 1986.
40. **Tagami, H., Kato, T., Terui, T., and Tadaki, T.**, Successful treatment of psoriasis with topical application of the active vitamin D_3 analog, 1α,24-dihydroxycholecalciferol (TV-02), in *Vitamin D: Molecular, Cellular and Clinical Endocrinology*, Norman, A. W., Schaefer, K., Grigoleit, H. G., and Herrath, D. v., Eds., Walter de Gruyter, Berlin, 1988, 958.
41. **Calverley, M. J.**, Synthesis of MC 903, a biologically active vitamin D metabolite analogue, *Tetrahedron*, 43, 4609, 1987.
42 **Binderup, L. and Bramm, E.**, Effects of a novel vitamin D analogue MC 903 on cell proliferation and differentiation *in vitro* and on calcium metabolism *in vivo*, *Biochem. Pharmacol.*, 37, 889, 1988.
43. **Kragballe, K., Beck, H. I., and Søgaard, H.**, Improvement of psoriasis by a topical vitamin D_3 analogue (MC 903) in a double-blind study, *Br. J. Dermatol.*, 119, 223, 1988.

44. **Thavarajah, M., Evans, D. B., Binderup, L., and Kanis, J. A.**, 1,25$(OH)_2D_3$ and calcipotriol (MC903) have similar effects on the induction of osteoclast-like cell formation in human bone marrow cultures, *Biochem Biophys. Res. Commun.*, 171, 1056, 1990.
45. **Marie, P. J., Connes, D., Hott, M., and Miravet, L.**, Comparative effects of a novel vitamin D analogue MC-903 and 1,25-dihydroxyvitamin D_3 on alkaline phosphatase activity, osteocalcin and DNA synthesis by human osteoblastic cells in culture, *Bone*, 11, 171, 1990.
46. **Murayama, E., Miyamoto, K., Kubodera, N., Mori, T., and Matsunaga, I.**, Synthetic study of vitamin D_3 analogues. VIII. Synthesis of 22-oxavitamin D_3 analogues, *Chem. Pharm. Bull.*, 34, 4410, 1986.
47. **Abe, J., Morikawa, M., Miyamoto, K., Kaiho, S., Fukushima, M., Miyaura, C., Abe, E., Suda, T., and Nishii, Y.**, Synthetic analogues of vitamin D_3 with an oxygen atom in the side chain skeleton, *FEBS Lett.*, 226, 58, 1987.
48. **Abe, J., Takita, Y., Nakano, T., Miyaura, C., Suda, T., and Nishii, Y.**, A synthetic analogue of vitamin D_3, 22-oxa-1α,25-dihydroxyvitamin D_3, is a potent modulator of *in vivo* immunoregulating activity without inducing hypercalcemia in mice, *Endocrinology*, 124, 2645, 1989.
49. **Brown, A. J., Ritter, C. R., Finch, J. L., Morrissey, J., Martin, K. J., Murayama, E., Nishii, Y., and Slatopolsky, E.**, The noncalcemic analogue of vitamin D, 22-oxacalcitriol, suppresses parathyroid hormone synthesis and secretion, *J. Clin. Invest.*, 84, 728, 1989.
50. **Wovkulich, P. M., Batcho, A. D., Baggiolini, E. G., Boris, A., Truitt, G., and Uskokovic, M. R.**, Synthesis of 1α,25S,26-trihydroxy-Δ^{22}-cholecalciferol, a potent inducer of cell differentiation, in *Vitamin D: A Chemical, Biochemical and Clinical Update*, Norman, A. W., Schaefer, K., Grigoleit, H. G., and Herrath, D. v., Eds., Walter de Gruyter, Berlin, 1985, 755.
51. **Holick, M. F., Uskokovic, M., Persons, K., Horst, R., Baggiolini, E. G., and Truitt, G. A.**, An evaluation of analogs of 1,25-dihydroxyvitamin D_3 on the inhibition of proliferation and induction of terminal differentiation in cultured human normal keratinocytes, in *Vitamin D: Molecular, Cellular and Clinical Endocrinology*, Norman, A. W., Schaefer, K., and Grigoleit, H. G., Eds., Walter de Gruyter, Berlin, 1988, 362.
52. **Zhou, J. Y., Norman, A. W., Lübbert, M., Collins, E. D., Uskokovic, M. R., and Koeffler, H. P.**, Novel vitamin D analogs that modulate leukemic cell growth and differentiation with little effect on either intestinal calcium absorption or bone calcium mobilization, *Blood*, 74, 82, 1989.
53. **Norman, A. W., Zhou, J. Y., Henry, H. L., Uskokovic, M. R., and Koeffler, H. P.**, Structure-function studies on analogues of 1α,25-dihydroxyvitamin D_3: differential effects on leukemic cell growth, differentiation, and intestinal calcium absorption, *Cancer Res.*, 50, 6857, 1990.
54. **Zhou, J. Y., Norman, A. W., Chen, D. L., Sun, G. W., Uskokovic, M., and Koeffler, H. P.**, 1,25-Dihydroxy-16-ene-23-yne-vitamin D_3 prolongs survival time of leukemic mice, *Proc. Natl. Acad. Sci. U.S.A.*, 87, 3929, 1990.
55. **Tanaka, Y., Sicinski, R. R., DeLuca, H. F., Sai, H., and Ikekawa, N.**, Unique rearrangement of ergocalciferol side chain *in vitro*: production of a biologically highly active homologue of 1,25-dihydroxyvitamin D_3, *Biochemistry*, 25, 5512, 1986.
56. **Osterm, V. K., Tanaka, Y., Prahl, J., DeLuca, H. F., and Ikekawa, N.**, 24- and 26-Homo-1,25-dihydroxyvitamin D_3: preferential activity in inducing differentiation of human leukemia cells HL-60 *in vitro*, *Proc. Natl. Acad. Sci. U.S.A.*, 84, 2610, 1987.
57. **Osterm, V. K., Lau, W. F., Lee, S. H., Perlman, K., Prahl, J., Schnoes, H. K., DeLuca, H. F., and Ikekawa, N.**, Induction of monocytic differentiation of HL-60 cells by 1,25-dihydroxyvitamin D analogs, *J. Biol. Chem.*, 262, 14164, 1987.
58. **Osterm, V. K. and DeLuca, H. F.**, The vitamin D-induced differentiation of HL-60 cells: structural requirements, *Steroid*, 49, 73, 1987.

59. **Hara, N., Eguchi, T., Ikekawa, N., Ishizuka, S., and Sato, J.**, Synthesis and biological activity of (22E,25R)- and (22E,25S)-22-dehydro-1α,25-dihydroxy-26-methylvitamin D_3, *J. Steroid Biochem.*, 35, 655, 1990.
60. **Koizumi, T., Nakao, Y., Ishizuka, S., Oshida, J., Hara, N., Ikekawa, N., and Fujita, T.**, Novel vitamin D_3 derivatives, 26-homo-Δ^{22}-dehydro-1α,25(S)-dihydroxyvitamin D_3 and 26-homo-Δ^{22}-dehydro-1α,25(R)-dihydroxyvitamin D_3: preferential activity in *c-myc* mRNA production and in induction of phenotypic differentiation of HL-60 cells, *Arch. Biochem. Biophys.*, 276, 310, 1990.
61. **Eguchi, T., Ikekawa, N., Sumitani, K., Kumegawa, M., Higuchi, S., and Otomo, S.**, Effect on carbon lengthening at the side chain terminal of 1α,25-dihydroxyvitamin D_3 for calcium regulating activity, *Chem. Pharm. Bull.*, 38, 1246, 1990.
62. **Sai, H., Takatsuto, S., Hara, N., and Ikekawa, N.**, Synthesis of 1α,25-dihydroxy-26,27-dimethylvitamin D_3, a highly active analogue of 1α,25-dihydroxyvitamin D_3, *Chem. Pharm. Bull.*, 33, 878, 1985.
63. **Ikekawa, N., Eguchi, T., Hara, N., Takatsuto, S., Honda, A., Mori, Y., and Otomo, S.**, 26,27-Diethyl-1α,25-dihydroxyvitamin D_3 and 24,24-difluoro-24-homo-1α,25-dihydroxyvitamin D_3: highly potent inducer for differentiation of human leukemia cells HL-60, *Chem. Pharm. Bull.*, 35, 4362, 1987.
64. **Eguchi, T., Sai, H., Takatsuto, S., Hara, N., and Ikekawa, N.**, Synthesis of 26,27-dialkyl analogues of 1α,25-dihydroxyvitamin D_3, *Chem. Pharm. Bull.*, 36, 2303, 1988.
65. **Gill, H. S., Londowski, J. M., Corradino, R. A., Zinsmeister, A. R., and Kumar, R.**, The synthesis and biological activity of 25-hydroxy-26,27-dimethylvitamin D_3 and 1,25-dihydroxy-26,27-dimethylvitamin D_3: highly potent novel analogs of vitamin D_3, *J. Steroid Biochem.*, 31, 147, 1988.
66. **Honda, A., Mori, Y., Ishizuka, S., and Ikekawa, N.**, Effects of novel 26,27-dialkyl analogs of 1α,25-dihydroxyvitamin D_3 on differentiation-inducing activity of human promyelocytic leukemia (HL-60) cells in serum-supplemented or serum-free culture, *Steroids*, 56, 142, 1991.
67. **Perlman, K., Kutner, A., Prahl, J., Smith, C., Inaba, M., Schnoes, H. K., and DeLuca, H. F.**, 24-Homologated 1,25-dihydroxyvitamin D_3 compounds: separation of calcium and cell differentiation activities, *Biochemistry*, 29, 190, 1990.
68. **Krisinger, J., Strom, M., Darwish, H. D., Perlman, K., Smith, C., and DeLuca, H. F.**, Induction of calbindin-D 9k mRNA but not calcium transport in rat intestine by 1,25-dihydroxyvitamin D_3 24-homologs, *J. Biol. Chem.*, 266, 1910, 1991.
69. **Tanaka, Y., DeLuca, H. F., Kobayashi, Y., and Ikekawa, N.**, 26,26,26,27,27,27-Hexafluoro-1,25-dihydroxyvitamin D_3: a highly potent, long-lasting analog of 1,25-dihydroxyvitamin D_3, *Arch. Biochem. Biophys.*, 229, 348, 1984.
70. **Inaba, M., Okuno, S., Nishizawa, Y., Yukioka, K., Otani, S., Matsui-Yuasa, I., Morisawa, S., DeLuca, H. F., and Morii, H.**, Biological activity of fluorinated vitamin D analogs at C-26 and C-27 on human promyelocytic leukemia cells, HL-60, *Arch. Biochem. Biophys.*, 258, 421, 1987.
71. **Inaba, M., Yukioka, K., Nishizawa, Y., Okuno, S., Otani, S., Morisawa, S., DeLuca, H. F., and Morii, H.**, Biological activities of 26,26,26,27,27,27-hexafluorinated vitamin D_3 analogs in human promyelocytic leukemia cells and human peripheral mononuclear cells, in *Calcium Regulation and Bone Metabolism: Basic and Clinical Aspects*, Vol. 9, Cohn, D. V., Martin, T. J., and Meunier, P. J., Eds., Elsevier, Amsterdam, 1987, 523.
72. **Yukioka, K., Otani, S., Matsui-Yuasa, I., Goto, H., Morisawa, S., Okuno, S., Inaba, M., Nishizawa, Y., and Morii, H.**, Biological activity of 26,26,26,27,27,27-hexafluorinated analogs of vitamin D_3 in inhibiting interleukin-2 production by peripheral blood mononuclear cells stimulated by phytohemagglutinin, *Arch. Biochem. Biophys.*, 260, 45, 1988.

73. **Tanaka, Y., Bush, K. K., Eguchi, T., Ikekawa, N., Taguchi, T., Kobayashi, Y., and Higgins, P. J.**, Effects on 1,25-dihydroxyvitamin D_3 and its analogs on butyrate-induced differentiation of HT-29 human colonic carcinoma cells and on the reversal of the differentiated phenotype, *Arch. Biochem. Biophys.*, 276, 415, 1990.
74. **Okumura, H., Yamamuro, T., Hibuchi, S., Harada, M., Takamura, T., Otomo, S., Aihara, H., Ikekawa, N., and Kobayashi, T.**, 26,27-Hexafluoro-1,25-dihydroxy-vitamin D_3 (F_6-1,25$(OH)_2D_3$) prevents osteoporosis induced by immobilization combined with ovariectomy in the rat, *Bone Miner.*, 9, 101, 1990.
75. **Nagata, A., Hamma, N., Katsumata, T., Sato, R., Komuro, S., Iba, K., Yoshitake, A., Kiriyama, T., Okamoto, S., Nagataki, S., and Morii, H.**, Mechanism of action of 26,27-hexafluoro-1,25-$(OH)_2D_3$, in *Vitamin D: Molecular, Cellular and Clinical Endocrinology*, Norman, A. W., Schaefer, K., and Grigoleit, H. G., Eds., Walter de Gruyter, Berlin, 1988, 141
76. **Kiriyama, T., Okamoto, S., Ejima, E., Kurihara, N., Hakeda, Y., Ito, N., Izumi, M., Kumegawa, M., and Nagataki, S.**, Effect of a highly potent fluoro analog of 1,25-dihydroxyvitamin D_3 on human bone-derived cells, *Endocrinology*, 128, 81, 1991.
77. **Inaba, M., Okuno, S., Inoue, A., Nishizawa, Y., Morii, H., and DeLuca, H. F.**, DNA binding property of vitamin D_3 receptors associated with 26,26,26,27,27,27-hexafluoro-1,25-dihydroxyvitamin D_3, *Arch. Biochem. Biophys.*, 268, 35, 1989.
78. **Evans, D. B., Thavarajah, M., Bunning, R. A. D., Uskokovic, M. R., and Kanis, J. A.**, Side chain modifications increase the potency of 1,25-dihydroxycholecalciferol (calcitriol) on human bone-derived osteoblast-like cells, in *Vitamin D: Molecular, Cellular and Clinical Endocrinology*, Norman, A. W., Schaefer, K., and Grigoleit, H. G., Eds., Walter de Gruyter, Berlin, 1988, 612.
79. **Evans, D. B., Thavarajah, M., Uskokovic, M. R., and Kanis, J. A.**, Increased potency of 1,25-dihydroxyvitamin D_3 on human osteoblast-like cells following structural side-chain modification, *Bone*, 11, 439, 1990.
80. **Koeffler, H. P., Amatruda, T., Ikekawa, N., Kobayashi, Y., and DeLuca, H. F.**, Induction of macrophage differentiation of human normal and leukemic myeloid stem cells by 1,25-dihydroxyvitamin D_3 and its fluorinated analogs, *Cancer Res.*, 44, 5624, 1984.
81. **Nakada, M., Tanaka, Y., DeLuca, H. F., Kobayashi, Y., and Ikekawa, N.**, Biological activities and binding properties of 23,23-difluoro-25-hydroxyvitamin D_3 and its 1α-hydroxy derivative, *Arch. Biochem. Biophys.*, 241, 173, 1985.
82. **Ikekawa, N.**, Chemical synthesis of vitamin D analogs with selective biological activities, in *Vitamin D: Molecular, Cellular and Clinical Endocrinology*, Norman, A. W., Schaefer, K., and Grigoleit, H. G., Eds., Walter de Gruyter, Berlin, 1988, 25.
83. **Gill, H. S., Londowski, J. M., Corradino, R. A., Zinsmeister, A. R., and Kumar, R.**, Synthesis and biological activity of novel vitamin D analogues: 24,24-difluoro-25-hydroxy-26,27-dimethylvitamin D_3 and 24,24-difluoro-1α,25-dihydroxy-26,27-dimethylvitamin D_3, *J. Med. Chem.*, 33, 480, 1990.
84. **Gill, H. S., Londowski, J. M., Corradino, R. A., and Kumar, R.**, The synthesis and biological activity of 22-fluorovitamin D_3: a new vitamin D analog, *Steroids*, 48, 93, 1986.
85. **Dauben, W. G., Kohler, B., and Roesle, A.**, Synthesis of 6-fluorovitamin D_3, in *Vitamin D: A Chemical, Biochemical and Clinical Update*, Norman, A. W., Schaefer, K., Grigoleit, H. G., and Herrath, D. v., Eds., Walter de Gruyter, Berlin, 1985, 767.
86. **Perlman, K. L., Sicinski, R. R., Schnoes, H. K., and DeLuca, H. F.**, 1α,25-Dihydroxy-19-nor-vitamin D_3, a novel vitamin D-related compound with potential therapeutic activity, *Tetrahedron Lett.*, 31, 1823, 1990.
87. **Okano, T., Tsugawa, N., Masuda, S., Takeuchi, A., Kobayashi, T., Takita, Y., and Nishii, Y.**, Regulatory activities of 2β-(3-hydroxypropoxy)-1α,25-hydroxyvitamin D_3, a novel synthetic vitamin D_3 derivative, on calcium metabolism, *Biochem. Biophys. Res. Commun.*, 163, 1444, 1989.

88. **Okamura, W. H., Aurrecoechea, J. M., Gibbs, R. A., and Norman, A. W.,** Synthesis and biological activity of 9,11-dehydrovitamin D_3 analogues: stereoselective preparation of 6β-vitamin D vinylallenes and a concise enynol synthesis for preparing the A-ring, *J. Org Chem.*, 54, 4072, 1989.
89. **Okamura, W. H., Shen, G. Y., Barrack, S. A., and Henry, H. L.,** Vitamin D analogues: potential inhibitors of vitamin D metabolism, in *Vitamin D: Molecular, Cellular and Clinical Endocrinology,* Norman, A. W., Schaefer, K., and Grigoleit, H. G., Eds., Walter de Gruyter, Berlin, 1988, 12.
90. **Barrack, S. A., Gibbs, R. A., and Okamura, W. H.,** Potential inhibitors of vitamin D metabolism: an oxa analogue of vitamin D, *J. Org. Chem.*, 53, 1790, 1988.
91. **Hayashi, T., Yamada, S., Miyaura, C., Tanaka, H., Yamamoto, K., Abe, E., Takayama, H., and Suda, T.,** Phagocytic cells metabolize 25-hydroxyvitamin D_3 to 10-oxo-19-nor-25-hydroxyvitamin D_3 and a new metabolite, 8α,25-dihydroxy-9,10-seco-4,6,10(19)-cholestatrien-3-one, *FEBS Lett.*, 218, 200, 1987.
92. **Thierry-Palmer, M., Richardson, K. A., Schnoes, H. K., Yamada, S., and Free, A. L.,** Identification of 8α-25-dihydroxy-9,10-seco-4,6,10(19)-cholestatrien-3-one as a product of metabolism of 25-hydroxycholecalciferol by liver microsomes, *Arch. Biochem. Biophys.*, 277, 277, 1990.
93. **Haussler, M. R., Terpening, C. M., Komm, B. S., Whitfield, G. K., and Haussler, C. A.,** Vitamin D hormone receptors: structure, regulation and molecular function, in *Vitamin D: Molecular, Cellular and Clinical Endocrinology,* Norman, A. W., Schaefer, K., Grigoleit, H. G., and Herrath, D. v., Eds., Walter de Gruyter, Berlin, 1988, 205.
94. **Pike, J. W., Kesterson, R. A., Scott, R. A., Kerner, S. A., McDonnell, D. P., and O'Malley, B. W.,** Vitamin D_3 receptor: molecular structure of the protein and its chromosomal gene, in *Vitamin D: Molecular, Cellular and Clinical Endocrinology,* Norman, A. W., Schaefer, K., Grigoleit, H. G., and Herrath, D. v., Eds., Water de Gruyter, Berlin, 1988, 215.
95. **Brown, T. A., Prahl, J. M., and DeLuca, H. F.,** Partial amino acid sequence of porcine 1,25-dihydroxyvitamin D_3 receptor isolated by immunoaffinity chromatography, *Proc. Natl. Acad. Sci. U.S.A.*, 85, 2454, 1988.

Chapter 9

MOLECULAR STRUCTURE AND BIOLOGICAL ACTIVITY OF BRASSINOLIDE AND RELATED BRASSINOSTEROIDS

Takao Yokota and Kenji Mori

TABLE OF CONTENTS

I. INTRODUCTION

Brassinosteroids (BRs) are steroidal growth-promoting plant hormones. Some reviews on BRs have appeared.[1-6] Brassinolide is the first BR isolated from rape (*Brassica napus* L.) pollen and elicits strong growth-promoting activity.[7] The structure is quite unique in that it contains vicinal diols in the A ring and side chain and a lactone (7-oxa-6-one) structure in the B ring. At present, more than twenty related natural BRs have been isolated from natural sources.

These are all 5α-cholestane derivatives and can be classified into C_{27}, C_{28}, and C_{29} steroids as is the case with typical phytosterols. These variations come from different substitution at C-24, i.e., no substitution in C_{27} steroids, a methyl or an exomethylene in C_{28} steroids, and an ethyl or an (*E*)-ethylidene in C_{29} steroids. In addition, recently C_{29} steroids which have an exomethylene at C-24 and an additional methyl at C-25 have been isolated.[8] The alkyl (methyl or ethyl) substituents at C-24 are *S*-oriented in most of natural BRs, but recently BRs with a 24*R*-methyl, namely 24-epicastasterone and 24-epi-brassinolide, have been characterized from a green alga [*Hydrodictyon reticulatum* (L.) Lagerheim] and pollens of broad bean (*Vicia faba* L.).[9,10] All BRs carry a 22*R*, 23*R*-vicinal glycol in the side chain. Most of natural BRs have another vicinal glycol system at C-2 and C-3 in A ring, and some natural BRs, instead, have a 2-deoxy structure. With respect to B-ring functionalities, BRs can be grouped into either a lactone (7-oxa-6-one), 6-ketone, or non-oxidized (deoxo) type.

Until now there have been known a number of synthetic BRs in which functionalities of the A ring, B ring, and side chain are modified.

The structure-activity relationships of these natural and synthetic BRs have been examined using bean second internode bioassay, bean first internode curvature bioassay, rice lamina inclination bioassay, or bioassays using tomato and radish seedlings. The structure-activity relationships of BRs tend to differ from one bioassay system to another, so that in this paper the structure-activity relationships of BRs are discussed in respective bioassay systems. Relative biological activities of various BRs shown in this review are generally reestimated by authors from the data in original papers or otherwise cited directly from sources.

TABLE 1
Effects of Four Naturally Occurring Brassinosteroids on the Elongation and Growth Response in the Bean Second Internode Bioassay

	% Elongation over controls and growth response[a] (μg)				
Compound	**50**	**10**	**1**	**0.1**	**0.01**
Brassinolide (**8**)	—	136 (5)	136 (5)	156 (5)	183 (4)
24-Epibrassinolide (**11**)	366 (5)	237 (5)	178 (5)	228 (4)	224 (3)
Castasterone (**24**)	—	152 (4)	120 (4)	105 (3)	—
24-Epicastasterone (**26**)	426 (4)	295 (3)	336 (3)	352 (2)	—

[a] Growth response on a 0 to 5 scale is given in parenthesis: 0, no elongation; 1, elongation only; 2, elongation with slight curvature and swelling; 3, elongation with good curvature and swelling; 4, elongation with excellent curvature and swelling; and 5, elongation with split internodes.

From Thompson, M. J., Meudt, W. J., Mandava, N. B., Dutky, S. R., Lusby, W. R., and Spaulding, D W , *Steroids*, 39, 89, 1982. With permission.

II. STRUCTURE-ACTIVITY RELATIONSHIP OF BRASSINOSTEROIDS IN THE BEAN SECOND INTERNODE BIOASSAY AND BEAN FIRST INTERNODE CURVATURE BIOASSAY

The bean second internode bioassay using pinto bean (*Phaseolus vulgaris* L. Pinto) is quite a historical one and was successfully used as a guide for the isolation of brasinolide.[7] Young second internodes extracted from the 6-day-old seedlings are treated with test compounds in lanolin paste.[11] After 4 days, BRs elicit morphological changes such as elongation, swelling, curvature, and splitting. Such morphological changes have been known to be due to elongation and cell division of plant cells.[12] The most reliable way to estimate the biological activity is based on the response codes shown in the footnote of Table 1. Data obtained for some BRs are shown in Table 1.[13] Elongation, namely increment in the length of the second internode, can be used for the evaluation of the biological activity. However, it should be kept in mind that elongation response does not tend to be dependent on the concentrations of test samples.[14] Another bioassay method using pinto bean is the first internode curvature bioassay in which the tissues are first treated with a BR then with indole-3-acetic acid and the resulting synergistic growth due to BR and indole-3-acetic acid is measured.[15] This bioassay, however, seems to be even less suitable for the structure-activity relationship study than elongation response of the bean second internode bioassay.

A. EVALUATION ON THE BASIS OF THE BEAN SECOND INTERNODE BIOASSAY (RESPONSE CODE) AND BEAN FIRST INTERNODE BIOASSAY

1. Effects of Functionalities in the Side Chain

Thompson et al.[13,14] examined using the bean second internode bioassay (response code) the biological activities of BRs of lactone type and 6-ketone type with various modifications in the side chain (Table 2). Among BRs examined, brassinolide (**8**) is the most active. In lactone-type (7-oxa-6-one) BRs, vicinal hydroxyls at C-22 and C-23, and an alkyl substituent at C-24 are quite important. Favorable alkyl substituent at C-24 is a methyl group and BRs with an ethyl group (**16**) are two orders of magnitude less active. BRs which lack either vicinal hydroxyl groups or C-24 substituent (**1, 2, 4** to **6**) are nearly inactive. Exceptionally compound **3** which has 22*S*, 23*S*-hydroxyls but lacks methyl groups (C-26 and C-28) is slightly active, the biological activity, however, being five orders of magnitude less than that of brassinolide. Stereochemistry of side chain functionalities also affect the biological activity. BRs with 22*R*, 23*R*-hydroxyls are more active than those with 22*S*, 23*S*-hydroxyls (**8** vs. **7**, **11** vs. **9**, **16** vs. **13**). Concerning the 24-methyl substituents, *S*-orientation (brassinolide) is more favorable than *R*-orientation in BRs with 22*R*, 23*R*-hydroxyls (**8** vs. **11**), but such relation is reversed in BRs with 22*S*, 23*S*-hydroxyls (**7** vs. **9**). In BRs with 6-ketone, the structure-activity relationship concerning the side chain is similar to that of lactone-type BRs. Thompson et al.[14] further evaluated the same BRs in the bean first internode curvature bioassay. It was found that the patterns of activity shown by the BRs were similar to those they showed in the bean second internode bioassay.

Kerb et al.[16] examined the biological activity of hexanorbrassinolide-22-ethers. Ethyl ether (**29**) was one tenth as active as 28-homobrassinolide (**16**). *n*-Propyl ether (**30**), methyl ether (**31**), and methoxyethyl ether (**32**) are all less active than the ethyl ether.

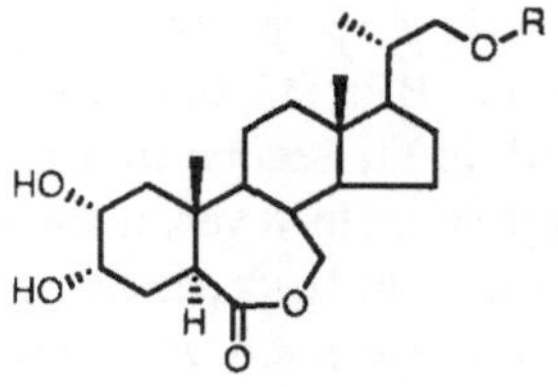

	R
29	Et
30	*n* - Pr
31	Me
32	$OCH_2CH_2OCH_3$

2. Effects of Functionalities in A/B Rings[14]

In bean second internode bioassay, BRs with 6-ketone are generally more than 100 times less active than lactone-type (7-oxa-6-one) BRs (Table 2). Isomeric lactone (6-oxa-7-one) of brassinolide (**10, 12, 15**) are inactive (Table 2). The acetates of highly active BRs are only marginally active. A 2β,3β-hydroxyl analog (**14**) is inactive. Similar biological activities are also observed in the first internode curvature bioassay.

TABLE 2
Relative Biological Activity of Brassinosteroids with Modified Side Chain in the Bean Second Internode Bioassay (Response Code)

Side chain	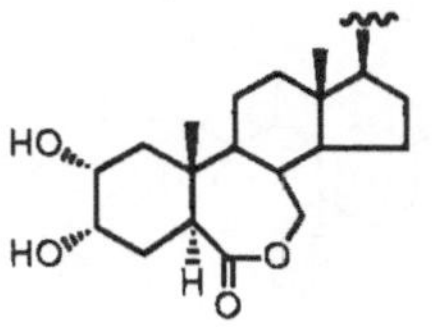		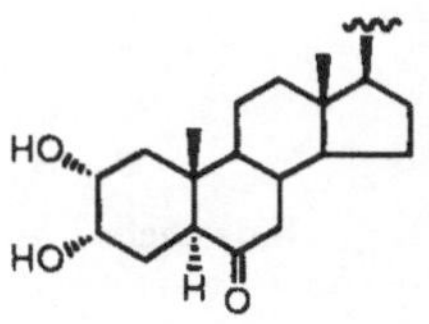	
	**1**	~0	**17**	-
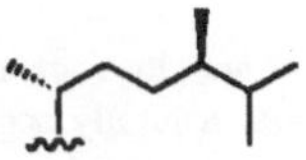	**2**	~0	**18**	-
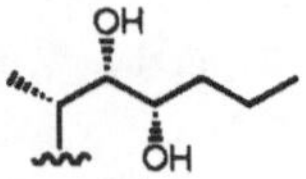	**3**	0.001	**19**	0.001
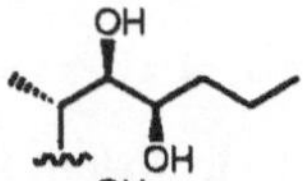	**4**	0	**20**	~0
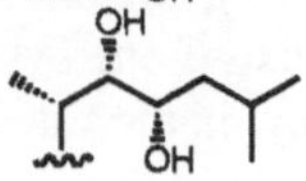	**5**	~0	**21**	~0
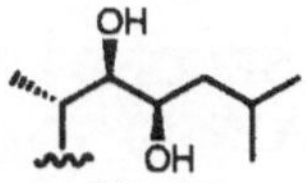	**6***	~0	**22***	~0
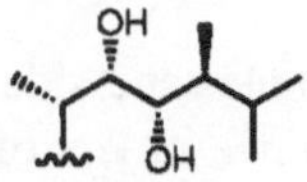	**7**	1	**23**	~0
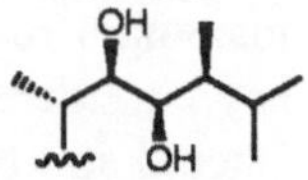	**8***	100	**24***	1
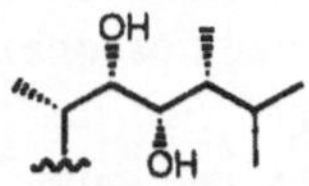	**9*** (Isolactone **10**, 0)	<10	**25**	0
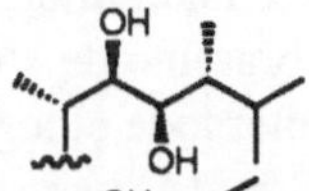	**11*** (Isolactone **12**, 0)	10	**26***	0.1
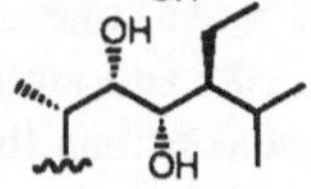	**13** (2,3-Diepimer **14**, 0) (Isolactone **15**, 0)	0.1	**27**	~0

TABLE 2 (continued)
Relative Biological Activity of Brassinosteroids with Modified Side Chain in the Bean Second Internode Bioassay (Response Code)

Side chain	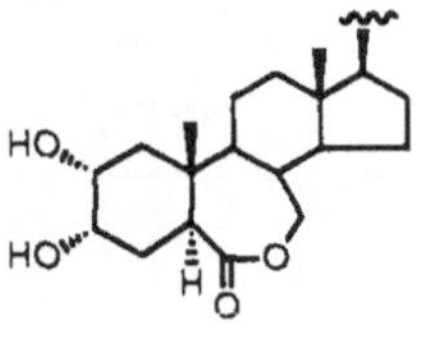		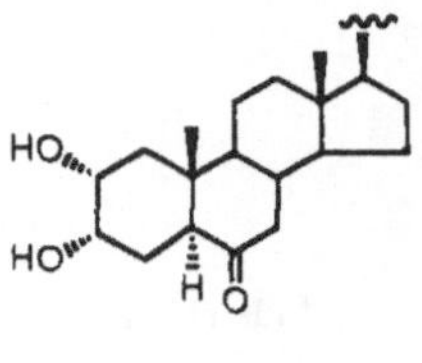	
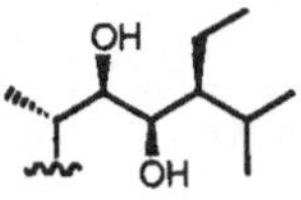	16	1	28*	0.001

Note: Biological activity of brassinolide = 100. Relative biological activities are roughly estimated on the basis of the data of Thompson et al. * Indicates naturally occurring.

From Thompson, M. J., Meudt, W. J., Mandava, N. B., Dutky, S. R., Lusby, W. R., and Spaulding, D. W., *Steroids*, 39, 89, 1982. With permission.

3. Conclusions

The structural requirements of BRs with trans-A/B ring system for the biological activity are 2α,3α-vicinal diol in A ring, a 7-oxa-6-one system in B ring, and 22*R*, 22*R*-diol or 22*S*, 23*S*-diol and a 24-methyl in the side chain.

B. EVALUATION ON THE BASIS OF THE BEAN SECOND INTERNODE BIOASSAY (ELONGATION RESPONSE)

1. Effects of Side Chain Modification

Kohout and Strnad[17] examined some BRs with a cholestane side chain (**1**, **33** to **35**). Their biological activities are more than three orders of magnitude less active than 24-epibrassinolide (**11**) and the maximum response observed was far less than that of 24-epibrassinolide. They considered that brassinolide side chain is not a prerequisite. However, it should still be said that 22, 23-hydroxyls and a 24-alkyl are important functionalities for exerting typical BR activity in the bean second internode bioassay.

Cerny et al.[18,19] prepared cholanic acid and their ester and amide derivatives (**36** to **42**), in which A and B ring are the same as brassinolde. Among them, an acid (**36**) shows the highest activity in second internode elongation. However, maximum elongation caused by the acid was as low as one seventh that caused by 24-epibrassinolide. Ester derivatives (**37**, **38**) and amide derivatives (**39** to **42**) are one or two orders of magnitude weaker than the acid (**36**). Compounds **43** to **46** without a 21-methyl show lower biological activity, indicating that 21-methyl is an important functionality.[19] Their data, however, seem to be less confirmatory to draw definite conclusions. They also examined these compounds by the bean first internode curvature bioassay. However,

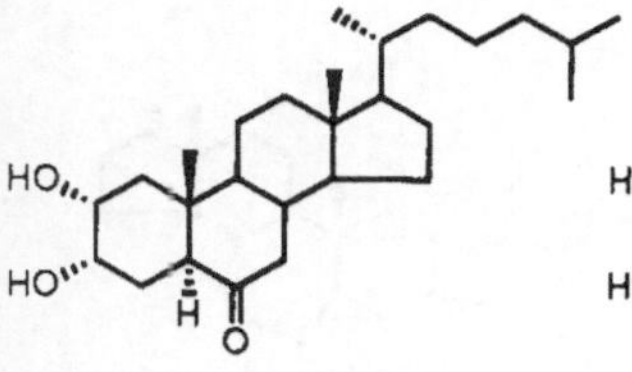

33

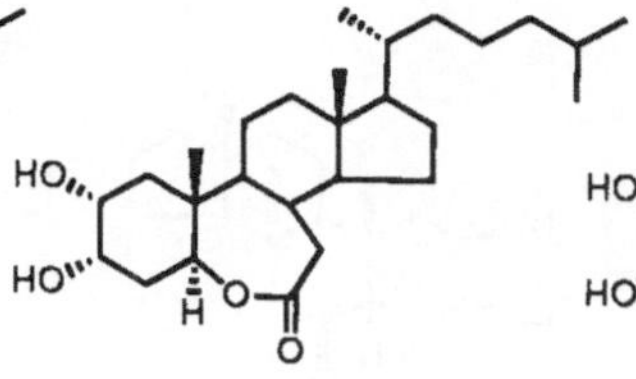

34

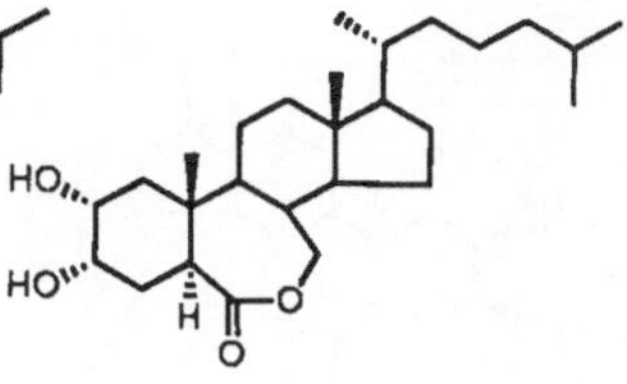

35

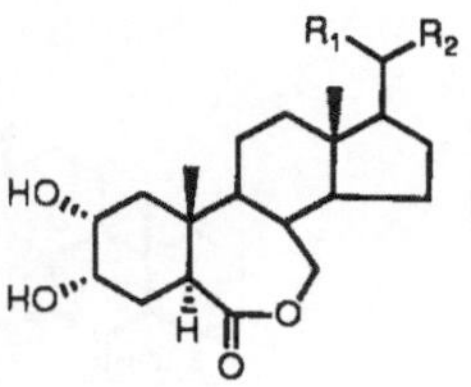

	R_1	R_2
36	CH_3	COOH
37	CH_3	COOMe
38	CH_3	COOCH(Me)-$CHMe_2$
39	CH_3	$CONH_2$
40	CH_3	$CONHC_2H_5$
41	CH_3	$CONHCHMe_2$
42	CH_3	$CONHCH_2CHMe_2$
43	H	COOH
44	H	COOMe
45	H	$CONHCHMe_2$
46	H	$CONHCH_2CHMe_2$

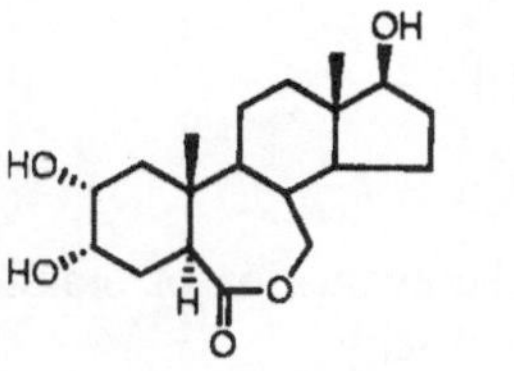

47

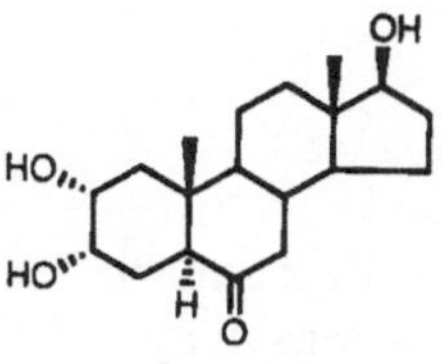

48

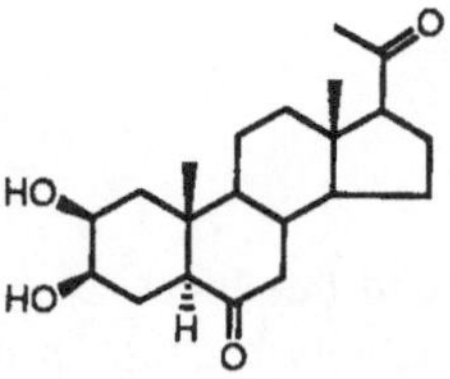

49

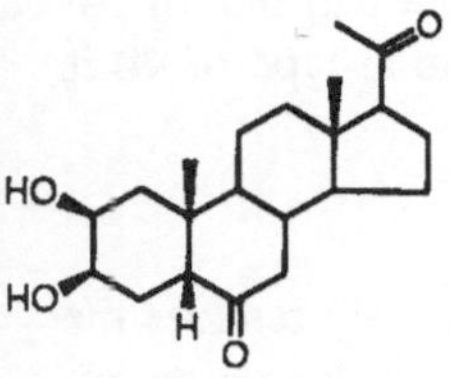

50

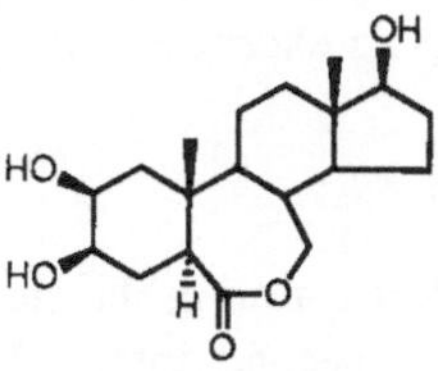

51

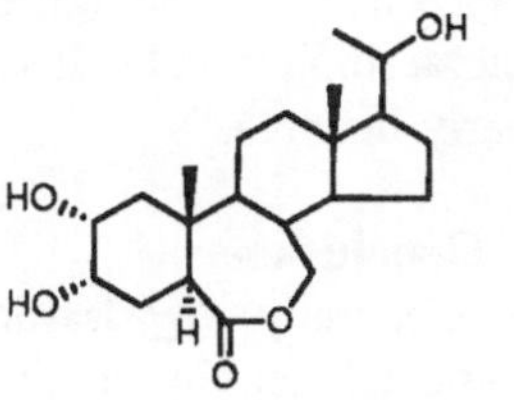

52

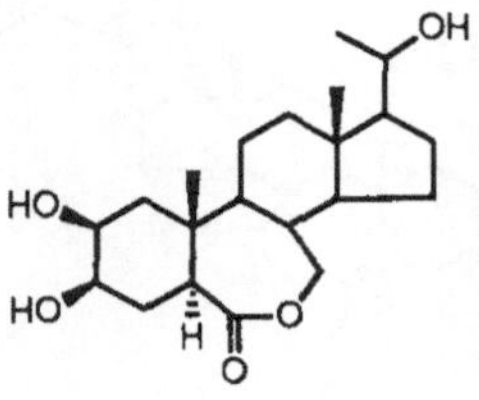

53

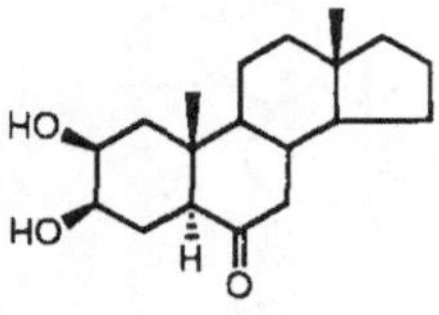

54

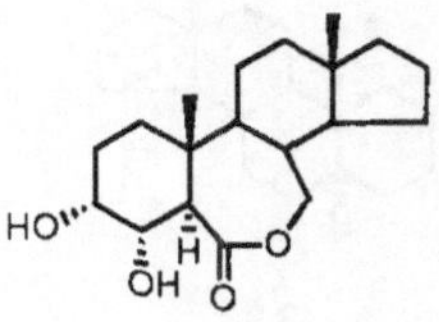

55

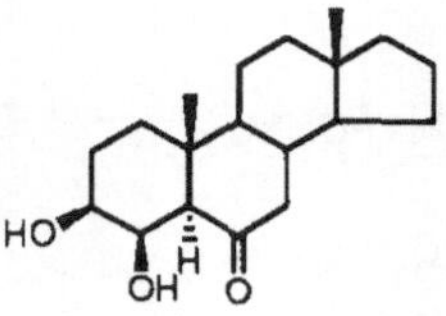

56

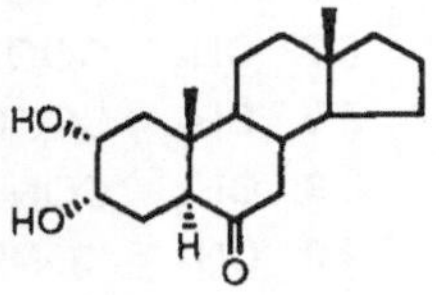

57

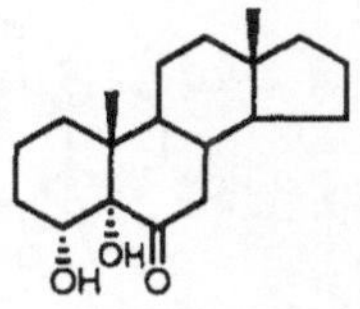

58

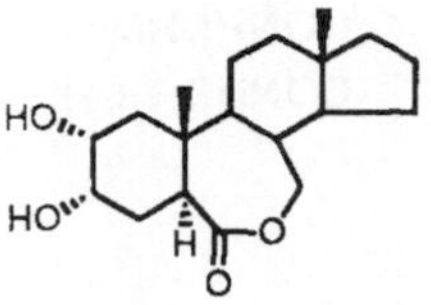

59

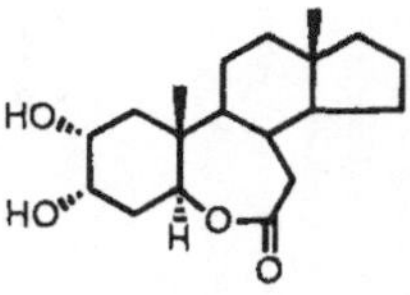

60

they found this bioassay system is not suitable for the evaluation of brassinolide activity because of the low sensitivity and specificity.

Kohout et al.[20] and Kohout and Strnad[21] examined biological activities of BRs with androstane and pregnane skeleton (**47** to **60**). Most of these compounds were essentially inactive. However, compound **48** is exceptionally very active, the biological activity being reported to be about half that of 24-epibrassinolide (**11**). It is also an exception that **47**, a lactone type of **48** is nearly inactive.[18]

2. Conclusions

Generally, any drastic modification in the side chain largely reduces the biological activity in the bean second internode bioassay (elongation response).

III. STRUCTURE-ACTIVITY RELATIONSHIP OF BRASSINOSTEROIDS IN THE RICE LAMINA INCLINATION BIOASSAY

The rice (*Oryza sativa* L.) lamina inclination bioassay was developed by Maeda in 1965,[22] far long before brassinolide was isolated from rape pollen. In 1968, Japanese scientists isolated three principles from leaf of *Distilium racemosum* Sieb. et Zucc., using this bioassay system,[23] although their structures remained undetermined, because, at that time, analytical methods such as high-resolution nuclear magnetic resonance spectrometry and mass spectrometry had not been developed. After the discovery of brassinolide from rape pollen, these compounds turned out to be brassinolide-related congeners.[24] Because of such historical reasons, this bioassay system has been successfully used by Japanese scientists to guide purification and isolation of BRs from various plant extracts and also to evaluate biological potency of natural and synthetic BRs.[25,26]

In this bioassay, explants comprising leaf blade, joint, and sheath are excised from etiolated rice seedlings and floated on test solutions. Cells aligned in the adaxial side of the joint between leaf blade and sheath are swollen by BRs, causing lamina bending.[27] This reaction is quite specific for BRs. Although indole-3-acetic acid is also active in this bioassay, the biological activity is five orders of magnitude weaker than brassinolide.[25] The response to BRs is linear to logarithm of the concentrations of BRs and, therefore, is suitable for quantitative evaluation of the biological potencies of BRs (Figure 1). Four cultivars, Arborio J1 (Indica), Kinmaze (Japonica), Koshihikari (Japonica), and Nihonbare (or Nipponbare, Japonica), have been used for this bioassay. Sensitivity and specificity of these cultivars to BRs seem to be similar, so that the names of the cultivars will not be specified in this discussion if unnecessary.

A. EFFECTS OF FUNCTIONALITIES IN A RING

An α-orientated diol at C-2 and C-3 is quite important for the biological activity (Tables 3 and 4). Biological activities of natural BRs with A-ring modification (Table 3) indicate that when the configurations of the 2α,3α-diol (**24**) are changed, the biological activity decreases in the order of 3-epimer (2α,3β-diol, **62**), 2-epimer (2β,3α-diol, **61**) and 2,3-diepimer (2β,3β-diol, **63**).[28] The biological activity of the 2,3-diepimer is extremely low.[28] Lack of the 2-hydroxyl (**64, 65, 69, 70, 78, 80, 86, 87**) reduces the biological activity by one order of magnitude (Tables 3 and 4).[28-31] By comparing the

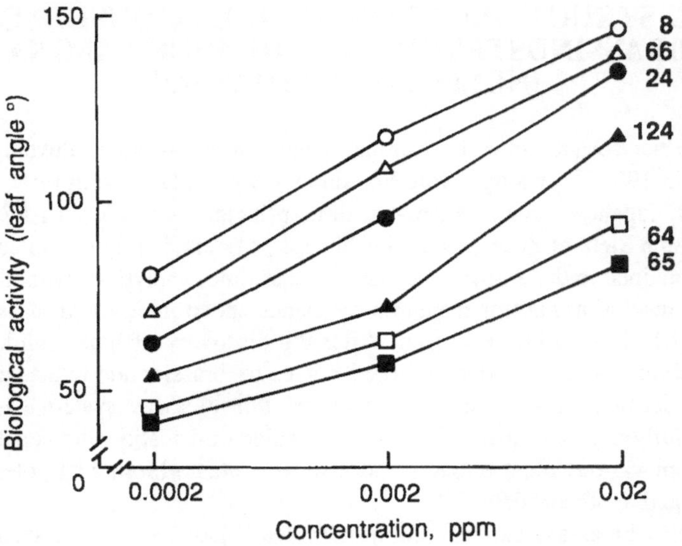

FIGURE 1. Biological activities of some naturally occurring brassinosteroids in the rice lamina inclination bioassay. Etiolated lamina explants obtained from a rice cultivar Koshihikari are used. Leaf angle in controls 42°.

biological activities of BRs shown in Table 3, it is found that 3α-hydroxyl has a greater growth-promoting activity than 3β-hydroxyl (**64** vs. **65, 24** vs. **62**) and that introduction of a hydroxyl at C-2α increases the activity about 5- to 10-fold (**64** vs. **24, 65** vs. **62**), but introduction of a hydroxyl at C-2β decreases the activity about 5- to 20-fold (**64** vs. **61, 65** vs. **63**). Acetylation or methylation of one of hydroxyls at C-2 and C-3 greatly reduces the biological activity (Tables 4 and 5).[29,31] Interestingly, a 4α-hydroxyl and 3-ketone group have promotive effects (Table 4).[30,31] Therefore, hydroxyls at C-2α, C-3α, C-3β, and C-4α and ketone at C-3 are promotive, but C-2β hydroxyl is inhibitory.

B. EFFECTS OF FUNCTIONALITIES IN B RING

Among the functionalities in B ring of natural BRs, lactone (7-oxa-6-one) is the most favorable one, while ketone is two to four times less effective than lactone, depending on the cultivars used in bioassay (Tables 3 and 4).[28,30,31] 6-Deoxo derivatives (**104** to **106**) show extremely low activity (Table 6).[33]

When the ether oxygen of 7-oxalactone is replaced by nitrogen (**96, 100**), the biological activities are reduced by two orders of magnitude (Table 6).[32] Replacement of the oxygen by sulfur (**98, 101**) reduces the biological activity by about two to three orders of magnitude (Table 6).[34] Removal of the carbonyl oxygen of 7-oxalactone (**99**) reduces the biological activity two orders of

TABLE 3
Relative Biological Activity of Naturally Occurring Brassinosteroids with A-Ring Modifications in the Rice Lamina Inclination Bioassay-1[1,28]

Compound	A ring	Activity	Compound	A ring	Activity
	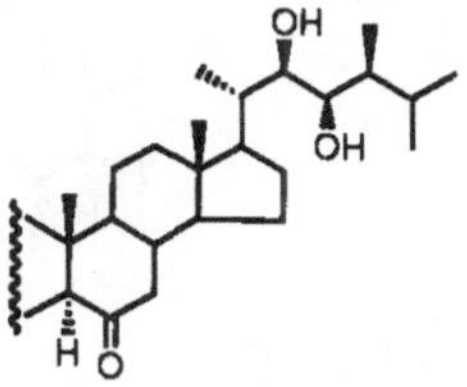			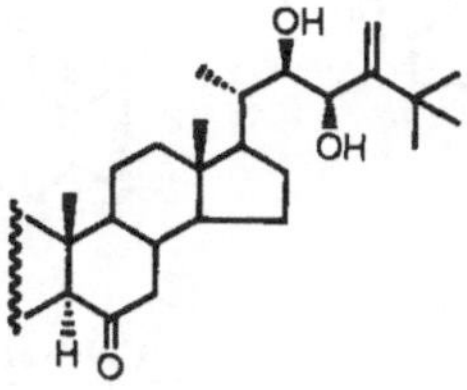	
24	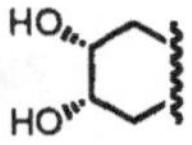	25	**66**	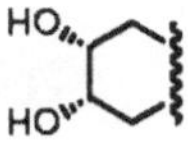	50
61	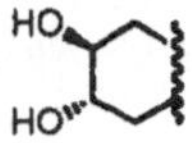	0.5	**67**	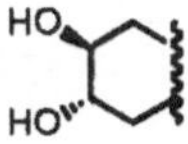	0.5
62	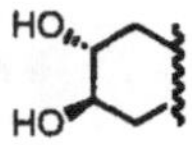	5	**68**	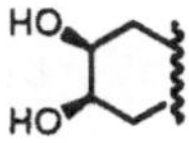	0.1
63	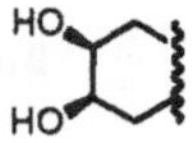	0.05	**69**	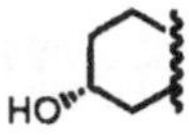	6
64	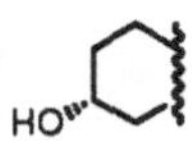	2.5	**70**	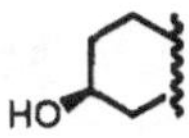	2
65	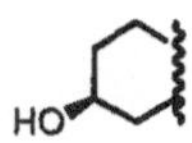	1			

Note: Rice cultivar, Koshihikari; biological activity of brassinolide = 100.

magnitude (Table 6).[34] These compounds have no synergistic or antagonistic activities.[34] Isomeric lactones (6-oxa-7-one) have 20- to 30-fold less activity than the corresponding 7-oxa-6-one BRs (Table 4).[30,31] This is the case in biologically very potent BRs, but, in some of less potent BRs, only little difference is observed. Isomeric lactam (**102**) and thiolactam (**103**) are inactive (Table 6).[35] Anastasia et al.[36] also stated that isomeric lactams (**102** and isomeric lactam of **97**) are inactive, although they did not specify the bioassay system.

TABLE 4
Relative Biological Activity of Brassinosteroids with A-Ring Modifications in the Rice Lamina Inclination Bioassay-2

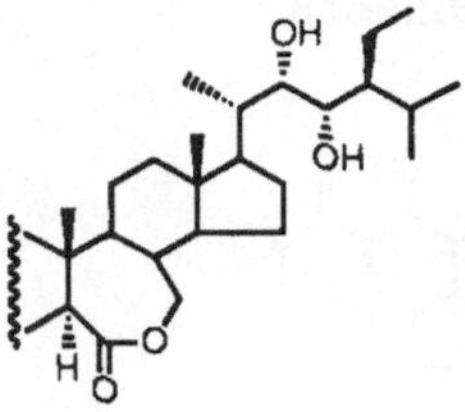

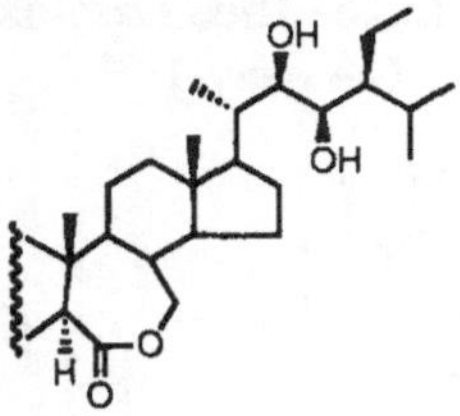

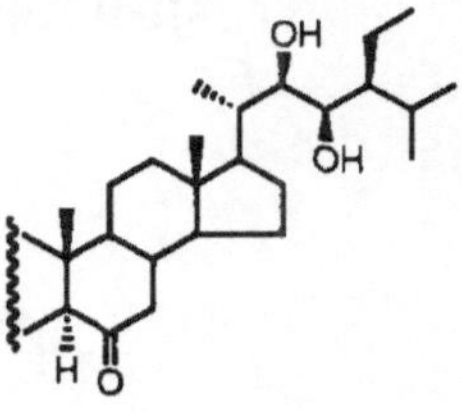

Cultivar used for bioassay								
Kinmaze[29]	A ring		Arborio J1[30,31]	A ring		Arborio J1[30,31]	A ring	
13	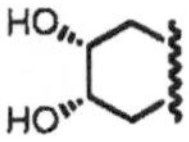	10	16	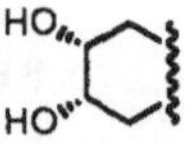	100 (Isolactone **77**, 5)	28*	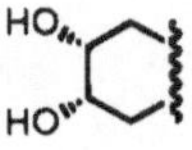	50
71	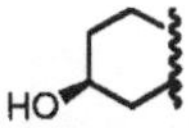	0.8 (Isolactone **15**, 0.01)	78	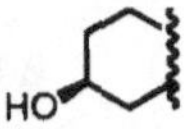	2 (Isolactone **79**, 2)	86	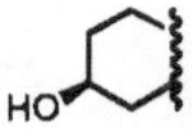	5
72		0.1	80	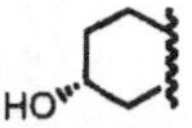	5 (Isolactone **81**, 2)	87	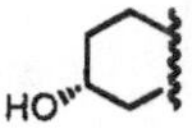	1.5
73		0.1	82	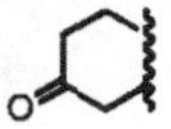	7	88	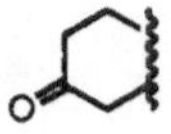	0.5
74	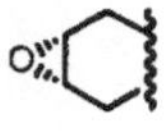	0.1	83	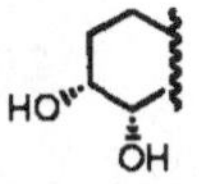	30 (Isolactone **84**, 1)	89	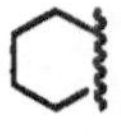	0.03
75	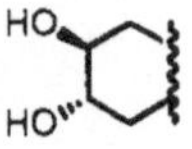	0.1	85		1			
76	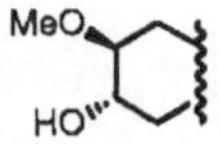	0.03						

Note: Biological activity of brassinolide = 100.

* Naturally occurring.

In 6-keto BRs the presence of a double bond at $C_{7\text{-}8}$ (**107, 109**) reduces the biological activity by one or two orders of magnitude, while introduction of a 5α-hydroxyl (**108, 110**) causes drastic inactivation, indicating B-ring structures typical of ecdysteroids are unfavorable against brassinosteroid activity (Table 7).[31]

TABLE 5
Relative Biological Activity of Brassinosteroids with A-Ring Modifications in the Rice Lamina Inclination Bioassay-3

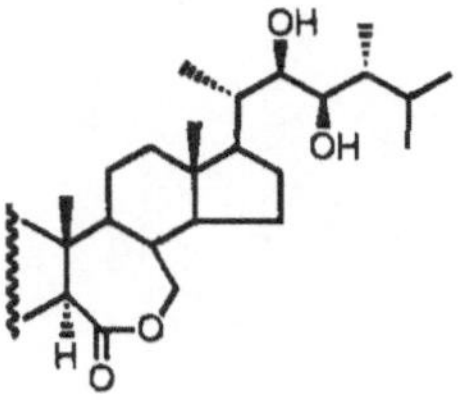

A ring

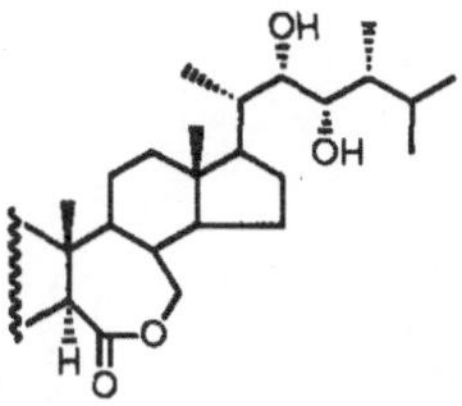

A ring

11*	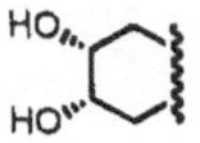	10
90	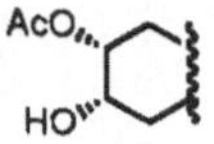	0.5
91	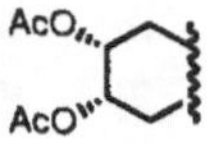	0.01
92	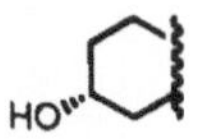	1
93	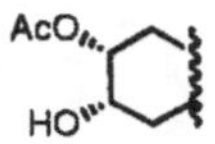	0.1
94	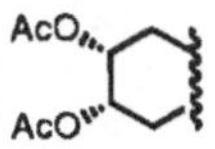	0.01
95	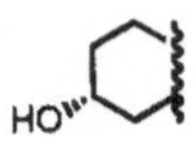	1

Note: Rice cultivar, Koshihikari, biological activity of brassinolide = 100.

* Naturally occurring.

From Takatsuto, S., Muramatsu, M , Ohya, K., Hayashi, S., and Shida, A., *Agric. Biol. Chem.*, 52, 2059, 1988. With permission.

C. EFFECTS OF FUNCTIONALITIES IN THE SIDE CHAIN

Takatsuto et al.,[37] Wada et al.,[38] and Kondo et al.[39] examined the biological activities of synthetic BRs while Yokota et al.[1,28] examined the biological activities of natural BRs (Tables 8 and 9). The most favorable alkyl substituents at C-24 are *S*-oriented methyl and ethyl groups, both being equally effective.[37-39] In the case of lactone-type BRs, the epimers with *R*-configuration are about tenfold less active (**8** vs. **11, 16** vs. **111, 13** vs. **112**).[37] The 28-nor derivatives, lacking an alkyl group at C-24, are further less active than *R*-alkylated derivatives (**6** vs. **8, 5** vs. **9**).[37]

BRs with 24-methylene (**119, 124**) have one sixth biological activity of 24*S*-methyl BRs.[1,28] BRs with 24-ethylidene (**120, 125**) have two- to threefold higher biological activity than the corresponding methylene derivatives.[1,28] Removal of 26- and 27-methyl groups (**113, 123**) does not affect the biological

TABLE 6
Relative Biological Activity of Brassinosteroids with B-Ring Modifications in the Rice Lamina Inclination Bioassay-1

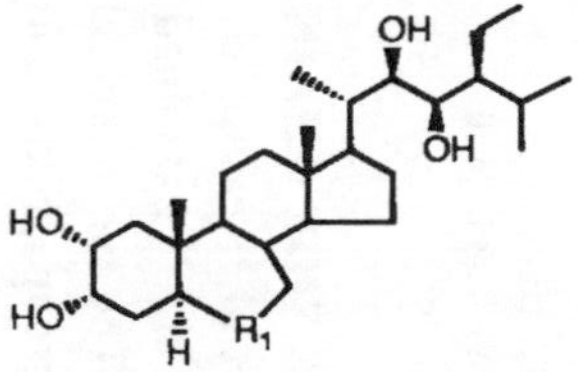
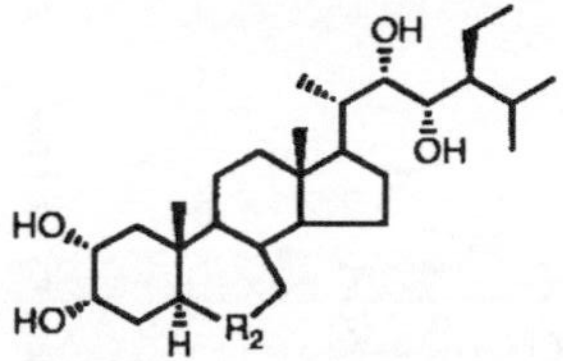
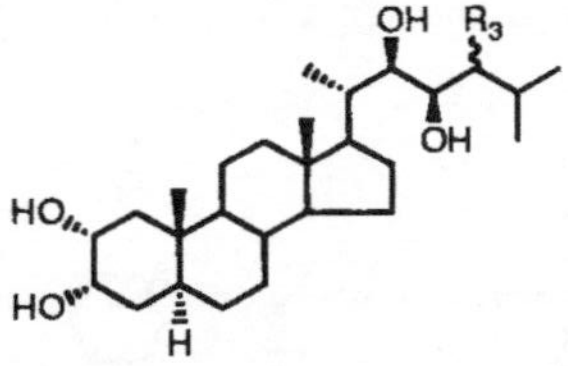

Cultivar used for bioassay											
Nihonbare[34]			**Nihonbare[34]**			**Koshihikari[35]**			**Koshihikari[33]**		
	R_1			**R_2**			**R_2**			**R_3**	
96	-CONH- (cholesterol side chain **97**, 0.01)	1	**100**	-CONH-	0.1	**27**	-CO-	<1	**104***	$-CH_3$	0.01
			101	-COS-	0.1	**102**	-NHCO-	0	**105***	$=CH_2$	0
98	-COS-	0.1				**103**	-NHCS-	0	**106***	$=CHCH_3$	0
99	$-CH_2O-$	1				**15**	-OCO-	<<1			

Note: Biological activity of brassinolide = 100.

* Naturally occurring.

TABLE 7
Relative Biological Activity of Brassinosteroids with B-Ring Modifications in the Rice Lamina Inclination Bioassay-2

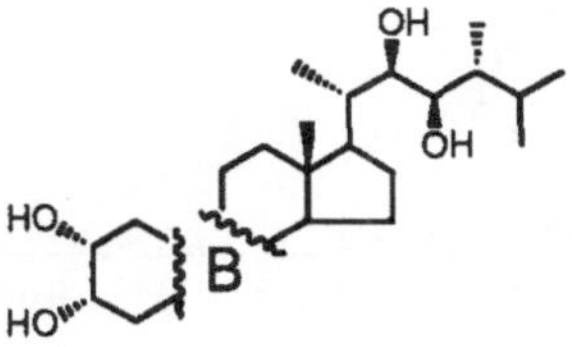

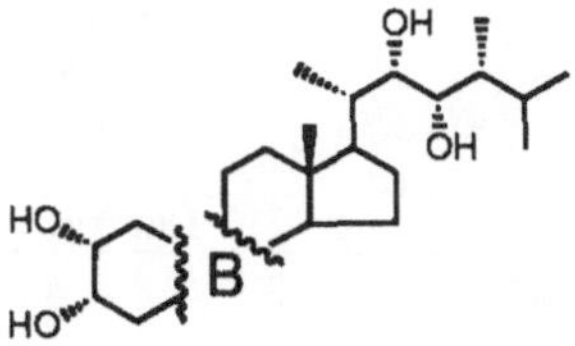

	B ring			B ring	
11*	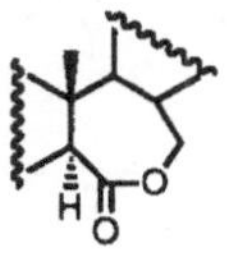	10	9	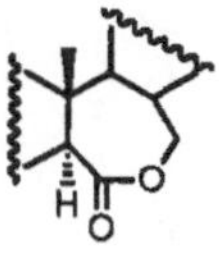	4
26*	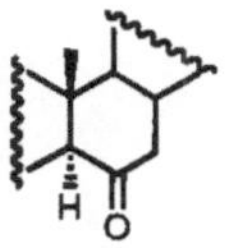	10	25	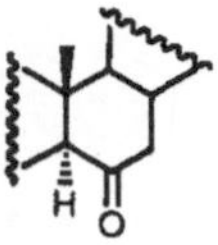	1
107	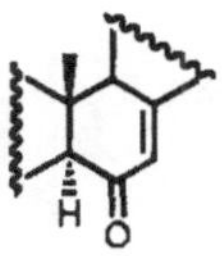	1	109	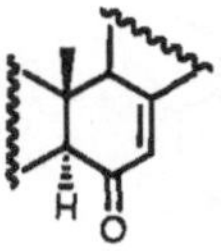	0.01
108	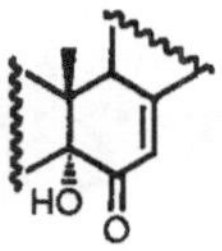	0.01	110	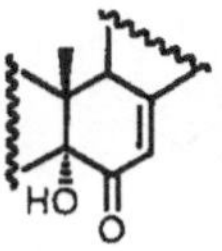	~0

Note: Rice cultivar, Arborio J1; biological activity of brassinolide = 100.

* Naturally occurring.

From Takatsuto, S., Ikekawa, N., Morishita, T., and Abe, H., *Chem. Pharm. Bull.*, 35, 211, 1987.

activity, indicating these methyls are not important for the biological activity.[40] Quite interesting is that a natural BR, 25-methyldolichosterone (**66**) which has a terminal tertiary butyl group exhibits six times higher activity than dolichosterone (**124**) (Table 9).[8] According to this observation, Mori and Takeuchi[41] synthesized 25-methylbrassinolide (**118**) and this compound was found to be three times as active as brassinolide (Table 8).[42] Thus, 25-methylbrassinolide (**118**) is the strongest BR among natural and synthetic BRs.

TABLE 8
Relative Biological Activity of Brassinosteroids with Modified Side Chain in the Rice Lamina Inclination Bioassay-1

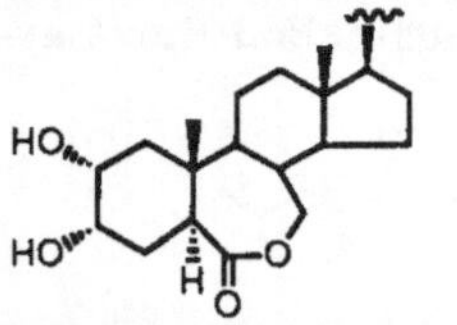

	Side chain	Cultivar used for bioassay: Arbo[37]	Kin[38,39]		Side chain	Kin[38,39]	Koshi[1,28]
8*	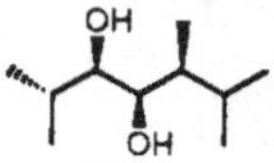	100	100	116	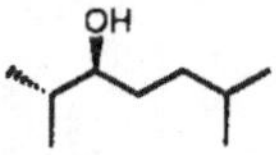	1	-
11*	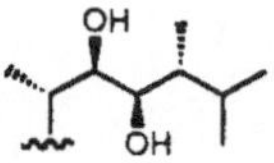	10	-	1		1	-
9	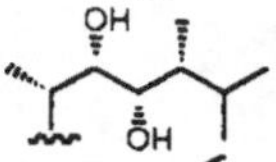	7	-	117	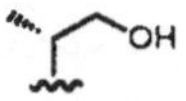	1	-
16	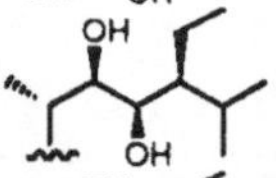	100	100	47		0.001	-
13	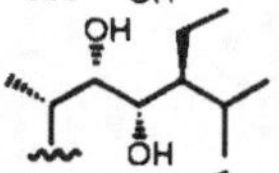	50	10	118	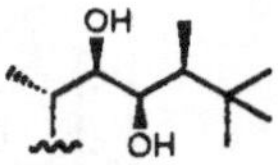	-	300[41,42]
111	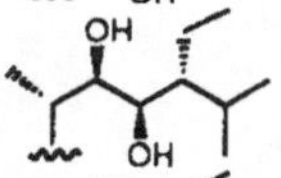	10	-	119*	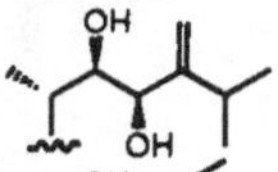	-	15
112	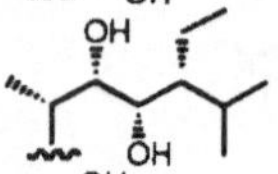	5	-	120*	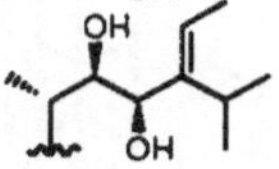	-	50
113	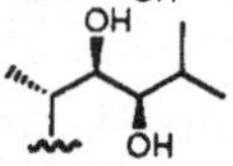	100[40]	-				
6*	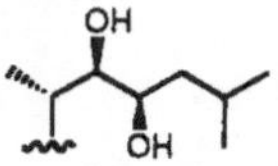	5	10				
5**	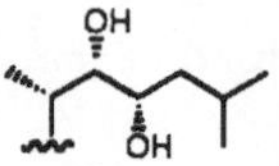	1	0.1				
114	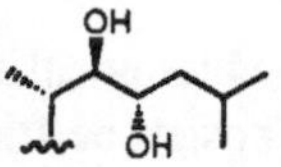	0.1	0.1				
115	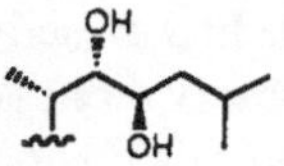	0.05	0.1				

Note: Abbreviation. Arbo, Arborio J1; Kin, Kinmaze; Koshi, Koshihikari. Biological activity of brassinolide = 100. * Indicates naturally occurring; ** indicates relative biological activity in Koshihikari = 1.[35]

TABLE 9
Relative Biological Activity of Brassinosteroids with Modified Side Chain in the Rice Lamina Inclination Bioassay-2

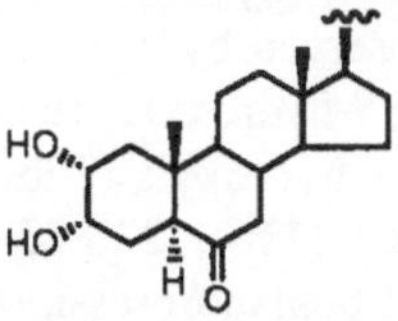

Cultivar used for bioassay		Arborio J1[37]			Koshihikari[1,28]
Side chain			Side chain		
24*	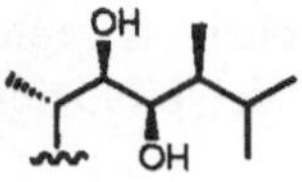	50	24*	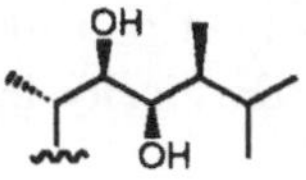	25
28*	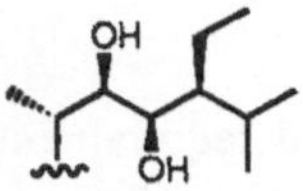	50	124*	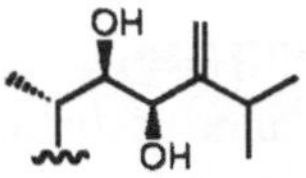	8
27**	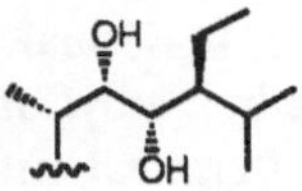	0.5	125*	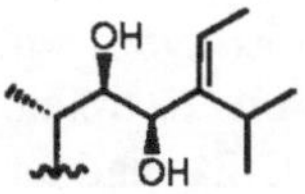	15
121	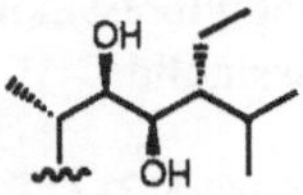	1	66*	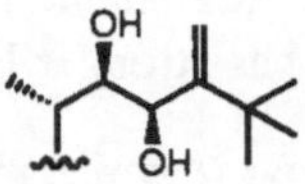	50
122	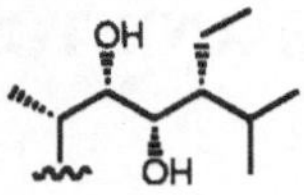	0.5			
22*[35]	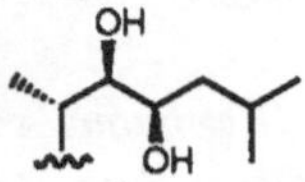	5			
21**	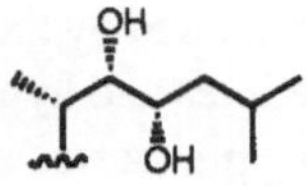	0.1			
123	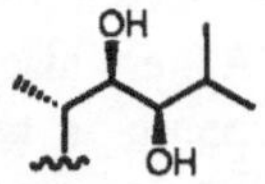	50[40]			

Note: Biological activity of brassinolide = 100. * Indicates naturally occurring; ** indicates similar relative biological activity is observed in Koshihikari.

With respect to the 22, 23-diol functionality, BRs with 22*R*, 23*R*-hydroxyls are the most active. The 22*S*, 23*S*-isomers are only slightly less active than 22*R*, 23*R*-hydroxyls in lactone-type BRs as determined by using Arborio J1 but larger differences are observed by using Kinmaze (Table 8). Some of 6-ketone type BRs with 22*S*, 23*S*-hydroxyls are two orders of magnitude less active than those with 22*R*, 23*R*-hydroxyls (Table 9).[37]

23-Deoxy-28-norbrassinolide (**116**), 22, 23-dideoxy-28-norbrassinolide (**1**, cholesterol side chain), and pentanorbrassinolide (**117**) are one order of magnitude less active than 28-norbrassinolide (**6**) (Table 8).[38,39] However, (22*R*, 23*S*)-28-norbrassinolide (**114**) and (22*S*, 23*R*)-28-norbrassinolide (**115**) are less active than the above three compounds.[37-39] This means 22*R*, 23*S*-hydroxyls and 22*S*, 23*R*-hydroxyls might have inhibitory effects on the biological activity. Compound **47** in which the side chain is replaced by a hydroxyl has drastically low (five orders of magnitude) biological activity (Table 8).[38,39]

D. CONCLUSIONS

In rice lamina inclination bioassay, the structural requirements are less stringent than in other bioassays. Favorable functionalities are a *trans*-A/B ring system (5α-hydrogen), 2α,3α-vicinal diol or 3α,4α-vicinal diol in A ring, 7-oxa-6-one or 6-ketone in B ring, as well as methyl, ethyl, methylene, ethylidene at C-24, and an additional methyl at C-25 (terminal tertiary butyl group) in the side chain. It is noteworthy that 25-methylbrassinolide is the first BR that has stronger biological activity than brassinolide.

IV. STRUCTURE-ACTIVITY RELATIONSHIP OF BRASSINOSTEROIDS IN THE RADISH AND TOMATO BIOASSAY[43]

A. EFFECTS OF FUNCTIONALITIES IN THE SIDE CHAIN AND A/B RINGS

In these bioassays, young seedlings of radish (*Raphanus sativus* cv. Tokinashi) and tomato (*Lycopersicon esculentum* cv. Giant cherry) are grown for 24 h in test solutions and the biological activity is evaluated by the resulting elongation and morphogenetic changes of cotyledon petioles and hypocotyls.[44] Brassinolide (**8**) at the concentration of as low as 0.01 ppm causes elongation and morphogenetic changes (curvature) of cotyledon petiole and hypocotyls. In these systems, indole-3-acetic acid and gibberellin (A_3) are also effective but the minimum concentration required for the response is as high as 10 ppm. Therefore, these tests are rather specific for BRs and are useful for the evaluation of their growth-promoting activities of BRs.

In the radish bioassay (Table 10), brassinolide (**8**) is the most active. (22*S*, 23*S*, 24*R*)-Brassinolide (**9**) and 28-nor-brassinolide (**6**) have the same activity as brassinolide. With respect to C-24 alkyl substituents, BRs with a methyl group or no alkyl group are more active than BRs with an ethyl group.

TABLE 10
Relative Biological Activity of Brassinosteroids in the Radish and Tomato Bioassay

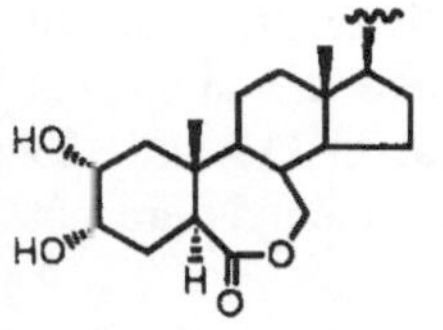

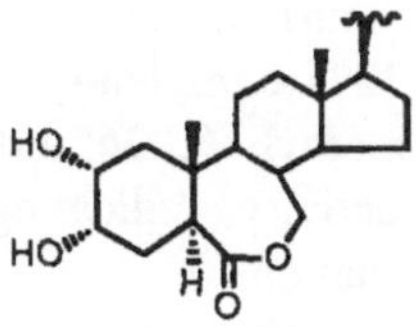

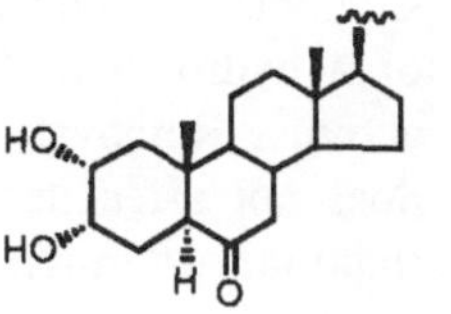

	Bioassay used										
	Side chain	R	T		Side chain	R	T		Side chain	R	T
8[a]	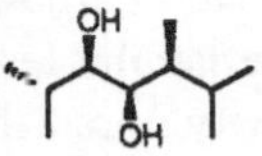	100	100	5	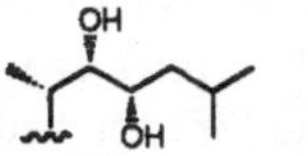	10	1	28	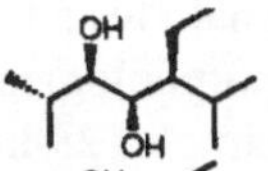	1	0.3
11[a]	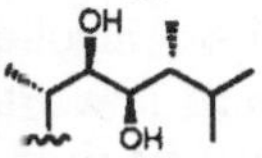	10	10	114	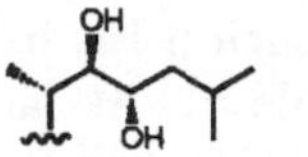	1	0.1	27	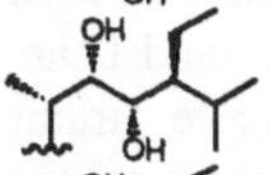	0.01	0
9	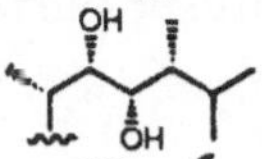	100	3	115	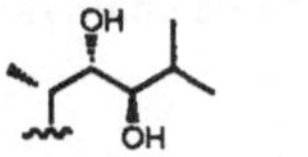	1	0.5	121	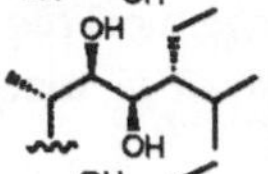	0.1	0
16	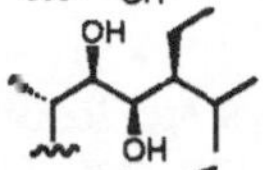	10	1	116	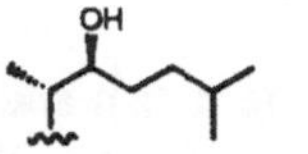	1	0.3	122	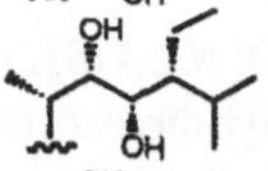	0.03	0
13	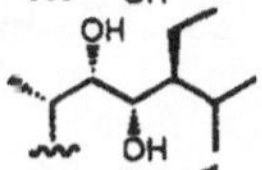	3	1	126	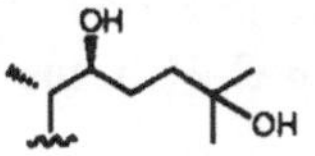	0	0	22	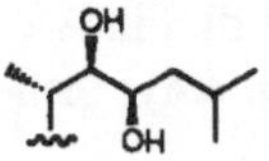	3	1
111	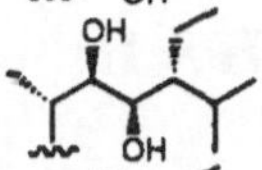	1	1	127	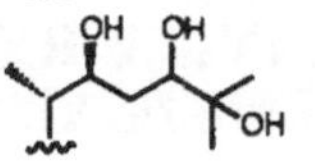	0	0	21	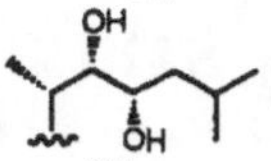	0.3	0.3
112	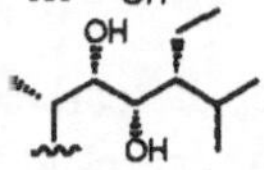	0.5	1	1	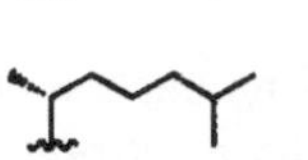	0	0	123	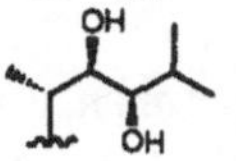	10[40]	-
6[a]	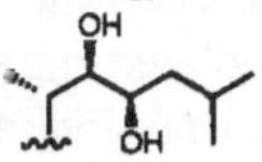	100	10	113	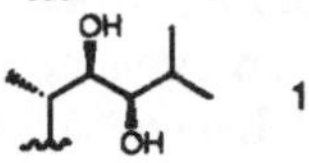	100[40]	-				

Note: Biological activity of brassinolide = 100. R: radish bioassay, T: tomato bioassay.

[a] Naturally occurring.

From Takatsuto, S., Yazawa, N., Ikekawa, N., Takematsu, T., Takeuchi, Y., and Koguchi, M., *Phytochemistry*, 22, 2437, 1983. With permission.

With an exception of (22*S*, 23*S*, 24*R*)-brassinolide (**9**), *S*-configuration of 24-alkyl group is better than *R*-configuration, and 22*R*, 23*R*-diol is better than 22*S*, 23*S*-diol. The biological activities of (22*R*, 23*S*)- or (22*S*, 23*R*)-28-norbrassinolide (**114, 117**), as well as 23-deoxy-28-norbrassinolide (**116**), are one hundredth that of brassinolide (**8**), namely even weaker than that of (22*S*, 23*S*)-28-norbrassinolide (**5**), the same relationship being observed in the rice lamina inclination bioassay (Table 8). Brassinolide without hydroxyls in the side chain (**1**, cholesterol side chain) has no biological activity. Introduction of a hydroxyl at C-25 (**126, 127**) completely extinguishes the biological activity. Removal of terminal methyls (C-26 and C-27) as in **113** and **123** does not affect the biological activity, indicating that the methyls are not important as in rice lamina inclination.

In the tomato bioassay (Table 10), structural requirements for the biological activity are basically similar to, but more stringent than, those observed in the radish bioassay.

6-Keto BRs are much less active than lactone-type BRs in both radish and tomato tests (Table 10). This fact is similar to the bean second internode test (morphological response) but is quite distinct from the rice-lamina test. For example, 28-homocastasterone (**28**) showed only slight activity in radish, tomato, and bean bioassay, although it has half the activity of brassinolide in the rice lamina bioassay (Table 4). As far as examined until now, the stereochemical requirements of the side chain functionalities seem to be similar to those of lactone-type BRs.

B. CONCLUSIONS

Favorable functionalities for the radish and tomato bioassays are 2α,3α-vicinal diol in A ring, 7-oxa-6-one in B ring, 22*R*, 23*R*-diol or 22*S*, 23*S*-diol and a methyl or no alkyl group at C-24 in the side chain.

V. STRUCTURE-ACTIVITY RELATIONSHIPS IN OTHER BIOLOGICAL SYSTEMS

BRs enhance auxin-mediated production of aminocyclopropane carboxylic acid, a direct precursor of the plant hormone ethylene. Arteca et al.[45] examined the effects of six BRs on ethylene production by etiolated mung bean (*Vigna radiata* L. Rwilcz cv. Berken) hypocotyl segments. Brassinolide (**8**) and (24*R*)-brassinolide (**11**) are the most effective. (22*S*, 23*S*, 24*R*)-brassinolide (**9**) is as active as or slightly less active than brassinolide. Inactive are 28-homocastasterone (**28**), brassinolide without hydroxyl groups in the side chain (**2**), and its 28-nor derivative (**1**).

Kakiuchi et al.[46] found that a 22, 23-epoxy-2,3-acetonide analog (**128**) is not biologically active in the rice lamina inclination bioassay but, in terms of long-term biological effects on field-grown radish (*Raphanus sativus* L. var. radicula DC) seedlings, is more effective than brassinolide.

Takatsuto et al.[47] reported that 22, 23-epoxy-2,3-diacetate analogs (**129, 130**) promote the growth of young soy bean (*Glycine max* Merrill cv. Orihime) plant grown in pots under greenhouse condition without any morphogenetic change and growth suppression characteristic of a high dose of brassinolide.

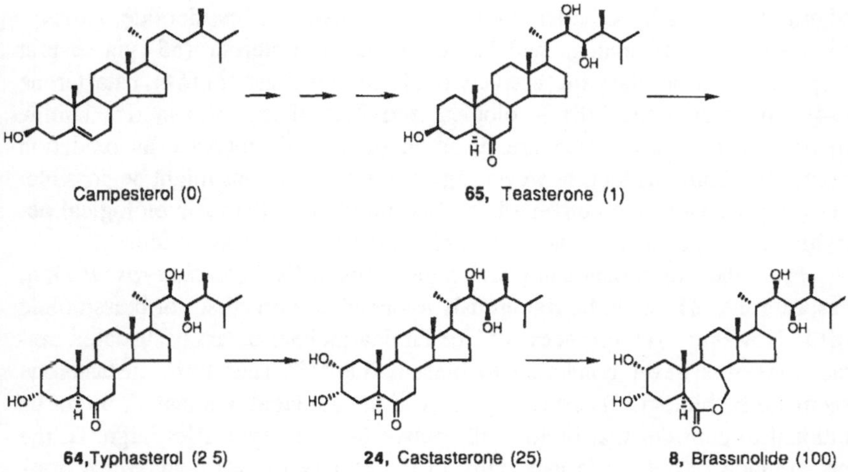

FIGURE 2. Hypothetical biosynthetic pathway of brassinolide in plant. Relative biological activities in the rice lamina inclination bioassay using a cultivar, Koshihikari are shown in parentheses.

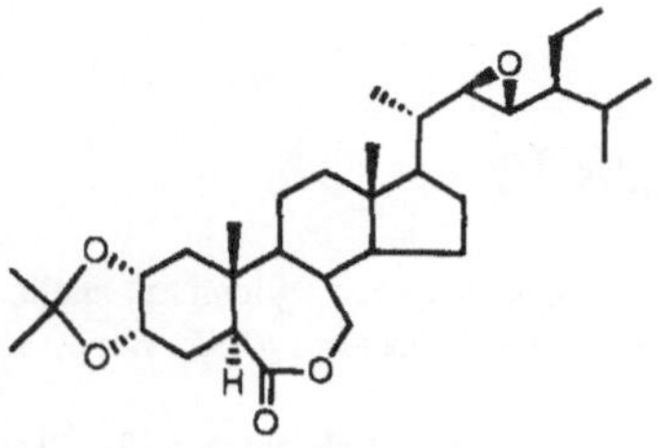

128

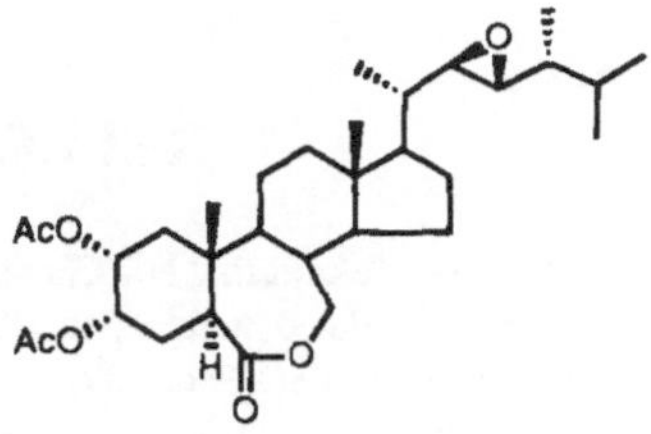

129

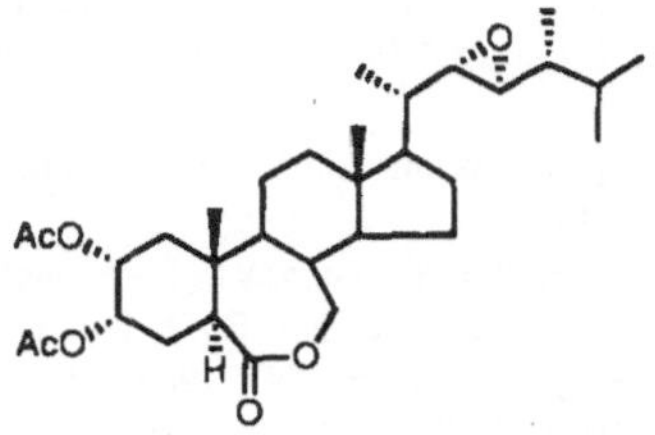

130

VI. CONCLUDING REMARKS

In the studies of structure-activity relationship of BRs, it has been in question whether 6-ketone type BRs fit the brassinolide receptor site(s) or are converted to the corresponding lactone (7-oxa-6-one)-type BRs.[14,31] Brassinolide (**8**) and castaterone (**24**) are lactone type and ketone type BRs, respectively, which have been frequently found in extracts of various plants.

Figure 2 shows a hypothetical biosynthetic pathway of brassinolide, namely, campesterol or its analogs will be converted to teasterone (**65**) via several steps of oxidation, then successively oxidized to typhasterol (**64**), catasterone (**24**), and brassinolide (**8**). Biological activities of the BRs in rice lamina inclination bioassay which are given in parenthesis increase as oxidation proceeds. From this fact, however, again two explanations might be possible: (1) the precursors are converted to brassinolide and (2) their biological activities are dependent on the structural similarities to brassinolide.

From the experiment using crown gall cells of *Catharanthus roseus* Don, castasterone (**24**) could be rigourously assigned as a precursor of brassinolide (**8**).[48] However, in the process of rice lamina inclination assay, tritiated castasterone was never converted to brassinolide.[49,50] Therefore, castasterone seems to be biologically active by itself in the etiolated rice leaf. It might be rational to consider that biologically active 6-ketone type BRs might fit the brassinolide receptor site especially in the etiolated rice tissue. Since brassinolide and castasterone, however, co-occur in various species of higher plants, it might be assumed that endogenous castasterone is convertible to brassinolide in plant tissues.

REFERENCES

1. **Yokota, T. and Takahashi, N.**, Chemistry, physiology and agricultural application of brassinolide and related steroids, in *Plant Growth Substances 1985*, Bopp, M., Ed., Sprnger-Verlag, Berlin, 1986, 129.
2. **Adam, G. and Marquart, V.**, Brassinosteroıds (review article number 19), *Phytochemistry*, 25, 1787, 1986.
3. **Marumo, S.**, The advance of brassinosteroid researches in Japan, in *Proc. 14th Annu. Plant Growth Regulator Socıety of America Meeting*, Cooke, A. R., Ed., Honolulu, 1987, 174.
4. **Mandava, N. B.**, Plant growth-promoting brassinosteroids, *Annu. Rev. Plant Physiol. Plant Mol. Biol.*, 39, 23, 1988.
5. **Meudt, W. J.**, Chemical and biological aspects of brassınolide, in *Ecology and Metabolism of Plant Lipids*, (AC Symp. Ser No. 325), Fuller, G. and Nes, W. D., Eds., American Chemical Society, Washington, DC, 1987, 53.
6. **Singh, H. S. and Bhardwaj, T. R.**, *Indian J. Chem.*, 25B, 989, 1986.
7. **Grove, M. D., Spencer, G. F., Rohwedder, W. K., Mandava, N., Worley, J. F., Warthen, J. D., Jr., Steffens, G. L., Flippen-Anderson, J. L., and Cook, J. C., Jr.**, Brassinolide, a plant growth-promoting steroid isolated from *Brassıca napus* pollen, *Nature*, 281, 216, 1979.
8. **Kim, S. K., Yokota, T., and Takahashi, N.**, 25-Methyldolichosterone, a new brassinosteroid with a tertiary butyl group from immature seed of *Phaseolus vulgarıs*, *Agric. Biol. Chem.*, 51, 2303, 1987.
9. **Yokota, T., Kim, S.-K., Fukui, Y., Takahashi, N., Takeuchi, Y., and Takematsu, T.**, Brassinosteroids and sterols from a green alga, *Hydrodictyon reticulatum:* configuration at C-24, *Phytochemistry*, 26, 504, 1987.

10. **Ikekawa, N., Nishiyama, F., and Fujimoto, Y.,** Identification of 24-epibrassinolide in bee pollen of the broad bean, *Vicia faba* L., *Chem. Pharm. Bull.*, 36, 405, 1988.
11. **Mitchell, J. W. and Livingston, G. A.,** Methods of studying plant hormones and growth-regulating substances, *Agricultural Handbook,* No. 336, Vol. 1, U.S. Department of Agriculture, Washington, D.C., 1968, 140.
12. **Worley, J. F. and Mitchell, J. W.,** Growth responses induced by brassins (fatty plant hormones) in bean plants, *J. Am. Soc. Hortic. Sci.*, 96, 270, 1971.
13. **Thompson, M. J., Mandava, N. B., Meudt, W. J., Lusby, W. R., and Spaulding, D. W.,** Synthesis and biological activity of brassinolide and its 22β,23β-isomer: novel plant growth-promoting steroids, *Steroids,* 38, 567, 1981.
14. **Thompson, M. J., Meudt, W. J., Mandava, N. B., Dutky, S. R., Lusby, W. R., and Spaulding, D. W.,** Synthesis of brassinosteroids and relationship of structure to plant growth-promoting effects, *Steroids,* 39, 89, 1982.
15. **Meudt, W. J. and Bennett, H. W.,** Rapid bioassay for auxin, *Physiol. Plant.*, 44, 422, 1978.
16. **Kerb, U., Eder, U., and Krämer, H.,** Synthesis of hexanorbrassinolide-22-ethers with plant growth-promoting activity, *Agric. Biol. Chem.*, 50, 1359, 1986.
17. **Kohout, L. and Strnad, M.,** Brassinosteroids with a cholestane side chain, *Coll. Czech. Chem. Commun.*, 51, 447, 1986.
18. **Cerny, V., Strnad, M., and Kaminek, M.,** Preparation of 2α,3α-dihydroxy-7-oxa-6-oxo-23,24-dinor-B-homo-5α-cholanic acid, its esters and amides as brassinolide analogues, *Coll. Czech. Chem. Commun.*, 51, 687, 1986.
19. **Cerny, V., Zajicek, J., and Strnad, M.,** 2α,3α-Dihydroxy-7-oxa-6-oxo-B-homo-5α-pregnan-21-oic acid and its derivatives as brassinolide analogues, *Coll. Czech. Chem. Commun.*, 52, 215, 1987.
20. **Kohout, L., Velgova, H., Strnad, M., and Kaminek, M.,** Brassinosteroids with androstane and pregnane skeleton, *Coll. Czech. Chem. Commun.*, 52, 476, 1987.
21. **Kohout, L. and Strnad, M.,** Brassinolide analogues without side chain, *Coll. Czech. Chem. Commun.*, 54, 1019, 1989.
22. **Maeda, E.,** Rate of lamina inclination in excised rice leaves, *Physiol. Plant.*, 18, 813, 1965
23. **Marumo, S., Hattori, H., Abe, H., Nonoyama, Y., and Munakata, K.,** The presence of novel plant growth regulators in leaves of *Distylium racemosum* Sieb et Zucc, *Agric. Biol. Chem.*, 32, 528, 1968.
24. **Ikekawa, N., Takatsuto, S., Kitsuwa, T., Saito, H., Morishita, T., and Abe, H.,** Analysis of natural brassinosteroids by gas chromatography-mass spectrometry, *J. Chromatogr.*, 290, 289, 1984.
25. **Wada, K., Marumo, S., Abe, H., Morishita, T., Nakamura, K., Uchiyama, M., and Mori, K.,** A rice lamina inclination test-a micro-quantitative bioassay for brassinosteroids, *Agric. Biol. Chem.*, 48, 719, 1984.
26. **Arima, M., Yokota, T., and Takahashi, N.,** Identification and quantification of brassinolide-related steroids in the insect gall and healthy tissues of the chestnut plant, *Phytochemistry,* 23, 1587, 1984.
27. **Maeda, E. and Saka, H.,** Establishment of a new method for study of the inclination rate of lamina in the excised rice leaves, *Proc. Crop Sci. Soc. Jpn.*, 37, 37, 1968.
28. **Yokota, T., Kim, S. K., Ogino, Y., and Takahashi, N.,** Various brassinosteroid from *Phaseolus vulgaris* seeds: structures and biological activity, in *Proc. 14th Annu. Plant Growth Regulator Society of America Meeting,* Cooke, A. R., Ed., Honolulu, 1987, 28.
29. **Wada, K. and Marumo, S.,** Synthesis and plant growth-promoting activity of brassinolide analogues, *Agric. Biol. Chem.*, 45, 2579, 1981.
30. **Takatsuto, S. and Ikekawa, N.,** Synthesis and activity of plant growth-promoting steroids, (22*R*, 23*R*, 24*S*)-28-homobrassinosteroids, with modifications in rings A and B, *J. Chem. Soc. Perkin Trans. 1,* 439, 1984.

31. **Takatsuto, S., Ikekawa, N., Morishita, T., and Abe, H.**, Structure-activity relationship of brassinosteroids with respect to the A/B-ring functional groups, *Chem. Pharm. Bull.*, 35, 211, 1987.
32. **Takatsuto, S., Muramatsu, M., Ohya, Y., Hayashi, S., and Shida, A.**, Synthesis of 24-epibrassinolide-related compounds with plant growth-promoting activity, *Agric. Biol. Chem.*, 52, 2059, 1988.
33. **Yokota, T., Koba, S., Kim, S.-K., Takatsuto, S., Ikekawa, N., Sakakibara, M., Okada, K., Mori, K., and Takahashi, N.**, Diverse structural variations of the brassinosteroids in *Phaseolus vulgaris* seed, *Agric. Biol. Chem.*, 51, 1625, 1987.
34. **Kishi, T., Wada, K., Marumo, S., and Mori, K.**, Synthesis of brassinolide analogs with a modified ring B and their plant growth-promoting activity, *Agric. Biol. Chem.*, 50, 1821, 1986.
35. **Okada, K. and Mori, K.**, Synthesis of brassinolide analogs and their plant growth-promoting activity, *Agric. Biol. Chem.*, 47, 89, 1983.
36. **Anastasia, M., Allevi, P., Brasca, M. B., Ciuffreda, P., and Fiecchi, A.**, Synthesis of (2*R*,3*S*,22*S*,23*S*)-2,3,22,23-tetrahydroxy-B-homo-6-aza-5α-stigmastan-7-one, an aza-analogue of brassinolide, *Gaz. Chim. Ital.*, 114, 159, 1984.
37. **Takatsuto, S., Yazawa, N., Ikekawa, N., Morishita, T., and Abe, H.**, Synthesis of (24*R*)-28-homobrassinolide analogues and structure-activity relationships of brassinosteroids in the rice-lamina inclination test, *Phytochemistry*, 22, 1393, 1983.
38. **Wada, K., Marumo, S., Mori, K., Takatsuto, S., Morisaki, M., and Ikekawa, N.**, The rice lamina inclination-promoting activity of synthetic brassinolide analogues with a modified side chain, *Agric. Biol. Chem.*, 47, 1139, 1983.
39. **Kondo, M. and Mori, K.**, Synthesis of brassinolide analogs with or without the steroidal side chain, *Agric. Biol. Chem.*, 47, 97, 1983.
40. **Takatsuto, S., Yazawa, N., and Ikekawa, N.**, Synthesis and biological activity of brassinolide analogues, 26,27-bisnor-brassinolide and its 6-oxo analogue, *Phytochemistry*, 23, 525, 1984.
41. **Mori, K. and Takeuchi, T.**, Synthesis of 25-methyldolichosterone, 25-methyl-2,3-diepidolichosterone, 25-methylcastasterone and 25-methylbrassinolide, *Liebigs Ann. Chem.*, 815, 1988.
42. **Yokota, T.**, unpublished data, 1987.
43. **Takatsuto, S., Yazawa, N., Ikekawa, N., Takematsu, T., Takeuchi, Y., and Koguchi, M.**, Structure-activity relationship of brassinosteroids, *Phytochemistry*, 22, 2437, 1983.
44. **Takematsu, T.**, Screening of plant growth regulators and the *Raphanus* test, *Chem. Regul. Plants*, 2, 132, 1967.
45. **Arteca, R. N., Bachman, J. M., Yopp, J. H., and Mandava, N. B.**, Relationship of steroidal structure to ethylene production by etiolated mung bean segments, *Physiol. Plant.*, 64, 13, 1985.
46. **Kakiuchi, T., Kamuro, Y., Takatsuto, S., and Kobayashi, K.**, A new brassinolide analog and its practical efficacy under field-cultivation conditions, *Agric. Biol. Chem.*, 52, 2381, 1988.
47. **Takatsuto, S., Futatsuya, F., Kobayashi, K., and Satoh, H.**, Synthesis and biological activity of 22, 23-epoxybrassinosteroid 2,3-diacetates, *Agric. Biol. Chem.*, 53, 263, 1989.
48. **Yokota, T., Ogino, Y., Takahashi, N., Saimoto, H., Fujioka, S., and Sakurai, A.**, Brassinolide is biosynthesized from castasterone in *Catharanthus* roseus crown gall cells, *Agric. Biol. Chem.*, 54, 1107, 1990.
49. **Yokota, T., Ogino, Y., Suzuki, H., Takahashi, N., Saimoto, H., Fujioka, S., and Sakurai, A.**, Metabolism and biosynthesis of brassinosteroids, in *Brassinosteroids — Chemistry, Bioactivity and Applications* (ACS Symp. Ser. 474), Cutler, H. G. C., Yokota, T., and Adam, G., Eds., American Chemical Society, Washington, D.C., 1991, 86.
50. **Yokota, T., Higuchi, K., Kosaka, Y., and Nobutaka, T.**, Transport and metabolism of brassinosteroids in rice, in *Plant Growth Substances 1991*, Karssen, C. M., Ed., Kluwer, Dordrecht, in press.

Chapter 10

STRUCTURE-ACTIVITY RELATIONSHIPS OF ANDROGENS AND ANABOLICS

Irena Urbánková, Zdeněk Peřina, and Oldřich Hanč

TABLE OF CONTENTS

I. INTRODUCTION

More than 55 years ago the structure of male sexual hormone — testosterone — was clarified and the first attempts to its synthesis were provided independently by Ruzicka and Butenandt. The androgenic effects have been known yet longer and they are characterized by a development of secondary male sexual features, they support a growth of sexual organs, and have a masculinic action. From the biochemical point of view, Kochakian[1] was the first who recognized the marked effects of androgens on protein metabolism in a body of higher vertebrates including man and explained the musculotropic testosterone effect by an increased protein formation (anabolism) and increased nitrogen retention in organism.

Shortly thereafter approaches were realized to prepare synthetically androgens and anabolics with more advantageous properties than testosterone alone. In newly synthetized compounds an interesting fact was observed that, in the various derivatives of basic androstane skeleton, the ratio of anabolic and androgenic effect was different even when both effects were present in all effective compounds.

II. QSAR APPROACHES IN THE FIELD OF ANDROGENS AND ANABOLICS

For years we have been collecting information on the above-mentioned problems partly of methodical orientation, partly of marginal data about androgens and anabolics. The preparations Demalon, Stenolon, and Superanabolon were prepared and commercialized as clinically used anabolics (Figure 1). From the world pharmaceutical literature a series of preparations with similar effect is known, i.e., with different ratio of androgenic and anabolic activity. Vida[2] gathered the empirical data about the chemistry and pharmacology of these compounds in the monograph published in 1969.

In the field of androgens and anabolics there are many more empirical findings about the structure-activity relationship than QSAR computer studies. The reason is very probably a lack of comparable biological preparations. In 1962 Saunders[3] tried to compare activities of 100 various anabolics known at that time on the empirical basis. In 1965, Neumann and Wiechert[4] from Schering Laboratories in Berlin compiled tables concerning the influence of some substituents on the anabolic and androgenic effect and also according to various applied doses. They especially noticed the substitutions in positions 1 and 2 in a steroid molecule. In a similar way they compiled empirical results for antiandrogenic activity,[5] where they paid attention to substitution changes in positions 1, 2, and 6 and in ring D.

The anabolic active esters were evaluated by the Hansch method[6] and further the correlations of testosterone esters activities were derived with regard to R_m values and hydrolysis rates.[7] In this case the acid component of

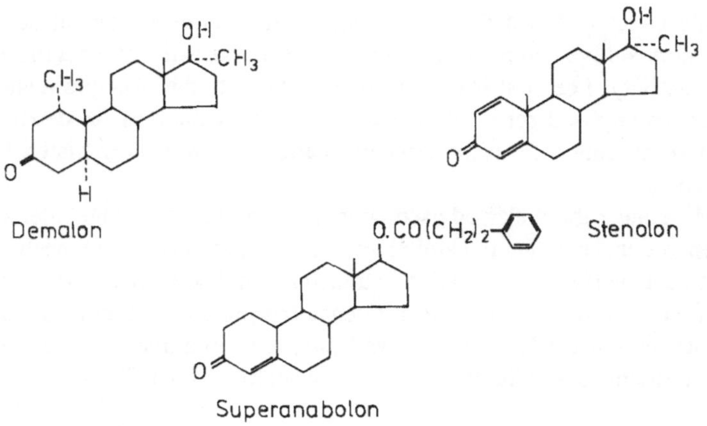

FIGURE 1. Demalon, Stenolon, and Superanabolon-CS commercial preparations.

appropriate esters was more important than steroid structure alone. Allinger et al.[8] tried at first to determine by quantum chemical calculations the charges distribution in the androsterone molecule as the structure characteristics.

A survey study on testosterone effect mechanism was published by Wilson.[9] Conformation aspects were followed especially in 5β-methyl-19-norsteroids by Glotter[10] and in testosterone and its 8β-methyl derivative in the course of roentgenographic analysis of 200 steroids in crystalline state by Wolff.[11] The parachor measurements regarding the steroid structure were realized by Ahmad[12,13] with the aim of these correlations for new structure design.

The hormone activity manifests after interaction of its molecules with receptor macromolecules in target tissues of its effect. Modeling of such interactions is based on a knowledge of hormone function groups rather than on a knowledge of tissue macromolecular receptors. Kier[14] supposes that the pharmacologically active molecule approaching the macromolecule of specific receptor has preferred conformation. Binding to the receptor occurs when there exists a complementarity between the receptor molecule field of force and steroid hormone, not only a complementary spatial in a sense "lock-key" as was schematized.

Natural steroids have a higher degree of conformation flexibility according to Delettre,[15] whereas further substitutions of this flexibility decrease in a detriment of receptor specificity. Duax[16] estimated various configurations and conformations of androgens by means of crystallography. From the models of steroid molecules it is evident why the reverse configuration on C-10 and C-13 completely suppresses the androgenic and anabolic activity of steroid. The reverse configuration on C-8 from β to α occurs without them suppressing. Cunningham[17,18] studied the differences of steroid specificity for androgen binding to cytoplasmic receptor. The decisive structures were those parts of molecule about C-3 and C-17. The author also followed the influence of various competitive compounds on androgen transport and total manifestation of their activity. Lee et al.[19] examined quantitatively the structure relation to receptor binding in a series of androstenones; the equations for dependence of relative binding affinity on structure parameters were formulated for 55 compounds.

Hobza and Zahradník[20] advised upon the importance of stage interactions between systems coupled in parallel levels, which are usually connected with a significant energetic stabilization in transfer complexes, as doubtless steroid and cytoplasmic receptor are. To influence the protein biosynthesis, the anabolic steroid must induce mRNA synthesis, e.g., in a muscle. The bond on histone proteins in a cellular nucleus is presupposed for it.[21]

III. SUBSTRUCTURAL ANALYSIS

With increasing information about newly prepared compounds, especially those which form a certain row according to their properties, a problem originated on the mutual relation of chemical structure and biological activity. The structural elements that would be necessary for activity were searched. The individual structure fragments were evaluated as increments of biological activity by definite numeral value.

In the 1970s, Berkoff et al.[22] developed a quite simple method for evaluating substituent influence on biological activity of drugs, known as "substructural analysis". By means of code the structure of a drug decomposes on fragments (substructures) and a contribution of these fragments to the total activity is characterized by SAF value (substructure activity frequency) and the following equation is accepted:

$$SAF_i = \frac{\text{number of active compounds with fragment}_i}{\text{total number of compounds with fragment}_i} \qquad (1)$$

The SAF rate has value of 0 to 1; when it is above 0.5, the contribution for activity is positive, below 0.5 is negative.

By means of these values a larger set of compounds with various degrees of activity followed; it is possible to forecast the expected activity of further not-yet tested compounds using the value MSAF (mean substructure activity frequency), which is defined as a arithmetical mean of SAF values of all fragments contained in a given compound. The procedure requires a large set of compounds, otherwise the substructures with low frequency could falsify the total MSAF value.

From 1970 to 1980 we have assembled by systematic excerption a set of 415 androstane and estrane steroids showing 6 types of activities: androgenic s.c. and p.o., anabolic p.o., according to the anabolic quotient Q, and antiandrogenic activities in the tests on rats and chickens. The structures were

TABLE 1
SAF Values of some Substructures According to Andro s.c.

	SAF in a set with criterion		
Element	Above 10 and below 10	Above 25 and below 10	Most active 10 and inactive 10
13β-Me	0.504	0.474	0.533
10β-Me	0.514	0.462	0.400
3-keto	0.597	0.590	0.563
Δ^4	0.573	0.575	0.600
17β-OH	0.584	0.578	0.767
17α-OH	0.000	0.000	0.000
17α-Me	0.653	0.654	0.750
17-keto	0.306	0.167	0.000

Note: Activity of testosterone standard = 100.

subjected to fragmentation into 151 groups both according to the properties of the substituents and to their position and steric conformation.

Sets were presented by Saunders[3] in 1962, Junkmann[23] in 1962, Dorfman[24] in 1963, Vida[2] in 1969, in antiandrogens by Wiechert[5] in 1967, Dorfman[25] in 1969, Stárka[26] in 1970, and Goto[27] in 1978. With regard to the anabolic quotient Q, the sets of Krüskemper[28] were confronted.

The activities were represented by numbers relative to the standards = 100 (testosterone, methyltestosterone, cyproterone); in some of the analyses the logarithms of these relative activities were used. Steroids were arranged into 10 basic sets (maximal number of substructural elements of the particular substances being 7 to 10) and for each of the sets the incidence tables of the substructures were given.

After providing the primary statistics of assembled data, the increments SAF for particular substructures and the respective activities were calculated, as well as the MSAF values for the particular steroids, those being correlated with the activities given (i.e. measured) (Tables 1 to 3).

In the substructural analysis a number of criteria for dividing into activity groups was applied in successive steps. For example, for anabolic quotients Q dividing into groups $Q \geqq 2$ and $Q \leqq 1$ has proved successful, otherwise the MSAF values were concentrated in the close range of 0.555 to 0.580 even with quite different Q (1 to 120).

For the particular activities the contributions of SAF were evaluated for each substructure. In the next step pairs of elements were evaluated and then after normalization the results were subjected to a 100-fold iterative procedure.

TABLE 2
SAF Values for some Substructures According to Anabolic p.o.

Element	Above 30 and below 30	Above 40 and below 10	Most active 10 and inactive 10
13β-Me	0.581	0.507	0.533
10β-Me	0.495	0.547	0.364
3-keto	0.556	0.653	0.563
Δ^4	0.547	0.664	0.500
17β-OH	0.628	0.744	0.714
17α-OH	0.000	0.000	0.000
17α-Me	0.796	0.894	0.778
17-keto	0.068	0.077	0.000
7α-Me	1.000	1.000	1.000

Note: Activity of 17α-methyltestosterone standard = 100.

TABLE 3
SAF Values for 16 Main Substituents of Skeleton

Element	>100/<8 andro s.c.[a]	>100/<30 anab. p.c.[b]	>2/<1 Q[c]
7α-Me	1.00	1.00	0.7143
Δ^9	1.00	1.00	
Δ^{11}	1.00	1.00	0
6α-F	1.00	1.00	1.00
17α-Me	0.826	0.857	0.885
17β-OH	0.6939	0.6406	0.8089
17α-Et	0.667	0.8182	0.75
Δ^4	0.65	0.4842	0.674
3-keto	0.625	0.5379	0.7564
5α-H	0.5385	0.5467	0.878
10β-Me	0.5	0.4241	0.7764
17α-OH	0	0	0
17β-Me	0	0	0
13α-Me	0	0	0
5β-H	0	0	1.00
10α-Me	0	0	0
Compounds	358	368	237

[a] Activity of testosterone standard = 100.
[b] Activity of 17α-methyltestosterone standard = 100.
[c] Q for 17α-methyltestosterone = 1.

The selection by computer allow dividing

1. Substructural elements according to the frequency
2. Steroid activities according to the intensity
3. Pairs of elements according to the frequency
4. Weight of each element for particular activity

Generally, it could be declared that the substructural analysis revealed results of approximative character. The SAF values for substructures with limitation Ø and 1 characterize the importance of appropriate element presence, but their sum MSAF does not offer a full assertion of interfering structure elements. The inactive compounds had MSAF values in the range of 0.225 to 0.450, quite broad. The statistical relation of structure element occurrence depends a great deal on criterion of limitation between activity groups, i.e., active and inactive (or weakly active) and on the data availability proportionally in both groups.

IV. PATTERN RECOGNITION, DISCRIMINATIVE ANALYSIS

The fragmentation processes could be used also in connection with the method known as "pattern recognition", where the algorithm for classifying with respect to the certain property by means of known features is determined. The conditions could be formed, e.g., for compound classification according to the order of their activity and also with a predictive significance for compounds not yet prepared. For compounds dividing into specific semiquantitative classes the discriminative analysis was used expressing nonelementary discriminative features as linear combinations of molecular parameters by means of discriminative function. Henry and Block[32,33] applied this principle for steroids dividing into five therapeutic classes using the molecular connectivity indexes.

For the "pattern recognition" method we have arranged smaller sets of steroids by dividing them into two groups according to known activities. Afterwards a discriminative analysis was applied. In a smaller group of antiandrogens with pregnane structure the activity criterion above 50 and below 50 was used at cyproterone activity = 100. It was verified that chlormadinone with given activity 80 belongs in group 2 and its activity was exaggerated.

The method of linear discriminative function can be applied favorably to the control results and for dividing steroids into activity groups.

V. MOLECULAR CONNECTIVITY INDEXES

On the basis of graph theory the quantitative topological descriptors were developed, the so-called molecular connectivity indexes. Structural formulae of compounds commonly used in organic chemistry are examples of nonoriented graphs where the top of a graph forms carbon resp. other atoms and the bonds are edges of a graph. The graph could be divided into subgraphs with a character of chains, branching (trees), cycles, or a combination.[34,35]

The top connectivity with a number of edges i is specified by the symbol δ_i, a valency connectivity δ_i^v, χ is the symbol for connectivity function either of the whole molecule or of particular structure elements with appropriate indexes. For connectivity function and index calculation all types and orders of subgraphs are considered; into them the original molecular graph can be decomposed. Usually it is not necessary to calculate indexes of all orders and all subgraphs; only the simplest fragments or parts of a molecule are indicated where the substitution occurs. Henry and Block[36] used this procedure for discriminative analysis of psychotropic drugs with eight positions in a basic skeleton.

We have used following subgraphs of steroidal skeleton:

a 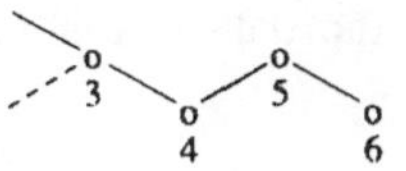

for structure alterations in position 3 and 5

b. 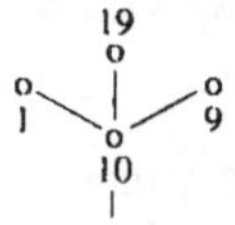

for alterations in position 10

c 

for alterations in position 17

χ_i^v values were calculated for positions 3 and 5:

3α-OH-5α-H	1.482	3-keto-5α-H	1.374
3-keto-Δ^5	1.201	3α-OH-Δ^5	1.309
3-keto-Δ^4	1.136		

Epimeric values for elements 5β-H (X) and 3β-OH (Y) calculated by variation analysis in a set of 31 steroids with a known androgenic and anabolic activity were −0.872 and −0.719, respectively. χ_i^v for positions 10 and 17:

10β-Me	1.143	10β-H	0.741
17β-OH	0.995	17-keto	0.808
17β-OH, α-Me	1.308		

Empimeric structures on C-17 were not considered; as to the empirical findings they lead to inactivation or strong decrease of steroid activity. The appropriate empirical regression equations for steroid activity were then derived:

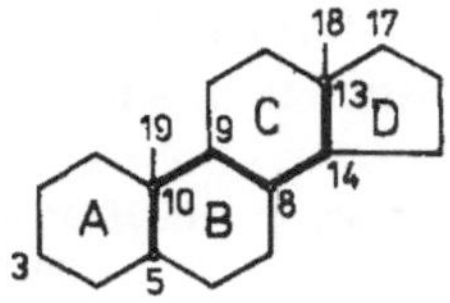

FIGURE 2. Configuration of androstane skeleton A-B-C-D rings position: *trans*-anti-*trans*-anti-*trans*.

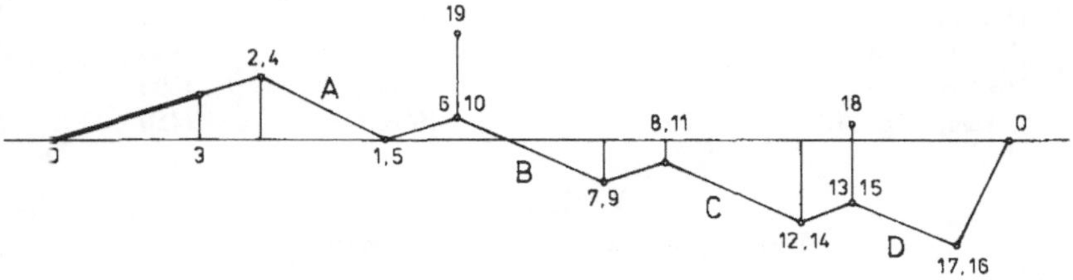

FIGURE 3. Projection of steroid skeleton into the plane perpendicular to cyclohexane rings in dihydrotestosterone.

$$\log A_{andro} = -0.872X - 0.719Y - 1.159\,\chi^{v}_{3-5} - 0.231\,\chi^{v}_{10} + 1.393\,\chi^{v}_{17} \tag{2}$$

$$\log A_{anab} = -1.065X - 1.304Y - 2.551\,\chi^{v}_{3-5} - 0.348\,\chi^{v}_{10} + 2.112\,\chi^{v}_{17} + 2.975 \tag{3}$$

The calculations were performed with 1 and 5% error permission and allowed prediction of activities of 41 remaining steroids with unknown activity. Thus, the fragments of molecular connectivity indexes can be utilized in correlations of substructure elements with androgenic and anabolic activity of steroids.

VI. TOPOLOGICAL ANALYSIS

Ghose and Crippen[37] explained a three-dimensional representation of molecular structure by the term "distance geometry". To understand the interaction between androgens and receptor macromolecules in the target organs, it is important to consider topographic and steric relations in the androstane skeleton.[38]

Crystallographic studies revealed that androgens have predominantly the following configuration: *trans*-anti-*trans*-anti-*trans*, which characterizes the A, B, C, D ring positions of steroid skeleton (Figure 2). Carbons 9-10 and 8-14 are in a *trans*-position, a chain of five bonds C_5–C_{13} is intensified, while the atoms arising in front of a projection plane are indicated by a ringlet. If the positions of particular C atoms are projected into the plane perpendicularly to the extrapolated planes of A to D rings, it is possible to signify substituents as α and β, i.e., above and below the projection of steroid skeleton planes (Figure 3). On this basis we have distinguished hydrophobic and hydrophile

TABLE 4
Relative Atom Distances in Androstane Skeleton

Atom distance	in 5α-androstane (pm)	in 5β-androstane (pm)	in 4-androstene (pm)
$[C_3]$-$[C_{17}]$	840	770	940
[3β-OH]-[17β-OH]	840	770	940
[3α-OH]-[17α-OH]	840	770	940
[3-keto]-[17-keto]	1200	1010	1190
[3β-OH]-[17α-OH]	800	670	840
[3α-OH]-[17β-OH]	970	940	1090
[3-keto]-[17β-OH]	1040	1000	1150

TABLE 5
Binding Distances in Testosterone

Position of graph top		Sums of binding distances	
7	11	65	64
2	16	85	88
1	15	72	77
5	14	68	62
6	12	69	69
8		57	
3	17	94	84
3–0	17–0	113	103

substituent configurations of various derivatives in an approximative scale for the bond lengths.[29]

By means of the distance matrices derived from a molecular graph with an usual atom numbering, the distances of particular atoms were determined in a minimal bond number among them together with a topological graph center (Table 4). By adding the minimal bond distances in testosterone we ascertain that the formal molecule center is C_8, with peripheries at C_3 and C_{17}. Symmetrical distances resemble the positions C_7 and C_{11}, C_2 and C_{16}, C_1 and C_{15}, C_5 and C_{14}, C_6 and C_{12}; likewise C_7 and C_9 toward C_3 and C_{17} and also toward oxygen atoms bound to them (Table 5).

The distance graph can indicate interactions of some substituents to steric relations in a molecule. Various testosterone conformations were derived crystallographically by Duax et al.[16] and deductions about distances C_3 and C_{17} with oxygen substituents are in agreement with our models of distance projection to one plane. Duax and Norton[39] published on the basis of X-ray studies *Atlas of Steroid Structure* with a conformation analysis of 200 steroids of various groups (estrogens, corticoids, etc.).

In 8α-testosterone with reversed configuration[40] the crystallographic analysis proved a warping of a relatively plain molecule (contrary to normal 8β-testosterone) with decreases in androgenic activity by 60% and anabolic by 90%. Simultaneously, the reverse configuration on C_{10} abolished the activity completely.

The relations between biologically active compounds and receptors are similar to interactions between substrate and enzyme. The X-ray studies of receptor macromolecules and steroid hormones revealed a marked receptor similarity for 3-keto-Δ^4 substructure and interference possibility for individual groups with a bond to various receptors.[41] The conformation influence for receptor affinity was emphasized in the hydrogen bond role of the 17β-OH group in androgen[42] together with a special importance of ring D for activity. The comparison of activities of 5α and 5β steroids showed that for s.c. androgenic activity is much more favorable the 5β configuration, while for anabolic activity the 5α configuration is more favorable.

For a forecast of geometrical receptor figures, the method was suggested based on the following minimal steric differences.[43] The compound affinity to the receptor must be directly proportional to the total nonsuperimposed volume and receptor cavity. According to this hypothesis more active methyltestosterone should be less active in comparison with testosterone, apart from the topography of antiandrogenic cyproterone, which could not bind competitively to testosterone receptor for its excessive voluminosity given by further substitutions of basic skeleton.

In an analogy to enzymological findings it is necessary to suppose a reciprocal influence in the substructure orientation of both interacting molecules with the conformation change caused by the proximity of other molecules (proximity effect). Interactions from the β-position of the androgenic molecule were considered with the conclusion that at least a three-point bond of two hydrophobic groups on C_{10} and C_{13} and at least one hydrophilic group, mostly in configuration β, are a necessary precondition for it.

VII. FREE-WILSON ANALYSIS AND ITS MODIFICATION

This multiple variant analysis also supposes additive contributions of particular substructure elements. These increments are calculated as deviations from an average of all biological effects estimated for a greater number of compounds in the same group. However, these increments are not transferable at other biological system. Free-Wilson[44] analysis was used for homological series of variously substituted compounds, where fragments consist of a basic skeleton and substituents in its particular positions. Presuming the increment additivity for individual molecular fragments, they participate in biological activity logarithm by a constant contribution and then can be accepted:

$$\log\Delta_i = \mu + \sum_{j=1}^{n} \sum_{k=1}^{m_j} b_{ijk} z_{jk} \tag{4}$$

where μ is an activity increment of basic skeleton, n is a number of substituted positions, m is a number of substituents in substituent position j, z_{jk} is an increment of kth substituent in jth position, and b_{ijk} is an auxiliary parameter.

The equation is completed by symmetrical conditions that the sums of all activity increments for every position must be equal to zero. Thus, for every series of compounds a set of linear equations originates and after resolving them by multiple regression analysis the activity increments z_{jk} and μ are obtained. By means of these values it is possible to forecast biological activity logarithm for all compounds of given fragmentation scheme. Some substituents combinations could cause intramolecular interactions and in this case it is necessary to use additional factors in considering interaction or the interacting fragment as a sole "pseudosubstituent" located in a sole "pseudoposition". The method according to Boček and Kopecký,[45] published nearly at the same time as Free-Wilson method, already assumes the influence of mutual interactions.

For Free-Wilson analysis of steroids divided into 4 groups we have chosen from each group 16 to 19 compounds with the activity factor expressed by the log of the relative activity value. Incidence maps were estimated for the following elements:

3α-H, 3β-H, 3α-OH, 3β-OH, 3-keto and Δ^3, Δ^4, and Δ^5

17α-OH, 17β-OH, 17α-H, 17β-H, 17α-Me, 17β-Me, and 17-keto and basic and symmetrical equations were formulated. However, more advantageous was a modification according to Fujita and Ban[46] with the elimination of symmetrical equations, which was applied on a matrix of 72 steroids substituted in positions 3, 5, 10 and 17, whereas 11 independently variable structure elements were reduced to 8 parameters. The regression coefficients were calculated and the respective empirical equations were formulated for the androgenic and anabolic activities:

$$\log A_{andro} = 0.856(8^*) + 0.195(9) + 0.790(15) + 0.745(16) + 1.368(30) + 0.731(54) - 0.172(76) - 0.089(105) - 0.354 \quad (5)$$

$$\log A_{anab} = 1.095(8) + 0.386(9) + 1.490(15) + 0.989(16) + 1.443(30) + 0.557(54) + 0.087(76) + 0.511(105) - 1.213 \quad (6)$$

(8*) = 3-keto (15) = Δ^{4-5} (30) = 17β-OH (76) = 5β-H

(9) = 3α-OH (16) = 5α-H (54) = 10β-Me (105) = 17α-Me

We have ascertained that for androgenic activity the highest weights have substructures 17β-OH, 3-keto, and Δ^4, whereas 5β-H has the lowest. For anabolic activity the highest weights have also substructures Δ^4, 17β-OH, 3-keto, and 5α-H, followed by 10β-Me and 17α-Me; 5β-H has for activity the lowest weight again.

This analysis is therefore applicable in that case, where a group of compounds has a common basic structure substituted at least in two various positions and at least by two various substituents in each of these positions. On the contrary, it is not applicable where the substituents alter only in one position. Another inevitable condition is that every substituent in a certain position must be represented in the analyzed compound series at least twice so that its contribution to the total activity could be estimated.[47] The predicted activity of a compound with a basic structure is expressed by an intercept, that is the value from which the activity predictions of other compounds are derived by adding the contribution predictions of individual substituents. The modification according to Fujita and Ban especially enables the prediction of activity of steroids substituted in particular positions. This is made possible because a skeleton substituted in several positions (3β-OH, $\Delta^{5\text{-}6}$, 17-keto) serves as a basic core. The importance of individual coefficients does not signify the replacement of hydrogen by a substituent, but the replacement of the mentioned substituent by the other, f.e. 3-keto instead of 3β-OH means the oxidation in a position 3, 5α-H instead of $\Delta^{5\text{-}6}$ means the hydrogenation, and 17β-OH instead of 17-keto the specific reduction in this position.

VIII. CONCLUSIONS

The chemical structure can be considered also as a property, so that structural characteristics are qualitative parameters and structure is a sum of substructure elements together with the type of characteristics common for all compounds of an estimated set. The activity similar to physical properties can be localized (pharmacophores) or be a function of the entire molecule.[48] The properties depend then on all elements or on particular structure elements unequally. For hormone bond to receptor a certain complementarity of properties in ligands set is anticipated, which represents a possibility of steric arrangement of receptor molecules into the "cavities".

In our series of androgens and anabolics we have suggested a modified model for the expression of activity coefficient, in which the member of interfering substructures is not added, but it is used as a multiplier:

$$K = (I_n + I_m + I_s) \cdot I_i \qquad (7)$$

where K is the activity coefficient, I_n is a set of increments for activity inevitable, I_m is a set of increments slightly modifying activity, I_s is an increment of skeleton, and I_i are increments interfering with activity. Steroids bind to receptors when they have (1) respective type properties and (2) all inevitable structure elements and no interfering ones.

A set of inevitable substructure elements models positive variables, whereas a set of interfering elements models its negation. Both of these types eventually with elements modifying or indifferent could offer an idea about geometric structure of binding receptor position, charges distribution, etc.

With some schematism we introduce the contribution of 40 substructure elements in estrane and androstane series to their respective activities (Table 6).

TABLE 6
Influence of Substituents in Various Positions on Androgen Activity (Position 1-19)

Activity	1α-Me	1β-Me	1-Me-Δ^1	1α-OH	1,2α-CH_2	1-keto	2α-Me	3-keto
Androgenic s.c.	−	−	−	+	−	−	+	+
Androgenic p.o.	−	−	?	?	−	−	?	+
Anabolic p.o.	+	+	−	+	+	−	+	+
Q	++	++	?	+	+	−	+	?

Activity	3α-OH	3β-OH	Δ^4	4-Cl-Δ^4	5α-H	5β-H	Δ^5	6α-Me	Δ^6
Androgenic s.c.	−	=	+	+	+?	=	−	−	?
Androgenic p.o.	?	?	+	−	−	−	−	−	?
Anabolic p.o.	+	−	+	+	+	−	−	−	+
Q	+	?	−	+	+	?	?	+	+

Activity	7α-Me	7β-Me	7-keto	9α-F	Δ^9	10β-Me	10α-Me	11α-OH	11β-OH
Androgenic s.c.	++	−	−	+	+	+	=	+	−
Androgenic p.o.	++	−	?	+	++	+	=	?	+
Anabolic p.o.	++	−	−	+	++	−	=	−	+
Q	+	?	?	?	+	−	=	?	+

Activity	11-keto	Δ^{11}	13β-Me	13β-Et	13α-Me	17β-OH	17α-OH	17-keto
Androgenic s.c.	−	?	+	+	=	++	=	−
Androgenic p.o.	?	++	+	?	=	+	=	−
Anabolic p.o.	+	++	+	+	=	+	=	−
Q	+	?	−	−	=	−	=	?

TABLE 6 (continued)
Influence of Substituents in Various Positions on Androgen Activity (Position 1-19)

Activity	17α-Me	17β-Me	17α-Et	19-nor	18-nor	17β-$COCH_3$
Androgenic s.c.	+	=	?	−	=	sl+
Androgenic p.o.	++	=	+	−	=	sl+
Anabolic p.o.	++	=	+	+	=	−
Q	+	=	+	+	=	−

Note: − influence negative, = strongly negative, + positive, + + strongly positive, ? uncertain, and sl = slightly.

REFERENCES

1. **Kochakian, Ch. D.**, Mechanism of anabolic action of androgens, in *Mechanism of Hormone Action*, Karlson, P., Ed., Thieme Stuttgart, 1965, 192; Historical review of androgenic steroids in *Mechanism of Androgen Action*, Mainwaring, W. J., Ed., Berlin, 1977, 1.
2. **Vida, J. A.**, *Androgens and Anabolic Agents, Chemistry and Pharmacology*, Academic Press, New York, 1969, 327.
3. **Saunders, F. J.**, Steroide anabolica, *Excerpta Med. Int. Congr. Ser.*, No. 51, 207, 1962.
4. **Neumann, F. and Wiechert, R.**, Über anabole und androgene Wirkung von C1 und C1C2 substituierten Steroiden, *Arzneim-Forsch.*, 15, 1168, 1965.
5. **Wiechert, R. et al.**, Wirkung und Struktur neuer antiandrogener Steroide, *Arzneim-Forsch.*, 17, 1103, 1967.
6. **Chaundry, M. A. Q. and James, K. C.**, A Hansch analysis of the anabolic activities of some nandrolone esters, *J. Med. Chem.*, 17, 157, 1974.
7. **James, K. C.**, Correlation of androgenic activities of lower testosterone esters in rat with R_m values and hydrolysis rates, *Eur. J. Med. Chem.*, 10, 55, 1975.
8. **Allinger, N. L. et al.**, Androstane, the structure by force-field calculations, *Steroids*, 26, 398, 1975.
9. **Wilson, J. D.**, Recent studies on the mechanism of action of testosterone, *N. Engl. J. Med.*, 287, 1284, 1972.
10. **Glotter, E. et al.**, Conformational aspects of some 5β-methyl-19-nor-steroids, *J. Chem. Soc. Perkin Trans 1*, 2104(21), 1975.
11. **Wolff, M. E. et al.**, Quantitation of chemical and biological effects in a steroid by Hansch type and molecular orbital calculations, in *Proc. Conf.*, Tichý, M., Ed., Praha, 1973, Akademiai; Kiadó, Budapest, 1976, 31.
12. **Ahmad, P. et al.**, Parachors in drug design, *Biochem. Pharmacol.*, 24, 1103, 1975.
13. **Ahmad, P.**, Parachors in drug design, structure-activity correlation study of steroids, National Library of Canada, Otawa, 1977; *Dis. Abstr. Int. B*, 39, 204, 1978.
14. **Kier, L. B.**, The conformation of 20-oxopregnane hormones from molecular orbital calculations and a consideration of the cortisol receptor, *J. Med. Chem.*, 11, 915, 1968.
15. **Delettre, J. et al.**, Steroid flexibility and receptor specificity, *J. Steroid Biochem.*, 13, 45, 1979.
16. **Duax, W. L. et al.**, Conformational studies of steroids. Correlations with biological data, *J. Steroid Biochem.*, 6, 195, 1975; *X-Ray Crystallography. Drug Action Course International*, Oxford University Press, New York, 1985, 405.
17. **Cunningham, G. R. et al.**, Differences in steroid specificity for rat androgen binding protein and the cytoplasmic receptor, *Steroids*, 33, 261, 1979.
18. **Cunningham, G. R. and Kemp, P. L.**, Studies of steroid structure and affinity for rat androgen binding protein, Am. Chem. Soc., 179 Meet. MEDI 6, 1980.
19. **Lee, D. L. et al.**, Quantitative relationships between steroid structure and binding to putative progesterone receptors, *J. Med. Chem.*, 20, 1139, 1977.
20. **Hobza, P. and Zahradník, R.**, Calculations of weak intramolecular (van der Waals) interactions in biologically significant systems, *Chem. Listy*, 76, 673, 1982.
21. **Sluyser, M.**, Binding of testosterone and hydrocortisone to rat tissue histones, *J. Mol. Biol.*, 22, 411, 1966.
22. **Berkoff, C. E. et al.**, Substructural analysis in drug design, in *Proc. Symp.*, Tichý, M., Ed., Akademiai Kiadó, Budapest, 1976, 41.
23. **Junkmann, K. and Suchowsky, G.**, Untersuchung anabol wirksamer Steroide, *Arzneim-Forsch.*, 12, 214, 1962.
24. **Dorfman, R. I. and Kincl, F. A.**, Relative potency of various steroids in an anabolic-androgenic assay using the castrated rat, *Endocrinology*, 72, 259, 1963.
25. **Dorfman, R. I.**, Antiandrogens, *Methods Hormone Res.*, 2A, 221, 1969.

26. **Stárke, L.**, Antiandrogens, *Čs. Farmacie,* 19, 116, 1970.
27. **Goto, G. et al.**, Synthesis and antiandrogenic activity of 16β-substituted 17β-hydroxysteroids, *Chem. Pharm Bull.*, 26, 1718, 1978.
28. **Krüskemper, H. L.**, *Anabole Steroide,* 2nd ed., Thieme Verlag, Stuttgart, 1965
29. **Urbánková, I.**, Structure and Biological Activity of Androgens, Dissertation, Prague, 1982, 133 p.
30. **Urbánková, I.**, The substitution influence on the activity of androgens and anabolics, *Čs. Farmacie,* 33, 331, 1984.
31. **Hanč, O. and Urbánková, I.**, Steroid hormones and QSAR, *Čs. Farmacie,* 37, 90, 1988.
32. **Henry, D. and Block, J. H.**, Pattern recognition of steroids using frequent molecular connectivity, *J. Pharm. Sci.*, 69, 1030, 1980.
33. **Henry, D. and Block, J. H.**, Steroid classification of discriminant analysis using fragment molecular connectivity, *Eur. J. Med. Chem.*, 15, 133, 1980.
34. **Kier, L. B. and Hall, L. H.**, *Molecular Connectivity in Chemistry and Drug Research,* Academic Press, New York, 1976.
35. **Kier, L. B. and Hall, L. H.**, The nature of structure-activity relationships and their relation to molecular connectivity, *Eur. J. Med. Chem.*, 12, 307, 1977.
36. **Henry, D. R. and Block, H. J.**, Classification of drugs by discriminant analysis using fragment molecular connectivity values, *J. Med. Chem.*, 22, 465, 1979.
37. **Ghose, A. K. and Crippen, G. M.**, Quantitative structure-activity relationship by distance geometry: quinazolines as dihydrofolate reductase inhibitors, *J. Med. Chem.*, 25, 892, 1982; Use of physicochemical parameters in distance geometry and related three-dimensional quantitative structure-activity relationship: a demonstration using *Escherichia coli* dihydrofolate reductase inhibitors, *J. Med. Chem.*, 28, 333, 1985.
38. **Urbánková, I.**, Distance geometry of androgens and receptors structure, *Čs. Farmacie,* 35, 207, 1986.
39. **Duax, W. L. and Norton, D. A.**, *Atlas of Steroid Structure,* Vol. 1, Plenum Press, New York, 1975.
40. **Chakrabarti, B. et al.**, The crystal and molecular structure of 17β-hydroxy-8α-androst-4-en-3-one, 8-iso-testosterone, *Steroids,* 37, 269, 1981.
41. **Ojasoo, T. and Raynoud, J. P.**, Unique steroid congeners for receptor studies, *Cancer Res.*, 38, 4186, 1978.
42. **Schmit, J. P. et al.**, The androgen receptor binding site, a conformation study of steroids in relation with their affinity, interaction with the D-ring, *J. Steroid Biochem.*, 13, 1387, 1980.
43. **Balaban, A. T.**, *Steric Fit in Quantitative Structure-Activity Relations,* Springer-Verlag, Berlin, 1980.
44. **Free, S. M. and Wilson, J. W.**, A mathematical contribution to structure-activity studies, *J. Med. Chem.*, 7, 395, 1964.
45. **Boček, K. et al.**, Chemical structure and biological activity of p-disubstituted derivatives of benzene, *Experientia,* 20, 667, 1964.
46. **Fujita, T. and Ban, T.**, Structure-activity study of phenethylamines, *J. Med. Chem.*, 14, 148, 1971.
47. **Peřina, Z. and Urbánková, I.**, Structure-activity relations of some steroids by Free-Wilson analysis, presented at 17th Conf. Čs. Pharm. Soc., Písek, Czechoslovakia, April 29, 1987.
48. **Exner, O. et al.**, QSAR in examples, House of Technique ČSVTS, Pardubice, Czechoslovakia, 1981; Correlations in organic chemistry, SNTL, Praha, 1981.

Chapter 11

STRUCTURE-ACTIVITY RELATIONSHIPS OF STEROIDAL NEUROMUSCULAR BLOCKING AGENTS

Zoltán Tuba and Sándor Mahó

TABLE OF CONTENTS

I. INTRODUCTION

The wide biological spectrum characterizing both the natural and the synthetic steroids has made this group of compounds one of the most important classes of biologically active materials. As a results of an almost unprecedented number of structural modifications, it has become possible to introduce highly active, specific, dependable, and last but not least economically producible derivatives into therapy. In many cases, the biological effects of the synthetic products have substantially surpassed those of the natural compounds both quantitatively and qualitatively. The literature shows that there is not one of the carbon atoms of the sterane structure that has remained unmodified in order to achieve a favorable biological effect.

The replacement of one or more carbon or hydrogen atoms of a steroid molecule with nitrogen or other heteroatoms often results in useful alterations to its biological properties.

For instance, various aza- and aminosteroids are reported to possess antibacterial, antifungal, hypocholesterolemic, and neuromuscular blocking activity and are known to inhibit the enzyme 5α-reductase in the conversion of testosterone to dihydrotestosterone.

The present survey covers a brief summary of the structure-activity relationships of quaternary amino- and azasteroids which possess high neuromuscular blocking activity and some of them are clinically useful agents.

II. MECHANISM OF ACTION OF THE NONSTEROIDAL TUBOCURARINE AND SUCCINYLCHOLINE

The first use of neuromuscular blocking agents (muscle relaxants) in anesthesiological practice in 1942 revolutionized the practice of anesthesia and started the modern era of surgery. The drug to be introduced as an adjuvant to anesthesia was a crude but standardized preparation called "Intocostrin" containing tubocurarine.

The chemical neurotransmission from nerve to muscle is mediated by acetylcholine which is synthetized and stored in the axon terminals of the motor nerve. Acetylcholine is released by nerve impulses; it diffuses across the junctional gap and stimulates the muscle fibers by acting on specific receptor sites on the postjunctional membrane of the motor endplates.[1] The action of acetylcholine at this site is traditionally classified, as nicotine-like or nicotinic receptor. Neuromuscular blocking agents are traditionally classified into two groups:

1. The depolarizing blocking agents (e.g., succinylcholine)
2. The nondepolarizing acetylcholine antagonists (also called competitive neuromuscular blocking drugs, e.g., tubocurarine).

Muscle relaxation as a result of block by depolarization became a practical reality in anesthetic practice. Succinylcholine is essentially the only neuromuscular blocking drug of the depolarizing type that is still in practice in Europe and the U.S. The drug is capable of producing complete muscle relaxation of rapid onset (one circulation time) and lasting only a few minutes. This characteristic makes it suitable for intubation and its use in anesthetic practice.

Nondepolarizing acetylcholine antagonists (e.g., tubocurarine) are also often described as competitive neuromuscular blocking drugs. These drugs combine with the same receptors as the endogenous agonist, acetylcholine, and inhibit the receptor-agonist interaction. Until recently, tubocurarine was believed to be a bisquaternary compound, but one of the nitrogens in the molecule is now known to be tertiary (1970). At body pH, the nitrogen is, however, protonated in most molecules so that tubocurarine does possess two positively charged centers. Very recently, it was observed that tubocurarine, besides its postsynaptic effect, has a presynaptic effect reducing the release of acetylcholine.[2,3]

According to Cecil Gray, the introduction of curare had the same importance for anesthesiology as the introduction of antisepsis had for surgery.[4]

Since the discovery of neuromuscular blocking properties of curare (D-tubocurarine), plenty of new nonsteroidal and steroidal compounds were synthetized and introduced for clinical use as muscle relaxant.

III. CLINICAL REQUIREMENTS

It is well known that muscle relaxants — neuromuscular blocking agents — with different duration of action from 5 to 10 min to several hours are needed in clinical practice, depending on the duration of manipulation.

Intubation and similar procedures should have rapid onset (1 to 2 min) giving complete paralysis for 5 to 10 min and complete recovery in 10 to 20 min. Somewhat longer duration of action is required for more general surgical work. Many operations are complete within 20 to 30 min so that a second group of compounds with a duration of some 15 to 20 min is required for general surgery. A third group of long-acting drugs similar to those presently used but devoid of cardiovascular effect are required for the relaxation of skeletal muscles in abdominal and thoracic surgery.

Other uses of neuromuscular blocking agents include the control of convulsions in tetanus and in electroconvulsive therapy of psychiatric disorders to prevent muscle and tone damage.

All neuromuscular blocking agents in clinical use act either by depolarizing the motor endplate or by competing with the chemical transmitter acetylcholine at the neuromuscular junction.

What are the main features (characteristics) of an ideal neuromuscular blocking agent?

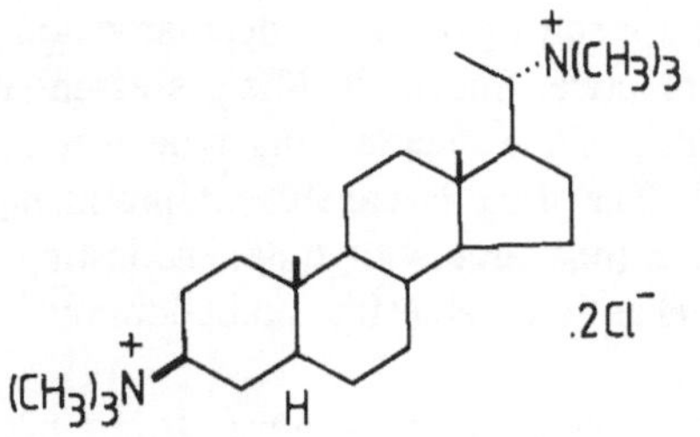

FIGURE 1. The chemical structure of malouétine chloride.

- Nondepolarizing (competitive) mechanism of action and hence it is capable of rapid reversal by anticholinesterases.
- Freedom from undue toxicity and side effects
  - No vagolitic activity
  - No histamine liberation
  - No ganglion blocking activity
- High potency
- High selectivity
- Absence of cumulative properties
- Lack of active metabolites
- Predictability of response
- Freedom from undesirable interaction with other anesthetics and substances used as adjuncts to anesthesia or surgery
- Optimum duration of action for the required use
- Stability in aqueous solution

IV. FROM MALOUÉTINE CHLORIDE TO CONESSINE DERIVATIVES

A. MALOUÉTINE CHLORIDE

The discovery of the curarizing properties of the bisquaternary steroidal alkaloid, malouétine chloride, 3β,20α-bisdimethylamino-5α-pregnane bismethochloride[5] prompted the synthesis of several quaternary ammonium analogues where the steroid nucleus has been used as the supporting moiety (Figure 1). Malouétine and its stereoisomers possess similar potency to tubocurarine but produce hypotension when injected.

Within the steroid group, it is known too that the trimethylammonium salts derived from the alkaloids funtumine (3α-amino-5α-pregnane-20-one) and funtumidine (3α-amino-5α-pregnan-20-ol) exhibit nondepolarizing neuromuscular-blocking activity.[6]

B. DIPYRANDIUM AND ITS ANALOGUES

Alauddin et al.[7] studied substitution of the quaternary ammonium heads at 3α and 17α positions of diamino-5α-androstane derivatives (Table 1).

TABLE 1
The Neuromuscular Blocking Potency of $3\alpha,17\alpha$-Diamino-5α-Androstane Derivatives in the Cat

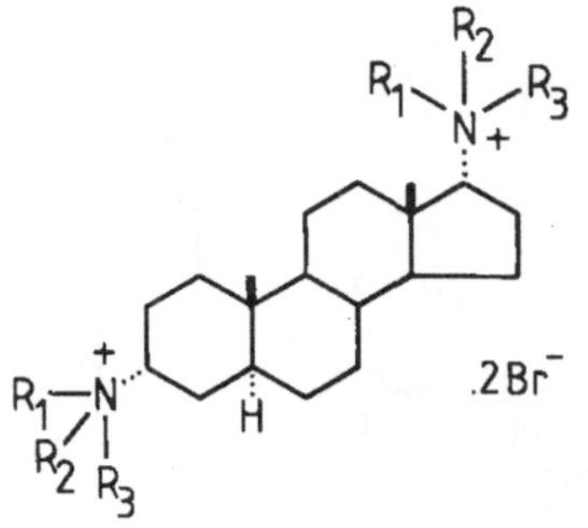

Compound no.	R_1	R_2	R_3	Molar blocking potency (tubocurarine = 1.0)
1	Me	Me	Me	0.27
2	Me	Et	Me	0.84
3	Et	Et	Me	0.87
4	Et	Et	Et	0.46

Note: The most potent compounds are (**2**) and (**3**).

The synthesis of $3\alpha,17\alpha$-dipyrrolidine-1′-yl-5α-androstane and its bisquaternary salts have been described by Davis and others;[8] neuromuscular blocking properties of stereoisomeric androstane-3,17-bisquaternary ammonium salts were studied by Bamford et al.[9] Relaxant action of dipyrandium chloride in man was described by Mushin and Mapleson[10] (Table 2).

The authors have studied eight 3,17-bisquaternary androstane derivatives forming a complete series of all possible stereoisomers at positions 3, 5, and 17 (Table 2). All four 5α-androstanes studied were very active, but the 3α-isomers were 1.5 to 2.3 times less potent than the corresponding 3β-compounds. These results agree well with those of Khoung and Pinto-Scognamiglio,[11] who synthesized a series of 5α-pregnanes based on the alkaloid malouétine. They observed that there was little difference in potency within their series, but that the 3α-isomers were 1.3 times less potent than the corresponding 3β compounds.

In the 5β-androstane series the 3α-members were approximately ten times less potent than the corresponding 3β-members and the 3β-members were 1.1 to 1.4 times less potent than the most active members of the 5α-series. Thus, in both 5α- and 5β-series, the most potent members possessed the 3β configuration and the least potent members possessed the 3α configuration.

In contrast, inversion of the basic center at the 17-position had little effect on potency in either the 5α- or the 5β-series, even when the two quaternary

TABLE 2
The Neuromuscular Blocking Action of the Stereoisomeric Androstane-3,17-Bisquaternary Ammonium Salts on the Sciatic Nerve Tibialis Muscle Preparation of the Cat or Monkey

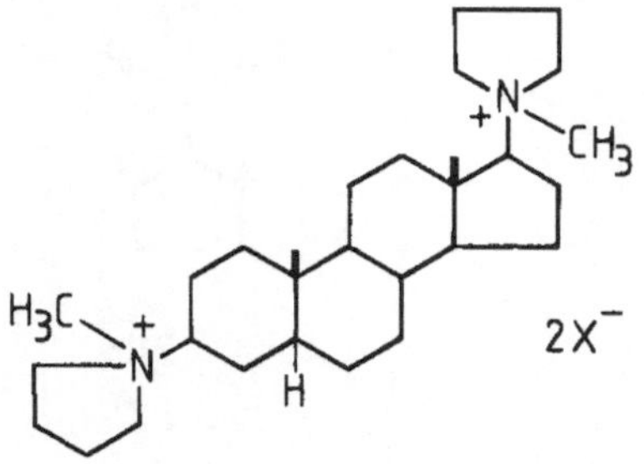

		Configuration			Relative molar potency (dipyrandium = 1)		
Compound no.	**X**	**3**	**17**	**5**	**Cat**	**Monkey**	**Interonium distance (pm)**
1	Cl	β	β	α	1.00	1.00	1100
2	I	β	α	α	1.00	0.92	1060
3	I	α	β	α	0.56	0.86	1040
4	I	α	α	α	0.43	0.76	960
5	I	β	β	β	0.91	1.25	1000
6	I	β	α	β	0.67	0.79	930
7	I	α	β	β	0.09	0.08	990
8	I	α	α	β	0.09	0.07	870
D-Tubocurarine	Cl				1.31	1.53	

The interonium distances were measured from Dreiding models of the stereoisomers.

From Bamford, D. G., Biggs, D. F., Davis, M., and Parnell, E. W., *Br. J. Pharmacol. Chemother.*, 30, 194, 1967. With permission.

heads were on opposite sides of the ring system and so hindered by the rigid skeleton that simultaneous attachment of both to the receptors seemed unlikely.

As far as the significance of interonium distance is concerned the authors found no general relationships between potency and interonium distance in these steroid derivatives. Thus, compounds (**5**) and (**7**), with a similar interonium distance, showed a tenfold difference in potency and compounds (**7**) and (**8**) where there was an interonium distance difference of 120 pm where equipotent. This would suggest that the configuration of the quaternary centers, and the shape and nature of the structure joining them, are at least as important as interonium distance in determining neuromuscular blocking potency.

The duration of action of the stereoisomers did not differ significantly from that of dipyrandium. This suggested that in the interaction with the

TABLE 3
The Relative Neuromuscular Blocking Activities of 3β,17β-Dipyrrolidin-1′-yl-5α-Androstane Bismethoiodide (1) and some Related Monoquaternary Salts in the Cat

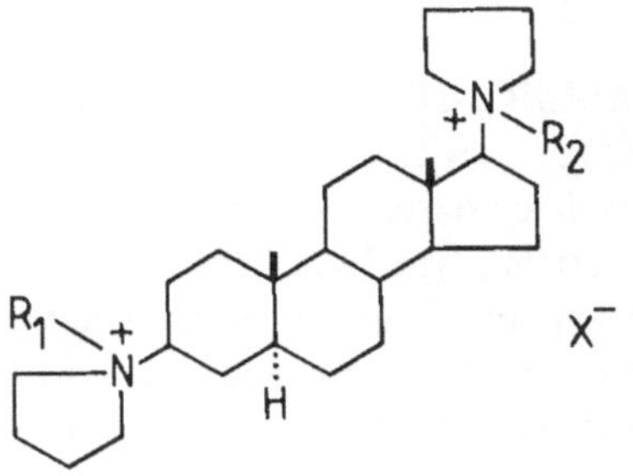

Compound no.	Configuration 3	17	R_1	R_2	X	Relative molar potency
1	β	β	Me	Me	2Cl	1.00
2	β	β	—	—	2Cl	0.04
3	β	β	Me	—	2Cl	1.07
4	β	β	—	Me	2I	0.22
5	β	=O	Me	—	I	0.025
6	=O	β	—	Me	I	0.0013
7	β	–OH	Me	—	I	0.0066
8	βOH	β	—	Me	I	0.0081
9	–H	β	—	Me	Cl	<0.007
10	–H	α	—	Me	Cl	<0.007
D-Tubocurarine						1.31

From Bamford, D. G., Biggs, D. F., and Parnell, E. W., *J. Pharm. Pharmacol.*, 23, 595, 1971. With permission.

receptor, position 3 was more important in determining potency in this series of steroidal neuromuscular blocking agents than position 17.

To test this hypothesis Davis et al.[12] synthetized several monoquaternary salts of 3- or 17-pyrrolidin-1′-yl-5α-androstane, including the 3β- and 17β-monoquaternary salts corresponding to dipyrandium chloride and compared their potency with dipyrandium chloride and with the corresponding nonquaternized dihydrochloride[13] (Table 3).

All the compounds had a nondepolarizing mechanism of action, with a similar duration of action in the cat.

A comparison of the results for compounds (**3**) and (**4**), and for (**5**) and (**6**) showed that the 3β-monoquaternary salts were more potent than the corresponding 17β-salts. This would be predicted from the study of the stereoisomers of the parent compound, dipyrandium chloride (**1**).

In addition, comparison of compounds (**3, 5**) and (**7**) with dipyrandium chloride demonstrates that, in this series of 3β-quaternary salts, potency is maximal when there is also a basic group in position 17, whether quaternized or not, and is reduced by a factor of 40 to 150 when there is a 17-keto or 17β-hydroxy group.

As would be expected the epimeric compounds (**9**) and (**10**), lacking any basic center in position 3, were both inactive.

C. QUATERNARY AMMONIUM DERIVATIVES OF CONESSINE

The pharmacological properties of quaternary ammonium compounds derived from conessine or related steroids which possess neuromuscular blocking activity of the nondepolarizing type have been examined by Busfield et al.[14] The activities of these compounds in cats are summarized in Table 4.

The conessine-type quaternary compounds are short-acting neuromuscular blocking agents of nondepolarizing type with rapid onset of action and low cumulation propensity.

In the 3N-monoquaternary series potency increased in the order of methyl, ethyl, butyl (**1, 2, 3**). Potency was markedly reduced, however, when the cation was differently constituted (**8**).

The bisquaternary compounds all possess an identical interonium distance (1010 pm, Dreiding stereomodels), an inflexible steroidal nucleus, and similar potencies and duration of action.

The similarity in activity of these particular mono- and bisquaternary steroids supports the one-point attachment theory of Loewe and Harvey.[15] In N,N′-dimethylconessine the cation head attached to the 3-position is in the same plane as rings A, B, C, and D of the steroid nucleus; both the double bond at 5, 6, and the possibility of steric hindrance between the 3*N*-methyl groups and the angular group at carbon-10 confer the chair conformation of rings A and B of the conessine skeleton. Ring E, which contains the second nitrogen at position 20, is at right angles to the plane of the cyclopentanoperhydrophenantrene nucleus and, if the one-point attachment theory is correct, it could be expected to point away from the receptor surface and therefore be without influence on activity.

In man N,N′-dimethylconessine caused rapid onset of nondepolarizing neuromuscular block without change in blood pressure or heart rate and without the development of secondary curarization, but the duration of effect was three times longer than with a corresponding dose of succinylcholine.

V. PANCURONIUM BROMIDE AND ITS ANALOGUES

The incorporation of biologically active group into a naturally occurring structure is an interesting approach to new pharmacological agents. Savage and co-workers[16,17] synthetized a series of 2,3-amino-alcohol-type androstane derivatives since such molecular fragments occur in endogenous neurotransmitter such as norepinphrine and acetylcholine (Figure 2).

TABLE 4
Structures and Neuromuscular Blocking Potencies of Conessine Derivatives Assessed on Anesthetized Cat Tibialis Preparation

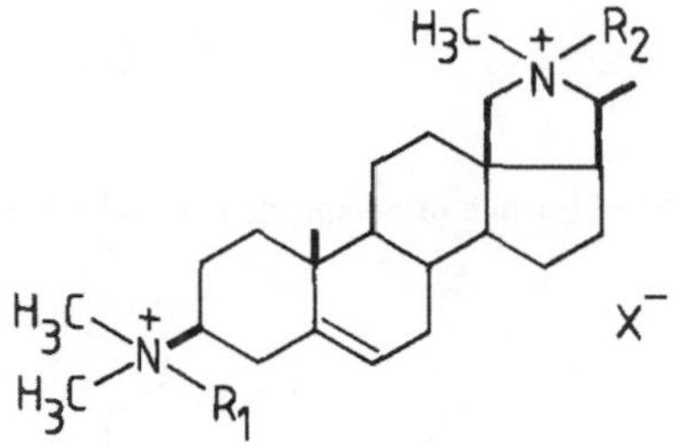

Compound no.	R_1	R_2	X	Relative potency (tubocurarine = 1)
1	Me		I	0.12
2	Et		I	0.33
3	Bu		I	0.64
4	MeCl		Br	0.22
5	CH_2COOEt		Br	0.10
6	CH_2Ph		Br	0.58
7	CH_2CH_2OPh		Br	0.45
8	O			0.02
9[a]	Me	Me	2I	0.62
10	Et	Et	2I	0.79
11	Pr	Pr	2I	0.47
12	i-Pr	Me	2I	1.61
13	Bu	Me	2I	1.54
14	Oc	Me	2I	0.25
15	Al	Al	2Br	0.67
16	Pg	Pg	2Br	0.56
17	Cr	Cr	2Br	1.00
18	H	Me	2Cl	0.35
19	OMe	Me	2I	0.61
20	MeCl	Me	Br I	0.71
21	CH_2OMe	H	2Cl	0.21
22	CH_2CH_2OMe	CH_2CH_2OMe	2I	0.45
23	CH_2COOEt	CH_2COOEt	2Br	0.97
24	CH_2Ph	CH_2Ph	2I	0.51
25	CH_2Ph	Me	2Cl	1.09
26	CH_2CH_2OPh	Me	Br I	1.37

Note: MeCl: chloromethyl; Oc: octyl; Al: allyl; Pg: propargyl; Cr: ceryl.

[a] *N,N'*-Dimethylconessine.

From Busfield, D., Child, K. J , Clarke, A. J., Davis, B., and Dodds, M. G., *Br. J. Pharmacol. Chemother.*, 33, 609, 1968. With permission.

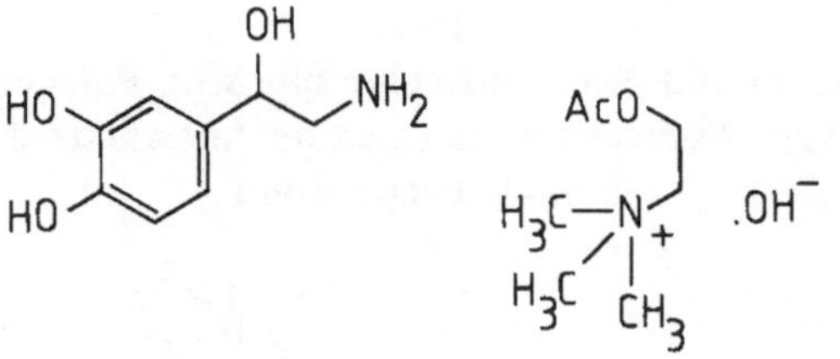

FIGURE 2. Structure of norepinephrine and acetylcholine.

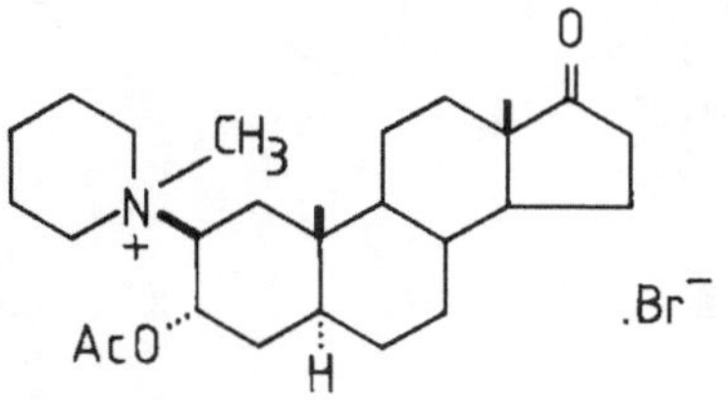

FIGURE 3. Molecular structure of 3α-acetoxy-2β-piperidino-5α-androstane-17-one methobromide.

Lewis et al.[18] studied the pharmacology of 2β- and 3α-steroidal monoquaternary salts having androstane or pregnane skeletons and related in structure to acetylcholine by possession of an oxygen function on the carbon atom next but one to the quaternary nitrogen atom. All compounds displayed typical nondepolarizing neuromuscular blocking activity, the duration of block being significantly less than that for d-tubocurarine in cat and in hen. The most active monoquaternary compound the 3α-acetoxy-2β-piperidino-5α-androstane-17-one methobromide being one sixteenth as active on a molar basis in the cat as d-tubocurarine (Figure 3).

Incorporation of an acetylcholine-like structural unit into the D-ring at 16- and 17-position enhanced the neuromuscular blocking activity of the monoquaternary compound. In 1965 Hewett and Savage[19] applied for a patent with title: "Improvements in or Relating to a New 2β,16β-Diamino-Androstanes".

The crystal and molecular structure of the most potent compound (3α,17β-diacetoxy-2β,16β-dipiperidino-5α-androstane dimethobromide) has been determined by a single-crystal X-ray analysis.[20]

In 1973 Buckett et al.[21] published a substantial paper about the synthesis and biological activity of 2,16-bis-amino-androstane derivatives which have acetylcholine-like structural fragments both in the A- and D-rings.

In the tables the potency and duration of action of the most potent compounds are summarized (Table 5 to 9).

For high potency it is essential to have two nitrogen atoms in the molecule, since the monoquaternary salts (**43**) and (**44**) which lack a tertiary nitrogen

TABLE 5
Bis- and Monoquaternary Salts of 2β,16β-Dipiperidino-3,17-Dioxy-Androstane Derivatives

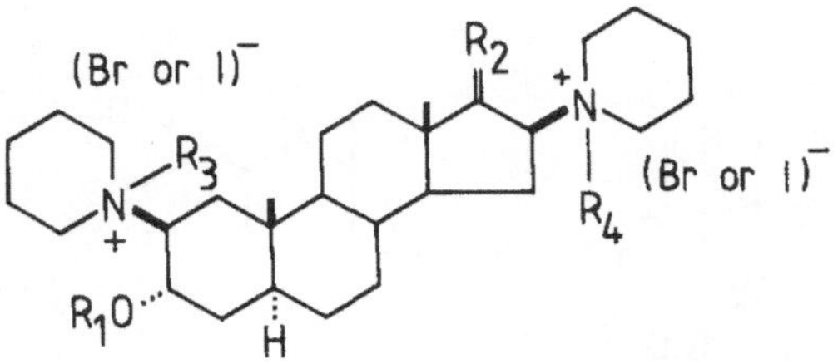

Compound no.	R_1	R_2	R_3	R_4	Anion	Potency[a,b]	Duration[a,b]
1	H	O	Me	Me	Br	0.67	0.47
2	H	β-OH,H	Me	Me	Br	0.55	0.53
3	H	β-OH,H	Al	Al	Br	2.50	1.00
4	H	β-OH,H	Pg	Pg	Br	0.77	0.44
5	H	β-AcO,H	Me	Me	Br	5.15	1.00
6	HCO	β-HCO,H	Me	Me	Br	1.30	0.46
7	Ac	O	—[c]	Me	Br	0.39	0.44
8	Ac	O	Me	Me	Br	1.53	0.37
9	Ac	β-OH,H	Me	Me	Br	1.72	0.49
10	Ac	β-OH,H	Me	Me	I	1.29	0.32
11	Ac	β-OH,H	—[c]	Me	Br	0.40	0.45
12	Ac	β-AcO,H	Me	Me	Br	9.41	1.15
13	Ac	β-AcO,H	Me	Me	I	15.0	1.1
14	Ac	β-AcO,H	—[c]	Me	Br	6.0	0.77
15	Ac	β-AcO,H	—[c]	Al	Br	3.44	0.89
16	Ac	β-AcO,H	Me	Al	Br	7.23	1.16
17	Ac	β-AcO,H	Pg	Pg	Br	6.31	0.81
18	Ac	β-EtCOO,H	Me	Me	Br	5.60	0.85
19	EtCO	β-EtCOO,H	Me	Me	Br	7.10	0.95
20	Me_3CCO	β-Me_3CCOO,H	Me	Me	I	3.80	1.47
21	Me_3CCO	β-OH,H	Me	Me	I	1.70	0.50
22	PhCO	β-OH,H	Me	Me	I	0.84	0.72
23	PhCO	β-PhCOO,H	Me	Me	Br	1.00	>3.0
24	Ac	H,α-AcO	Me	Me	Br	6.15	0.75

[a] Intravenously in anesthetized cat sciatic-gastrocnemius preparation.
[b] Tubocurarine = 1.
[c] Not quaternized.

atom are almost inactive. At least one of these nitrogen atoms should be quaternized. The dimethobromide (**12**) and monomethobromide (**14**) are very potent.

The potency was decreased when the 16β-piperidino substituent of (**12**) replaced with dimethylamino group (**29**). The compound (**42**) has two acetylcholine-like units, and also has a low potency. Comparing the morpholino

TABLE 6
Dimethohalides of other 2β,16β-Diamino-3,17-Dioxy-5α-Androstane Derivatives

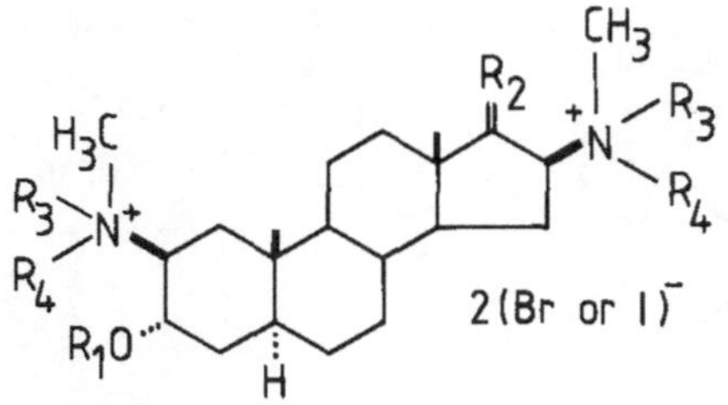

Compound no.	R_1	R_2	NR_3R_4	Potency[a,b]	Duration[a,b]
25	H	O	Morpholino	0.06	0.80
26	H	β-OH,H	Morpholino	0.01	
27	H	β-OH,H	Pyrrolidino	0.17	0.32
28	Ac	β-OAc,H	Morpholino	3.44	0.96
29	Ac	β-OAc,H	—[c]	0.40	1.05
30	H	H,α-OH	Morpholino	<0.10	
31	Ac	H,α-OAc	Pyrrolidino	5.9	1.0
32	Ac	β-OAc,H	Morpholino[d]	0.95	0.95

[a] Intravenously in anesthetized cat sciatic-gastrocnemius preparation.
[b] Tubocurarine = 1.
[c] 2β-Piperidino-16β-dimethylamino.
[d] 2α,3β.

TABLE 7
Dimethohalides of 2β,17β-Diamino-5α-Androstane-3α,16β-Diols and Diacetates

Compound no.	Substituents 2β	17β	3α,16β	Potency[a,b]	Duration[a,b]
33	Pyrrolidino	Pyrrolidino	OAc	0.13	1.5
34	Piperidino	Pyrrolidino	OAc	0.70	1.1
35	Piperidino	Piperidino	OH	0.15	0.62
36	Piperidino	Piperidino	OAc	0.59	0.73
37	Piperidino	Dimethylamino	OAc	0.57	1.0

[a] Intravenously in anesthetized cat sciatic-gastrocnemius preparation.
[b] Tubocurarine = 1.

derivatives (**25, 26, 28**) with the corresponding piperidino derivatives (**1, 2, 12**) the latter are significantly more potent. The 2β-ammonio substituent may confer greater potency than one of a 2α-configuration as (**32**) is less active than its 2β-epimer (**28**).

TABLE 8
Dimethohalides of 2β,17β-Diamino-5α-Androstane-3α-ol Acetates and 3α,17β-Diamino-5α-Androstane-2β-ol Acetates

Compound	Substituents 2β	3α	17β	Potency[a,b]	Duration[a,b]
38	Piperidine	OAc	Pyrrolidine	0.20	0.70
39	Piperidine	OAc	Piperidine	<0.10	
40	OAc	Pyrrolidine	Pyrrolidine	0.48	0.65
41	OAc	Piperidine	Pyrrolidine	1.15	0.75

[a] Intravenously in anesthetized cat sciatic-gastrocnemius preparation.
[b] Tubocurarine = 1.

TABLE 9
Miscellaneous

Compound no.	Name	Potency[a,b]	Duration[a,b]
42	3β,17β-Di-(methyl-2′-acetoxyethylamino)-5α-androstane dimethobromide	0.37	0.44
43	17β-Acetoxy-16β-piperidino-5α-androstane methobromide	<0.10	
44	3α-Acetoxy-2β-piperidino-5α-androstane methobromide	<0.10	

[a] Intravenously in anesthetized cat sciatic-gastrocnemius preparation.
[b] Tubocurarine = 1.

From Buckett, W. R., Hewett, C. L., and Savage, D. S., *J. Med. Chem.*, 16, 1116, 1973. With permission.

The diesters yielded the most potent compounds with the diacetate (**12**) > dipropionate (**19**) > dipivalate (**20**) > dibenzoate (**23**). Potency appears to decrease with increasing liphophilicity. In the less potent alcohol series the 17-monoacetate (**5**) which have half the potency of (**12**) is > 3-monoacetate (**9**) > 3,17-diol (**2**) = the 3α-ol-17-one (**1**) which has less than 10% the potency of (**12**) and the 3-monopivalate (**21**) = 3-monoacetate (**9**) which is > 3-monobenzoate (**22**). With respect to duration of action the long-acting dibenzoate (**23**) ⩾ diacetate (**12**) > 17-monoacetate (**5**) > 3-monobenzoate (**22**) > 3-monoacetate (**9**) = 3,17-diol (**2**). The potency and prolonged action are associated with the presence of a 17-ester.

How do the quaternizing groups influence the activity? In the 3,17-diacetates of this series the most potent are dimethyl (**12**) > methyl, allyl (**16**) > dipropargyl (**17**) = monomethyl (**14**) > monoallyl (**15**). In the corresponding diols of this series the order is the diallyl (**3**) > dipropargyl

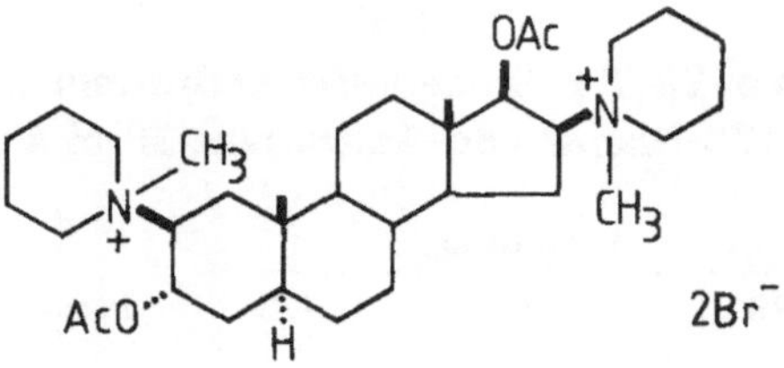

FIGURE 4. Molecular structure of pancuronium bromide.

(**4**) = dimethyl (**2**). This shows that the optimal N-alkyl group in the 17-ester series is methyl, whereas in the 17-alcohol series the most potent is compound (**3**).

As far as the position of ammonio-groups is concerned the most potent derivatives belong to the 2β,16β-series in comparison with that of 2β,17β-series.

As it is seen from the biological data the most potent compound is the 3α,17β-diacetoxy-2β,16β-dipiperidino-5α-androstane dimethobromide (pancuronium bromide) (Figure 4).

Pancuronium bromide has proved a clinically useful agent of medium duration of action with moderate cardiovascular side effects, e.g., a vagolytic effect and a noradrenaline reuptake block in the clinical dose range. The nondepolarizing pancuronium elevates arterial blood pressure and increases heart rate.[22-41]

The neuromuscular blocking activities of pancuronium bromide and other analogues were studied in more detail by Durant et al.[42]

The pharmacological data are given in Table 10.

The salient member of the series is 3α,17β-diacetoxy-2β,16β-dipiperidino-5α-androstane-16*N*-monomethobromide,[43-65] ORG NC-45, (vecuronium bromide) which is highly selective in blocking neuromuscular transmission in that a dose approximately 60 times greater than the neuromuscular blocking dose was required to block responses to vagal stimulation (Figure 5).

In contrast, in doses sufficient to produce neuromuscular block pancuronium simultaneously blocked responses to vagal stimulation.

The minor molecular modification compared to pancuronium produced an increased cardiovascular stability and shortened the duration of neuromuscular blocking action. However, it lacks the rapid onset desirable for an intubating agent.

In the 1980s vecuronium bromide was introduced into anesthesiological practice.

A further step forward was the development of the experimental compound ORG-9426.[66] This compound was recently released for clinical testing[67-74] (Figure 6).

A glance at the molecular structure reveals the loss of both the acetylcholine-like fragments which characterize the molecular structure of pancu-

TABLE 10
Pancuronium, Vecuronium, and Analogues

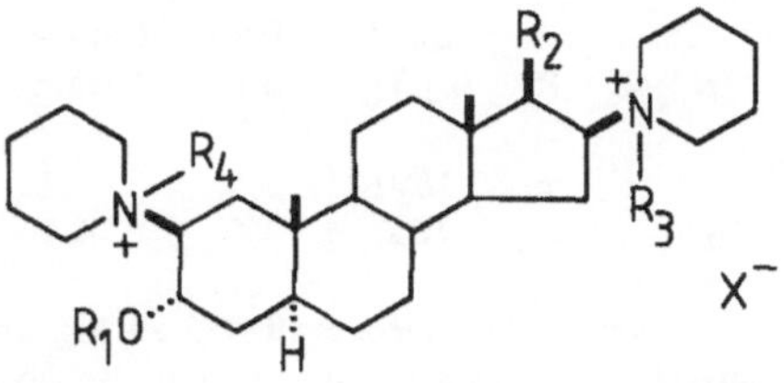

Compound no.	X-	R_1	R_2	R_3	R_4	ED_{50}[t] (μg/kg)	Onset[t] (min)	Recovery[t] (min)	ED_{50}[v] (μg/kg)	V/N
1	2Br	a	b	c	c	34	4.3	8.8	62	1.82
2	Br	a	b	c	—	38	3.5	3.9	2145	56.4
3	2Br	d	b	c	c	59	1.7	5.1	263	4.45
4	Br	d	b	c	—	39	2.8	4.1	315	8.00
5	Br	d	b	e	—	17	4.6	4.7	280	16.4
6	Br	d	b	f	—	73	2.5	6.0	383	5.2
7	Br	d	b	g	—	232[a]	1.0	2.2	528	2.3[a]
8	2Br	h	i	c	c	87	2.4	6.6	564	6.5
9	Br	h	i	c	—	520	1.1	2.8	251	0.48
10	2Br	d	i	c	c	544	1.3	2.5	296	1.83
11	Br	d	i	c	—	544	0.5	2.7	226	0.4
12	2Br	j	k	c	c	50	2.0	4.3	203	4.1
13	Br	j	k	c	—	405	1.5	3.6	265	0.61
14	2Br	d	k	c	c	250	1.4	1.4	155	0.44
15	Br	d	k	e	—	260	1.3	3.5	215	0.82
16	Br	d	k	f	—	670	1.7	2.1	480	0.71
17	Br	d	k	g	—	790	1.7	2.4	376	0.47
18	2Br	l	m	c	c	161	0.9	1.9	328	2.03
19	Br	l	m	c	—	2000			347	0.17[a]
20	2Br	d	m	c	c	511	2.6	3.1	220	0.43
21	2Br	a	n	c	c	230	1.5	4.7	250	1.1
22	2I	a	o	c	c	240	1.1	3.8	230	0.95
23	2Br	a	d	c	c	121	1.6	3.7	125	1.03
24	Br	d	n	c	—	1850	1.4	3.0	250	0.13
25	Br	a	o	c	—	1100	1.4	5.1	400	0.36

TABLE 10 (continued)
Pancuronium, Vecuronium, and Analogues

Compound no.	X-	R_1	R_2	R_3	R_4	ED_{50}^{t} (μg/kg)	Onset^{t} (min)	Recovery^{t} (min)	ED_{50}^{v} (μg/kg)	V/N
26	2I	d	o	c	c	1650	3.5		280	0.16
27	I	d	o	c	—	2200	1.0	4.4	620	0.28

Note: a: acetyl; b: acetoxy; c: methyl; d: hydrogen; e: ethyl; f: allyl; g: propargyl; h: propionyl; i: propionyloxy; j: butyryl; k: butyryloxy; l: isobutyryl; m: isobutyryloxy; n: hydroxy; o: oxygen; t: tibialis; v: vagal blockade; V/N: vagal block/neuromuscular block ratio, tibialis; s: soleus.

From Durant, N. N , Marshall, I. G., Savage, D. S., Nelson, D. J., Sleigh, T., and Carlyle, I. C., *J. Pharm. Pharmacol.*, 31, 831, 1979. With permission.

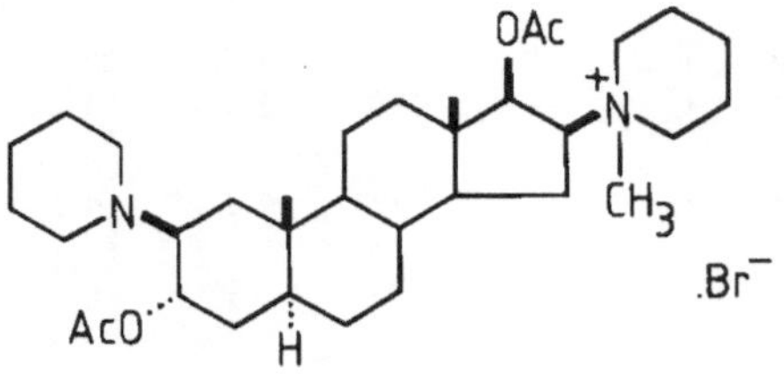

FIGURE 5. Molecular structure of vecuronium bromide.

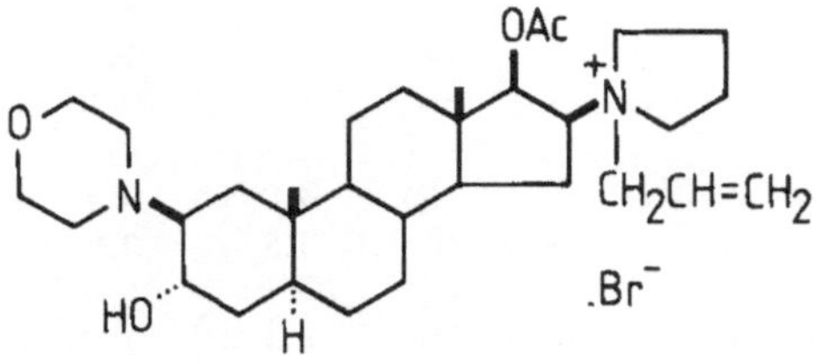

FIGURE 6. Molecular structure of ORG-9426.

ronium. The effects of ORG-9426 were studied in anesthetized cat's and pig's isolated nerve muscle preparations.[69] ORG-9426 was approximately 20% as potent as vecuronium, its onset of action was twice as rapid as that of vecuronium in the cat, and its duration of action was similar to that of vecuronium in both cats and pigs. It blocked the bradycardia produced by vagal stimulation only in doses greater than those necessary to produce neuromuscular block. Ganglion block was seen only at doses several times those producing vagal block. In general the effects of ORG-9426 on the cardiovascular system were slight, whereas a small depressor effect occurred at high doses in the cat. However, recovery and duration values for ORG-9426 were similar to those for vecuronium in both the cat and pig. Thus, the overall time course profile

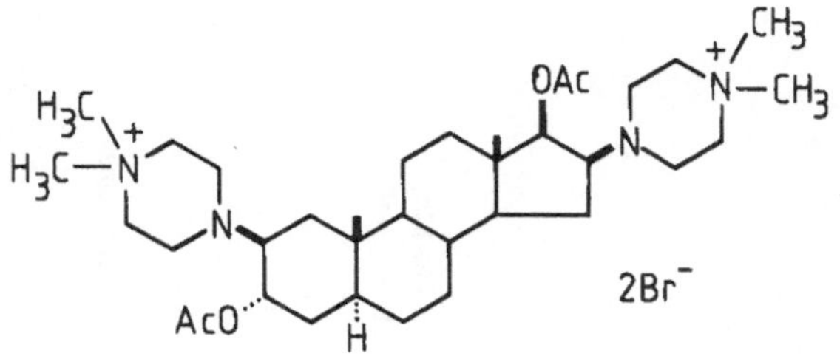

FIGURE 7. Molecular structure of pipecuronium bromide.

of ORG-9426 is that of a drug with medium duration of action, similar to vecuronium, but with a faster onset of action. ORG-9426 may have advantages over existing nondepolarizing neuromuscular blocking agents with respect to rate of development of good intubating conditions, and is stable in aqueous solution.

If the rapid onset is also seen in humans ORG-9426 represents a potential advance in this field.

VI. PIPECURONIUM BROMIDE (RGH-1106) AND ITS ANALOGUES

In 1971 we began a search for pharmacologically active nonhormonal agents based upon readily accessible steroid intermediates. A series of mono- and bisquaternary ammonio-steroids having androstane skeletons has been prepared,[75-82] some of which possessed high neuromuscular blocking activity. One of the series 2β,16β-bis-(4′,4′-dimethyl-1′-piperazino)-3α,17β-diacetoxy-5α-androstane dibromide (pipecuronium bromide) has proved to be a clinically useful agent of medium-long duration of action without cardiovascular side effects (Figure 7).

By changing the dose the duration of action can be regulated within rather wide time intervals from 25 to 90 min. Within these intervals different types of surgical interventions can be performed; therefore, pipecuronium bromide can be considered as a well applicable "practical" muscle relaxant. In the doses mentioned above in balanced anesthesia and in cases of normal renal function even multiple pipecuronium doses do not produce prolongation of effect. According to the clinico-pharmacological investigations performed so far 20 to 30% lower doses of pipecuronium have similar or shorter effect than pancuronium without its cardiovascular side effects. Its application field is identical with that of pancuronium. It can well be applied in the anesthesia of "poor risk" patients and of those having cardiovascular diseases.[83-96]

In this part of the chapter the structure-activity relationships of pipecuronium bromide and its analogues are described (Tables 11 to 21).

The most potent quaternary compounds synthetized in our laboratory are those having piperazino groups at position 2 and 16 (**1, 16, 17**) of the an-

TABLE 11
Comparison of Neuromuscular Vagolytic and Ganglion Blocking Effects of Pipecuronium Bromide and Its Three Metabolites in Anesthetized Cat's Tibialis Anterior Muscle

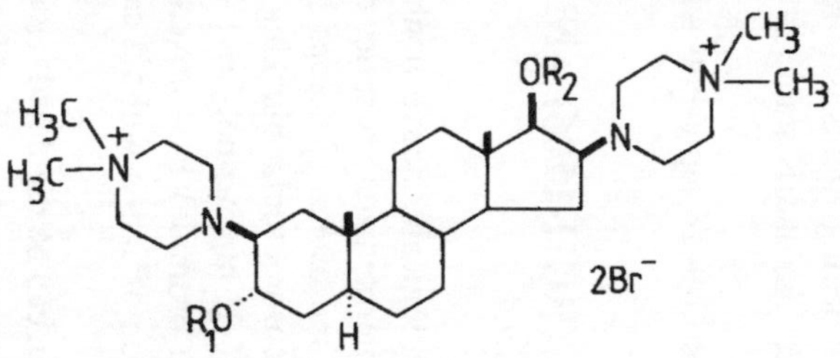

Compound no.	R_1	R_2	Neuromuscular blocking activity: ED_{50} (μg/kg)	ED_{90} (μg/kg)	OT (min)	RR (min)	TDT (min)	Vagolytic effect ED_{50} (μg/kg)	V/N	Ganglion blocking activity at −3 mg/kg
1	Ac	Ac	4.6	8.00	8.0	8.1	30.2	623	135	28
2	H	Ac	11.2	19.0	4.8	7.9	27.3	727	65	41
3	Ac	H	204	374	2.4	7.8	21.8	535	2.6	17
4	H	H	133	330	2.1	4.8	13.3	778	5.9	21

Note: OT: onset time; RR: recovery rate; TDT: total duration time; overall at 90% neuromuscular blockade. V/N: vagal block/neuromuscular block ratio.

TABLE 12
Effects of 2β,16β-Bis Piperazino-Androstane Derivatives

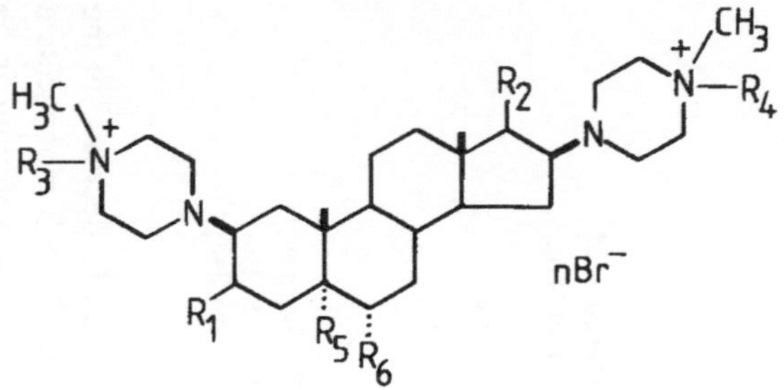

Compound no.	R_1	R_2	R_3	R_4	R_5	R_6	n	Neuromuscular blocking activity: ED_{50} (μg/kg)	OT (min)	D (min)	Vagolytic effect ED_{50} (μg/kg)	V/N
5	αOAc	βOAc	—	Me	H	H	1	27	5.4	17	1200	44
6	αOAc	βOAc	Me	—	H	H	1	100[a]	2.2	16		
7	βOAc	βOAc	Me	Me	H	H	2	34	7.6	28	5500	161
8	αOAc	αOAc	Me	Me	H	H	2	880	3.0		3100	~3.5
9	αOPi	βOPi	Me	Me	H	H	2	600[b]	4.8	24		
10	H	βOAc	Me	Me	H	H	2	230	1.6	11	>2500	~11
11	βBr	βOAc	Me	Me	H	H	2	170	5.2	31	>4000	~24
12	αOAc	βOAc	Pr	Pr	H	H	2	18	9.3	35	1200	66
13	αOAc	αOAc	Al	Al	H	H	2	25	6.3	19	1800	72
14	αOAc	βOAc	Me	Me	OH	H	2	19	6.5	20	2100	110
15	αOAc	βOAc	Me	Me	H	OH	2	45[b]	3.6	24		

Note: OT. onset time; D: duration of block; V/N: vagal block/neuromuscular block ratio; Pi: pivaloyl; Pr: propyl; Al: allyl. [a] = ED_{100}; [b] = ED_{96}.

TABLE 13
2β,6β,16β-Tris-Piperazino-Androstane Derivatives

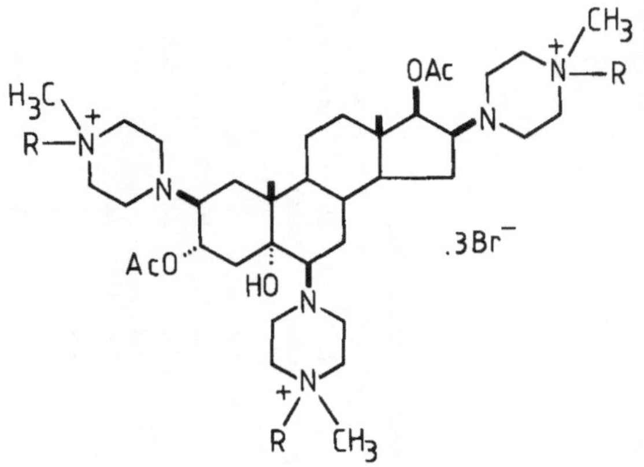

Compound no.	R	ED_{100} (μg/kg)	Duration of action (min)
16	Me	6.8	180
17	Pr	7.1	220

Note: Pr: propyl.

From Tuba, Z., Marsai, M., Bíró, K., Szporny, L., Kárpáti, E., and Szeberényi, Sz., U.S. Patent 4,071,515, 1978; British Patent 1,533,454, 1979.

drostane skeleton. Within the series potency appears to decrease in the following order 3,17-diacetate (**1**) > 17-monoacetate (**2**) > 3-monoacetate (**3**). There is no significant difference in the action of the compounds having quaternary groups such as methyl (**1, 16**), allyl (**13**), or propyl (**12, 17**). The configuration of the piperazino group at position 16 is important namely if it is α (**49**) potency has decreased in comparison with that of β. The configuration of acetoxy group at position 3 does not influence the biological activity (**1, 7**) but if the acetoxy group's configuration at position 17 is a α potency has decreased (**1, 8**).

The monoquaternary salts have the same range of action as the bisquaternary ones. The duration of activity has decreased in the order of tris- (**16, 17**) > bis- (**1**) > monoquaternary derivatives (**5, 6**).

Changing the piperazino group's position attached at ring A, from 2β to 3α (**47**) or 2β to 4β (**48**) potency has decreased.

Changing the 16β-piperazino group's position to 17β, in the D-ring the activity has also decreased.

TABLE 14
2β-Piperazino-16β-Piperidino-Androstane Derivatives

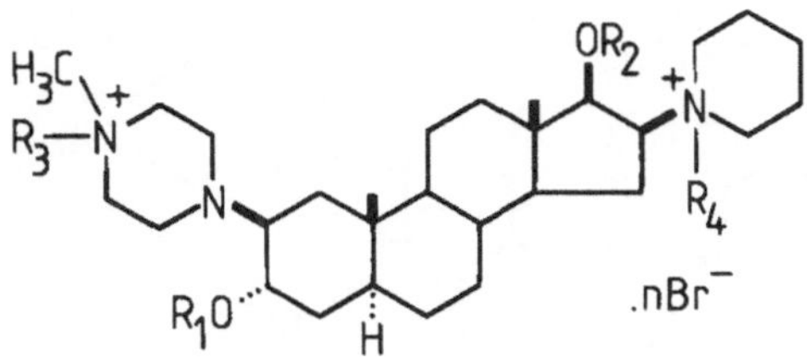

Compound no.	R_1	R_2	R_3	R_4	n	Neuromuscular blocking activity ED_{50} (μg/kg)	OT (min)	D (min)	Vagolytic effect ED_{50} (μg/kg)	V/N
18	Ac	Ac	Me	Me	2	18	6.8	15	1100	61
19	Ac	Ac	—	Me	1	42	6.2	20	700	16.6
20	Ac	Ac	Me	—	1	120	10	33	510	4.25
21	H	H	Me	—	1	1600[a]	2[a]	9[a]		
22	Ac	Ac	Al	—	1	21	11	40	1500	71.4
23	Ac	Ac	Pr	—	1	31	7.5	28	2700	87

Note: OT: onset time; D: duration of block; V/N: vagal block/neuromuscular block ratio.

[a] ED_{90}.

From Tuba, Z., Marsai, M., Bíró, K., Szporny, L., Kárpáti, E., and Szeberényi, Sz., U.S. Patent 4,101,545, 1978; British Patent 1,504,404, 1979.

The safety factor (V/N value) is extremely high in the case of pipecuronium bromide (**1**) and some analogues (**5, 7, 12, 13, 14**).

Attached to the N-methyl-piperazino groups at position 6 and 16 the interonium distance shortened from 1601 pm (pipecuronium bromide determined by X-ray diffraction analysis) to 1185 pm (determined by Dreiding stereomodel). (The latter is not exactly correct caused by the 1,3-diaxial interaction of the 6β-piperazino and the angular methyl groups.) The range of the biological activity has not changed.

How can we explain the high activity of the 6,16-bisquaternary-piperazino derivatives in comparison with that of pipecuronium bromide?

The common feature of the two different quaternary derivatives — 2,16- and 6,16- — is that the onium centers are on the same side — in β-position — of the androstane skeleton. That is why stereochemically there is not much difference between the two potential parts of the molecules. The hydrophility of the two compounds differs, since the 6,16-bis-piperazino derivatives contain a free-hydroxy function at position 5α. As we have seen earlier the interonium distance is variable to a certain extent.

TABLE 15
2β-Piperidino-16β-Piperazino-Androstane Derivatives

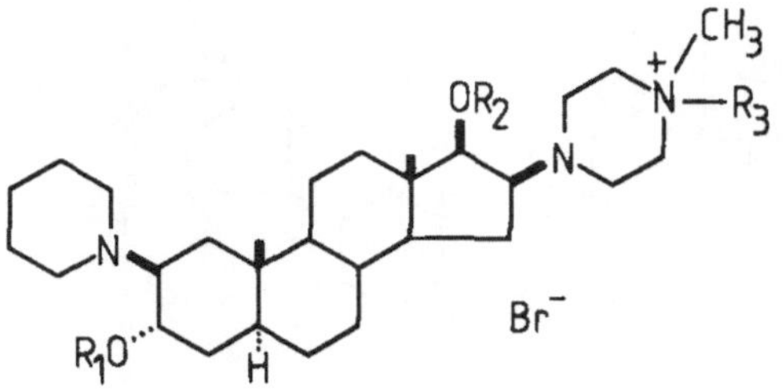

Compound no.	R_1	R_2	R_3	Neuromuscular blocking activity: ED_{50} (μg/kg)	OT (min)	D (min)	Vagolytic effect ED_{50} (μg/kg)	V/N
24	Ac	Ac	Me	51	6.3	18.8	5300	104
25	Ac	Ac	Pr	330	4.0	23	5900	17.8
26	Ac	Ac	Al	220	2.4	10	1000	4.5
27	H	H	Me	700[a]	3[a]	15[a]		

Note: OT: onset time; D: duration of block; V/N: vagal block/neuromuscular block ratio.

[a] ED_{90}.

Later on we have synthetized more 5α-androstane derivatives substituted at position 2 and 16 with different amines (**33, 37**) to get further proof of the significance of interonium distance.

In the course of 1990 we succeeded in synthetizing the "5β-vecuronium bromide" (**43**) and the "5β-pipecuronium bromide" (**44**). Surprisingly they have remarkable effects. The biological data are given in the table. Probably the X-ray diffraction analysis gives some help to explain the high activity; shall we have more information about the receptor site too?

VII. STRUCTURE-ACTIVITY RELATIONSHIPS OF THE TWO MOST POTENT NEUROMUSCULAR BLOCKING AGENTS, PANCURONIUM BROMIDE AND PIPECURONIUM BROMIDE

As it is seen from the X-ray diffraction analysis, the pancuronium bromide ring A conformation changed due to the quaternary center attached to position 2 (Figure 8). The 2β-quaternary-piperidino and the 3α-acetoxy groups both are almost pseudoequatorial due to the twisted-boat conformation of ring A. Buckett et al.[21] assumed that in this preferred conformation, which may be rigid due to the steric compression of ring-A substituents, the acetylcholine-like fragments resemble a specific molecular conformation of the neurohor-

TABLE 16
6,16-Bis-Piperazino-Androstane Derivatives

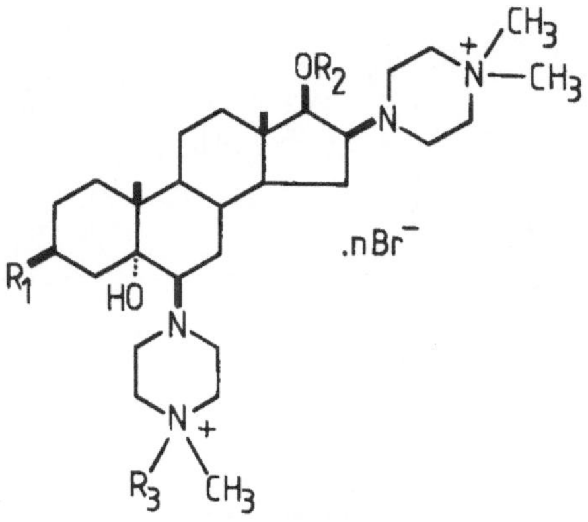

Compound no.	R_1	R_2	R_3	n	Neuromuscular blocking activity: ED_{50} (μg/kg)	OT (min)	D (min)	Vagolytic effect ED_{50} (μg/kg)	V/N
28	OAc	Ac	Me	2	61	4.4	20	6000	98
29	OAc	Ac	—	1	130	4.8	23	>10000	>76
30	H	Ac	Me	2	85	5.2	25	3700	43
31	H	Ac	—	1	410	3.6	16	3800	9.2
32	H	H	Me	2	9000	3.5	17	3100	0.34

Note: OT: onset time; D: duration of block; V/N: vagal block/neuromuscular block ratio.

From Tuba, Z., Horváth, J., Lovasnee-Marsai, M., Riesz, M., Szporny, L., and Kárpáti, E., European Patent 208,497, 1987; European Patent 345,471, 1989.

monal transmitter, acetylcholine. The acetylcholine molecule is known to be released at nerve endings and to be highly specific in inducing muscle contractions. Therefore, it is reasonable to expect that the rigid acetylcholine-like substance such as pancuronium bromide would occupy the transmitters site of action and block neuromuscular transmission.

Starting from this theory how could we explain the structure-activity relationships of the pipecuronium bromide?

Pipecuronium bromide has no direct acetylcholine-like structural units attached to the ring A and D.

The ring-A conformation has not changed, so the configuration of 2β-piperazino and 3α-acetoxy groups are in the original axial positions. The remote nitrogens are quaternized. (The interonium distance is much longer, i.e., 1600 pm, than that of pancuronium bromide which is about 1100 pm.) The potent parts of the molecule closed to the onium centers are more flexible. The structures of the two active compounds are deeply different.

TABLE 17
6β-Piperidino-16β-Piperazino-Androstane Derivatives

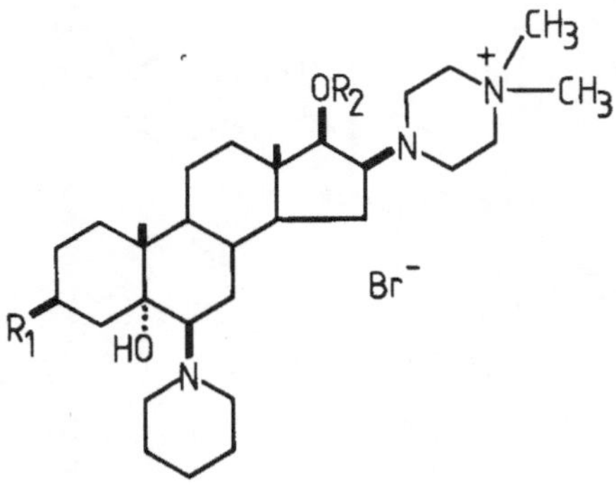

Compound no.	R_1	R_2	Neuromuscular blocking activity ED_{50}	OT	D	Vagolytic effect ED_{50}	V/N
33	OAc	Ac	560	2.3	14	2500	4.5
34	OH	H	10000	25	13	710	0.07
35	H	Ac	870	3.6	12	960	1.1
36	H	H	18000	4.6	15	430	0.02

Note: OT: onset time; D: duration of block; V/N: vagal block/neuromuscular block ratio.

As far as the biological activity is concerned pipecuronium bromide has higher activity than that of pancuronium bromide. The safety factor — the V/N value — as for pipecuronium bromide 131.6 on the contrary is about 2 in pancuronium bromide.

The human applications have proved these advantages and that is why pipecuronium bromide can be applied in the anesthesia of "poor risk" patients and of those having cardiovascular diseases.

More recently we have succeeded in synthetizing such a "pipecuronium bromide analogue" in which the N-atom attached at position 16 is substituted by a carbon atom. We wanted to have a better proof for the necessity of the special D-ring 1,2-amino-alcohol type arrangement, since according to our recent knowledge it has a fundamental role for the high and selective activity.

The pharmacological data have proved undoubtedly that the nitrogen — at position 16 — more exactly the 1,2-amino-alcohol arrangement, is not inevitable for the high activity and selectivity. The most potent neuromuscular blocking agents belong to the derivatives of androstan series, substituted at position 2 and 16 with piperidino or N-methyl-piperazino groups. Pancuronium bromide, pipecuronium bromide, and vecuronium bromide are widely used in clinical practice.

TABLE 18
6β-Piperazino-16β-Piperidino-Androstane Derivatives

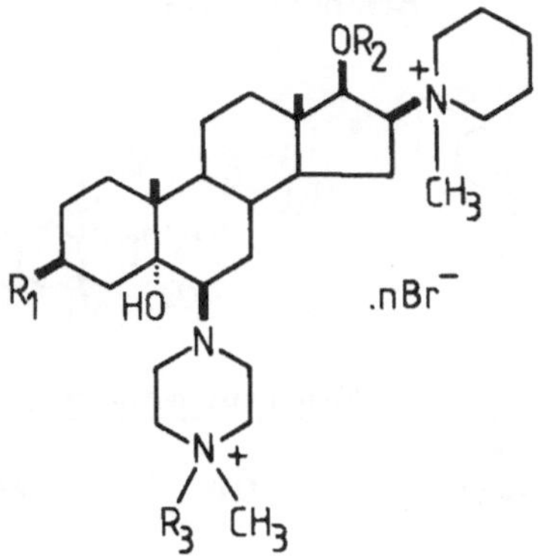

Compound no.	R_1	R_2	R_3	n	Neuromuscular blocking activity ED_{50} (μg/kg)	OT (min)	D (min)	Vagolytic effect ED_{50} (μg/kg)	V/N
37	OAc	Ac	Me	2	160	3.4	17	640	4
38	OH	H	Me	2	8800			1300	0.14
39	OAc	Ac	—	1	330	4.7	16	1800	5.4
40	H	Ac	—	1	1070	2.8	27	810	0.7
41	OAc	Ac	Pr	2	160	3.9	13	730	4.56
42	H	H	Me	2	7300	5.0	22	1200	0.16

Note: OT: onset time; D: duration of action; V/N: vagal block/neuromuscular block ratio.

VIII. CHANDONIUM AND ITS ANALOGUES

As we have seen previously the steroid skeleton has often been used as a nucleus upon which to hang onium groups in the hope of producing useful neuromuscular blocking drugs. Such compounds include pancuronium, pipecuronium, and vecuronium. All of these compounds belong to the amino steroid series, in which the amino or quaternary amino groups are attached at position 2 and 16.

Chandonium is a new addition to this group, being one of a novel series of nitrogen-containing steroids in which the quaternary nitrogens are part of the ring skeleton. Parts of these compounds were synthesized by Singh and co-workers[106] in Chandigarh (hence the name of chandonium) and pharmacologically studied by Marshall and co-workers.[104,109,110,118]

Singh and co-workers[102] first developed the 4,17a-diaza-D-homosteroid system.

HS-342 exhibited activity approximately equal to that of tubocurarine,[103,104] but of shorter duration and with rapid onset of action; it also showed

TABLE 19
"5β-Vecuronium Bromide" (43) and "5β-Pipecuronium Bromide" (44)

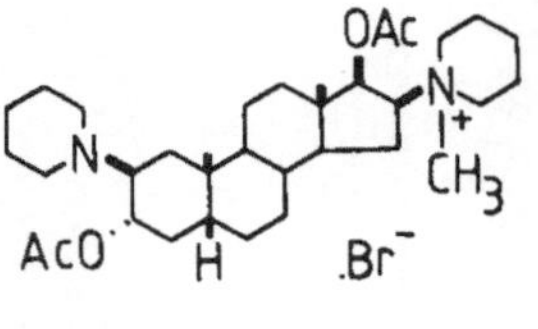
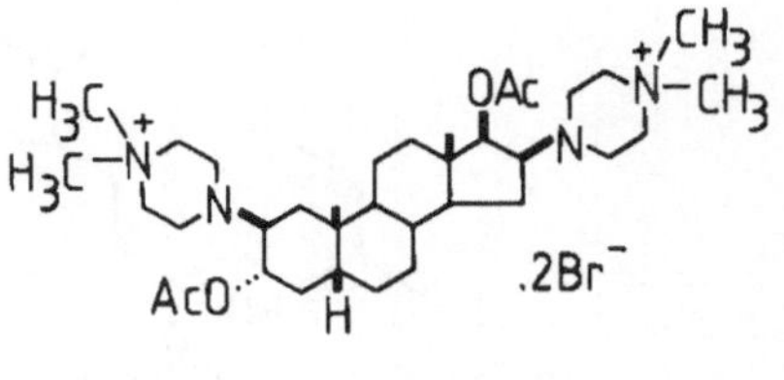

Compound no.	Neuromuscular blocking activity ED$_{99}$ (μg/kg)	OT	RR	TDT
43	200	2.6	4.2	18.6
44	20	3.6	9	22

Note: OT: onset tıme; RR: recovery rate; TDT: total duration time.

TABLE 20
C-Analogues of Pipecuronium Bromide

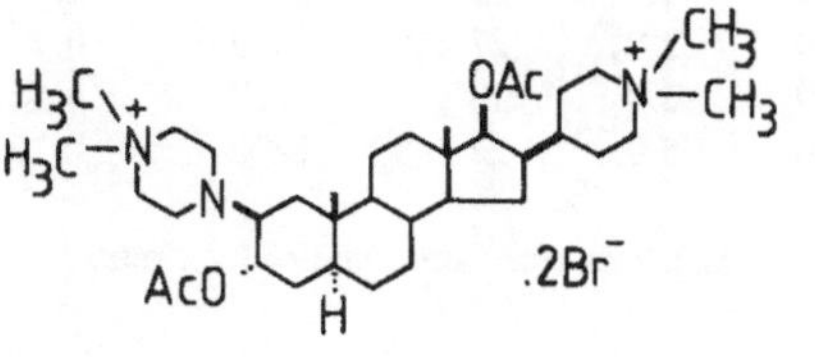
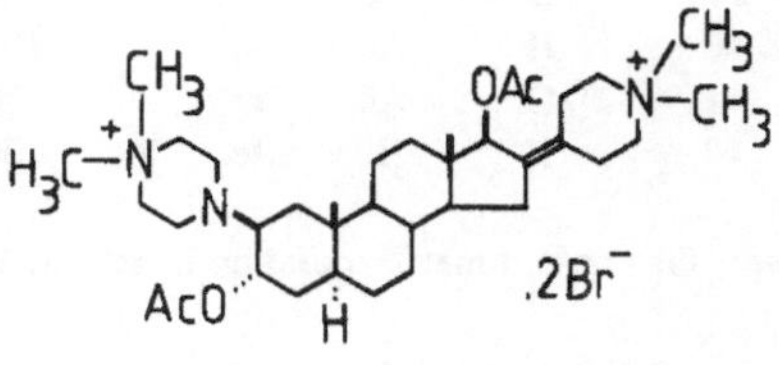

Compound no.	Neuromuscular blocking activity ED$_{99}$ (μg/kg)	OT	RR	TDT
45	35	5.0	5.8	17
46	60	4.2	3.6	15

Note: OT: onset time; RR: recovery rate; TDT: total duration time.

ganglion blocking activity (Figure 9). The interonium distance 817 pm falls only just outside the range of values exhibited by tubocurarine.[105] The drug was not investigated clinically.

Continuing the work towards the synthesis of different onium-azasteroids Singh and co-workers[106-109] prepared two more potent compounds chandonium iodide (HS-310) and HS-467 (Figure 10).

Both the bisquaternary compounds HS-310 and HS-467 of the series exhibited neuromuscular blocking activity of short duration and rapid onset

TABLE 21
Miscellaneous

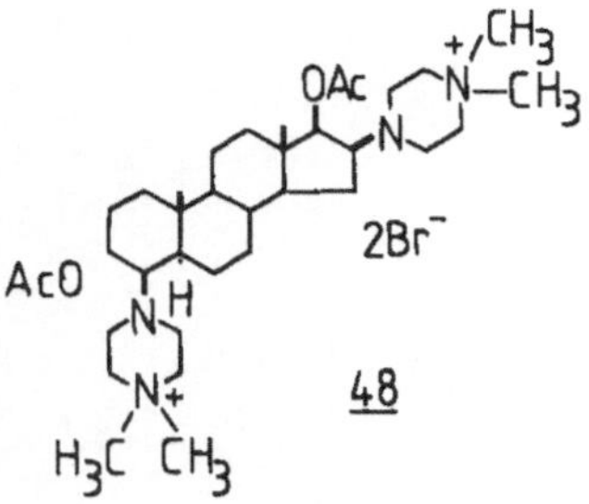

48

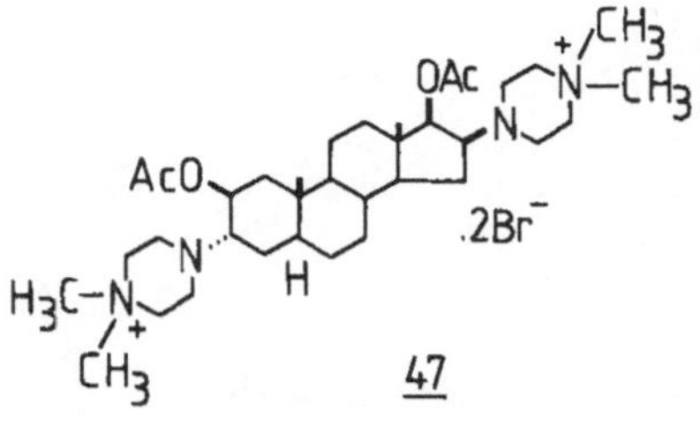

47

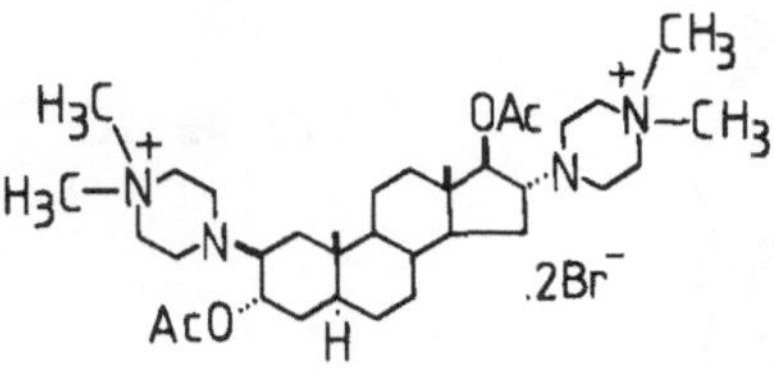

49

Str.R

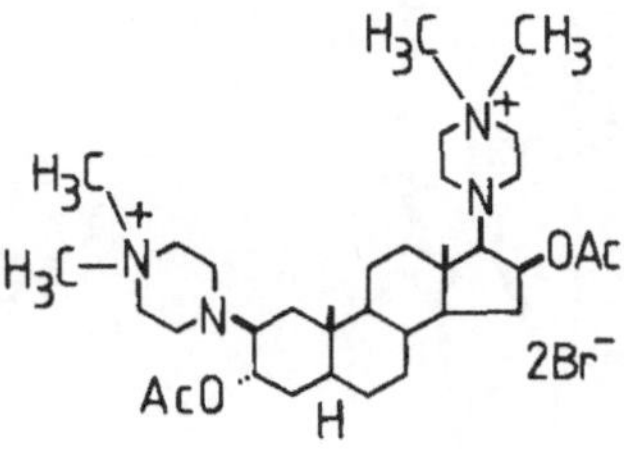

50

Compound no.	Neuromuscular blocking activity ED$_{50}$ (μg/kg)	OT (min)	TDT (min)	Vagolytic effect ED$_{50}$ (μg/kg)	V/N
47	1800	2.4	17	4200	2.3
48	520	2.9	19	2000	3.8
49	180	7.7	40	410	2.2
50	780	4.7	30	600	0.76

Note: OT: onset time; TDT: total duration time; V/N: vagal block/neuromuscular block ratio.

of action.[110-112] HS-467 was approximately equipotent with tubocurarine, whereas HS-310 was four to five times more active. Both compounds exhibited weak ganglion blocking activity. HS-310 has relatively powerful vagal blocking activity and blocks norepinephrine uptake.[113] In accordance with this its neuromuscular blocking action in humans is accompanied by pronounced tachycardia (Bowman, 1980). The monoquaternary salts of the series were weak neuromuscular blocking agents, their main action being to produce ganglion blockade and a fall in blood pressure.

Using chandonium as a prototype further chemical modifications have been carried out. Four new close analogues of chandonium have been synthesized[114] in an attempt to gain insight into the structural requirements for neuromuscular blocking potency and the undesirable autonomic blocking

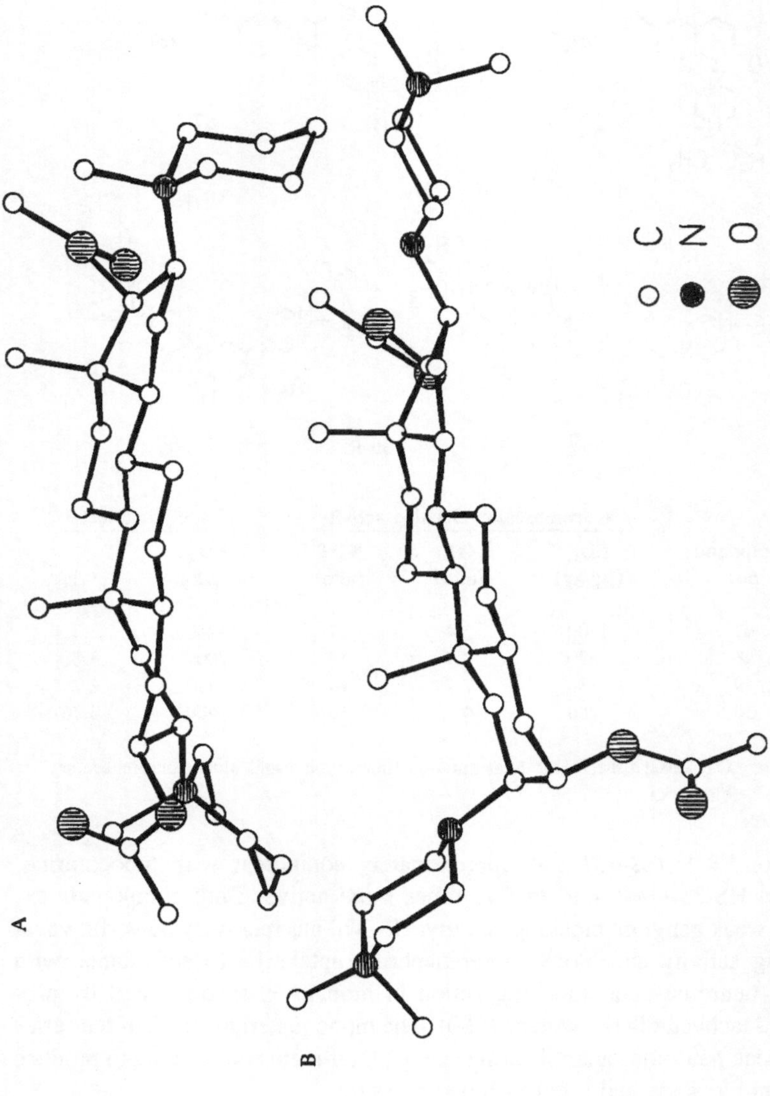

FIGURE 8. The structure of (A) pancuronium[20] and (B) pipecuronium[100,101] determined by single-crystal X-ray diffractometry.

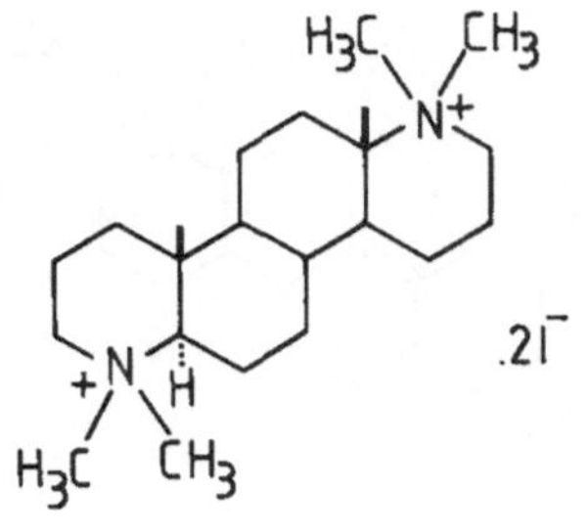

FIGURE 9. Chemical structure of HS-342.

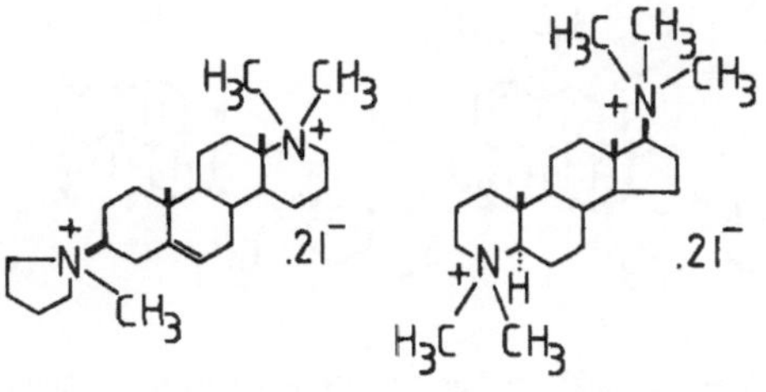

FIGURE 10. Chemical structures of HS-310 and HS-467.

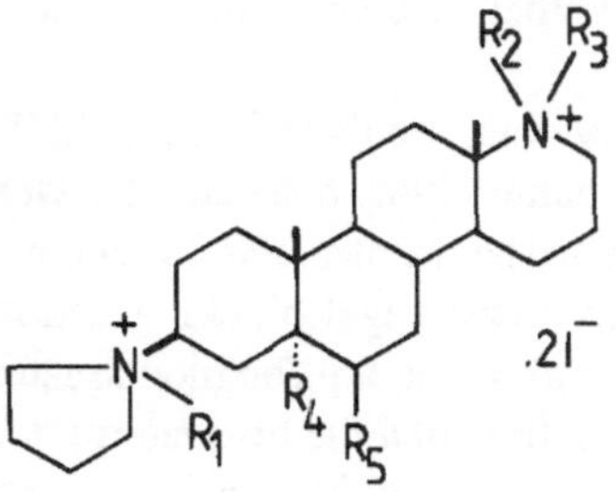

FIGURE 11. Chemical structures of HS-692, HS-693, HS-705, and HS-704.

actions in chandonium-like compounds. The bisquaternary analogues HS-692, HS-693, HS-705, and HS-704 have been prepared (Figure 11).[114-116]

The order of the potencies relative to chandonium (1.0) was HS-692 (0.5) > HS-693 (0.25) > HS-704 (0.1) = HS-705, characteristics such as onset and duration being indistinguishable from those of chandonium. All compounds were antagonized by neostigmine. At neuromuscular blocking doses none of the compounds exhibited any ganglion blocking activity and no marked changes in blood pressure or heart rate were observed, but these compounds depressed the response of the heart rate vagal stimulation: HS-693 (86%) > HS-705 (55%) = HS-704 (55%) > HS-692 (28%) > chandonium (9%). HS-692, the dihydro derivative of chandonium, possessed the most desirable spectrum of activity.[117,118]

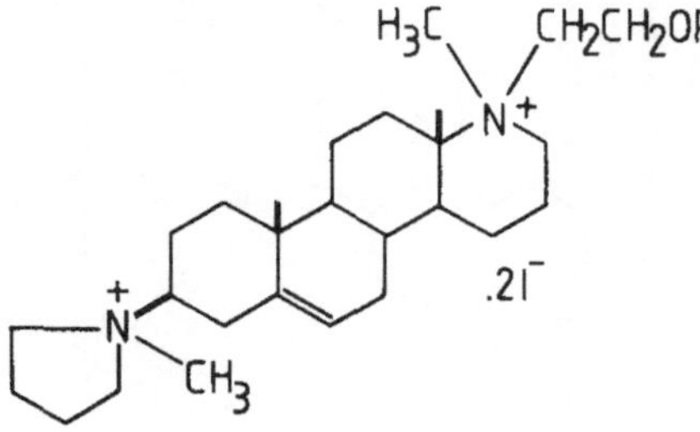

FIGURE 12. Chemical structures of HS-627 (R = Ac) and HS-626 (R = H).

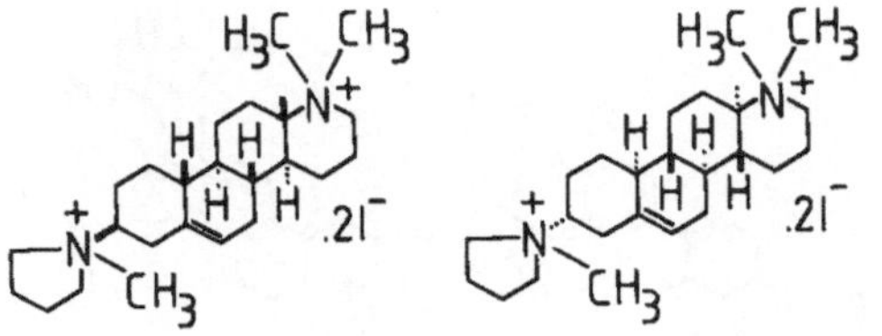

FIGURE 13. The chemical structures of 19-norchandonium and its enantiomer.

The high neuromuscular blocking activity of pancuronium has given Singh et al.[119,120] the idea to synthetize compounds which have acetylcholine-like structural units.

The two bisquaternary compounds HS-626 and HS-627, choline and acetylcholine analogues of chandonium, respectively, were approximately equipotent with chandonium as neuromuscular blocking agents in the cat, but HS-627 possessed slightly more vagal blocking action (Figure 12). No evidence was obtained for ganglion block. The monoquaternary compounds were more than 100 times less active than the bisquaternary ones as neuromuscular blocking agents, and produced vagal blockade at subneuromuscular blocking doses.[121]

To study the steric effects in neuromuscular blocking series Zeelen et al., have synthetized the enantiomer form of the 19-norchandonium iodide and compared their pharmacological effects (Figure 13).[122]

Both enantiomers were approximately three to four times less active than chandonium with short duration and rapid onset of action. A side effect is vagolytic activity.

The effect of the complete change in steric configuration is therefore insignificant, compared to that resulting from the change in liphophilicity caused by the removal of the 10-methyl group of chandonium iodide. The steric effects in these analogues of chandonium iodide are not a major factor in determining the relationship between neuromuscular blocking and vagolytic activities and chemical structure. The steroid nucleus is probably important for contributing to a sufficiently high liphophilicity.

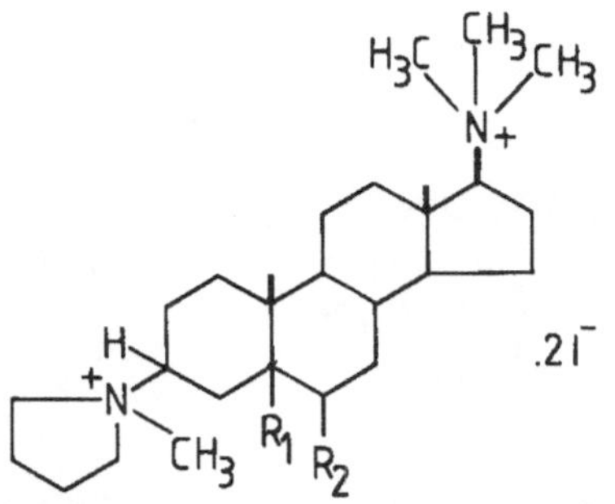

FIGURE 14. Chemical structures of HS-854, HS-1046, HS-944, and HS-892.

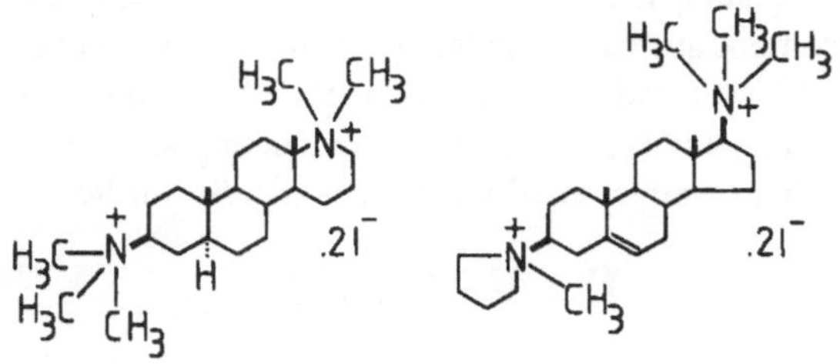

FIGURE 15. Chemical structures of HS-777 and HS-873.

None of the chemical modifications has lead to a compound more potent than chandonium (HS-854, HS-1046, HS-944, and HS-892) (Figure 14).[123]

The neuromuscular blocking activity of HS-777 and HS-873 in rat phrenic nerve diaphragm preparation is one half and one fifth of that of chandonium (Figure 15).[124]

The need for a short-acting nondepolarizing neuromuscular blocking agent prompted us to prepare new steroidal quaternary derivatives which have the onset time and duration of action in the same range as suxamethonium.

A new series of azasteroids has been developed. One of the series, 3α-pyrrolidino-17a-methyl-17a-aza-D-homo-5α-androstane dimethobromide (RGH-4201), was studied pharmacologically in detail (Figure 16).[125]

RGH-4201 exhibited a nondepolarizing neuromuscular block that was considerably faster in onset and shorter in duration of action than the reference substances suxamethonium and chandonium.[126] RGH-4201 blocked cardiac muscarinic receptors like all other chandonium analogues. The recovery rate was more rapid of those compounds in rat phrenic nerve hemidiaphragm preparation. Small doses of RGH-4201 possessed vagolytic action. The vagolytic action was more than 50% at effective neuromuscular blocking doses and lasted 35 min.

RGH-4201 has been extensively studied in clinical anesthesia; however, it has unwanted effects on the cardiovascular system.[127]

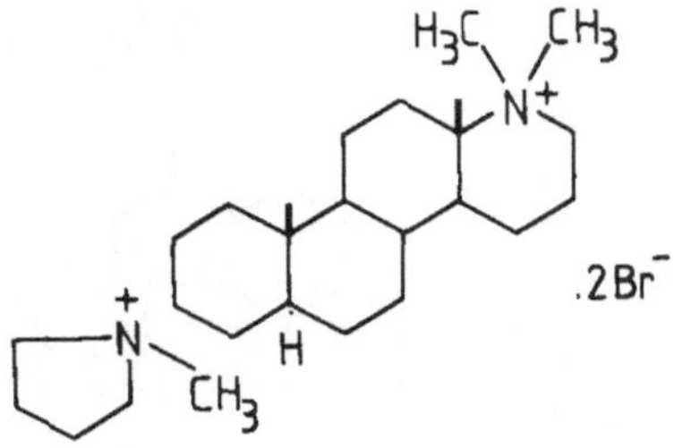

FIGURE 16. Chemical structure of RGH-4201.

Our next experiments were related with the synthesis of aza-aminosteroids in which we have changed the positions of the onium centers (Figure 17).

All of the compounds have nondepolarizing neuromuscular blocking activity, with rapid onset and short duration of action; however, all of them have vagal blocking activity, too. Since in human application it causes tachycardia, none of them were studied pharmacologically further.

IX. SUMMARY

The steroid skeleton possessing onium centers at different position and configuration has given a good chance to study the structure-activity relationships within the series of neuromuscular blocking agents. These studies resulted in the last 25 years being very successful in producing highly active and selective drugs. Pancuronium bromide, pipecuronium bromide, and vecuronium bromide are widely used in clinical practice. Pipecuronium bromide which has an exceptional high cardiovascular stability can well be applied in the anesthesia of "poor risk" patients and of those having cardiovascular diseases. Vecuronium bromide which has shorter duration, and much higher V/N ratio, than that of pancuronium bromide, is useful for shorter manipulations.

In spite of these results a new nondepolarizing neuromuscular blocking agent with an even rapid onset, shorter duration, and with faster recovery rate will be required in the future to replace the depolarizing suxamethonium.

A number of question remains to be answered with respect to structure-activity relationships, but since the introduction of the purified d-tubocurarine into the anesthesia plenty of them has been clarified. The different types of structural modification of most potent neuromuscular blocking agents having performed so far will help to supply us with more information about the receptor site, too, and then to design more active and selective derivatives for the different types of manipulation.

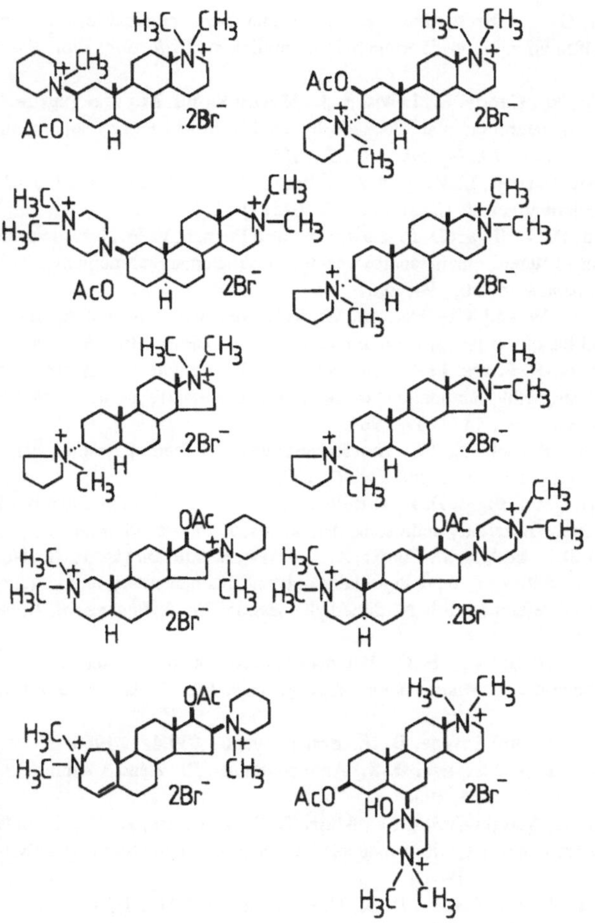

FIGURE 17. Chemical structures of different quaternary aza-aminosteroids.

REFERENCES

1. **Bowman, W. C.,** *Pharmacology of Neuromuscular Function,* John Wright and Sons, Bristol, 1980.
2. **Vizi, E. S., Somogyi, G. T., Nagashima, H., Duncalf, D., Chaudry, I. A., Kobayashi, O., and Földes, F. F.,** d-Tubocurarine and Pancuronium inhibit evoked release of acetylcholine from the mouse hemidiaphragm preparation, *Br. J. Anaesth.*, 59, 226, 1987.
3. **Vizi, E. S. and Somogyi, G. T.,** Prejunctional modulation of acetylcholine release from the skeletal neuromuscular junction: link between positive (nicotinic) — and negative (muscarinic) — feedback modulation, *Br. J. Pharmacol.*, 97, 65, 1989.
4. **Földes, F. F.,** The impact of neuromuscular blocking agents on the development of anaesthesia and surgery, presented at Harold Griffith Memorial Lecture, Montreal, 1986.
5. **Quevauviller, A. and Lainé, F.,** Sur la toxicité et le pouvoir curarisant du chlorure de malouétine, *Ann. Pharm. Fr.*, 18, 678, 1960.

6. **Blanpin, O. and Bretaudeau, J.,** Etude pharmacodynamique des sulfomethylates de triméthylfuntuminium et de triméthylfuntumidinium, *C. R. Soc. Biol. Paris*, 155, 878, 1961.
7. **Alauddin, M., Caddy, B., Lewis, J. J., Martin-Smith, M., and Sugrue, M. E.,** Non-depolarising neuromuscular blockade by 3α,17α-bis-quaternary ammonium-5α-androstanes, *J. Pharm. Pharmacol.*, 17, 55, 1965.
8. **Davis, M., Parnell, E. W., and Rosenbaum, J.,** Steroid amines. IV. 3,17-Diaminoandrostane derivatives. *J. Chem. Soc. C*, 1045, 1967.
9. **Bamford, D. G., Biggs, D. F., Davis, M., and Parnell, E. W.,** Neuromuscular blocking properties of stereoisomeric androstane-3,17-bisquaternary ammonium salts, *Br. J. Pharmacol. Chemother.*, 30, 194, 1967.
10. **Mushin, W. W. and Mapleson, W. W.,** Relaxant action in man of dipyrandium chloride (a steroid bis-quaternary ammonium salt), *Br. J. Anaesth.*, 36, 761, 1964.
11. **Khuong, H.-L. F. and Pinto-Scognamiglio, W.,** Activité curarisante du dichlorure de 3β,20α bistrimethyl-ammonium 5α-prégnane (malouétine) et de ses stéréoisoméres, *Arch. Int. Pharmacodyn.*, 147, 209, 1964.
12. **Davis, M., Parnell, E. W., and Rosenbaum, J.,** Steroid Amines. III. 3-Aminoandrostane derivatives, *J. Chem. Soc. C*, 1983, 1966.
13. **Bamford, D. G., Biggs, D. F., and Parnell, E. W.,** The neuromuscular blocking activity of some monoquaternary androstane derivatives, *J. Pharm. Pharmacol.*, 23, 595, 1971.
14. **Busfield, D., Child, K. J., Clarke, A. J., Davis, B., and Dodds, M. G.,** Neuromuscular blocking activities of some steroidal mono and bis-quaternary ammonium compounds with special reference to N,N′-dimethylconessine, *Br. J. Pharmacol. Chemother.*, 32, 609, 1968.
15. **Loewe, S. and Harwey, S. C.,** Equidistance concept and structure-activity relationship of curarizing drugs, *Naunyn-Schmiedebergs Arch. Exp. Pathol. Pharmakol.*, 214, 214, 1952.
16. **Hewett, C. L. and Savage, D. S.,** British Patent 1,039,441, 1966.
17. **Hewett, C. L. and Savage, D. S.,** Amino-steroids. III. 2 and 3-Amino-5α-androstane, *J. Chem. Soc. C*, 1134, 1968.
18. **Lewis, J. J., Martin-Smith, M., Muir, T. C., and Ross, H. H.,** Steroidal monoquaternary ammonium salts with nondepolarising neuromuscular blocking activity, *J. Pharm. Pharmacol.*, 19, 502, 1967.
19. **Hewett, C. L. and Savage, D. S.,** U.S. Patent 3,553,212, 1971.
20. **Savage, D. S., Cameron, A. F., Ferguson, G., Hannaway, C., and Mackay, I. R.,** Molecular structure of pancuronium bromide (3α,17β-diacetoxy-2β,16β-dipiperidino-5α-androstane dimethobromide), a neuromuscular blocking agent. Crystal and molecular structure of the water: methylene chloride solvate, *J. Chem. Soc. B*, 410, 1971.
21. **Buckett, W. R., Hewett, C. L., and Savage, D. S.,** Pancuronium bromide and other steroidal neuromuscular blocking agents containing acetylcholine fragments, *J. Med. Chem.*, 16, 1116, 1973.
22. **Bonta, I. L. and Goorissen, E. M.,** Different potency of pancuronium bromide on two types of skeletal muscle, *Eur. J. Pharmacol.*, 4, 303, 1968.
23. **Taylor, A. F.,** New neuromuscular blocking agent, *Organorama*, 6, 4, 1969.
24. **Buckett, W. R.,** Pharmacology of pancuronium bromide. Nondepolarizing neuromuscular blocking agent, *Ir. J. Med. Sci.*, 1, 565, 1968.
25. **Dick, W., Droh, R., Frey, R., Hadjidimos, M., Halmagyi, M., Heymer, G., and Oettel, P.,** Experimental and clinical investigations of the muscle relaxant pancuronium bromide, *Anaesthesist*, 19, 248, 1970.
26. **Loh, L.,** Cardiovascular effects of pancuronium bromide, *Anaesthesia*, 25, 356, 1970.
27. **Smith, G., Proctor, D. W., and Spence, A. A.,** Comparison of some cardiovascular effects of tubocurarine and pancuronium in dogs, *Br. J. Anaesth.*, 42, 923, 1970.
28. **Varma, Y. S. and Sharma, P. L.,** Clinical experience with pancuronium bromide (NA97), a new steroid neuromuscular blocking agent, *Indian J. Med. Res.*, 59, 69, 1971.

29. **Vourc'h, G., Lecharny, B., and Madre, F.,** Pancuronium bromide, a new curarizing drug, *Anesth. Analg. Reanim.*, 28, 1, 1971.
30. **Kelman, G. R. and Kennedy, B. R.,** Cardiovascular effect of pancuronium in man, *Br. J. Anaesth.*, 43, 365, 1971
31. **Telivuo, L., Ertama, P., Aalto-Setala, M., Appelquist, R., Kaljunen, A., Karhunen, U., Nuuttila, K., and Welling, I.,** Clinical study of pancuronium bromide as a neuromuscular blocking agent in anesthesia, *Acta Anaesthesiol Scand.*, 15, 261, 1971.
32. **Ogura, Y. and Furuta, H.,** Pharmacological properties of pancuronium bromide a new potent steroidal neuromuscular blocking agent, with reference to d-tubocurarine, *Oyo Yakuri*, 5, 715, 1972.
33. **Gergis, S. D., Dretchen, K. L., Sokoll, M. D., and Long, J. P.,** Effect of pancuronium bromide on acetylcholine release, *Proc. Soc. Exp. Biol. Med.*, 139, 74, 1972.
34. **Bowman, W. C. and Webb, S. N.,** Neuromuscular blocking and ganglion blocking activities of some acetylcholine antagonists in the cat, *J. Pharm. Pharmacol.*, 24, 762, 1972.
35. **Barzu, T., Cuparencu, B., and Cardan, E.,** Anticholinesterase activity of pancuronium bromide (Pavulon), *Biochem. Pharmacol.*, 23, 166, 1974.
36. **Reitan, J. A. and Warpinski, M. A.,** Cardiovascular effects of pancuronium bromide in mongrel dogs, *Am. J. Vet. Res.*, 36, 1309, 1975.
37. **Duke, P. C., Fung, H., and Gartner, J.,** Myocardial effects of pancuronium, *Can. Anaesth. Soc. J.*, 22, 680, 1975.
38. **Bodman, R. I.,** Pancuronium and histamine release, *Can. Anaesth. Soc. J.*, 25, 40, 1978.
39. **Son, S. L. and Wauld, D. R.,** A vagolytic action of neuromuscular blocking agents at the pacemaker of the isolated guinea pig atrium, *Anesthesiology*, 48, 191, 1978.
40. **Su, P. C., Su, W.-H. L., and Rosen, A. D.,** Pre- and postsynaptic effects of pancuronium at the neuromuscular junction of the mouse, *Anaesthesiology*, 50, 199, 1979.
41. **Bowman, W. C., Rodger, I. W., Houston, J., Marshall, R. J., and McIndewar, I.,** Structure-action relationships among some desacetoxy analogues of pancuronium and vecuronium in the anesthetized cat, *Anesthesiology*, 69, 57, 1988.
42. **Durant, N. N., Marshall, I. G., Savage, D. S., Nelson, D. J., Sleigh, T., and Carlyle, I. C.,** The neuromuscular and autonomic blocking activities of pancuronium, Org. NC 45, and other pancuronium analogues, in the cat, *J. Pharm. Pharmacol.*, 31, 831, 1979.
43. **Carlyle, I. C., Sleigh, T., and Savage, D. S.,** European Patent Application 8,824, 1980.
44. **Bowman, W. C.,** A new non-depolarizing neuromuscular blocking drug, *Trends Pharmacol. Sci.*, 1, 263, 1980.
45. **Durant, N. N., Houwertjes, M. C., and Crul, J. F.,** Comparison of the neuromuscular blocking properties of Org NC 45 and pancuronium in the rat, cat and rhesus monkey, *Br J. Anaesth.*, 52, 723, 1980.
46. **Marshall, I. G., Ágoston, S., Booij, L. H. D. J., Durant, N. N., and Földes, F. F.,** Pharmacology of Org NC 45 compared with other non-depolarizing neuromuscular blocking drugs, *Br. J. Anaesth.*, 52, 118, 1980.
47. **Marshall, R. J., McGrath, J. C., Miller, R. D., Docherty, J. R., and Lamar, J. C.,** Comparison of the cardiovascular actions of Org NC 45 with those produced by other non-depolarizing neuromuscular blocking agents in experimental animals, *Br. J. Anaesth.*, 52, 21S, 1980.
48. **Savage, D. S., Sleigh, T., and Carlyle, I.,** The emergence of Org NC 45, 1-[2β,3α,5α,16β,17β-3,17-bis(acetyloxy)-2-(1-piperidinyl)-androstan-16-yl]-1-methylpiperidinium bromide, from the pancuronium series, *Br. J. Anaesth.*, 52, 3S, 1980.
49. **Booij, L. H. D. J., Edwards, R. P., Sohn, Y. J., and Miller, R. D.,** Cardiovascular and neuromuscular effects of Org NC 45, pancuronium metocurine, and d-tubocurarine in dogs, *Anaesth. Analg. (Cleveland)*, 59, 26, 1980.

50 **Docherty, J. R. and McGrath, J. C.**, A comparison of the effects of pancuronium bromide and its monoquaternary analog, Org NC 45, on autonomic and somatic neurotransmission in the rat, *Br. J. Pharmacol.*, 71, 225, 1980.
51. **Son, S. L., Waud, B. E., Waud, D. R., and Phil, D.**, A comparison of the neuromuscular blocking and vagolytic effects of ORG NC45 and pancuronium, *Anesthesiology*, 55, 12, 1981.
52. **Booij, L. H. D. J., Vree, T. B., and Crul, J. F.**, Org-NC45: a new steroidal nondepolarizing muscle relaxant, *Pharm. Weekbl.*, 4, 1, 1982.
53. **Barnes, P. K., Brindle Smith, G., White, W. D., and Tennant, R.**, Comparison of the effects of Org NC 45 and pancuronium bromide on heart rate and arterial pressure in anesthetized man, *Br. J. Anaesth.*, 54, 435, 1982.
54. **Durant, N. N.**, Norcuron, a new nondepolarizing neuromuscular blocking agent, *Semin. Anesth.*, 1, 47, 1982.
55. **Marshall, I. G., Gibb, A. J., and Durant, N. N.**, Neuromuscular and vagal blocking actions of pancuronium bromide, its metabolites, and vecuronium bromide (Org NC45) and its potential metabolites in the anesthetized cat, *Br. J. Anaesth.*, 55, 703, 1983.
56. **Savage, D. S.**, The invention and discovery of vecuronium, *Curr. Clin. Pract. Ser.*, 11(Clin. Exper. Norcuron), 9, 1983.
57. **Miller, R. D., Rupp, S. M., Fisher, D. M., Cronelly, R. F., Fahey, M. R., and Sohn, Y. J.**, Clinical pharmacology of vecuronium and attracurium, *Anesthesiology*, 61, 444, 1984.
58. **Miller, R. D.**, Vecuronium: a new nondepolarizing neuromuscular blocking agent. Clinical pharmacology, pharmacokinetics, cardiovascular effects and use in special clinical situations, *Pharmacotherapy*, 4, 238, 1984.
59. **Jones, R. S.**, Neuromuscular blocking action of vecuronium in the dog and its reversal by neostigmine, *Res. Vet. Sci.*, 38, 193, 1985.
60. **Bencini, A. F., Scaf, A. H. J., Ágoston, S., Houwertjes, M. C., and Kersten, U. W.**, Disposition of vecuronium bromide in the cat, *Br. J. Anaesth.*, 57, 782, 1985.
61. **Conil, J. M., Richard, J. P., Conil, C., Antonini, A., Bareille, P., and Bertrand, J. C.**, Vecuronium bromide pharmacology, *Convergences Med.*, 4, 227, 1985.
62. **Ogura, Y., Kikuchi, T., Kisara, K., Tadano, T., Sakurada, S., Sakurada, T., Fujita, K., Inoguchi, K., Kogaku, H., et al.**, An investigation of the properties of vecuronium bromide, a new short-acting non-depolarizing neuromuscular blocking agent, *Oyo Yakuri*, 32, 465, 1986.
63. **Savarese, J. J. and Wastila, W. B.**, Current research in relaxant development, *Semin. Anesth.*, 5, 304, 1986.
64. **Liu, X. and Zhang, L.**, Clinical pharmacology of vecuronium, *Zhongua Mazuixue Zazhi*, 6, 308, 1986.
65. **Liu, S., Jiang, Y., Jiang, J., Gui, B., and Wu, F.**, Pharmacology and toxicology of OrgNC45, *Zhongguo Yaoke Daxue Xuebao*, 19, 22, 1988.
66. **Savage, D. S., Sleigh, T., and Carlyle, I. G.**, European Patent Application 287,150, 1988.
67. **Miller, R. D., Brady, K., Ágoston, S., Castagnoli, K., and Claver, C.**, Pharmacokinetics of Org 9616 and Org 9426 the new short acting nondepolarizing muscle relaxants, presented at 9th World Congress of Anaesthesiologist, Washington, Abstract A0535, 1988.
68. **Booij, L. H. D. J., Marshall, I. G., Grul, J. F., and Muir, A. W.**, Pharmacology of four steroidal muscle relaxants, presented at 9th World Congress of Anaesthesiologists, Washington, Abstract A0533, 1988.
69. **Muir, A. W., Houston, J., Green, K. L., Marshall, R. J., Bowman, W. C., and Marshall, I. G.**, Effects of a new neuromuscular blocking agent (Org 9426) in anaesthetized cats and pigs and in isolated nerve-muscle preparations, *Br. J. Anaesth.*, 63, 400, 1989.

70 **Wierda, J. M. K. H., DeVit, A. P. M., Kuizenga, K., and Ágoston, S.,** Clinical observations on the neuromuscular blocking action of Org 9426, a new steroidal non-depolarizing agent, *Br. J. Anaesth.*, 64, 521, 1990.

71. **Witkowski, T. A., Barkowski, R. R., Azad, S. S., Epstein, R. H., Marr, A., and Lessin, J.,** Org 9426, a new short-acting muscle relaxant: onset and duration of action, *Anesth. Analg.*, 70 (suppl.), 437, 1990.

72. **Cason, B., Baker, D. G., Hickey, R. F., Miller, R. D., and Ágoston, S.,** Cardiovascular and neuromuscular effects of three steroidal neuromuscular blocking drugs in dogs (ORG 9616, ORG 9426, ORG 9991), *Anesth. Analg.*, 70, 382, 1990.

73. **Khuenl-Brady, K., Castagnoli, K. P., Canfell, P. C., Caldwell, J. E., Ágoston, S., and Miller, R. D.,** The neuromuscular blocking effects and pharmacokinetics of ORG 9426 and ORG 9616 in the cat, *Anesthesiology*, 72, 669, 1990.

74. **Golpariani, M., Ohta, Y., Nagashima, H., Goldiner, P. L., and Földes, F. F.,** ORG 9426 potentiates neuromuscular blocking effects of other nondepolarizing muscle relaxants in rats, *Anesth. Analg.*, 70, 131, 1990.

75. **Tuba, Z., Bor nee Szabó, M., Szeberényi, Sz., Kárpáti, E., Szporny, L., and Szepesi, G.,** British Patent 1,398,050, 1975; German Patent 2,337,882, 1978.

76. **Tuba, Z.,** Synthesis of 2β,16β-bis-(4′-dimethyl-1′-piperazino)-3α,17β-diacetoxy-5α-androstane dibromide and related compounds, *Arzneim.-Forsch.*, 30(2A), 342, 1980.

77. **Tuba, Z. and Marsai, M.,** Synthesis of biologically active amino and aza steroids and some of their new chemical reactions, presented at First International Conference on Chemistry and Biotechnology of Biologically Active Natural Products, Varna, Conference Proceedings 2, 135, 1981.

78. **Tuba, Z., Mahó, S., Gálik, Gy., Balogh, G., and Csehi, A.,** Synthesis of biologically active amino steroids and some of their new chemical reactions, presented at 16th International Symposium on the Chemistry of Natural Products, Kyoto, Abstract 434, 1988.

79 **Tuba, Z., Mahó, S., and Gálik, Gy.,** Synthesis of Pipecuronium Bromide (ARDUAN®) Structure-Activity Relationships, presented at 4th Symposium on Biochemical Aspects of Steroid Research, Holzhau, Abstract 18, 1988.

80. **Tuba, Z., Gálik, Gy., and Mahó, S.,** Synthesis of Pipecuronium Bromide Haptens, presented at 4th Symposium on Biological Aspects of Steroid Research, Holzhau, Abstract 49, 1988.

81. **Tuba, Z., Mahó, S., and Gálik, Gy.,** Synthesis and Structure-Activity Relationships of Nonhormonal Nitrogen Containing Steroids, presented at 5th International Conference on Chemistry and Biotechnology of Biologically Active Natural Products, Varna, Conference Proceedings 3, 169, 1989.

82. **Gazdag, M., Szepesi, G., Varsányi-Riedl, K., and Tuba, Z.,** HPLC of pipecuronium bromide (Arduan) and related compounds, *Anal. Chem. Symp. Ser.*, 23(Adv. Steroid Anal. '84), 431, 1985.

83 **Kárpáti, E. and Bíró, K.,** Pharmacological study of a new competitive neuromuscular blocking steroid, pipecuronium bromide, *Arzneim.-Forsch.*, 30(2A), 346, 1980.

84. **Alyautdin, R. N., Buyanov, V. V., Fisenko, V. P., Lemina, E. Yu., Muratov, V. K., Samoilov, D. N., and Shorr, V. A.,** On some properties of a new steroid curare-like compound pipecuronium bromide, *Arzneim.-Forsch.*, 30(2A), 355, 1980.

85. **Pulay, I., Alánt, O., Darvas, K., Weltner, J., and Zetény, Z.,** Respiration paralyzing and circulatory effects of a new non-depolarizing relaxant, pipecuronium bromide, in anesthetized dogs, *Arzneim-Forsch.*, 30(2A), 358, 1980.

86. **Simon, G., Bíró, K., Kárpáti, E., and Tuba, Z.,** The effect of the steroid muscle relaxant pipecuronium bromide on the acetylcholinesterase activity of red blood cells *in vitro*, *Arzneim.-Forsch.*, 30(2A), 360, 1980.

87. **Vereczkey, L. and Szporny, L.,** Disposition of pipecuronium bromide in rats, *Arzneim.-Forsch.*, 30(2A), 364, 1980.

88. **Bodrogi, L., Fehér, T., Váradi, A., and Vereczkey, L.,** Pharmacokinetics of pipecuronium bromide in the rat, *Arzneim.-Forsch.*, 30(2A), 366, 1980.
89. **Bowman, W. C.,** Comparative animal pharmacology of atracurium, vecuronium, pipecuronium and pancuronium, *Exerpta Med. Int. Congr. Ser.*, 637 (Anaesthesia), 261, 1984.
90. **Ágoston, S. and Richardson, F. J.,** Pipecuronium bromide (Arduan) — a new long-acting non-depolarizing neuromuscular blocking drug, *Clin. Anaesthesiol.*, 3, 361, 1985.
91. **Kobielski, J., Wysocki, M., Orlicz, P., Trzaski, M., and Gajewski, K.,** Pipecuronium bromide (Arduan) — a new steroid non-depolarizing muscle relaxant, *Anest. Reanim. Intensywna Ter.*, 18, 5, 1986.
92. **Jones, R. S.,** Observations on the neuromuscular blocking action of pipecuronium in the dog, *Res. Vet. Sci.*, 43, 101, 1987.
93. **Bíró, K.,** Effects of respiratory and metabolic alkalosis and acidosis on pipecuronium neuromuscular block, *Eur. J. Pharmacol.*, 154, 329, 1988.
94. **Deam, R. K. and Soni, N.,** Effects of pipecuronium and pancuronium on the isolated rabbit heart, *Br. J. Anaesth.*, 62, 287, 1989.
95. **Khuenl-Brady, K. S., Sharma, M., Chung, K., Miller, R. D., Ágoston, S., and Caldwell, J. E.,** Pharmacokinetics and disposition of pipecuronium bromide in dogs with and without ligated renal pedicles, *Anesthesiology*, 71, 919, 1989.
96. **Törocsik, A., Chaudhry, I. A., Bíró, K., Nagashima, A., Kinjo, M., Duncalf, D., Nagashima, R., Földes, F. F., Goldiner, P. L., and Vizi, E. S.,** *In vitro* comparison of the neuromuscular antinicotinic and intestinal antimuscarinic effects of different non-depolarizing muscle relaxants, *Arch. Int. Pharmacodyn. Ther.*, 299, 1989.
97. **Tuba, Z., Marsai, M., Bíró, K., Szporny, L., Kárpáti, E., and Szeberényi, Sz.,** U.S. Patent 4,071,515, 1978; British Patent 1,533,454, 1979.
98. **Tuba, Z., Marsai, M., Bíró, K., Szporny, L., Kárpáti, E., and Szeberényi, Sz.,** U.S. Patent 4,101,545 1978; British Patent 1,540,404, 1979.
99. **Tuba, Z., Horváth, J., Lovasnee-Marsai, M., Riesz, M., Szporny, L., and Kárpáti, E.,** European Patent 208,497, 1987; European Patent 345,471, 1989.
100. **Tuba, Z., Mahó, S., Párkányi, L., and Czugler, M.,** Chemistry and Structure-Activity Relationships of Neuromuscular Blocking Agents, Presented at 3rd Cyprus Conference on New Methods in Drug Research, Limassol, Abstract 44, 1987.
101 **Tuba, Z., Mahó, S., Párkányi, L., and Czugler, M.,** Pipecuronium Bromide and Pancuronium Bromide: Structure and Activity, presented at Steric Aspects of Biomolecular Interactions, International Symposium, Sopron, Abstract 39, 1985.
102. **Singh, H., Paul, D., and Parashar, V. V.,** Steroids and related studies. XX. 4,17a-Diaza-D-homo-steroids, *J. Chem. Soc. Perkin Trans. 1*, 1204, 1973.
103. **Marshall, I. G., Paul, D., and Singh, H.,** The neuromuscular and other blocking actions of 4,17a-dimethyl-4,17a-diaza-D-homo-5α-androstane dimethoiodide in the anaesthetized cat, *Eur. J. Pharm.*, 22, 129, 1973.
104. **Marshall, I. G., Paul, D., and Singh, H.,** Some actions of 4,17a-dimethyl-4,17a-diaza-D-homo-5α-androstane dimethoiodide (HS-342) a new neuromuscular blocking drug, *J. Pharm. Pharmacol.*, 25, 441, 1973.
105. **Palmer, R. A., Kalam, M. A., Singh, H., and Paul, D.,** Steroids and related studies. LIII. Structure and function of synthetic bisquaternary aza steroidal neuromuscular blocking agents, *J. Cryst. Mol. Struct.*, 10, 31, 1980.
106. **Singh, H. and Paul, D.,** Steroids and related studies. XXV. Chandonium iodide (17a-Methyl-3β-pyrrolidino-17a-aza-D-homo-androst-5-ene dimethoiodide) and other quaternary ammonium steroid analogues, *J. Chem. Soc. Perkin Trans. 1*, 1475, 1974.
107. **Singh, H. and Paul, D.,** Indian Patent 145,657, 1977.
108. **Chauhary, A. K., Gombar, V. K., and Singh, H.,** Steroids and related studies LXII. Molecular orbital calculations on some neuromuscular blocking azasteroids, *Indian J. Chem.*, 24B, 2, 1985.

109. **Ghandiha, A., Marshall, I. G., Paul, D., and Singh, H.**, Neuromuscular and other blocking actions of a new series of mono and bisquaternary aza steroids, *J. Pharm. Pharmacol.*, 26, 871, 1974.
110. **Ghandiha, A., Marshall, I. G., Paul, D., Rodger, I. W., Scott, W., and Singh, H.**, Some actions of chandonium iodide, a new short-acting muscle relaxant, in anaesthetized cats and on isolated muscle preparations, *Clin. Exp. Pharmacol. Physiol*, 2, 159, 1975.
111. **Harvey, A. L., Paul, D., Rodger, I. W., and Singh, H.**, Actions of the muscle relaxant chandonium iodide on guinea-pig ileum and vas deferens preparations, *J. Pharm. Pharmacol.*, 28, 617, 1976.
112. **Singh, H. and Chaudhary, A. K.**, Pharmacokinetics and disposition of chandonium iodide in rat, *Indian J. Exp. Biol.*, 23, 253, 1985.
113. **Marshall, R. J. and Ojewole, J. A. O.**, Comparison of the automic effects of some currently used neuromuscular blocking agents, *Br. J. Pharmacol.*, 66, 77, 1979.
114. **Singh, H., Bhardwaj, T. R., Ahuja, N. K., and Paul, D.**, Steroids and related studies. XLIV. 17a-Methyl-3β-(N-pyrrolidinyl)-17a-aza-D-homo-5α-androstane bis(methoiodide) and certain other analogues of chandonium iodide, *J. Chem. Soc. Perkin Trans.*, *1*, 305, 1979.
115. **Singh, H., Bhardwaj, T. R., and Paul, D.**, Indian Patent 148, 376, 1977.
116. **Husain, J., Tickle, I. J., and Palmer, R. A.**, Steroids and related studies. LIV. The structure of 17a-methyl-3β-pyrrolidinyl-17a-aza-D-homo-5α-androstane (HS-691), *Acta Cryst.*, B38, 130, 1982.
117. **Teerapong, P., Marshall, I. G., Harvey, A. L., Singh, H., Paul, D., Bhardwaj, T. R., and Ahuja, N. K.**, Neuromuscular and autonomic actions of four chandonium analogues, presented at British Pharmaceutical Conference, Sheffield, 1977, *J. Pharm. Pharmacol. Suppl.*, 80, 1977.
118 **Teerapong, P., Marshall, I. G., Harvey, A. L., Singh, H., Paul, D., Bhardwaj, T. R., and Ahuja, N. K.**, The effects of dihydrochandonium and other chandonium analogues on neuromuscular and autonomic transmission, *J. Pharm. Pharmacol.*, 31, 521, 1979.
119 **Singh, H., Bhardwaj, T. R., and Paul, D.**, Steroids and related studies. XLVIII. A chandonium iodide analogue possessing an acetylcholine-like moiety, *J. Chem. Soc. Perkin Trans.*, *1*, 2451, 1979.
120 Council of Scientific and Industrial Research. Indian Patent 147,560, 1978.
121 **Marshall, I. G., Harvey, A. L., Singh, H., Bhardwaj, T. R., and Paul, D.**, The neuromuscular and autonomic blocking effects of azasteroids containing choline and acetylcholine fragments, *J. Pharm. Pharmacol.*, 33, 451, 1981.
122. **Marshall, R. J., McIndewar, I., Peters, J. A. M., van Vliet, N. P., and Zeelen, F. J.**, An enantiomeric pair of steroidal neuromuscular blockers, *Eur. J. Med. Chem.-Chim. Ther.*, 19, 43, 1984.
123. **Singh, H., Gupta, R. K., and Bhardawaj, T. R.**, Steroids and related studies. LXXXII. Chandonium related azasteroidal neuromuscular blockers. *Indian J. Chem.*, 27B, 508, 1988.
124. **Bhardwaj, T. R., Kapoor, S., Shekhar, C., Paul, D., and Singh, H.**, Steroids and related studies. LXXXI. Potential azasteroidal neuromuscular blockers, *Indian J. Chem.*, 27B, 209, 1988.
125. **Tuba, Z., Marsai, M., Görög, S., Bíró, K., Kárpáti, E., and Szporny, L.**, British Patent 2,003,479, 1982.
126. **Bíró, K. and Kárpáti, E.**, The pharmacology of a new short acting nondepolarizing muscle relaxant steroid (RGH-4201), *Arzneim.-Forsch.*, 31, 674, 1982.
127. **Fitzal, S., Ilias, W., Kalina, K., Schwarz, S., Földes, F. F., and Steinbereitheimer, K.**, Neuromusculäre und Kardiovasculäre Effecte von Duador, einem neuen kurz Wirksamen nicht depolarisierenden Muskelrelaxans, *Anaesthesist*, 31, 674, 1982.

Chapter 12

MOLECULAR BIOLOGY OF CARDIOTONIC STEROIDS

Richard E. Thomas

TABLE OF CONTENTS

I. INTRODUCTION

The cardiotonic steroids include a diverse group of compounds that act on the heart to produce a positive inotropic effect (increase in force of contraction) and, in higher concentrations, toxic effects including arrhythmias and cardiac arrest. Representative examples of cardiotonic steroids are shown in Figure 1.

Cardiotonic steroids occur naturally in a wide variety of plants and in certain species of toad (Figure 1A). In plants, these substances usually occur

A

BUTENOLIDES

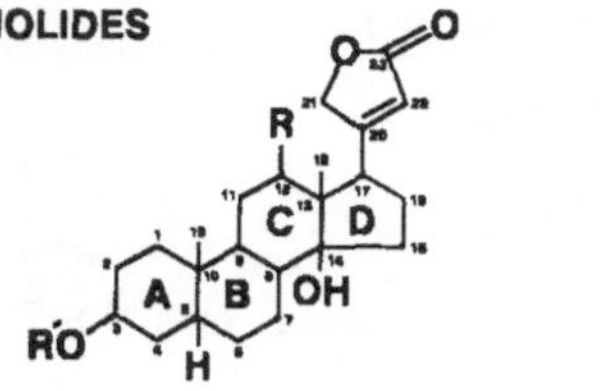

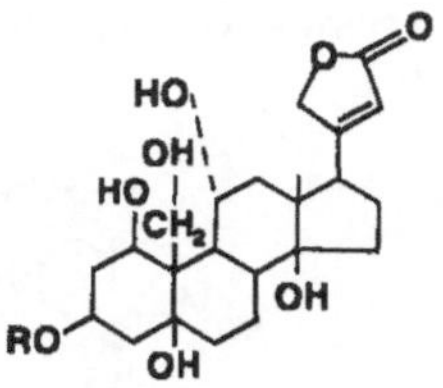

I Digitoxigenin (R, R' = H)
II Digitoxin (R = H; R' = digitoxose$_3$)
III Digoxigenin (R = OH; R' = H)
IV Digoxin (R = OH; R' = digitoxose$_3$)

V Ouabain (R = rhamnose)

PENTADIENOLIDES

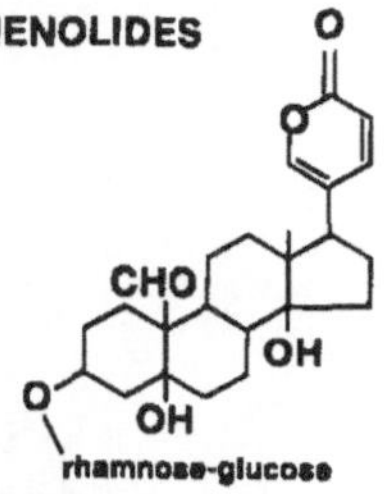

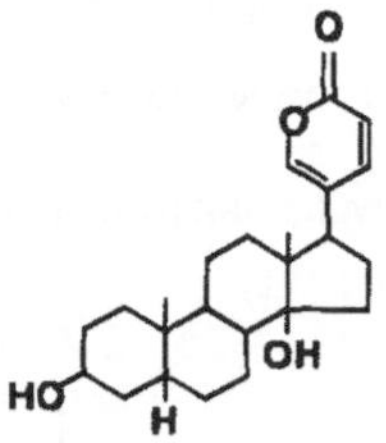

VI Hellebrin

VII Bufalin

ERYTHROPHLEUM ALKALOIDS

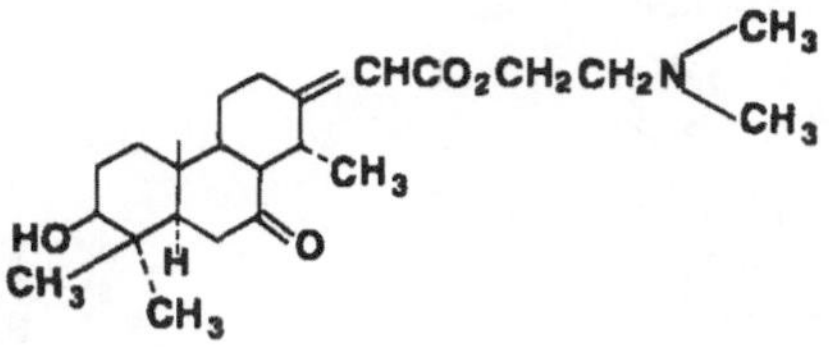

VIII Cassaine

FIGURE 1. (A) Representative selection of naturally occurring cardiotonic steroids. (B) Representative selection of synthetic and semisynthetic substances with digitalis-like activity.

as glycosides (''cardiac glycosides'') with the sugar moiety attached at position 3 (in some glycosides, the sugar is bi-linked through positions 2 and 3). Removal of the sugar residue gives the corresponding genin (aglycone) which is usually less potent than the parent glycoside. The best-known cardiac glycosides are those from digitalis (foxglove) species (I to IV in Figure 1A). The expression ''digitalis'' is often used as a generic term to cover all cardiotonic steroids. In particular, it is usual practice to refer to the receptor for cardiotonic steroids as the ''digitalis receptor''.

Recently, there has been extensive speculation that an endogenous factor exists in man and other animals that has pharmacological properties in common with the cardiotonic steroids. This ''endogenous digitalis factor'' is

B SYNTHETIC AND SEMISYNTHETIC ANALOGS

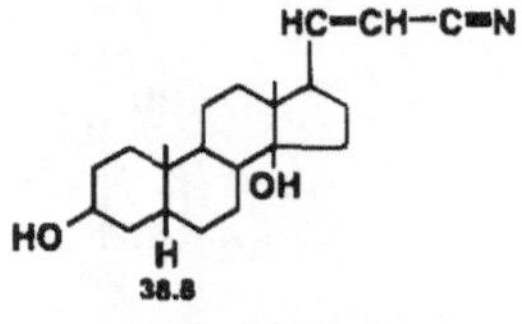

IX α,β-unsaturated nitrile

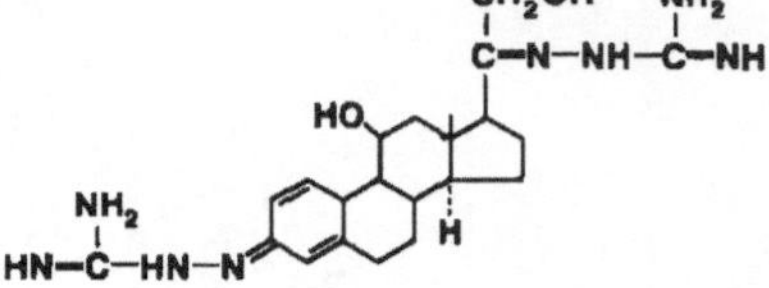

X Prednisolone 3,20-bisguanylhydrazone

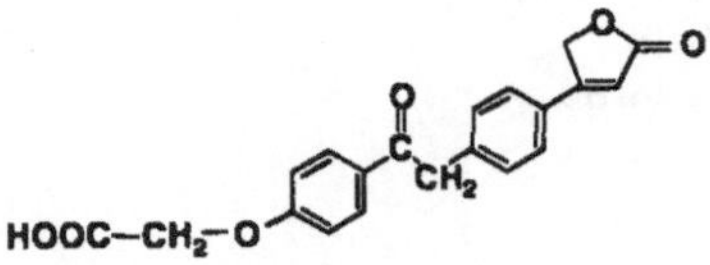

XI "Pseudo" steroid

PREGNANE DERIVATIVES

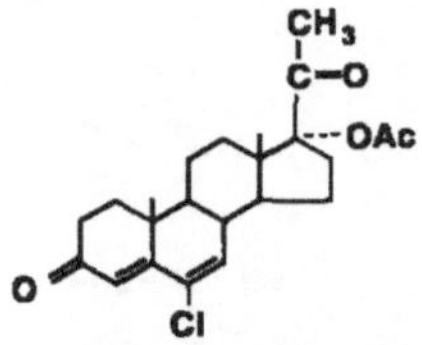

XII Chlormadinone acetate (CMA)

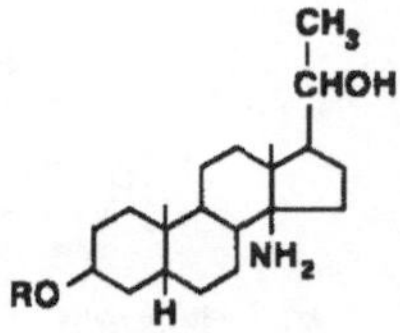

XIII LND 623 (R = rhamnose)

N-CONTAINING 17β HETEROCYCLICS

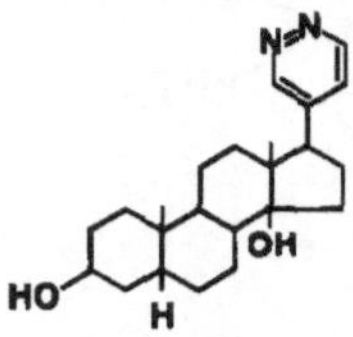

XIV Pyridazine derivative

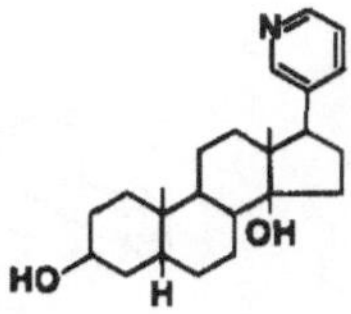

XV Pyridine derivative

FIGURE 1B.

believed to play a role in the renal excretion of Na^+ and may contribute to the development of essential hypertension in certain individuals. Evidence for the existence and role of this factor is controversial.[1-3] For a general review, see Reference 4.

Digitalis has long been used for the treatment of congestive heart failure (CHF) and for certain cardiac arrhythmias. Originally, it was used in the form of the dried plant or crude plant extract but, since the second world war, pure glycosides (usually digoxin) have been used. During the last 2 decades, the value of digitalis in the treatment of CHF has been extensively challenged.[5] However, recent studies have restored its reputation to some degree and it should now be regarded as one of the drugs of choice in the treatment of

CHF and certain arrhythmias.[6,7] There seems no doubt that positive inotropic agents are required for the optimal treatment of certain types of chronic heart failure.[8] Although β-agonists have a role in the treatment of acute heart failure, Erdmann and Böhm[9] have shown that their value in chronic heart failure is greatly diminished because of the down regulation of β-adrenergic receptors that occurs in chronic heart failure, particularly if severe. This, together with the disappointing results that have been obtained clinically with the phosphodiesterase inhibitor class of inotropic agents,[6] leaves digitalis as the sole inotropic agent of proven value in the treatment of chronic heart failure.

Cardiotonic steroids have very narrow therapeutic to toxic ratios, and one of the objectives of SAR studies has been to develop glycosides with improved therapeutic ratios. However, the main scientific interest in the study of cardiotonic steroids arises from their ability to inhibit Na^+,K^+-ATPase. This enzyme is involved in the active transport of Na^+ and K^+ across cell membranes and plays a key role in maintaining the transmembrane potential difference which is the basis for membrane excitability and conductance. Sequencing and other studies of Na^+,K^+-ATPase have now provided a firm basis for speculation about the nature of the digitalis receptor and the way in which the formation of the drug receptor complex mediates its effect on the activity of the enzyme. The possibility that an endogenous digitalis-like factor may play a role in the pathogenesis of essential hypertension by combining with Na^+,K^+-ATPase in vascular smooth muscle is a another reason for studying the molecular biology of cardiotonic steroids.

The mode of action, receptor, and structure-activity relationships (SARs) of cardiotonic steroids have recently been extensively reviewed by Thomas et al.[4] In this chapter, only a brief overview of the subject will be given with the object of outlining its scope and stimulating further interest in its study. The reader is referred to the earlier review[4] for a more-detailed treatment of the subject.

Lack of space prevents me from citing the seminal papers that formed the milestones in digitalis research and I hope that the authors concerned will forgive this omission. However, I would like to acknowledge the work of Professor K. Repke who was the first to establish the metabolic pathways of cardiac glycosides[10] and the first to suggest that the digitalis receptor was located on Na^+,K^+-ATPase.[10,11] Professor Repke's work has also included extensive SAR studies which have recently focused on quantitative SARs.[12]

II. PHARMACOLOGY

A. DIRECT EFFECTS OF DIGITALIS ON THE HEART

The cardiotonic steroids act on the heart by direct and indirect mechanisms. The direct effects are best demonstrated in isolated heart muscle preparations where the drugs produce both inotropic and toxic effects. However, cardiotonic steroids also affect the heart by indirect mechanisms, and, according to some investigators, it is the indirect rather than the direct effects that are responsible for the therapeutic and toxic effects seen clinically.[13,14]

Certainly, one of the major difficulties in interpreting studies of digitalis on isolated heart muscle is that the concentrations reported as being needed to produce a therapeutic (positive inotropic) effect are one to two orders of magnitude greater than required in the whole animal.[4,15] Grupp et al.[16] reported that concentrations of cardiac glycosides that apply clinically have no positive inotropic effects on isolated myocardium from several species, including the cat, which has sensitivity similar to that of man.

However, as will be discussed later, isoforms of Na^+,K^+-ATPase have now been identified and selective interaction with one of these may explain the concentration paradox.

B. INDIRECT EFFECTS OF DIGITALIS ON THE HEART

In whole animals, therapeutic concentrations of cardiotonic steroids can affect the heart by a number of indirect mechanisms, including reflexogenic effects and direct CNS effects. These mechanisms will be discussed briefly below. For a more detailed review see Reference 4.

1. Reflexogenic Effects

The reflexogenic effects of digitalis include excitation of the baroreceptors of the carotid sinus and aortic arch and stimulation of the mechanoreceptors of ventricular muscle. Activation of these receptors produces afferent responses that lead to activation of efferent parasympathetic fibers and inhibition of efferent sympathetic fibers. These actions are claimed to produce the following useful effects:

1. A decrease in sinus rate (slowing of the heart).[17]
2. Antiarrhythmic effects due to (1) slowing of the conduction in the AV node, (2) decreased automaticity in supraventricular tissue, (3) decreased atrial refractory period, and (4) suppression of ventricular arrhythmia.[18]
3. Reduction in preload and afterload due to vasodilation consequent to inhibition of the sympathetic nervous system.[19,20]
4. Increased blood flow to the kidneys due to inhibition of the renin/angiotensin/aldosterone system consequent to inhibition of the sympathetic nervous system.[21]

According to Gillis and Quest,[13,14] a significant part of the improved hemodynamics produced by therapeutic doses of digitalis, and all of the useful antiarrhythmic effects are due to the indirect actions of the drug. These claims await clinical verification but, if true, are likely to be more significant in heart failure patients than in normals.

2. CNS Effects

Although therapeutic concentrations of cardiotonic steroids are said to reduce sympathetic outflow due to reflex-mediated effects, toxic concentrations have been claimed to increase CNS sympathetic outflow thereby lowering the threshold for cardiac arrhythmia.[13,14] The site of action of cardiotonic steroids in the CNS is controversial, as is its clinical significance. Several investigators have concluded that digitalis acts somewhere within the hypothalamus,[22] possibly in the área postrema[23,24] (a part of the brain not protected by the blood brain barrier). However, it is also possible that the increased sympathetic activity associated with toxic doses of digitalis may be due to indirect (reflex-mediated) effects on the CNS.

III. BIOCHEMICAL PHARMACOLOGY

There is a growing appreciation that the biochemical basis of the therapeutic and toxic effects of cardiac glycosides is far more complex than previously realized. Apart from indirect effects it is possible that the direct effects of digitalis on the heart may involve multiple mechanisms, and that different mechanisms may be dominant under different testing conditions.

When myocardial cells are depolarized, a small amount of extracellular Ca^{2+} enters the cell via the slow calcium channel and triggers the release of stored calcium which then activates the contractile process, the extent of which being proportional to the amount of Ca^{2+} released from the internal stores. Following contraction, released calcium is returned to the intracellular stores and the "excess" calcium that entered the cell via the slow calcium channel is removed from the cell via the calcium pump (Ca^{2+},Mg^{2+}-ATPase) and the Ca^{2+}/Na^{+} antiport. However, if these transporting mechanisms are slowed, more Ca^{2+} will be taken up by the stores and this will lead to an increase in the force of contraction when the cell is next depolarized.

There is now general agreement that the direct inotropic effects of digitalis are due to an effect that leads to an increase in the amount of calcium stored in the sarcoplasmic reticulum of the myocardium. There are several mechanisms whereby digitalis could produce such an increase. These include inhibition of Na^{+},K^{+}-ATPase,[11] inhibition of "basal" Ca^{2+},Mg^{2+}-ATPase,[25] stimulation of the Na^{+},K^{+},Cl^{-} co-transporter,[26] and release of membrane-bound stores of Ca^{2+}.[27] There is good evidence supporting the first three of these mechanisms and weaker evidence in support of the last.

Two of the above mechanisms, namely inhibition of Na^{+},K^{+}-ATPase and stimulation of the Na^{+},K^{+},Cl^{-} co-transporter, lead to a transient increase in the level of intracellular Na^{+} which slows the outward transport of Ca^{2+} via the Ca^{2+}/Na^{+} antiport. Inhibition of "basal" Ca^{2+},Mg^{2+}-ATPase slows the outward passage of Ca^{2+} while the cell is in diastole. All three mechanisms can thus slow the outward transport of Ca^{2+} with a resulting increase in the amount of stored intracellular calcium.

It is possible that several of the mechanisms mentioned contribute to the inotropic effects of cardiotonic steroids. It is also possible that different mechanisms are expressed under different test conditions.

The therapeutic and toxic effects of digoxin in man are associated, respectively, with concentrations of the order of 1 and 2 $\times$ 10^{-9} *M*, whereas concentrations several magnitudes higher are used to produce inotropic responses in isolated myocardium from animals of similar sensitivity to man. It is thus possible that "clinical" concentrations of glycosides produce positive inotropy by one mechanism (e.g., stimulation of the Na^+,K^+,Cl^- co-transporter[26]) and laboratory studies by another (e.g., inhibition of Na^+,K^+-ATPase). Alternatively, as suggested by Gillis and Quest,[13,14] the therapeutic effects of digitalis seen clinically may be due mainly to its indirect effects.

A further complication is that Na^+,K^+-ATPase can exist in a number of isoforms whose affinities for digitalis vary by up to 1000-fold. (This will be discussed later.)

VI. NA$^+$,K$^+$-ATPase

Although the mechanism of action of digitalis is controversial,[4] there is no doubt that a highly specific receptor for digitalis exists on the extracellular face of the enzyme Na^+,K^+-ATPase. In fact, it is the only such receptor for digitalis that has been identified. There is also a considerable body of evidence which indicates that Na^+,K^+-ATPase plays a key role in the therapeutic and toxic effects of cardiotonic steroids.[4] For this reason, many workers in the field believe that inhibition of Na^+,K^+-ATPase mediates the inotropic and toxic effects of digitalis under both clinical and experimental conditions. For a more-detailed discussion of the evidence for and against this hypothesis, see Reference 4.

Na^+,K^+-ATPase is a phosphohydrolase (ATP-splitting enzyme) that is found in the plasma membranes of all eukaryotic cells. It extends across the full width of the plasma membrane and its role is to actively transport Na^+ out of the cell in return for the coupled inward transport of K^+. The pump is electrogenic and leads to a net loss of cations from the cell. This arises because three Na^+ ions are transported out of the cell in return for two K^+ ions. This contributes significantly to the transmembrane potential difference that exists across cell membranes. The enzymology of Na^+,K^+-ATPase has been extensively reviewed.[4,28-45]

Cardiotonic steroids inhibit Na^+,K^+-ATPase and this is generally considered to be the primary event that gives rise to their biological activity.[29] However, some studies have indicated that therapeutic doses of cardiotonic steroids stimulate the enzyme[10,25,40,47-50] and others that the therapeutic effects are unrelated to effects on Na^+,K^+-ATPase, or at least are unrelated to its inhibition.[51,52] At the basis of the controversy is the fact that most experimental work is done at concentrations at least an order of magnitude greater than those that apply clinically.[4]

Na^+,K^+-ATPase shows a remarkable degree of homology across the phyla.[53] Sequence homology between Na^+,K^+-ATPase and K^+-ATPase from prokaryotes has led to the suggestion that Na^+,K^+-ATPase may have evolved from the fusion of single-cation transport systems that exist in more primitive life forms.[39] It is possible that transmembrane cation transport systems evolved, initially, to deal with the problem of excess osmotic pressure in primitive cells. They may then have become adapted to other specialized roles, such as the maintenance of the transmembrane potential difference necessary for excitation and conduction in muscle and nerve cells. As well as regulating the environment of intracellular fluids, Na^+,K^+-ATPase also plays a key role in controlling the amount and composition of many extracellular fluids including cerebrospinal fluid, lung cavity fluids, and the aqueous humour of the eye. To quote from a recent review: "The importance of this enzyme cannot be overstated, and it has been estimated that approximately 25% of the ATP catabolized by an individual at rest is used by this enzyme."[54]

A. THE EFFECTS OF CARDIOTONIC STEROIDS ON NA$^+$,K$^+$-ATPASE: THE SODIUM PUMP LAG HYPOTHESIS

The most widely accepted model of digitalis activity is the Na^+ pump lag hypothesis.[4] According to this theory, therapeutic concentrations of cardiotonic steroids inhibit a small proportion of a cell Na^+,K^+-ATPase ("Na^+ pump units"). Because of this, intracellular Na^+ levels begin to rise, but, since the enzyme is stimulated by intracellular Na^+, the noninhibited pump units will cycle faster. Even so, there is a lag before all excess Na^+ is cleared. There is thus a transient increase in intracellular Na^+ and this leads to a transient inhibition of the Na^+/Ca^{2+} antiport. Although this antiport can act in both directions, its normal mode is to transport Ca^{2+} out of the cell in exchange for extracellular Na^+. Under these conditions, rising levels of intracellular Na^+ slow down the antiport leading to an increase in intracellular Ca^{2+} stores which leads to an increase in contractility when the cell is depolarized.

Toxic concentrations of cardiotonic steroids, according to this model, cause extensive inhibition of Na^+,K^+-ATPase such that the increased levels of intracellular Na^+ cannot be cleared before the next membrane depolarization. Toxic concentrations thus lead to a sustained increase in intracellular Na^+ and excessive retention of Ca^{2+}, thus upsetting the transmembrane potential as well as disrupting mitochondrial function.

Although there has been much support for the Na^+ pump lag hypothesis, new evidence about the nature of Na^+,K^+-ATPase and its isoforms suggests that the theory, as stated above, may need to be modified.

B. ENZYMOLOGY OF NA$^+$,K$^+$-ATPASE

Part of the basis of the contention about the nature of Na^+,K^+-ATPase and its interaction with digitalis relates to the drastic procedures usually used to isolate the enzyme, and to the highly artificial conditions under which it

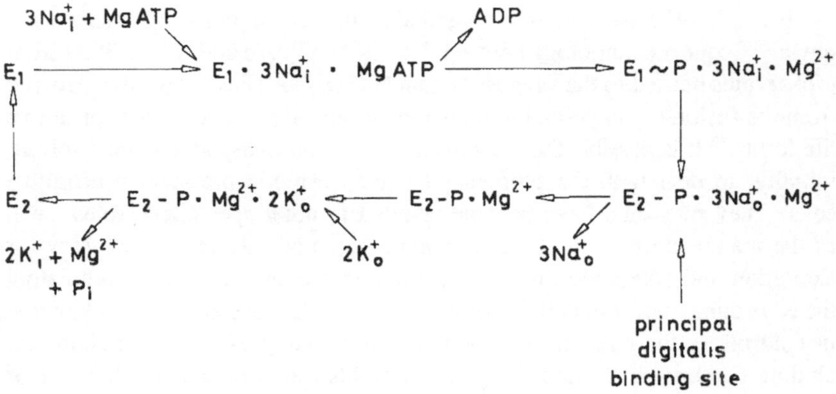

FIGURE 2. Modified version of the Albers-Post model for the cycling of Na^+,K^+-ATPase (see text for an explanation). (From Thomas, R., Gray, P., and Andrews, J., *Advances in Drug Research*, Vol. 19, Testa, B., Ed., Academic Press, New York, 1990, 311. With permission.)

is studied. In the intact cell, the enzyme is held within the phospholipid matrix of the plasma membrane which, apart from influencing the enzyme's conformation, maintains different electrolyte environments on each side of the macromolecular complex.

The above notwithstanding, there is much that can be said with certainty about the nature of Na^+,K^+-ATPase. The enzyme extends across the full width of the plasma membrane and consists of two polypeptides designated α and β. The polypeptides are associated with phospholipids which must be present for phosphohydrolase and pump activity. The α polypeptide has a relative molecular mass (M_r) of about 110,000 and is the ATP splitting subunit. The β subunit has an M_r of about 55,000 and is a glycoprotein. Its role has not been established, although it is essential for activity. The enzyme has high- and low-affinity binding sites for ATP. The latter have been called regulatory sites but this is unlikely since intracellular concentrations of ATP are very constant, except in ischemia. The digitalis binding site is located on the extracellular surface of the α subunit.

The enzyme is known to cycle through a series of conformations. According to the most widely quoted model, that of Albers[55] and Post,[56] the α subunit can exist in at least four main states designated E_1, E_1~P, E_2P, and E_2, where E is a particular conformation of the α subunit, P is a phosphate moiety, and ~ a high-energy bond (Figure 2). According to the model, cardiotonic steroids bind to the extracellular face of the E_2P state. Recent studies indicate that the Albers-Post model is an oversimplification and that there are, probably, at least three phosphorylated states, and possibly epicycles within the main cycle.[57]

The Na^+,K^+-pump probably exists as a dimer of Na^+,K^+-ATPase: [(α,β)(α,β)]. Various suggestions for how the enzyme units operate as a pump have been made, with the flip-flop model being the most favored:[58,59]

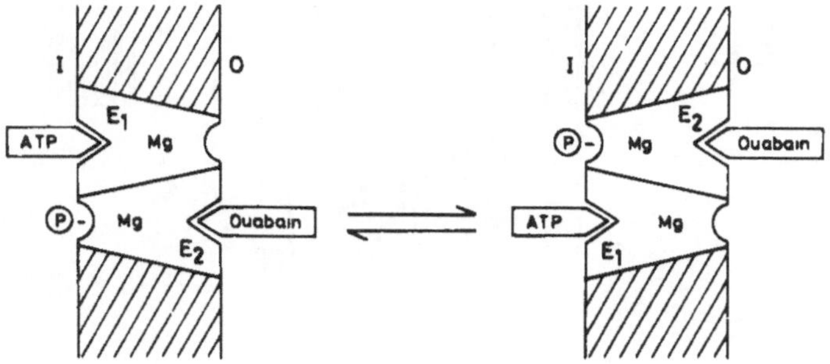

FIGURE 3. Possible coordinated conformational changes in Na^+,K^+-ATPase based on the Albers-Post model (Figure 2). It is suggested that when one catalytic unit is in the E_1 conformation, the other must, of necessity, take the E_2 conformation. Coordinated changes from the E_1-E_2 to the E_2-E_1 arrangement open a pore through which Na^+ and K^+ are transported. These changes are associated with the binding of ATP to the E_1 conformation and subsequent phosphorylation to produce the E_2P conformation. Only the latter contains the ouabain-binding site (i.e., the digitalis receptor). The binding of digitalis freezes the system so that further conformational changes are prevented and ion transport ceases. I and O refer to the inside and outside of the membrane. (From Schoner, W., Pauls, H., and Patzelt-Wenczler, R., *Myocardial Failure*, Riecker, B., Weber, A., and Goodwin, J., Eds., Springer-Verlag, Berlin, 1977, 115. With permission.)

$$E_1 - E_2 \rightleftharpoons E_2 - E_1$$
(E = conformation of the α subunit of Na^+,K^+-ATPase)

A suggested[60] relationship between the above states is shown in Figure 3, which depicts the two α subunits oscillating between the E_1 and E_2 conformations. The figure shows only the α subunits of the [(α,β)(α,β)] dimer. The conformational changes are such that, when the α subunit of one member of the dimer is in the E_1 state, the other is in the E_2 form. As the two enzyme units undergo coordinated change from the E_1 to E_2 states, pores are opened that allow for the coupled transport of Na^+ and K^+. The binding site for cardiotonic steroids is located on the extracellular surface of the low-energy phosphorylated from of the E_2 state (E_2P — refer to Figures 2 and 3). When the site is occupied, the conformation of the E_2P state is frozen, the cycle of conformation changes stops, and cation transport ceases.

Schoner's group[61,62] has recently shown that cooperativity exists between the two α subunits: occupation of the low-affinity ATP binding site causes the high-affinity site to shift to the Na^+ form.

The phosphorylation of Na^+,K^+-ATPase takes place through an aspartate residue, as it does with H^+,K^+-ATPase and Ca^{2+},Mg^{2+}-ATPase. These enzymes are therefore known as P-type ATPases. Formerly, they were called E_1-E_2 enzymes, but this term does not discriminate from other ATPases which also undergo E_1-E_2 conformational changes.

Purified membrane-bound Na^+,K^+-ATPase has been crystallized and major conformational differences have been shown for two forms of the α subunit corresponding to the E_1P and E_2P states.[63]

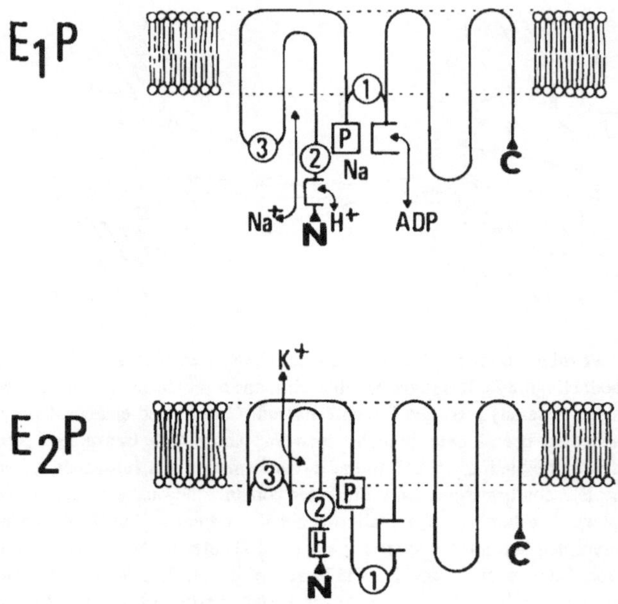

FIGURE 4. Model for the arrangement of the α-subunit in the E_1 and E_2 forms of Na^+,K^+-ATPase. The circled numbers mark the sites of primary tryptic cleavage in KCl (1) and (2) or NaCl (3) and chymotryptic cleavage in NaCl (3). In the E_1P conformation, bonds 2 and 3 are exposed to cleavage while bond 1 is protected. Transition to E_2P is accompanied by protonation of an ionizable group close to the NH_2 terminus. In the E_2P conformation, bond 3 is protected while bond 1 is exposed to trypsin; the position of bond 2 is such that it is cleaved secondary to cleavage of bond 1 within the same α-subunit. It is proposed that transition from E_1P to E_2P is accompanied by movement of a part of the segment containing the aspartyl phosphate from a relatively hydrophilic to a more hydrophobic environment. This segment between bonds 2 and 3 is proposed to engage in cation binding and formation of the pathway across the membrane. (From Jørgensen, P. L., Skriver, E., Hebert, H., and Maunsbach, A. B., *Ann. N.Y. Acad. Sci.*, 402, 207, 1982. With permission.)

Controlled proteolysis combined with specific chemical labeling enabled Jørgensen and co-workers[64] to propose the model for Na^+,K^+-ATPase shown in Figure 4. The figure depicts the α subunit in the E_1P and E_2P states. Circled numbers mark the spot of tryptic or chymotryptic cleavage. A major feature of the model is the proposal that the α subunit loops through the membrane eight times.

Herbert et al.[63] used electron microscopy of crystalline Na^+,K^+-ATPase to obtain Fourier-filtered projection structures. A three-dimensional model of the dimeric form of the enzyme based on these studies is shown in Figure 5 and is significant for the deep cleft between the two members of the dimer. The model is useful in that it shows that, even though the α subunit may loop through the membrane, the loops are nevertheless closely associated to give a compact structure.

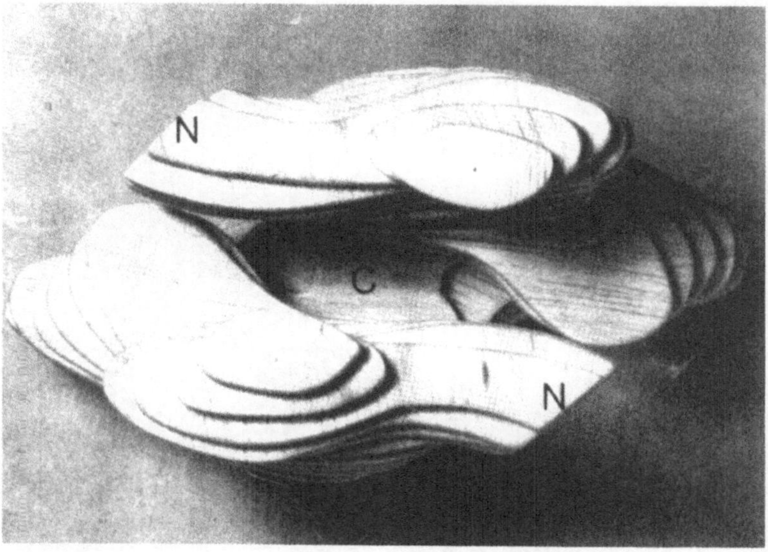

FIGURE 5. 3-D model of the Na⁺,K⁺-ATPase dimer based on electron microscopy studies When observed in the direction of the c-axis (perpendicular to the membrane) (above panel) there is a deep cleft between the symmetry-related units of the dimer, which are connected a c (bottom of cleft) and relate to neighboring units at N. The panel on following page shows the model in a direction parallel to the membrane The vertical line to the right indicates the level where the contrast variation has a minimum, which may correspond to the level of the lipid bilayer. (From Herbert, H , Skriver, E., and Maunsbach, A. B., *FEBS Lett.*, 187, 182, 1985. With permission.)

C. THE SEQUENCING OF NA^+,K^+-ATPASE

Sequencing studies of the α and β subunits of Na^+,K^+-ATPase from various sources have been carried out in a number of laboratories using complementary DNA (cDNA). This technique involves the preparation of a labeled DNA probe which is then used to identify appropriate colonies from a DNA library prepared using recombinant DNA technology. Once the appropriate DNA molecule has been identified, it is sequenced, and from this the amino acid sequence is inferred.

Essential to the success of such studies is the availability of suitable probes. These are prepared by synthesizing a DNA sequence corresponding to a known amino acid sequence from the protein under study. Typically, such a probe would correspond to a sequence of about eight amino acids. Ideally several probes should be available to confirm the identity of the suspected gene. Also, the amino acid sequence inferred for the protein should show properties established by other (classical) biochemical studies. These conditions have been met in the sequencing studies of Na^+,K^+-ATPase, thanks largely to the basic investigations of workers such as Collins et al.[65] and Jørgensen et al.[64] A typical sequencing procedure is shown in Figure 6.

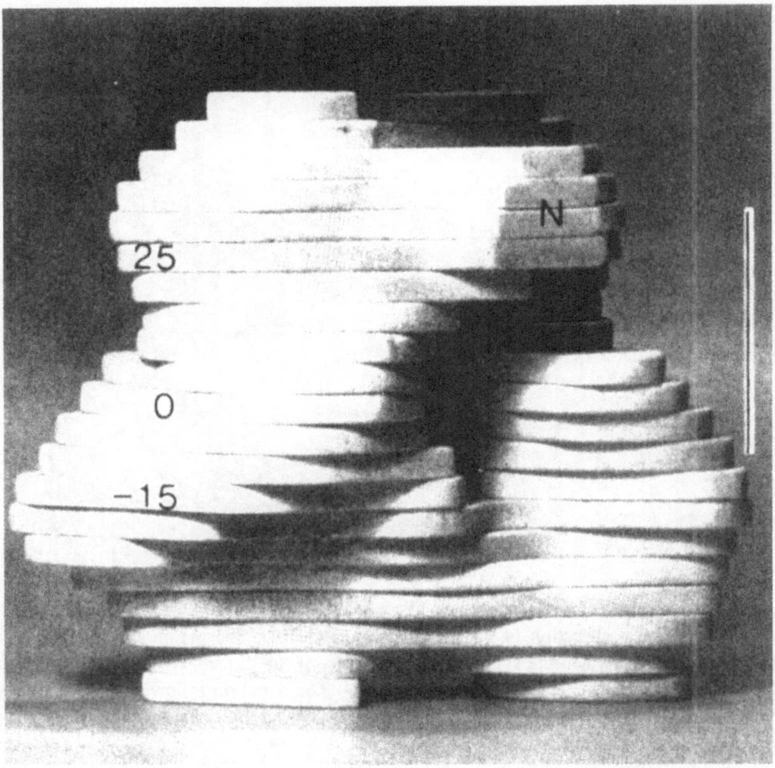

FIGURE 5 continued.

The amino acid sequences of the α and β subunits have been determined by several groups using cDNA [reviewed by Lingrel et al.,[54] whose group (Shull et al.) were the first to report the sequencing of a Na^+,K^+-ATPase polypeptide].

1. The α Subunit

Shull et al.[66-69] sequenced the α subunit from sheep kidney Na^+,K^+-ATPase and found it to contain 1016 amino acids (M_r = 112,177). Eight major hydropathic regions were identified and it was suggested that these represented the eight transmembrane domains proposed by Jørgensen et al.[64] There is some ambiguity in the interpretation of the hydropathy profile data and this has led to six to ten transmembrane regions being suggested for the α subunit (discussed by Lingrel et al.[54]).

At least three isoforms of the α subunit have been identified and designated α_1, α_2, and α_3 (older nomenclature: α, α^+, and αIII).

A hypothetical representation of the findings of Shull et al. for the sheep α_1 subunit is given in Figure 7. As can be seen, the bulk of the polypeptide lies on the intracellular side. There are four extracellular domains which are

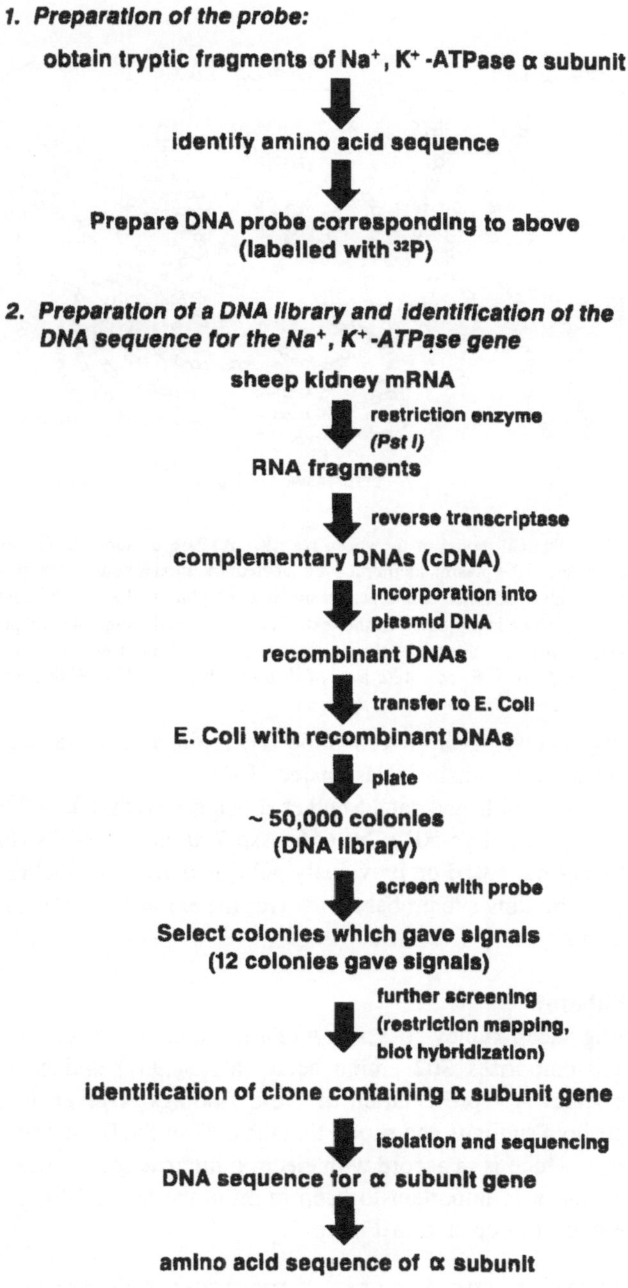

FIGURE 6. Flow chart showing the essential steps in the sequencing of the α subunit of Na^+,K^+-ATPase.

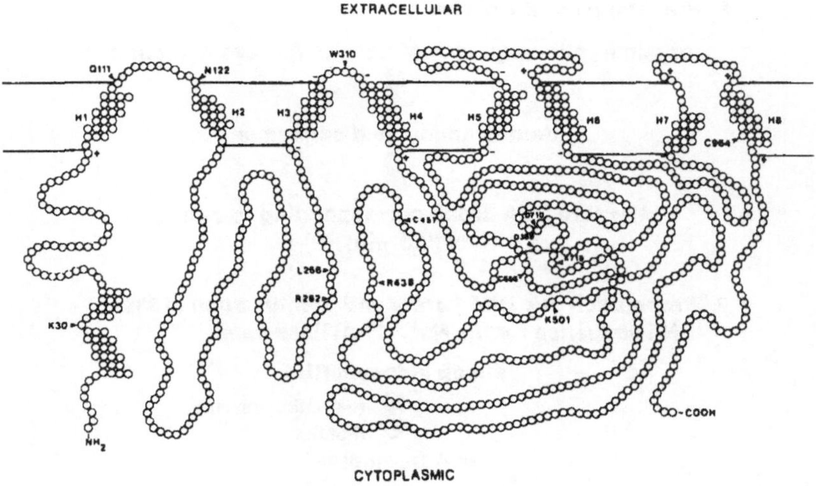

FIGURE 7. Hypothetical model of the sheep Na^+,K^+-ATPase α_1 subunit. The extracellular and intracellular sides of the plasma membrane are labeled; the model depicted here shows eight transmembrane domains. Specific amino acid residues are highlighted and numbered. The significance of these residues is discussed in the text. Certain charged amino acids that define the transmembrane domains are indicated by the appropriate charge. (From Lingrel, J. B., Orlowski, J., Shull, M. M., and Price, E. M., *Acid Res. Mol. Biol.*, 38, 37, 1990. With permission.)

referred to by specifying the hydropathic (H) regions on either side. Thus, the first extracellular domain is designated H_1-H_2.

Based on other published data, Shull et al. suggested that the ATP binding site involves Asp-369, Lys-501, Cys-656, Asp-710, and Lys-719 (Figure 7). They also suggested, based on previously published affinity labeling studies, that the digitalis binding site probably involves the extracellular H_3-H_4 domain of the α subunit.

2. The β Subunit

Also using sheep kidney Na^+,K^+-ATPase, Shull et al.[69] established that the β subunit comprises 302 amino acids (M_r 34,937) and contains one hydropathic zone. A representation of these findings is given in Figure 8 which depicts both subunits and shows that the bulk of the β subunit is located extracellularly, which is in accord with electron microscopy studies. In viewing such models, it is important to keep in mind the space-filling nature of the coiled protein as depicted in Figure 5.

D. LOCATION OF THE DIGITALIS BINDING SITE ON NA$^+$,K$^+$-ATPase

1. Labeling Studies

Photoaffinity labeling and other forms of chemical modification have been used extensively to locate the ATP and digitalis binding sites on Na^+,K^+-ATPase (reviewed by Pedemonte and Kaplan[70]). Although the results of these

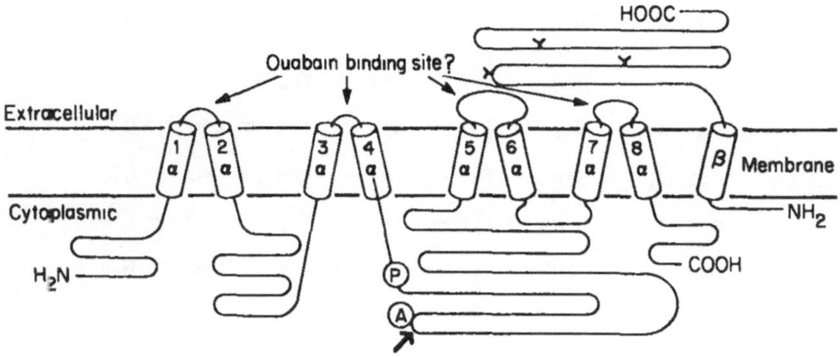

FIGURE 8. Hypothetical model showing the α (to the left) and β (to the right) subunits of Na^+,K^+-ATPase. Unlike the α subunit which probably makes eight traverses through the membrane, the β unit is considered to make only one. The β unit also differs from the α in that its major domain lies on the extracellular side of the membrane. P and A represent the locations of the phosphorylation and putative ATP-binding regions. (From Price, E. M. and Lingrel, J. B., *Biochemistry*, 27, 8400, 1988. With permission.)

studies have led to some conflicting interpretations, it would appear from this work that the digitalis receptor is located exclusively on the α subunit, or at least, that this unit contains the main locus of the receptor. A photoaffinity label attached to the steroid or sugar residue of a monoglycoside will label only the α subunit,[71-73] whereas those with the label attached to the terminal sugar of a triose glycoside label the β subunit.[74]

Photoaffinity studies also indicate that a tryptophan residue is part of the digitalis binding region.[75] Extensive studies have shown that the digitalis receptor is located in an extracellular portion of Na^+,K^+-ATPase, and the only tryptophan residue so located is in the H_3-H_4 domain (Figure 9).

2. The Problem of Species Differences

A major problem with the suggested location for the digitalis receptor is that it does not appear to explain species differences. The amino acid sequences of the H_1-H_2 and H_3-H_4 domains for various species are shown in Figure 9 (A to F).

Na^+,K^+-ATPase containing the rat α_1 isoform is about two to three orders of magnitude less sensitive than that from the other sources shown. This reduced sensitivity is reflected by reduced affinity, so that species differences are generally considered to be due to reduced binding strength rather than reduced intrinsic activity. However, as can be seen (Figure 9), the amino acid sequence in the H_3-H_4 domain of the low-sensitivity rat α_1 isoform is the same as those in the H_3-H_4 domains of the sensitive sheep and pig α_1 forms. Furthermore, the H_3-H_4 regions of the digitalis-sensitive α_2 and α_3 isoforms differ in amino acid content compared with the other sensitive isoforms. The amino acid sequence common to the H_3-H_4 domains of all isomers is tryptophan-leucine-glycine. This may, thus, be the region that binds digitalis.

(a) H_3-H_4 domains of α-subunit from rat, sheep and pig.

Rat α_1	Glu	Tyr	Thr	Trp	Leu	Glu
rat α_2	Gly	Tyr	Ser	Trp	Leu	Glu
rat α_3	Gly	Tyr	Thr	Trp	Leu	Glu
sheep α_1	Glu	Tyr	Thr	Trp	Leu	Glu
pig α_1	Glu	Tyr	Thr	Trp	Leu	Glu

(b) H_3-H_4 domain of rat, sheep and pig α_1

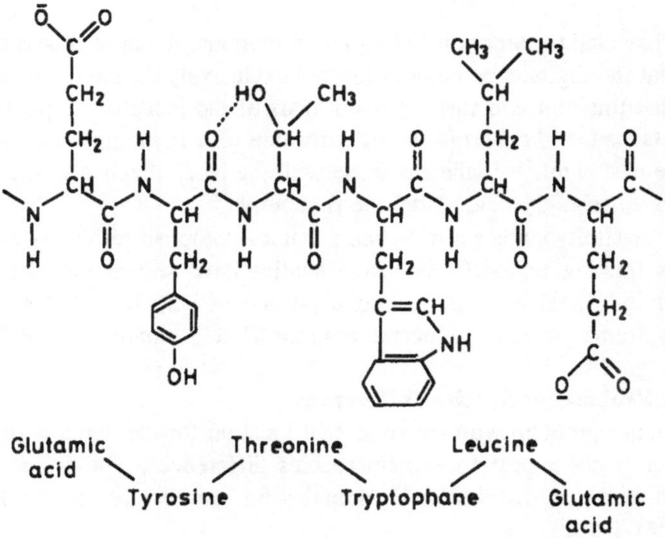

FIGURE 9. The amino acid sequences of the H_1-H_2 and H_3-H_4 domains of Na^+,K^+-ATPase α subunits from several species (based on the findings of Shull and co-workers — see text for references). The amino acids underlined in (e) are those in which the rat α_1 subunit differs from that of other species. (From Thomas, R., Gray, P., and Andrews, J., *Advances in Drug Research*, Vol. 19, Testa, B., Ed., Academic Press, New York, 1990, 311. With permission.)

How then to explain species differences? It is possible that interaction with the H_3-H_4 domain may be only the first step in binding the glycoside to the receptor. The initial binding step may be followed by conformational changes which intensify binding in high-sensitivity but not in low-sensitivity isoforms. There have been many suggestions but no hard evidence as to what

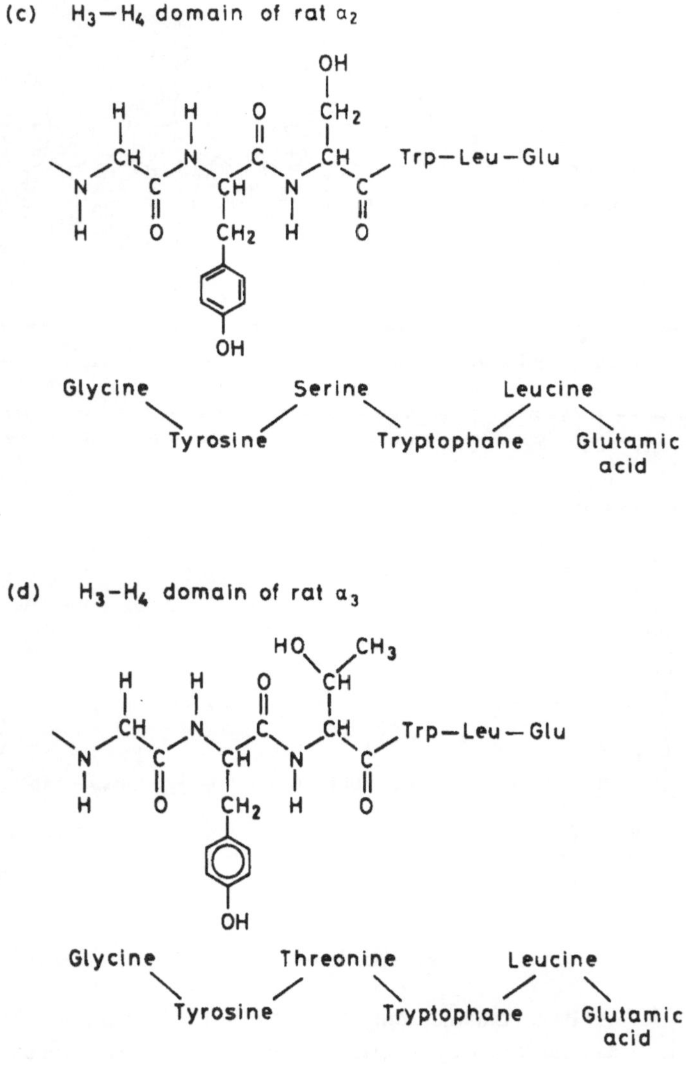

FIGURE 9c and d.

these changes may be. Another possibility is that the H_3-H_4 domain does not form part of the digitalis receptor. A hypothetical model for explaining species differences is given in Figure 11. Before discussing this, some other aspects of Na^+,K^+-ATPase need to be considered.

E. MULTIPLE BINDING SITES FOR CARDIOTONIC STEROIDS AND THE EXISTENCE OF ISOFORMS OF NA^+,K^+-ATPASE

During the first 15 years in which the binding of radioactive isotopes of digitalis was studied, it was generally concluded from Schatchard plot analysis

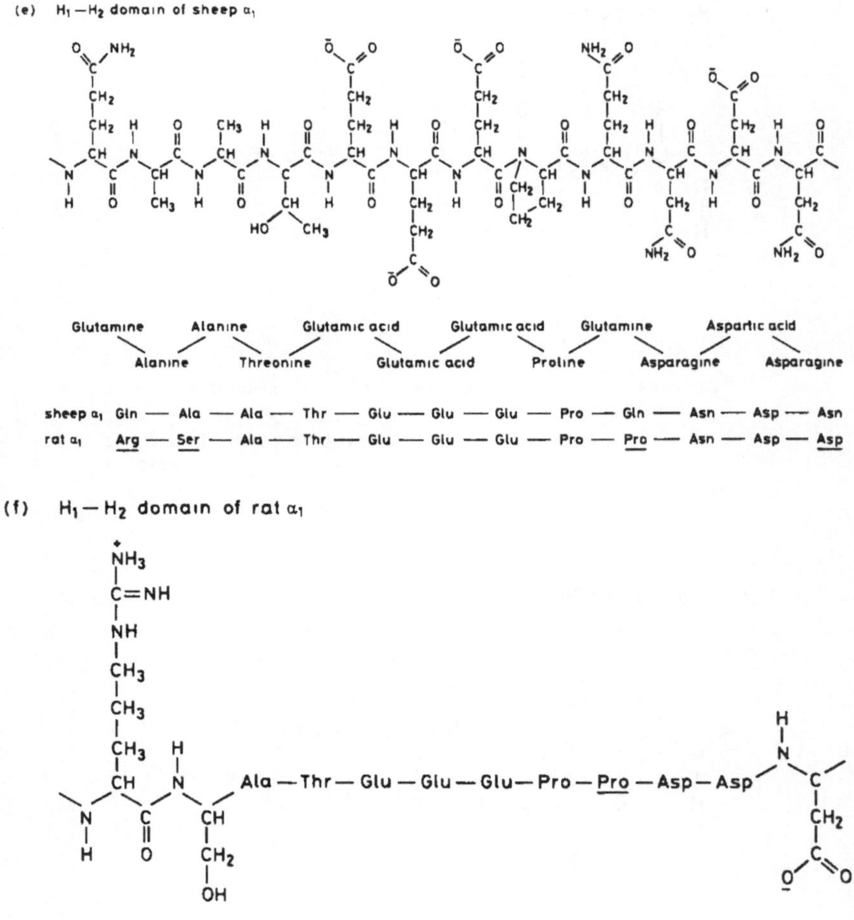

FIGURE 9e and f.

that Na$^+$,K$^+$-ATPase contained only one specific binding site for digitalis. This belief was overturned by a series of studies in the early 1980s which showed, at least in some species, that digitalis produced separate high- and low-dose inotropic effects. This was correlated with high- and low-binding affinities to Na$^+$,K$^+$-ATPase, and this in turn was correlated with the co-existence of separate isoforms of the enzyme.

Such isoforms were first reported by 1979 Sweadner,[76] who found that the ouabain-binding affinity of rat brain Na$^+$,K$^+$-ATPase containing the α_2 isoform was 100 times greater than that containing the α_1 obtained from rat heart, and that the isomers differed in their antigenic determinants. Subsequent studies showed that the isomers differed in their N-terminal sequences and were coded for by different genes which differed in their chromosomal location (reviewed by Sweadner[53] and Lingrel et al.[54]).

The human genome contains three (possibly five) genes corresponding to α isoforms which suggests a distinct biological role for these entities. This is supported by evidence for tissue-specific expression of Na^+,K^+-ATPase isoforms in lungs, kidney, heart, and brain.[53,54]

An important step in determining whether the digitalis receptor on Na^+,K^+-ATPase mediates the inotropic and/or toxic effects of cardiotonic steroids was to establish whether a correlation exists between pharmacological activity and affinity for the various isoforms and binding sites on Na^+,K^+-ATPase.

One of the most elegant studies of this type was that of Ng and Akera,[77] who studied the effects of the cardiotonic steroid ouabain on ferret heart. These workers found

1. That ouabain showed two binding affinities for ferret heart Na^+,K^+-ATPase (K_D = 10 and 230 n*M*).
2. That ferret heart Na^+,K^+-ATPase could be separated into two isoforms showing the above K_Ds for ouabain binding.
3. That, on isolated ferret heart, ouabain showed a biphasic inotropic effect at concentrations that correlated with the K_Ds for binding to Na^+,K^+-ATPase.
4. That the ratios of α_2 (high digitalis sensitivity) to α_1 (low sensitivity) isoforms in ferrets of different ages were, respectively

5-day-old ferret:	1:7
10-day-old ferret:	1:4
Adult ferret:	1:1

5. That the above corresponded to reduced sensitivity of the neonate to the inotropic effects of ouabain.

Similar correlations were found by Sweadner and Farshi[78] for rat myocardium.

Earlier in this review it was mentioned that one of main reasons for doubting that the effects of digitalis seen in patients were due to inhibition of Na^+,K^+-ATPase was the "concentration problem", namely that the clinical effects (therapeutic and toxic) are associated with concentrations of drug several orders of magnitude lower than those used to demonstrate correlations between enzyme inhibition and inotropy in laboratory animals. The discovery of the co-existence of Na^+,K^+-ATPase isoforms with greatly differing affinities for cardiotonic steroids in the same tissue may offer an explanation of the "concentration problem".

Maixent et al.[79,80] suggested that the inotropic and toxic effects of cardiotonic steroids are correlated, respectively, with binding to high- and low-affinity binding sites (i.e., with high- and low-sensitivity isoforms of Na^+-K^+-ATPase).

F. THE MOLECULAR BASIS OF SPECIES DIFFERENCES IN SENSITIVITY TO CARDIOTONIC STEROIDS

1. Transfection Studies

Although there is persuasive evidence that the digitalis receptor on Na^+,K^+-ATPase is located on the H_3-H_4 domain, the amino acid sequence of this region is the same for Na^+,K^+-ATPase with high and low sensitivity to cardiotonic steroids. A possible explanation is that other Na^+,K^+-ATPase domains are also involved in binding cardiotonic steroids.

Lingrel's group[54,81] has used chimeric DNA constructs (expressed in HeLa cells) to provide evidence for the involvement of the H_1-H_2 domain in the expression of digitalis sensitivity. (In Greek mythology, the Chimera was a fire-breathing monster with a lion's head, a goat's body, and a serpent's tail). Lingrel's group found that the construct consisting of the amino-terminal half of rat α_1 and the carboxylic half of sheep α_1 conferred digitalis resistance on HeLa cells, whereas the reverse construct did not.

These workers also found that replacement of the sheep α_1 H_1-H_2 domain with that of the rat α_1 conferred resistance. Even more startling, transfection of the two terminal amino acids from the rat α_1 H_1-H_2 domain (arginine and aspartic acid) also conferred digitalis resistance on sheep Na^+,K^+-ATPase. This, together with evidence previously cited, suggests that both the H_1-H_2 and H_3-H_4 domains are involved in the digitalis receptor.

Kinetic studies comparing the binding of cardiotonic steroids to high- and low-sensitivity forms of Na^+,K^+-ATPase have established that κ_{+1} is similar for both forms, where κ_{-1} is much greater for low-sensitivity Na^+,K^+-ATPase.[29] This suggests that the initial binding event is the same for both types of the enzyme, but that some event (such as a conformational change in the enzyme) takes place after binding which greatly stabilizes the drug-receptor complex in high-sensitivity forms of the enzyme.

In view of the previous discussion, one can imagine that binding of digitalis takes place initially with the H_3-H_4 domain, followed, in sensitive forms of the enzyme, by additional binding to the H_1-H_2 domain.

2. A Hypothesis Based on Structure-Activity Relationships (SAR)

In 1974, based on SAR studies, I and my co-workers suggested that the lactone of cardiac glycosides interacted with the receptor by means of a hydrogen bond reinforced by a charge transfer complex formed between the lactone and a full or partial negative charge in the receptor surface. The reasons for making this suggestion will be discussed later.

As can be seen from the sequencing studies (Figure 9) there are many negative charges in the H_1-H_2 and H_3-H_4 domains.

However, taking into account the fact that the initial interaction between digitalis and its receptor on Na^+,K^+-ATPase seems to involve the sequence tryptophan-leucine-glycine, we have recently published[4] a hypothesis that suggests that the lactone interacts with tryptophan. The heterocyclic nitrogen

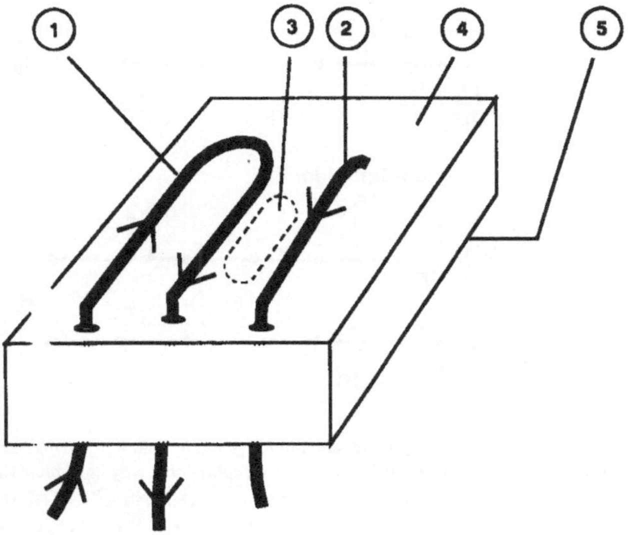

FIGURE 10. Hypothetical representation of the H_1-H_2 and H_3-H_4 domains of the α subunit of Na^+,K^+-ATPase. It is suggested that the domains run parallel to each other with the H_1-H_2 domain looping to form a U-shaped structure. The H_3-H_4 is shown as running prograde to the adjacent segment of the H_1-H_2 domain. The digitalis-binding site is shown as being located between the H_1-H_2 and H_3-H_4 domains. The arrangement shown is entirely speculative. (1) H_1-H_2 domain, (2) H_3-H_4 domain, (3) digitalis-binding site, (4) extracellular surface of plasma membrane, and (5) intracellular surface of plasma membrane. Refer to the text and Figure 11 for further details.

of the tryptophan ring system has a significant partial negative charge even in the nonionized state, and, if ionized, has a charge significantly greater than -1.

As will be discussed later, we have also proposed, based on our SAR studies, that there is a full negative charge in the region that binds the first sugar moiety.

On this basis, we have suggested the mechanism depicted in Figure 11 as a possible explanation of how digitalis binds to its receptor and why the presence of arginine and aspartic acid in the H_1-H_2 domain is so important with respect to species differences in sensitivity, and hence to binding affinity. As originally published, the model had an error in that the additional carboxylic acid residue present in the H_1-H_2 domain of the rat α_1 polypeptide was not shown. This has been added to the revised version of the figure.

Before considering the model, I would like to speculate, based on the model of Herbert et al. (Figure 5) that the extracellular domains run approximately parallel and that the long H_1-H_2 domain doubles back on itself as shown in Figure 10. Our model (Figure 11) suggests, for digitalis-sensitive forms of Na^+,K^+-ATPase, that the glycoside binds to both the H_3-H_4 domain and one of the suggested limbs of the H_1-H_2 domain. If this is so, an important

(A)

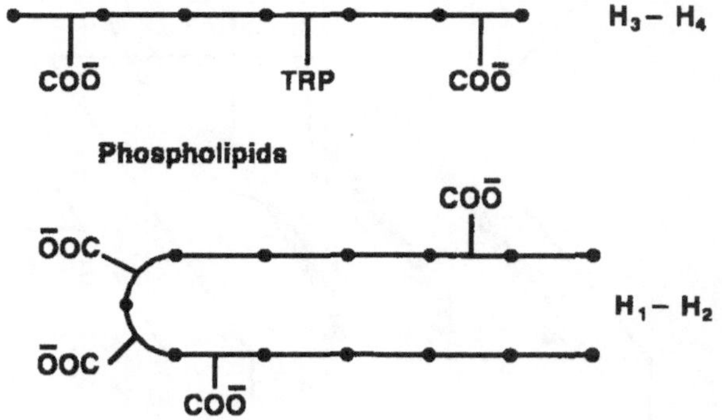

FIGURE 11. This is a purely hypothetical model that has been suggested in the hope of stimulating further thought. The model attempts to explain the way in which the binding of digitalis to the receptor induces major conformational changes in Na^+,K^+-ATPase and why the rat α_1 isozyme has low sensitivity to digitalis. In (A) the two domains that appear to be involved in binding digitalis, namely the H_3-H_4 and H_1-H_2 domains, are shown lying side by side separated by a zone of phospholipid. The two domains are prevented from close association by the repulsive forces of the ionized carboxylic acid residues. In (B) a digitalis molecule is shown associating first with the H_3-H_4 domain and then with the H_1-H_2 domain. It is proposed that interaction takes place first between the lactone and the tryptophan ring system. It is further suggested, although this is not essential to the model, that during this interaction a proton is transferred from the tryptophan nitrogen to the neighboring carboxylic acid which twists down into a water-free cleft where the protonated form is stable. Simultaneously with this, the lactone forms an ion dipole interaction with the tryptophan anion. The protonation and movement of the carboxylic acid residue overcomes the repulsive force between the H_1-H_2 and H_3-H_4 domains allowing the two to move together further enclosing the digitalis molecule and inhibiting dissociation, thereby increasing potency. It is suggested that the carboxylate anion of the H_1-H_2 domain forms part of the sugar-binding site (as discussed in the text, there is good reason for believing that such an anion exists in the sugar-binding site). The situation that applies in the rat α_1 subunit is depicted in (C). As discussed in the text, the feature that appears to give the rat α_1 subunit digitalis resistance is the presence of two amino acids in the H_1-H_2 domain, one of which is the highly basic arginine residue which is protonated (positively charged) at biological pH. The model suggests that a salt bridge forms between the arginine and the Asp-121 carboxylic acid group which fixes the conformation of the H_1-H_2 domain, preventing it from moving toward the H_3-H_4 domain as shown in (B). This decreases the stability of the complex with digitalis but does not prevent high concentrations of drug from completely inhibiting rat Na^+,K^+-ATPase, since the inhibition step is associated with the initial conformational changes that involve only the H_3-H_4 domain. (Modified from Thomas, R., Gray, P., and Andrews, J., *Advances in Drug Research*, Vol. 19, Testa, B., Ed., Academic Press, New York, 1990, 311.)

question is whether these two binding areas emerge from the membrane to run in a retrograde or prograde fashion. In Figure 10, I have shown the two regions as running prograde. However, if the H_3-H_4 domain runs in the opposite sense, it does not invalidate the model, since the H_3-H_4 domain has carboxylic side chains at each end with tryptophan roughly in the middle.

The key concept in the model shown in Figure 11 is that there are two ways in which cardiac glycosides can bind to Na^+,K^+-ATPase. In enzymes

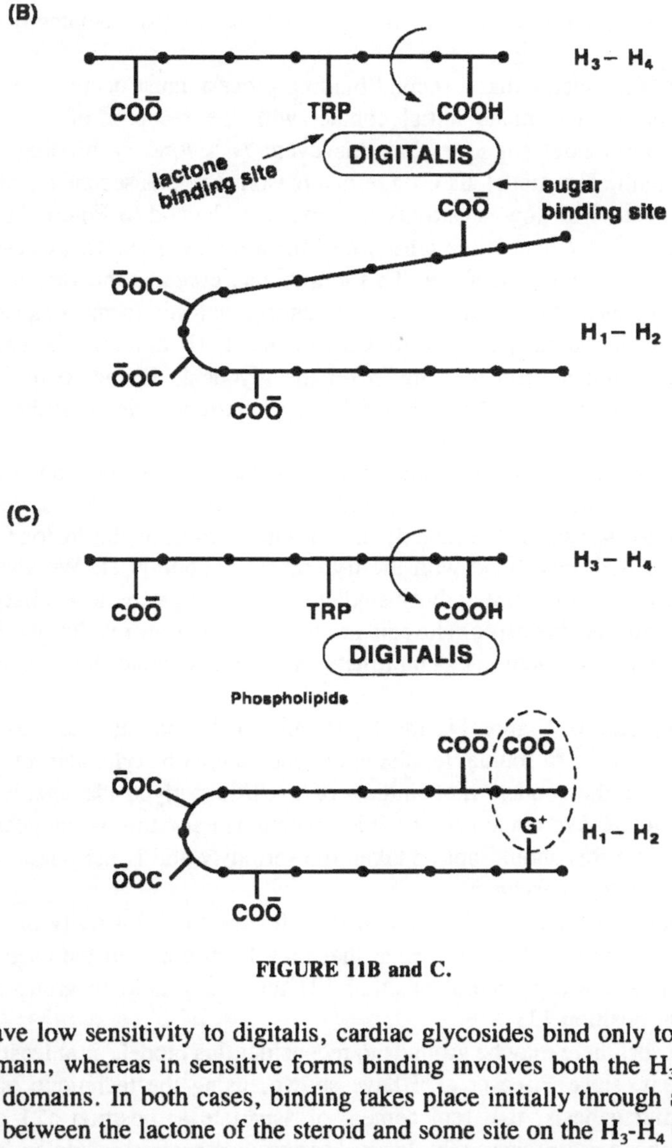

FIGURE 11B and C.

that have low sensitivity to digitalis, cardiac glycosides bind only to the H_3-H_4 domain, whereas in sensitive forms binding involves both the H_3-H_4 and H_1-H_2 domains. In both cases, binding takes place initially through an interaction between the lactone of the steroid and some site on the H_3-H_4 domain. This event then leads to a conformational change in this region which leads to inhibition of the enzyme. It should be noted that even in so-called "digitalis-insensitive" forms of the enzyme, complete inhibition of the enzyme is possible if enough drug ($\sim 10^{-4}$ *M*) is added. In other words, binding to the H_3-H_4 is all that is needed to inhibit the enzyme, but additional binding is needed to achieve high sensitivity. It should also be noted that the association rate constants for a particular glycoside are much the same for high- and low-sensitivity enzymes, indicating that the initial binding event is much the same for both types of enzyme. High-sensitivity enzymes owe their low dissociation rate constants mainly to the slow rate at which the drug dissociates from the receptor.

It therefore seems that a second binding process must occur once a cardiotonic steroid has made initial contact with the receptor of a sensitive enzyme. Our model suggests that this event is secondary binding to the H_1-H_2 domain. The model also suggests a reason why this secondary binding does not occur with low-sensitivity enzymes (see legend to Figure 11). According to our model, it is the presence of the arginine group that causes low sensitivity in the rat α_1 isoform. This amino acid contains the very strongly basic group, guanidine, and it is our contention that this forms a salt bridge with one of the carboxylic acid moieties of the H_1-H_2 domain, thereby preventing the domain from closing in on the glycoside bound to the H_3-H_4 domain. Recent work by Price et al.[82,83] may require a revision of the model and this will be discussed below.

The model, as originally published, suggests that the lactone overlays and forms a charge transfer complex with the tryptophan of the H_3-H_4 domain. Although not shown in the model, the carbonyl group of the lactone could also form a hydrogen bond with the tryptophan peptide N-H. We also suggested, based on structure activity studies, that the sugar moiety adjacent to the steroid binds to the carboxylic acid group of the aspartate residue at position 121 of the H_1-H_2 domain (the penultimate amino acid residue shown in Figure 9E).

As depicted in Figure 11, the H_1-H_2 and H_3-H_4 domains are normally kept apart because of mutual repulsion between the carboxylic acid residues. According to the model, the binding of a cardiac glycoside enables this repulsion to be overcome. A possible mechanism for this is suggested in Figure 11, but this should not be taken too seriously and is peripheral to the basic ideas being presented.

We have thus suggested an explanation for the low sensitivity of the rat α_1 isoform in terms of a salt bridge that may form between the carboxylic acid group of the aspartate at position 121 and the guanidine group of the arginine at position 111.

However, there may be grounds to reconsider this model, or at least some details of it, since Price et al.[82] have shown, using the technique of site-specific mutagenesis, that replacement of aspartate at position 121 of the sensitive sheep α isoform with alanine, serine, asparagine, or glutamic acid gave enzymes of low sensitivity when the DNA was expressed in HeLa cells. In other words, the arginine that we have posited as being the key factor in conferring low digitalis sensitivity in the rat α_1 isoform is not an absolute requirement for low sensitivity. This may invalidate the role we have suggested for arginine or it may simply have introduced another factor that influenced the structure of the H_1-H_2 domain in such a way that it would either not move towards the H_3-H_4 domain, or, if it did, it would not bind to the glycoside. Of the four mutants only one, that containing glutamate, had a free carboxylic acid. According to our hypothesis, a carboxylic group at position 121 is desirable for optimum binding to the sugar moiety. However, as will be shown later, the interaction of the sugar with its binding site is highly distance specific. It is possible that the low sensitivity of the Glu-121 mutant arises because the longer side chain of glutamic acid places the carboxylic group in an unfavorable position for binding to the glycoside and may even set up a point of steric repulsion [see Section V.D.3 for a discussion of the sensitivity of the sugar-binding site to unfavorable steric factors and note that acetylation of the 2′-OH of digitoxigenin rhamnoside reduced potency by 98% (Table 4)].

Also, the original aspartate β carboxylic group (pK_a 3.7) is about four times stronger an acid than the glutamate γ carboxylic group (pK_a 4.3), although I would not expect that to account for the difference in sensitivity.

I must emphasize that our proposed model is entirely hypothetical and is presented merely to stimulate thought and further work. For example, it would be most interesting to conduct an SAR study in which a series of glycosides was tested using the Glu-121 mutant. Glycosides which showed low activity against the natural enzyme may show higher potency with the mutant because of a more favorable alignment of binding forces.

In another study, Price et al.[83] prepared mutants in which the border amino acids (positions 111 and 122) of the H_1-H_2 domain of the sensitive sheep α_1 isoform of Na^+,K^+-ATPase were replaced with other amino acids as shown in Table 1.

In this series, the aspartic acid at 121 was conserved in all mutants (this is the acid that we suggest binds to the sugar of the glycoside). The strong base referred to in Table 1 is the amino acid, arginine (pK_a 12.5). The weak bases contain amine side chains (pK_a about 10.5). However, both types of base will be almost completely ionized when in contact with biological fluids at pH 7.4, as will all types of free carboxylate groups. Differences in pK_a values become significant in anhydrous pockets within the protein matrix, as may apply to the digitalis binding sites on the H_1-H_2 and H_3-H_4 domains.

As shown, all the mutants had low ouabain sensitivity. The model in Figure 11 can account for some, but not all, of the activities shown by the mutants. The model assumes that the salt bridge forms if an acid and a base are at either end of the border region of the H_1-H_2 domain and that the domain is folded as shown in Figures 10 and 11. The model thus readily explains the low activity of mutants DR and RD. The results for KE are also compatible: that lysine is a much weaker base than arginine could account for the higher sensitivity seen with this mutant. Mutant R- is also compatible with the model since, although there is no acid group at 122, the strength of the guanidine base is such that it could exert some attractive influence on the aspartate acid group at 121. The model does not explain the activity of mutants KK and EE. It is possible, in the case of KK, that the lysine at 111 becomes associated with the aspartate at 121. All that could be said about -D is that perhaps the

TABLE 1
Ouabain Sensitivity of Position 111 and 122 Mutants of the α_1 Isoform of Sheep Na^+,K^+-ATPase

	Position			Na^+,K^+-ATPase activity	
	111	121	122	Ouabain I_{50} (*M*)	Relative sensitivity
Natural enzyme	Gln (N)	Asp (A)	Asn (N)	8×10^{-8}	1000
Mutant DR	Asp (A)	Asp (A)	Arg (SB)	$\sim 4 \times 10^{-3}$	0.02
Mutant KK	Lys (WB)	Asp (A)	Lys (WB)	4×10^{-4}	0.2
Mutant RD	Arg (SB)	Asp (A)	Asp (A)	1×10^{-4}	0.8
Mutant EE	Glu (A)	Asp (A)	Glu (A)	8×10^{-6}	10
Mutant KE	Lys (WB)	Asp (A)	Glu (A)	5×10^{-6}	16
Mutant -D	Gln (N)	Asp (A)	Asp (A)	1×10^{-6}	80
Mutant R-	Arg (SB)	Asp (A)	Asn (N)	1×10^{-6}	80

Note: A = acid; WB = weak base; SB = strong base; N = neutral.

repulsive forces of the two carboxylic acid groups at 121 and 122 distort the shape of the peptide such that the group at 121 is moved away from its optimal position for binding to the sugar of the glycoside.

One aspect of our model not yet addressed is its applicability to the activities of genins (cardiotonic steroids from which the sugar has been removed).

Although genins are usually significantly less active than their corresponding glycosides, the fall in activity that accompanies removal of the sugar is nowhere near that which was found for the mutant Na^+,K^+-ATPases prepared by Price et al. This can be accounted for by the model in two ways. First, the genin would still be capable of producing the conformational change in the H_3-H_4 domain (see Figure 11b) thereby overcoming the repulsion between the two carboxylic acid groups and allowing the H_1-H_2 and H_3-H_4 domains to come together. Second, certain sugar derivatives destabilize the drug receptor complex (see Table 4). Thus, the H_1-H_2 may bind to the steroid as well as to appropriate sugar moieties.

V. SAR STUDIES

A. RANGE OF STRUCTURES WITH DIGITALIS-LIKE ACTIVITY

The literature on SAR studies of cardiotonic steroids is vast and has been reviewed recently by Thomas et al.[4] In the present review no attempt will be

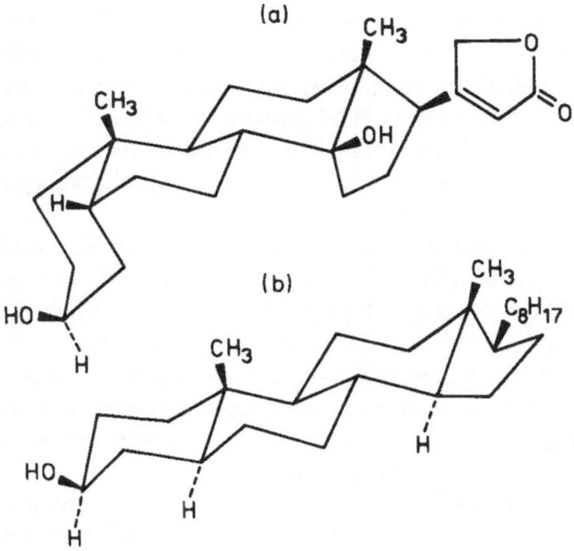

FIGURE 12. Stereochemistry of (a) digitoxigenin and (b) cholestanol. The ring junctions in digitoxigenin are *cis/trans/cis*, whereas those of cholestanol are all *trans*. (From Thomas, R. E., *Berger's Medicinal Chemistry*, 4th ed., Vol. 3, Wolff, M. E., Ed., John Wiley & Sons, New York, 1980, chap. 38. With permission)

made to repeat the systematic coverage given in the earlier publication. Instead, I will select some features that I think are appropriate to the themes being developed in this review. The range of substances that have "digitalis-like" activity are depicted in Figure 1.

The naturally occurring cardiotonic steroids (Figure 1A) have a *cis*-C/D ring junction, a 14β-hydroxyl group, and a 17β-unsaturated lactone. The A/B ring junction is usually *cis*, although there are some very active glycosides that have *trans*-A/B junctions. The steroids have 3β-hydroxyl groups to which sugars are attached when in the form of a glycoside. The 17β-unsaturated lactone can be five-membered (a butenolide) or six-membered (a pentadienolide) (see Figure 1 for examples).

In addition to the naturally occurring cardiotonic steroids, a large number of synthetic and semisynthetic derivatives and analogues have been studied (some examples are given in Figure 1B). Finally, there are a number of nonsteroidal substances that have "digitalis-like" activity (i.e., produce a positive inotropic effect and inhibit Na^+,K^+-ATPase). Examples include the naturally occurring erythrophleum alkaloids (e.g., VIII in Figure 1A) and the synthetic "pseudosteroids" (e.g., XI in Figure 1B).

1. Classical Cardiotonic Steroids: Is There a "Lead Structure"?

The largest group of cardiotonic steroids is that congeneric with digitoxigenin (Figure 12a, where 5H can be α or β). The biological activity of these compounds is often greatly affected by relatively minor changes in structure, which raises the question of whether these compounds and other, structurally very different compounds, such as the pregnanes and bisguanylhydrazones (which compete with ^{3}H-ouabain for its binding site on Na^{+},K^{+}-ATPase), are all based on the one type of lead structure, defined as the minimum structural requirement for powerful and specific recognition by the receptor.

This question was considered by Repke and co-workers who conducted a Free-Wilson type analysis of 73 digitalis-like compounds covering most of the compound types shown in Figure 1.[84] This type of study systematically examines the effects of a series of chemical modifications on biological activity to see if the changes in activity consistently reflect what seems to be the same set of binding features in the receptor's topography. In this study, the biological activity measured was inhibition of human myocardial Na^{+},K^{+}-ATPase, expressed in terms of ΔG values.

The study concluded that the bisguanylhydrazones, the erythrophleum alkaloids, and the pregnanes, such as the chlormadinone esters, were not interacting with the receptor in the same way as the congeners of digitoxigenin and hence should not be considered with these compounds when using SAR studies to define the topography of the digitalis receptor.

As for the digitoxigenin congeners, the study concluded that the minimum structural element that had to be present for specific recognition by the receptor was 5β,14β-androstane-3β-14-diol, and that this, therefore, was the lead structure for digitoxigenin congeners. This compound has the *cis/trans/cis* configuration of the steroid ring system depicted in Figure 12a.

However, we have found that some 5α analogues of digitoxigenin and its derivatives can be very potent inotropes and inhibitors of Na^{+},K^{+}-ATPase.[85,86] In addition, Shigei et al.[87] found that replacement of the 14β-OH of digitoxigenin with βH gave a compound with about one tenth the potency of the parent genin, indicating that the OH was an important but not essential substituent.

Of great interest are the amino steroids described by Jarreau et al.[88] These compounds are 5β,14β (i.e., *cis/trans/cis*) steroids with an amino group in place of the 14β-OH, and a 17β hydroxyethane [-CH(OH)CH_3] side chain (Figure 1B: XIII). The compounds are potent inhibitors of Na^{+},K^{+}-ATPase, produce pronounced inotropic effects and have wider therapeutic ratios than digoxin.[91] The compounds will be discussed further under pregnane derivatives. Repke and co-workers believe that the digitalis-like activity of these compounds is due to ion pair formation between the 14β-amino group and an anionic group present in the receptor. As will be discussed later, these compounds have complex activities in which inhibition of Na^{+},K^{+}-ATPase is probably only one aspect.[91]

It is thus possible that the digitalis receptor has some degree of flexibility in its ability to bind to cardiotonic steroids, or else that certain structural deviations from the identified lead structure are acceptable, provided they correctly position the key binding groups. This will be considered later, but first a brief review will be made of the so-called noncongeneric digitalis-like compounds.

2. Guanylhydrazones

The biological activity of the bisguanylhydrazones (such as X, Figure 1B) is not greatly influenced by changes in the structural features of the steroid ring system (reviewed by Thomas et al.[28]). Instead, the steroid appears to act simply as a bridging group to position the two side chains. In 1974 we suggested that drugs of this class were bound to the receptor by ionic attraction between the strongly basic guanylhydrazone side chains and two, appropriately separated, anionic groups that we suggested were present in the digitalis receptor.[28] This suggestion is consistent with the model proposed in Figure 11 (note: the H_1-H_2 and H_3-H_4 domains of the α subunit of sheep Na^+,K^+-ATPase together contain six free carboxylic groups which would give several binding possibilities for what, in essence, are bisonium-type compounds).

In 1974, Thomas et al.[90] reported that the lactone of digitoxigenin could be replaced with a 17β-formyl guanylhydrazone moiety to produce a compound with about one fifth the inotropic potency of the parent genin when tested against guinea pig atria. The compound also inhibited Na^+,K^+-ATPase, although it was less potent in this respect and produced cardiac arrhythmia in the atria at concentrations that produced less inhibition of Na^+,K^+-ATPase (tested separately) than was associated with the onset of atrial toxicity for other compounds tested.

In 1978, Gelbart and Thomas[91] reported the synthesis and biological activity of 17β-formyl guanylhydrazones of a series of some nondigitalis-like steroids based on the pregnane ring system, namely 3β-hydroxy, 17β-formylpregn-4-ene; 3β-hydroxy,17β-formyl-5β-pregnane; 3α-hydroxy,17β-formyl-5β-pregnane; and 3β-hydroxy,17β-formyl-5α-pregnane. These compounds inhibited Na^+,K^+-ATPase whereas open-chain lactone analogues (active when used to replace the lactone of digitoxigenin) produced no significant effect on the enzyme or on contractility in concentrations ranging up to 10^{-4} *M*.[92] These results indicated that the interaction of the guanylhydrazones with Na^+,K^+-ATPase was based on SAR characteristics that were very different from those that applied to the digitalis-type steroids. Even more startling, at the time, was the observation that the pregnane-based guanylhydrazones were negative inotropes, thus suggesting a separation of positive inotropy from inhibition of Na^+,K^+-ATPase. Subsequently, it was shown that the pregnane guanylhydrazones acted as inhibitors of mitochondrial respiration and that negative inotropy was preceded by a short period of positive inotropy.[31]

Recently, Ng et al.[93] showed that prednisolone-3,20-bisguanylhydrazone (PBG) differed from ouabain in that it did not show high and low affinities for α subunit isoforms of Na^+,K^+-ATPase. These workers compared the activity of ouabain and PBG on several tissues from guinea pigs, ferrets, and dogs. When tested on guinea pig heart containing only the α_1 isoform, PBG was about six times more potent than ouabain as an inhibitor of Na^+,K^+-ATPase. This could suggest a particular affinity for the α_1 isoform but this was not found when other sources of enzyme were used. For all other sources of enzyme, the guanylhydrazone was significantly less active than ouabain. Differences in the potency of PBG as an inhibitor of Na^+,K^+-ATPase reflected tissue and species differences and was not related to isoform content or ratio. However, Ng et al. did find that PGB inhibited the binding of tritiated ouabain. From this it would seem that the guanylhydrazone probably binds to the same region of the receptor as do "classical" cardiotonic steroids but that the topographical requirements for binding are different. This is consistent with the conclusions drawn in the previous paragraphs.

3. Progesterone Derivatives

These compounds are of interest because they may provide leads as to the nature of the putative endogenous digitalis-like substance mentioned earlier, and have been extensively studied by Labella, Templeton, and co-workers.[94-96,98] These workers used ability to compete with ^{3}H-ouabain for binding to Na^+,K^+-ATPase as a screening test for a variety of endogenous substances (e.g., prostaglandins, steroid hormones) and related synthetic analogues.[94] Of the compounds tested, only derivatives of hydroxyprogesterone having a 17α-acetate substituent were found to be effective competitors for ^{3}H-ouabain.

Although 17α-acetoxyprogesterone was a very weak inhibitor of ouabain binding, activity was greatly increased by modification of ring B. Substitution at C6 with an α-methyl, chloro, or bromo group plus unsaturation at C6 increased activity significantly.[95] Chlormadinone acetate (CMA) (Figure 1A: XII) was the most active compound tested, having about 2 to 5% the activity of ouabain in the test system.[96] It was suggested[95] that the ring-B substituents were acting sterically rather than electronically and were distorting ring A towards a digitalis-like conformation.

Fullerton et al.[97] used X-ray crystallography, molecular graphics, and molecular mechanics computer programs to compare the structures of digitoxigenin and CMA.[97] They found that, when rings B and C of both compounds were superimposed, the 20 carbonyl of CMA approximates very closely the position of the lactone carbonyl group of digitoxigenin. Both CMA and digitoxigenin were approximately equipotent in their ability to inhibit ^{3}H-ouabain binding.

Although effective as an inhibitor of Na^+,K^+-ATPase, CMA, when applied to the isolated perfused guinea pig heart, produced a profound decrease in the force of contraction.[96] This is similar to the results obtained by Gelbart and Thomas (discussed above) for a series of pregnane 17β-formyl guanylhydrazones.[91] In both cases, it is likely that the drugs act on several sites:

extracellularly to inhibit Na^+,K^+-ATPase and intracellularly to depress metabolism.

In a recent report, Templeton et al.[98] described the synthesis and biological activity of a series of 14β-hydroxy,5β,14β-pregnanes. The steroid ring systems of these compounds resemble digitoxigenin in all respects except that the 17β substituent was $-CH(OH)CH_3$ instead of an unsaturated lactone. The most active compound in the series was the glucoside, 14β,20β-dihydroxy-3β-(β-D-glucopyranosyloxy)-5β,14β-pregnane, which had 5% the potency of ouabain as an inhibitor of ^3H-ouabain binding to Na^+,K^+-ATPase. Unlike earlier compounds in this series, its effect on contractility was typical of digitalis, i.e., it did not act as a cardiodepressant but produced a marked inotropic response leading, at higher doses, to an increase in diastolic tension. The lack of cardiodepressant activity could have been due to the presence of the glucose moiety which would have limited the ability of the steroid to enter the cell by passive diffusion.

The importance of the work of Labella and Templeton, with respect to the search for an endogenous digitalis-like substance and its possible relationship to essential hypertension, is that it focuses attention on the type of compound that could be involved and shows that it is possible to have a digitalis-like inhibitor of Na^+,K^+-ATPase, such as CMA, that does not increase myocardial contractility.

Among the most interesting digitalis-like compounds with a pregnane-type side chain at the 17β position is 3β-rhamnosyloxy-14β-amino-5βH-pregnan-20β-ol (LND 623) (Figure 1B: XIII) and the related compound LND 796. LND 623 has the same steroid ring system as digitoxigenin but has an amino group instead of OH at the 14β position, and an hydroxyethane chain at the 17β position. An important feature is the 3β rhamnose moiety.

On isolated guinea pig ventricular muscle, LND 623 and LND 796 are powerful inotropic agents, LND 623 being two to three times more potent than digoxin and LND 796 being one third as potent; both LND 796 and digoxin produced negative inotropic effects at low concentrations whereas LND 623 did not.[89] Both LND compounds inhibited Na^+,K^+-ATPase.[99] A notable feature about both compounds was their very significant improvement in therapeutic ratios.[89] Adamantidis et al. examined the basis of the inotropic and toxic effects of both LND compounds and concluded that their cardiotoxic effects are due to inhibition of Na^+,K^+-ATPase, but that their inotropic effects are mediated by some different mechanism. It was suggested that stimulation of the slow calcium channel could account for part, at least, of the inotropic action of LND 623, and release of endogenous catecholamines for part of the inotropic effects of LND 796. The latter mechanism was not applicable to LND 623. It was also suggested that the negative inotropic effect seen with low doses of LND 796 may have been due to stimulation of Na^+,K^+-ATPase.[89]

Both LND compounds raise significant SAR questions. As mentioned, Repke's group has categorized these compounds as being "noncongeneric" with classical cardiotonic steroids as represented by digitoxigenin and its derivatives.[84] However, the potentiating effect of the 3β-rhamnosyloxy group, which we found to be one of the most effective sugar moieties in enhancing the potency of classical cardiotonic steroids,[100] suggests that both groups of compounds may be fitting the same binding site on Na^+,K^+-ATPase.

4. Erythrophleum Alkaloids

The erythrophleum alkaloids, of which cassaine (Figure 1A: VIII) is the best known example, are diterpene derivatives. They are potent inotropic agents and inhibitors of Na^+,K^+-ATPase. In 1974 Thomas et al.[28] suggested that these compounds may owe their digitalis-like properties to their α,β-unsaturated ester side chain:

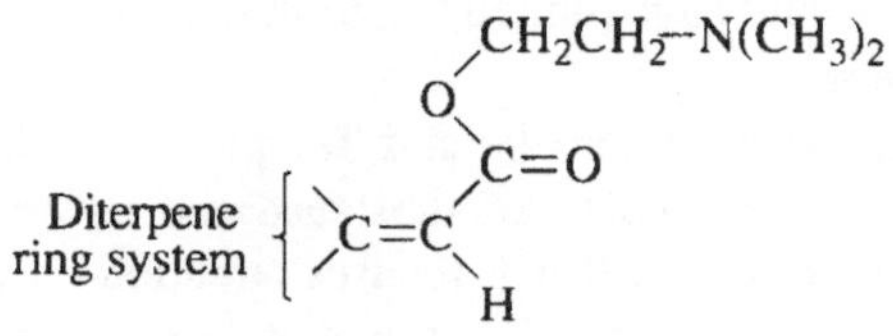

which has some resemblance to the α,β-unsaturated carbonyl system of the cardenolide lactone. This conclusion was supported by the work of Beard et al.,[101] who found that replacement of the lactone of digitalis-type steroids with the above basic ester side chain gave potent inotropic agents which seemed to have an improved therapeutic ratio, although they were more emetogenic than the parent cardenolide.

However, Repke's group (in the study cited above[84]) concluded that cassaine and related compounds were binding to the receptor by an ion-pair interaction and not by forces analogous to those that bound the cardenolide lactone. In view of the abundance of anionic groups in the receptor region (Figure 9) this is a likely possibility.

B. DO SPECIES AND ISOFORM DIFFERENCES WITH RESPECT TO DIGITALIS SENSITIVITY PRECLUDE THE FORMULATION OF GENERALLY APPLICABLE SARs?

The biological potency of cardiotonic steroids is usually determined using isolated sections of myocardium or partially purified preparations of Na^+,K^+-ATPase. Since there are marked differences in species and isoform sensitivity to digitalis, it raises the question of how generally applicable are SAR data obtained from any individual test system. In particular, most of the SAR studies reported in the literature, including our own, have not taken into account the existence of Na^+,K^+-ATPase isoforms.

In 1983 we published a comparison of K_D values for 27 glycosides tested for ability to displace ^{3}H-ouabain from beef heart with ability to produce

TABLE 2
Biological Activity of Reference Glycosides and Semisynthetic Genins and Glycosides Synthesized According to Figure 12

Compound[a]	Relative potencies[b] (guinea pig)		Binding to beef heart Na^+,K^+-ATPase[c]		
	Atria[d]	Na^+,K^+-ATPase[e]	Relative potency	K_D ($\times 10^{-9}$ *M*)	$\Delta G°$ (kJ mol^{-1})
Reference Compounds					
Digitoxigenin (1)	100	100	100	10.27	−47.3
Digoxin	182	324	288	3.57	−49.8
Digitoxin	750	1446	1268	0.81	−53.6
Ouabain	413	232	343	2.99	−50.3
Compounds Synthesized According to Figure 13					
Genins					
Nitrile (4a)	69				
Methyl ester (4b)	29				
Methyl ketone (4h)	20				
Acid (7)	Inactive				
Amide (8)	Almost inactive[f]				
Glucosides					
Digitoxigenin (2)	300	255	406	2.53	−51.1
Nitrile (6a)	240	324	138	7.42	−48.2
Methyl ester (6b)	170	249	222	4.63	−49.4
Ethyl ester (6c)	21	35	81	12.69	−46.9
n-Propyl ester (6d)	9	14	54	19.14	−45.6
i-Propyl ester (6e)	22	47	63	16.32	−46.1
n-Butyl ester (6f)	2	4	13	79.02	−42.0
i-Butyl ester (6g)	2	3	12	88.90	−41.7
Methyl ketone (6h)	80	139	74	13.95	−46.5

[a] The numbering of compounds corresponds to the numbering in Figure 12.
[b] Relative potencies were calculated by giving digitoxigenin a potency of 100 in each of the separate sets of data (contractility, enzyme inhibition, and binding studies). Each data set was then compared internally with digitoxigenin.
[c] Data from Brown et al.[102]
[d] Data for guinea pig left atria from Smith, et al.[105] and Fallon and Thomas.[106]
[e] Data for guinea pig myocardial Na^+,K^+-ATPase obtained by Fallon and Thomas.[106]
[f] Data from Smith et al.[105] A small (<20%) increase in force of contraction was obtained for the amide at 10^{-4} *M*.

inotropic effects in the much less sensitive guinea pig heart (Table 2 includes some of these data).[102] When the two sets of data were compared in terms of relative potencies, a correlation coefficient of 0.95 was obtained for comparison of the 22 5β-glycosides and 0.96 for the five 5α-glycosides. We separated 5β and 5α glycosides because they trended towards different regression lines, although the difference was not statistically significant. In the same paper we reported a strong correlation between ability to bind to guinea pig Na^+,K^+-ATPase and positive inotropy in guinea pig myocardium (r = 0.98 for the 5β series and 0.95 for the 5α series). Thus, relative inotropic activity in guinea pig atria correlated strongly with binding to myocardial Na^+,K^+-ATPase from guinea pig and beef, the former having low intrinsic sensitivity and the latter high. The glycosides tested in this study included a diverse group of steroid and sugar moieties.

Fullerton and co-workers have also addressed the problem of the general applicability of SAR data obtained from any one species or tissue. In 1983 they published a study which showed that relative potencies for a range of cardiotonic genins, determined using Na^+,K^+-ATPase from several species, were very similar, in spite of great differences in intrinsic sensitivity.[103] More recently, this group examined relative potencies of 12 glycosides of digitoxigenin using Na^+,K^+-ATPase from hog kidney, rat brain, cat brain, and rabbit kidney.[104] Regression analysis of cross comparisons of relative potencies obtained using the various enzyme preparations indicated that all enzyme preparations showed similar relative sensitivities to the structural features that appear to be the major contributors to the potentiating effects of sugars on the activity of the genin. These features include the axial C3′-OH and the equatorial C4′OH. However, there were significant species differences with respect to the less important determinants of activity such as the C6′-OH.

The conclusion reached by Fullerton's group was that the genin-binding site was probably homogenous across species and the sugar-binding site less so but still capable of recognizing the major sugar-binding groups.

According to the model presented in Figure 11 of the present review, and further elaborated in the text, the genin binds first to the H_3-H_4 domain of the α subunit of Na^+,K^+-ATPase and then undergoes secondary binding to the H_1-H_2 subunit. This secondary binding is, according to the model, greatly potentiated by an appropriate sugar. The conclusion of Fullerton et al. that the genin binding site is homogenous across species suggests that there should be a sequence of amino acids in the H_3-H_4 domain common to the Na^+,K^+-ATPase of all species and, presumably, all isoforms. As shown in Figure 9a, such a sequence does exist and comprises tryptophan-leucine-glutamic acid. Our model suggests that this is the region that initially binds the genin.

The reason for the lesser degree of across-species homogenity with respect to sugar SARs could be due to (1) variability in the amino acids that constitute the other three residues of the H_3-H_4 domain (see Figure 9a); (2) variations in the amino acid content of the H_1-H_2 domain, or (3) variations in other regions of the α subunit that may affect the ability of the polypeptide to make the types of accommodating conformational changes suggested in Figure 11. Such variations may also explain the differences in intrinsic sensitivity to cardiotonic steroids of Na^+,K^+-ATPases from different species or isoform type.

A limitation that applies to all of the SAR studies discussed above is that the Na^+,K^+-ATPase used appeared to contain only one isoform. As suggested by Fullerton et al., this may have been due to the severe conditions used to isolate and partially purify the enzyme, leading to selective loss of one or more isoforms.

There is clearly a need to re-evaluate digitalis SAR with respect to different Na^+,K^+-ATPase isoforms and to determine, if possible, the basis for differences in intrinsic sensitivity. The rat α_1 isoform is about 100 to 1000 times less sensitive to ouabain than is the rat α_2, but quite different orders of sensitivity may, perhaps, be found if the isoforms were studied using other digitalis analogues.

Another limitation to the interpretation of SAR data is the influence of variations in experimental technique used in different laboratories. These seem to be responsible for some of the very divergent values found in the literature for the relative potencies of particular cardiotonic steroids.[100]

C. HOW STRONG ARE THE FORCES THAT BIND CARDIOTONIC STEROIDS TO THE RECEPTOR SITE?

1. Reversible Binding to Drug Receptors

Hydrophobic forces are usually the main factor causing drug molecules to leave an aqueous environment and bind to specific regions of a protein. Water molecules tend to associate with each other by a random process whereby hydrogen bonds are continually being formed and broken. A hydrophobic solute disrupts this process, thereby decreasing its entropy. Because of this, hydrophobic molecules tend to leave an aqueous environment and aggregate with molecules of similar properties. The energy released when this happens is often spoken of as hydrophobic bonding, although it is not bonding in a chemical sense. The body has abundant regions capable of accepting hydrophobic molecules, including receptors that have evolved to selectively bind natural substrates that the drug may resemble.

The forces that bind drugs reversibly to receptors are electrostatic and include ion-ion, ion-dipole, dipole-dipole, hydrogen bonds, and London dispersion forces (better known as van der Waals forces).

The strength of electrostatic forces is inversely proportional to some power of the interatomic distance and the dielectric constant of the medium. Since the latter at the receptor surface is ambiguous and impossible to measure, estimates of the likely contribution of electrostatic effects to the strength of the drug receptor interaction can only be approximations. Usually electrostatic forces between drugs and receptors are only significant when they take place in some water-free cleft in the receptor.

Ion-ion interactions are the strongest electrostatic forces and can contribute up to 20 kJ mol^{-1} of binding energy. Ion-dipole and dipole-dipole interactions are weaker and their power falls off, respectively, with the second and third power of the interatomic distance, whereas that of an ion-ion interaction falls with the first power. Ion-dipole and dipole-dipole interactions are thus far more specific than ion-ion interactions since they depend on a much closer fit of the binding surfaces. This is relevant to the understanding of the SAR of the bisguanylhydrazones which probably bind by ion-ion interactions. Hydrogen bonds are a special type of dipole-dipole interaction in which a hydrogen atom (covalently bound to a small electronegative atom such as O, N, or F), bonds electrostatically to another electronegative atom such as O or N. The bonds are short range and can be considered negligible if the O···O distance is much greater than about 3 Å. Hydrogen bonds are also highly directional and their strength is greatly reduced if the hydrogen atom points more than 30° away from the acceptor atom. Under optimal conditions, a hydrogen bond can contribute 8 to 20 kJ mol^{-1} to the binding energy.

Charge transfer bonds occur when a good electron donor comes in close contact with a good electron accepter. Among organic molecules, such complexes form most readily between planar conjugated systems. The energy of such bonds is dependent on the ionization potentials and geometry of the interacting substances. Significant binding occurs only if the respective molecular orbitals overlap. They are thus very specific.

Van der Waals forces can take place between any two atoms and are due to the fact that all atoms are continually producing transient dipoles. As two atoms approach, their respective dipoles become synchronized so that the negative pole of one tends to cancel the positive pole of the other. The maximum binding strength between any two atoms due to these "dispersion" forces is about 4 kJ mol^{-1} and is inversely proportional to the seventh power of the interatomic distance. Such bonds are thus highly specific, depending on a geometry which allows for very close association of particular atoms.

For large molecules, fitting closely together, the total van der Waals binding energy can be considerable. For example, the van der Walls binding energy between an antigenic determinant (epitope) and the binding site of an antibody consisting of a few amino acids can be as high as 120 kJ mol^{-1}.

The overall energy changes that drive chemical reactions, including drug receptor interactions, is given by:

$$\Delta G = \Delta H - T\Delta S$$

where ΔG is the change in Gibbs free energy; ΔH is the enthalpy or change in internal energy; ΔS the change in entropy; and T is the absolute temperature.

Reactions proceed spontaneously when ΔG decreases and can be enthalpy driven, entropy driven, or driven by both these changes. As will be discussed later, it has been claimed that the interaction of cardiotonic steroids with their receptor is entropy driven.

Bond energies are expressed in terms of the energy required to break the bond and thus equal $-\Delta G$ for the formation of the bond. This is related to the dissociation constant (k_D) as follows:

$$\Delta G = RT\ln K_D$$

R being the universal gas constant, T the absolute temperature.

2. Bonding of Substrates to the Digitalis Receptor

a. Thermodynamic Approaches

Cardiotonic steroids bind to the digitalis receptor by reversible bonds. The most potent glycosides are considered to be relatively strongly bound, at least compared with most other drugs. However, as shown in Table 2, the total binding energy, even for very potent glycosides binding to very sensitive enzymes, is only of the order of 50 kJ mol^{-1}. By comparison, the bond strength of a typical covalent bond is about 400 kJ mol^{-1}.

Furthermore, the difference in binding energies between the most potent glycoside shown in Table 1 (digitoxin) and the least potent (compound 6g) is a mere 11.9 kJ mol^{-1} (roughly equivalent to one weak hydrogen bond or three van der Waal's bonds) and yet they differ in potency by a factor of about 100.

These small energy differences do not seem to explain the selectivity of cardiotonic steroids for their receptor. Although the binding strength of one hydrogen bond can increase the affinity (i.e., equilibrium constant, K) by a factor of up to 4000, this does not explain *selectivity*. In the biophase, there exists a wide range of binding opportunities, especially for hydrogen bonding which, as we will see, has been suggested as being one of the key binding interactions between digitalis and its receptor.

This apparent anomaly arises because the "binding energy" ($-\Delta G$) is not a measure of the actual forces that bind the drug to the receptor but of the total energy changes in the system. When a drug interacts with a receptor protein, energy is usually required to bring about certain conformational changes in the protein as well as to make other changes such as the shedding of structural water. In addition, there is the contribution of the hydrophobic forces mentioned earlier. Thus, ΔG is the final energy balance after all the energy changes have been summed.

In many cases, it would seem that part of the energy changes that take place when a drug binds to a receptor are used to "pay" for unfavorable energy changes associated with changes in the receptor molecule. This concept was expressed by Franklin[107] as follows:

$$\Delta G_{obs} = \Delta G_{int} + \Delta G_D - T\Delta S_{int}$$

where ΔG_{obs} = observed free energy change; ΔG_{int} = the intrinsic energy of binding (i.e., the summation of all electrostatic attractions that take place between the drug and the receptor surface); ΔG_D "destabilization" energy changes (i.e., the changes needed to bring the receptor from its ground state to that which binds the drug); and $T\Delta S_{int}$ = the intrinsic entropy changes associated with the interaction of the drug with receptor surface.

In many cases, ΔG_D is positive, that is, it inhibits the formation of the drug-receptor complex, since the more negative the value of ΔG_{obs} the stronger is the binding of the ligand.

Franklin's equation expresses the possibility that portion of the energy liberated by the formation of electrostatic bonds between the drug and receptor (ΔG_{int}) can be used to pay for the unfavorable allosteric changes that occur in the receptor protein as a consequence of the drug receptor interaction.

This concept helps to explain the specificity of drug action: only certain drug molecules are capable of binding to a receptor in such a way that the energy released can be used to bring about the conformational changes in the receptor protein that are needed for the expression (or inhibition) of the protein's biological activity.

The concept of intrinsic energy of binding also emphasizes the fact that ΔG_{obs} (and hence K_D or any other measure of potency) cannot be used to deduce the strength of the electrostatic interactions between the drug and receptor. The $-\Delta G$ values given in Table 1 are thus measures of the overall energy balance of the binding event (i.e., of potency) but tell us little about the nature of the electrostatic interactions between drug and receptor.

The thermodynamics of the binding of cardiotonic steroids to the digitalis receptor have been extensively studied by Repke and co-workers. In a recent paper this group analyzed the enthalpy, entropy, and activation energy changes associated with the inhibition of Na^+,K^+-ATPase by a series of cardiotonic steroids.[108] The authors found that the drug receptor interaction was an endothermic process that was entropy driven. Activation energies ranged from 45 to 59 kJ mol^{-1} for the forward reaction and 84 to 99 kJ mol^{-1} for the dissociation reaction.

The endothermic nature of the reaction would inhibit the reaction process but this is more than offset by the very large increase in entropy which arises not only from hydrophobic bonding but mainly, according to the authors, from a relaxation in the conformational energy of Na^+,K^+-ATPase. This is compatible with the model presented in Figure 11 where the binding of digitalis is depicted as overcoming the charge repulsion between the H_1-H_2 and H_3-H_4 domains.

b. Physical Properties in Relation to Binding

Several authors have attempted to explain the interaction of cardiotonic steroids with their receptor in terms of certain physical properties. Repke[109] has suggested that the digitalis molecule has an overall dipole character with the lactone operating as the negative pole and potency being related to favorable dipole vectors which guide the molecule to its site of action. This theory now seems unlikely in view of recent sequencing studies which show that there are no strongly positive charges in the H_3-H_4 domain of the receptor (Figure 9). The theory has also been criticized on conformational grounds by Fullerton et al.[110]

Bohl used a technique called the minimal topological difference (MTD) to determine the shape of the digitalis binding site. The method interprets SAR in terms of ability to fit the receptor cavity.[111] Molecular modifications that lead to increased activity are considered to improve fit for a particular part of the cavity, modifications that lead to falls in activity are considered to produce unfavorable steric interactions with the receptor wall, and modifications that have no effect on activity are considered to be sterically irrelevant. The technique is very similar to a Free-Wilson analysis with the effect of each structural modification being regarded as independent of the influence of other structural modifications. The analysis indicated that the steroid fits into a very tight cavity with strong steric repulsions occurring in regions of the receptor corresponding to the 11β, 12β, and 15β positions of the steroid. Receptor cavities were identified that corresponded to the 17β side chain, the steroid ring system, and the sugar moiety. A point of strong repulsion was identified in the sugar-binding site. Overall, the study emphasized that the whole molecule fits into a very tight cavity. This could mean that van der Waals forces play a major part in binding the molecule.

Dzimiri and Fricke have attempted to correlate lipophilicity with inotropic activity as determined in isolated guinea pig heart muscle preparations. The correlation was not good when glycosides of the digoxin series were included ($r = 0.20$) but was significantly improved when these and the cardanoldes (saturated lactone derivatives) were omitted ($r = 0.54$). The authors concluded that lipophilicity made a major contribution to the tendency of cardiotonic steroids to bind to the digitalis receptor. They considered that the reduction in activity associated with the 12β-OH (digoxin) and 16β-OH (gitoxin) series was due to reduction in hydrophobicity and not steric factors. This is supported by the fact that derivatization of the 16β-OH increases activity but is not consistent with the fact that activity falls when the 12β-OH is derivatized.

Kamernitzky et al. have determined the heteropolarity or biphilicity moments of a series of cardiotonic steroids and related these to activity. They concluded that these values depended on the location of substituent groups but were independent of the hydrophilicity of the molecule as a whole.

D. THE ROLE OF THE FUNCTIONAL GROUPS IN THE ACTIVITY OF CARDIOTONIC STEROIDS

For the purpose of SAR studies, cardiac glycosides are considered to have three distinct structural regions: the 17β side chain, the steroid ring system, and the 3β side chain. There is abundant evidence that all three regions can be involved in binding to the receptor.

1. The 17β Side Chain

One of the characteristic features of the naturally occurring cardiac glycosides is the unsaturated lactone attached to the 17β position of the steroid (Figure 12). It was long thought that the unsaturated lactone was an essential feature for cardiotonic activity, although it was also known that the lactone by itself, or attached to simple carbocyclic systems, was inactive.[114]

a. Replacement of the Lactone with Open-Chain Structures

In 1966 this author showed that it was possible to replace the lactone of cardiotonic steroids with an α,β-unsaturated ester and still retain biological activity.[115] This work was further extended by Thomas and co-workers,[28,90,105] who established that biological activity was retained when the lactone was replaced by open-chain structures of the general type:

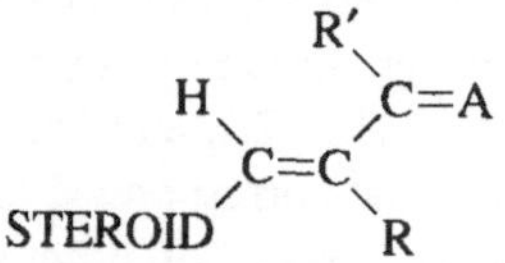

where A is a heteroatom. The synthesis of some of these analogues is shown in Figure 13 and their activities are given in Table 2. The most active compounds were the nitriles and methyl esters (Figure 13: 4a, b and 6a, b) which had about half the activity of the corresponding lactone. Activity was lost if R′ was larger than methyl or R was larger than H. Reduction of the ethylene group led to loss of most of the biological activity, as did extension of the conjugated system as shown below:

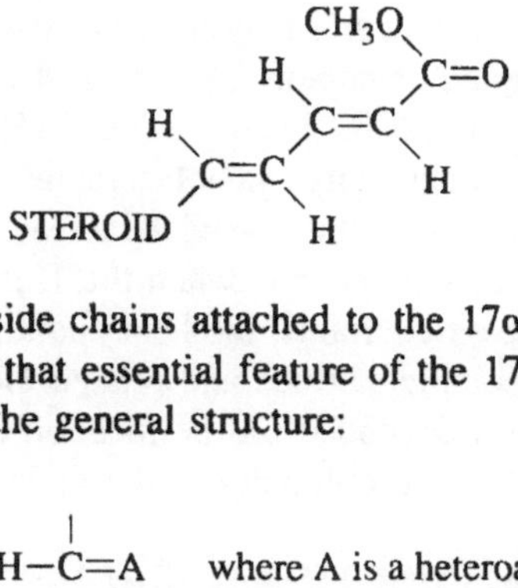

Analogues with side chains attached to the 17α position were inactive. The results indicated that essential feature of the 17β side chain was a conjugated system with the general structure:

$$-\text{C}{=}\text{CH}-\overset{|}{\text{C}}{=}\text{A} \qquad \text{where A is a heteroatom.}$$

The fall in activity as the chain was made more bulky implied that steric constraints were imposed by the receptor.

The contribution of the unsaturated lactone, or its isosteres, to biological activity is shown by data obtained by Repke's group using human myocardial Na^+,K^+-ATPase.[84] These workers found that saturation of the lactone of digitoxigenin gave a compound with only 0.9% the activity of the parent molecule and replacement of the lactone with H reduced activity by a factor of about 3000. Repke and co-workers estimated that the binding energy of the lactone was about 20 kJ mol^{-1} or roughly equivalent to the energy of a very strong hydrogen bond. This is compatible with the earlier suggestion of

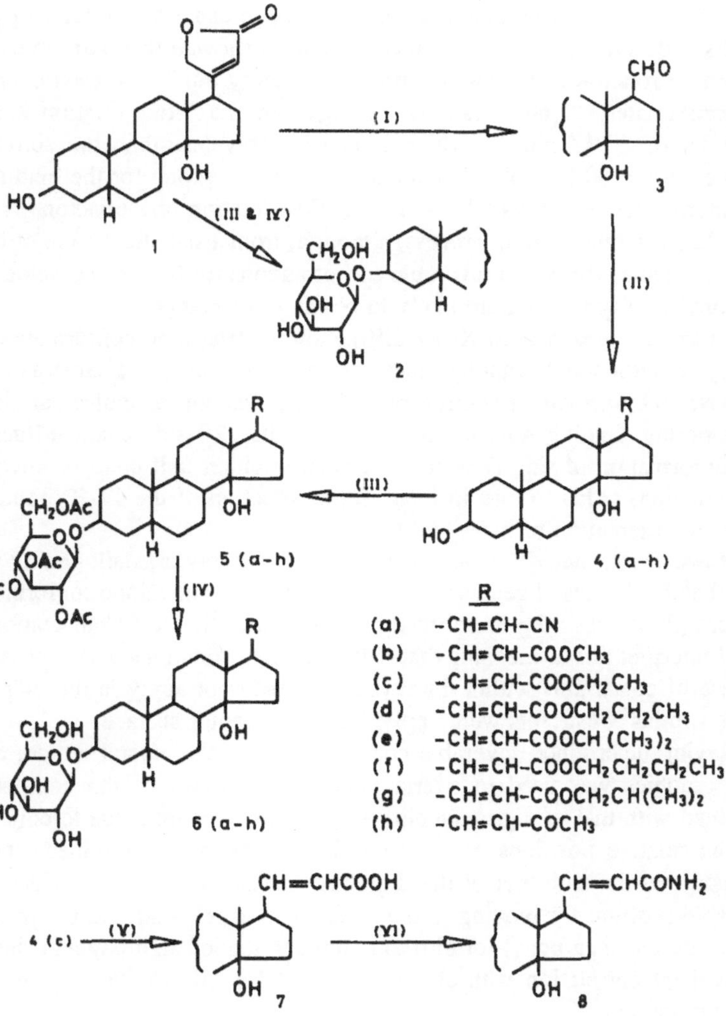

FIGURE 13. Synthesis of C17β-modified derivatives of digitoxigenin and the subsequent conversion of these to glucosides. I = (a) $(Ac)_2O$/pyridine, (b) O_3, (c) $NaBH_4$, (d), NaOH, (e) $NaIO_4$; II = $(R'O)_2P(O)CH^-R''$; III = Ag_2CO_3 (Celite)/$HgBr_2$/$Hg(CN)_2$; VI = Et_3N/MeOH/H_2O; V = OH^-; VI = Et_3N, $ClCOOCH_2CH(CH_3)_2$, NH_4OH. (From Smith, P., Brown, L., Boutagy, J., and Thomas, R., *J. Med. Chem.*, 25, 1222, 1982. With permission.)

Repke and Portius[116] that the lactone is bound to the receptor by hydrogen bonding through its carbonyl group.

b. Evidence that the 17β Side Chain Binds to the Receptor by Hydrogen Bonding

Fullerton and co-workers used a combination of X-ray crystallography, computer graphics, and molecular mechanics to determine the spatial orientation of the 17β side chains of a series of cardiotonic steroids including those with saturated and unsaturated lactones and open-chain structures (for a review, see Reference 110). Molecular mechanics showed that barriers existed to the free rotation of the lactone ring with energy minima at torsion angles of approximately ± 60°. Maximum energy barriers to free rotation were of the order of 20 kJ mol^{-1}, which is considerably less than the activation energies (50 to 60 kJ mol^{-1}) calculated by Repke's group for the binding of cardiotonic steroids to Na^+,K^+-ATPase. Free rotation of the lactone is thus possible during the binding process, although, from a statistical point of view, those molecules for which favorable conformations for binding coincide with rotational energy minima are likely to be the most potent.

Fullerton's group used X-ray diffraction to determine conformations in the crystal lattice and, although these are not necessarily the same as in the biophase, they provide a starting point for comparison of molecular shape. An important finding was that the nature of the 17β side chain influenced the conformation of ring D which can exist in either half-chair or envelope conformations. This in turn influenced the direction of the 17-20 bond and hence the orientation of the 17β side chain.

However, as stated by Go and Bhandary,[117] X-ray crystallography indicates that the D rings of genins and monosides have the envelope conformation whereas glycosides with more than one sugar have the half-chair conformation. I interpret this as meaning that half-chair conformations occur, possibly, because of constraints within the crystal and may not apply in the biophase unless similar constraints were applied by the receptor surface.

Having determined favorable conformations in the crystal lattice, computer graphics was used to overlay the conformations of the compounds examined with that of digitoxigenin. Rings B and C were made to coincide, and the relative positions of the side chain carbonyls determined. It was assumed that the carbonyl of the digitoxigenin lactone was in a reasonably favorable position for binding to the receptor. It was found that the position of the side-chain carbonyl (or nitrile) relative to that of digitoxigenin showed an excellent correlation with ability to inhibit Na^+,K^+-ATPase from three different sources.[110]

As shown in Figure 14, the carbonyl oxygens of the three most active compounds were considered to lie 1 to 2 Å closer to the putative H donor than did the carbonyl group of digitoxigenin. Assuming that the most potent compounds were at the optimal distance for H bonding, perhaps a little less than 3 Å (O····O), compounds lying on the top half of the correlation line (Figure 14: VIII to XIV) would have H bonds 6 to 10 Å long which is far too great a distance for any form of hydrogen bonding.

It is possible that the receptor is capable of some degree of flexibility which would bring the H donor closer to the carbonyl group, thus allowing some degree of bonding, even for the least potent compounds. However, this would destroy the basis for the linear relationship shown in Figure 14.

In a recent paper, Go and Bhandary[117] showed that crystal type can influence the orientation of the 17β side chain, thus casting doubt on the

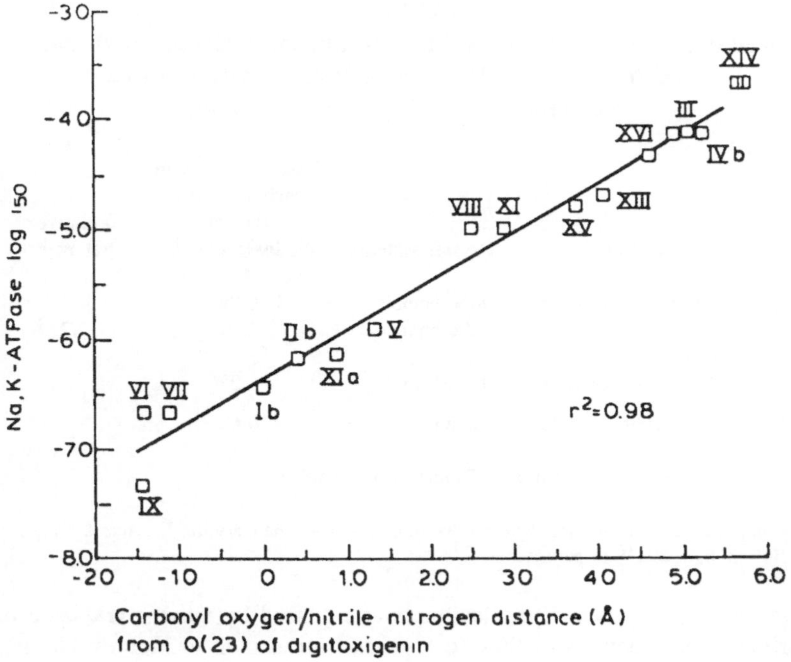

FIGURE 14. Relationship of log I_{50} (rat brain Na^+,K^+-ATPase inhibition) to C17 side chain group carbonyl oxygen (or nitrile nitrogen) position. (From Fullerton, D. S., Ahmed, K., From, A. H. L., McParland, R. H., Rohrer, D. C., and Griffin, J. F., *Molecular Graphics and Drug Design*, Burgen, A. S. V., Roberts, G. C. K., and Tute, M. S., Eds., Elsevier, Amsterdam, 1986, 257, With permission.)

value of X-ray crystallography as a means of estimating conformations in the biophase. These workers determined the conformations of the bisdigitoxosides of digitoxigenin, gitoxigenin, and digoxigenin in the crystal lattice. They then used computer graphics and energy minimization procedures to compare the relative positions of the lactone carbonyl groups using digitoxigenin as the reference compound. The distances obtained are given in Table 3 together with some relative potencies measured in my laboratory (the authors did not measure biological activity themselves).

The most relevant feature of Go and Bhandary's data is that which relates to the two crystal forms of digoxigenin bisdigitoxoside. The distances between lactone carbonyl groups and that of the reference compound (digitoxigenin) was 1.25 Å if the crystal type was orthorhombic and 0.67 Å if triclinic. Furthermore, the distance values obtained for all compounds do not compare particularly well with biological activity. The same can be said for the more potent compounds included in Fullerton's series (Figure 14).

Bohl and Süssmilch[118] have determined the conformation, charge distribution, and electrostatic potentials of digoxigenin, bufalin, and two digitoxi-

TABLE 3
Relative Distances Between Lactone Carbonyl Groups and that of Digitoxigenin for Biosides of Digitalis Glycosides as Determined by X-Ray Crystallography

Compound	Crystal system	Distance of lactone carbonyl from that of digitoxigenin (Å)	Relative potency[a]
Digitoxigenin bisdigitoxoside	Orthorhombic	0.89	4
Gitoxigenin bisdigitoxoside	Orthorhombic	1.09	0.7
Digoxigenin bisdigitoxoside	Orthorhombic	1.25	1
Digoxigenin bisdigitoxoside	Triclinic	0.67	

[a] Relative potencies obtained by Thomas and co-workers.

Compiled from the author's data and that of Go, K. and Bhandary, K. K., *Acta Cryst.*, B45, 306, 1989. With permission.

genin analogues in which the lactone was replaced by either pyridazine or pyridine. The structures of these compounds are shown in Figures 1A (III, IV) and B (XIV, XV). Note that bufalin is a pentadienolide.

The net charges on the atoms of the four side chains studied by Bohl and Süssmilch are shown in Figure 15, together with relative potencies. (Note: the potency given by these authors for the cardenolide derivative II in Figure 15 was that for digoxigenin, whereas the other three analogues were derivatives of digitoxigenin. I have therefore increased the relative potency of II by a factor of 4 to make it comparable with the digitoxigenin-based analogues.) Since there are great variations between laboratories with respect to potencies, small differences between relative potencies are not significant. On that basis, I and II of Figure 15 can be considered roughly equipotent and III and IV about one tenth as potent.

The charge distributions shown are very interesting. Apart from the low charge on C20, which I will return to later, it can be seen that the fractional negative charges on the ether and carbonyl oxygens of both lactones (I and II) are about the same and, hence, both could equally well take part in hydrogen bonding interactions. This is shown by the potential energy maps given in Figure 16. The N-containing side chains (III and IV) have extremely small negative charges on the hetero atoms which could imply only a slight tendency to form hydrogen bonds. This fact could explain the loss of 90% of their biological activity. However, when a theoretical positive charge is advanced toward the ring systems, potential energy wells of roughly equal magnitude develop for all four side chains (Figure 16). The wells represent the most favored position for the placement of a hydrogen atom in hydrogen

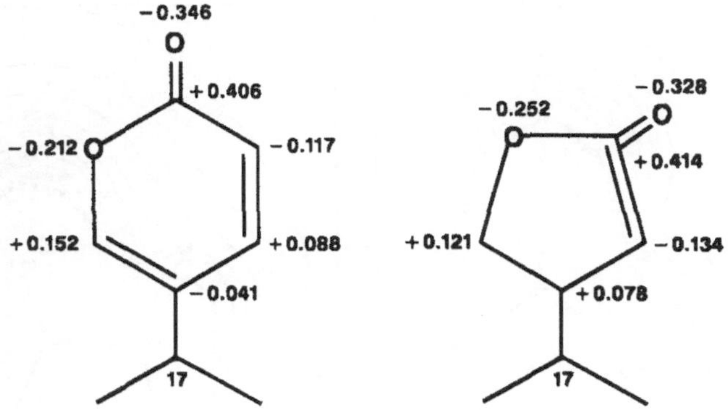

I
(RP = 1.0)

II
(RP = 0.7)

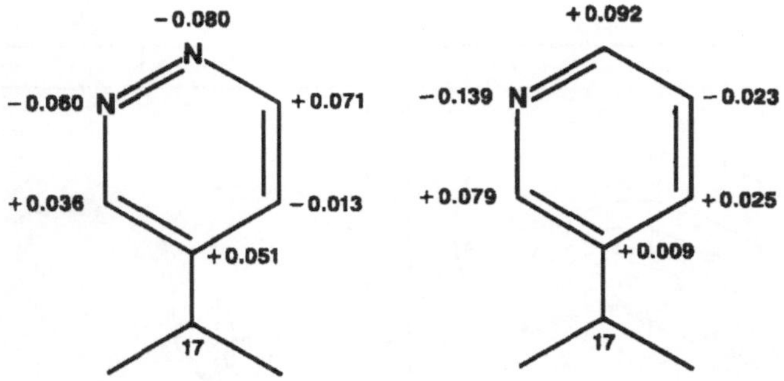

III
(RP = 0.1)

IV
(RP = 0.08)

FIGURE 15. 17β Side chains of (I) bufalin (structure VII in Figure 1A), (II) digitoxigenin (structure I in Figure 1A), (III) pyridazine analogue (structure XIV in Figure 1B), and (IV) pyridine analogue (structure XV in Figure 1B). Shown are relative potencies (RP) and net atomic charges as calculated by the CNDO/2 method (a semiempirical quantum-chemical procedure). Data from Reference 118. Note: in the source quoted II was digoxigenin. To make the comparison more meaningful, this reviewer has substituted the RP of digitoxigenin for that of digoxigenin on the assumption that the net charges would be very similar.

bond formation, and occur, as would be expected, in positions facing the heteroatoms (O or N). Where there are two such atoms, two wells are found. The reason the wells are of similar magnitude for all side chains, whereas the electronegativity in the ground state of the N-containing compounds is very low (Figure 15), is due to the polarizability of the conjugated system which allows electron density to flow in the direction of the positive charge.

The orientations shown for the side chains in Figures 15 and 16 were determined by CNDO/2 geometry optimization. As shown, the wells corresponding to the carbonyl oxygens of the lactones do not correspond with

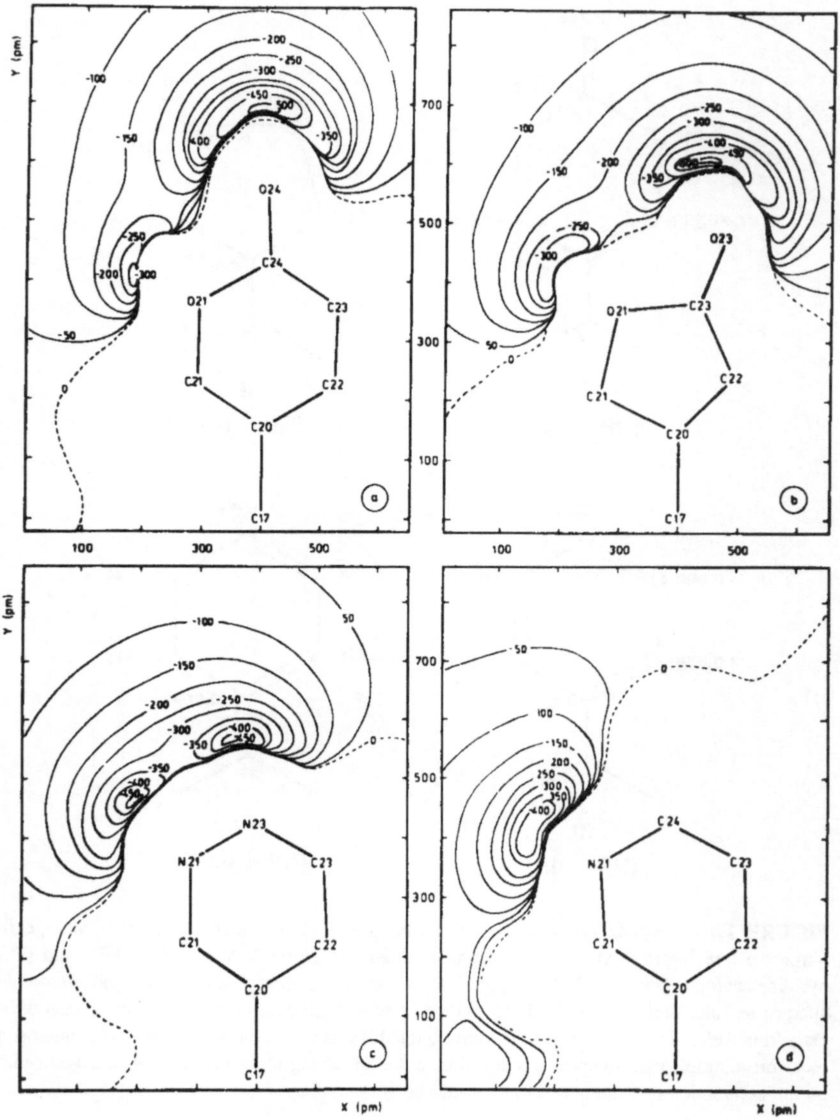

FIGURE 16. Electrostatic isopotential maps (energy values for a positive unit charge in kilojoules per mole (a) bufalin (plane C22-C23-021), (b) digoxigenin (plane C22-C23-021), (c) pyridazine derivative (plane C22-N21-N23), and (d) pyridine derivative (plane C22-C23-N21). (From Bohl, M. and Süssmilch, R., *Eur. J. Med. Chem.*, 21, 193, 1986. With permission.)

those for the N-containing derivatives, particularly that for IV. However, the wells which correspond to the ether oxygens of the lactones do correspond with one of the wells for III. From this, the authors concluded that biological activity could be associated with different geometrical arrangements of steroid and receptor, and, hence, that receptor models based on rigid geometries would have limited applicability.

c. *Other Bonding Possibilities for the 17β Side Chain*

All investigators of the SAR of the 17β side chain, including the present author, have identified the carbonyl group as a likely binding moiety. The work of Repke, Fullerton, and Bohl (quoted above) implied that the side chain binds to the receptor by a single electrostatic attraction (most likely a hydrogen bond) acting through the electronegative heteroatom (oxygen or nitrogen). However there are some limitations to this conclusion. The analysis of Fullerton et al. requires the existence, in some cases, of extremely long hydrogen bonds. Bohl's analysis, while supporting the single hydrogen bond concept, required a measure of flexibility in the drug receptor interaction which is not compatible with the hypothesis of Fullerton and co-workers. The analysis of Repke et al. required the rejection of certain compounds, particularly the bases, from SAR correlations on the grounds that these compounds were binding to the receptor in a way that was different from compounds congeneric with digitoxigenin.

In spite of the above, the excellent correlation obtained by Fullerton and co-workers (Figure 14) requires an explanation. Hydrogen bonds are a form of dipole-dipole interaction and are very distance specific. It is not possible for hydrogen bonds to form over the distances that are implied for the compounds shown on the upper half of the correlation line in Figure 14, and yet these compounds fit the correlation line better than the more potent compounds. A possible explanation is that the carbonyl group is simply acting as a suitable marker for the orientation of the side chain as a whole, since as the side chain moves so does the carbonyl group. It could be that features of the side chain other than, or as well as, the carbonyl group are involved in binding to the receptor.

In my own musings on this subject, I have been puzzled by certain compounds that did not seem to fit the hydrogen bond model. However, it could be that these were special cases where activity, or lack of it, was due to some feature peculiar to the compound in question. For example, we found that the replacement of the lactone of digitoxigenin with a 17β-formyl guanylhydrazone group gave a compound with about one fifth the activity of the parent compound, whereas the corresponding semicarbazone was inactive.[90]

Guanylhydrazone: $R{-}CH{=}N{-}NH{-}\overset{\overset{\large NH_2^+}{|}}{C}{-}NH_2$ (active)

Semicarbazone: $R{-}CH{=}N{-}NH{-}\overset{\overset{\large O}{||}}{C}{-}NH_2$ (inactive)

The guanylhydrazone is a strong base that is protonated in biological media and, as such, would not act as a proton accepter in hydrogen bonding, whereas the semicarbazone is neutral and can form hydrogen bonds through its carbonyl group. In order to explain the activity of the guanylhydrazone, we suggested that a negative charge was located in the receptor. This seemed to be confirmed when we found that hydrolysis of the active methyl ester side chain (Figure 13: 4b) to give the corresponding acid

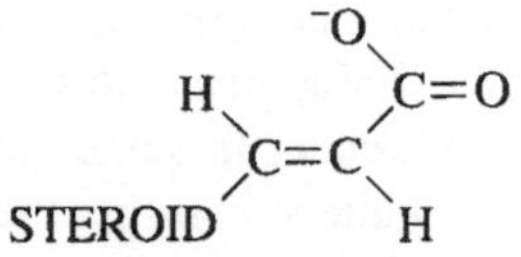

gave an inactive compound. Although such acids can form strong hydrogen bonds, lack of activity may have been due to mutual repulsion between the acid and the carboxylic groups that we now know are found in abundance in the receptor region. Less easy to understand was the very low activity of the amide:[105]

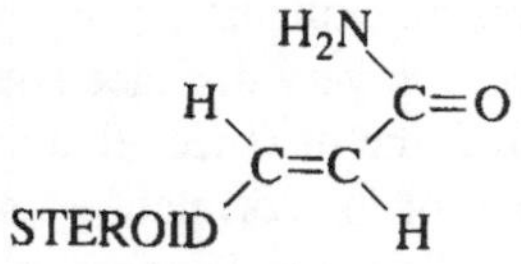

This compound is capable of forming strong hydrogen bonds due to charge transfer from the nitrogen to the carbonyl oxygen as shown in Figure 17. Megges et al.[119] also reported that the lactam analogue of the lactone was inactive. This compound would also be expected to form strong hydrogen bonds.

A possible explanation for the lack of activity of the amides (open chain and lactam) is a conformational factor that may affect the orientation of the side chain so that the carbonyl oxygen is not appropriately positioned for H bonding to the receptor. This possibility has not, to my knowledge, been examined. Alternatively, it could be due to an inappropriate charge distribution as shown in Figure 17.

Because of the need to explain the activity of the above compounds, as well as those of the bisguanylhydrazones and erythropleum alkaloids, I and my colleagues suggested that charge distribution along the side chain could be an important factor in determining interaction with the receptor. In particular, we focused attention on a possible fractional positive charge on C20 (Figure 17).

As shown in Figure 15, the charge on C20 of digoxigenin is very small (+0.078) and in the case of bufalin is negative (−0.041). However, these are conjugated systems and charge density will shift from the ground state if

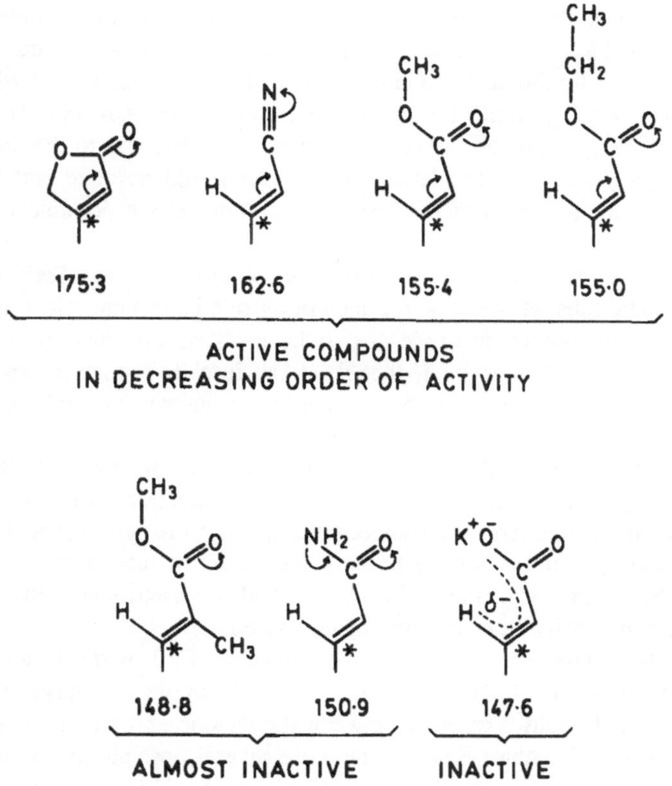

FIGURE 17. Charge distribution patterns due to resonance effects in a selection of 17β open-chain analogues of the butenolide lactone and the relationship of these to biological activity. The numbers refer to the ^{13}C-NMR shifts for C20 (*) in ppm. The higher the value, the greater is the fractional positive charge. The fractional charge is negligible at values below 150 ppm. (From Thomas, R., Gray, P., and Andrews, J., *Advances in Drug Research*, Vol. 19, Testa, B., Ed., Academic Press, New York, 1990, 311. With permission.)

another group with appropriate charges or polarizabilities is brought into close association. This is illustrated in Figures 15 and 16 where the negative charges on the N atoms of III and IV were extremely small in the ground state but polarized to give significant potential energy wells when a positive charge was advanced toward the N atom.

One way to shift electron density away from C20 would be to form a hydrogen bond through the carbonyl group which would enable the latter to draw more electron density from the conjugated system. It is possible that the binding of the lactone to the receptor involves hydrogen-bonding reinforced by an electrostatic interaction involving other atoms in the conjugated system.[90]

Earlier, I pointed out that the amino acid sequence common to all H_3-H_4 domains was tryptophan-leucine-glutamic acid (Figure 9). This sequence could well be the prime binding site for cardiotonic steroids and could contain the site that binds the 17β side chain. The glutamic acid would be ionized (unless folded into a hydrophobic cleft) and thus would not hydrogen bond with the lactone. However, it could be the site that binds to the aminoethanol side chain of the erythrophleum alkaloids (Figure 1A: VIII) and to the strongly basic guanylhydrazones. The leucine side chain would not take part in any electrostatic interactions which leaves the tryptophan as a possible binding site for the lactone.

The five-membered pyrrole ring of tryptophan is the most likely source of electrostatic interactions. The nitrogen of pyrrole contributes electron density to all four carbon atoms of the ring so that each has a net negative charge. Pyrrole can also act as an acid, that is, it can lose its H to form an anion with the negative charge spread over the entire conjugated system, including the fused benzene ring.

There are thus several possible ways in which the lactone could interact with the tryptophan side chain: it could form a hydrogen bond with the pyrrole N-H, it could form a charge transfer complex in which the pyrrole ring donated electron density to the lactone, a process that would be more likely to occur if a hydrogen bond with the carbonyl group also existed since this would draw electron density from the conjugated system.

In drawing attention to the positive charge on C20 of the lactone, we may have lost sight of the other possible dipole-dipole or charge transfer interactions that could take place between the lactone and a system such as the tryptophan side chain. Because there are several possible forms of electrostatic interaction, different kinds of association could form between tryptophan and diverse side chains such as those depicted in Figure 15.

On the other hand, the lactone could hydrogen bond to the OH of tyrosine (also present in all forms of the H_3-H_4 domain) and the sugar could interact with the glutamic acid residue that is four amino acids upstream. The tryptophan and leucine could then play a role in binding the steroid which would have to align closely with these groups for maximum activity. Cardiotonic steroids with basic 17β side chains could form ion pairs with the glutamic acid residue that is downstream from tyrosine.

2. The Steroid Ring System

a. Stereochemistry

As discussed in Section V.A, it is not clear that all compounds with digitalis-like activity react with the receptor in the same way. In particular, compounds with basic side chains seem to bind in a different and less constrained manner than cardiotonic steroids congeneric with digitoxigenin. In this section I will deal only with the latter group of compounds. For a more detailed discussion see Reference 4.

The essential feature of the steroid ring system is a carbocyclic system that is capable of fitting a constrained receptor cavity and providing sufficient

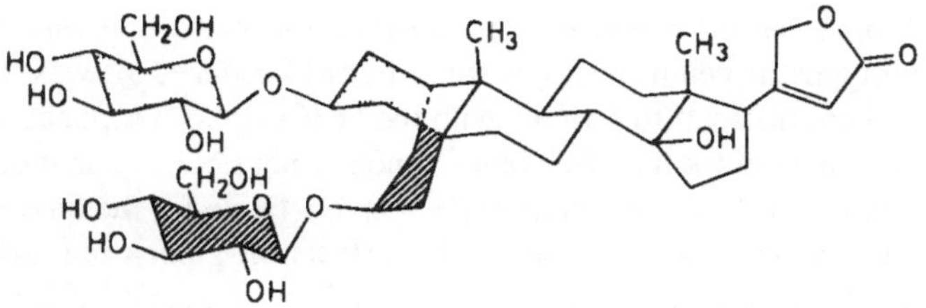

FIGURE 18. Schematic representation of the differences between uzarigenin glucoside and digitoxigenin glucoside. The nonshaded areas are similar for both compounds, the speckled area applies to the uzarigenin glucoside, and the cross-hatched to digitoxigenin glucoside. Although both genins are approximately equipotent, glycosidation of the 3-OH reduces potency in the case of the 5α-steroid, uzarigenin, but increases it in the case of the 5β-steroid, digitoxigenin. (From Brown, L. and Thomas, R. E., *Arzneim.-Forsch.*, 34, 572, 1984. With permission.)

binding energy to bring about the conformational changes needed for biological activity. As previously mentioned, Schönfeld et al.[84] concluded that the minimum requirement for the steroid was 5β,14β-androstane-3β,14-diol (i.e., digitoxigenin with the lactone replaced by H) which had 0.03% the activity of the parent genin.

Although the C/D *cis*-ring junction seems essential, our studies indicate that the A/B *cis*-(5βH)-ring junction is not a requirement for potent cardiotonic activity. We found that 5αH-digitoxigenin (uzarigenin) was roughly equipotent with digitoxigenin when tested for inotropy using guinea pig atria[85] and in binding studies using beef heart Na^+,K^+-ATPase.[102] We also found that gomphoside, a glycoside of a uzarigenin-type steroid, had 23 times the potency of digitoxigenin or almost 5 times the potency of ouabain.[86]

Although digoxigenin and its 5αH analogue uzarigenin are approximately equipotent, attachment of sugars to the 3β-OH of both compounds produced very dissimilar results. For example, glucosidation of digitoxigenin increased activity by almost 300% but reduced the activity of uzarigenin by 60% for the monoglucoside, and by 80 to 90% for the di- and triglucosides.[85] On the other hand, we found that the 6-deoxyalloside and rhamnoside of uzarigenin were, respectively, 2.2 and 7.8 times more potent than the genin, clearly showing that the A/B *cis*-ring junction is not an essential requirement for potent cardiotonic activity.

These results led us to conclude that ring A was probably not important for binding the steroid but influenced activity by its effect on the orientation of the sugar moiety (Figure 18). This will be discussed further in the next section.

We also found that the onset and washout times for the effects of uzarigenin on guinea pig atria were significantly slower than for digitoxigenin: 19 and 32 min compared with 4 and 5 min, respectively.[85] Usually genins have fast onset and washout times, and the much slower times for uzarigenin suggested to us that the nature of the drug-receptor interaction was different for the two types of genin: possibly some form of induced fit was involved which was more difficult to achieve with the 5αH steroid, but, once formed, was more stable than for the 5βH genin. Another possible explanation is that the approach to and from the binding site was less facile for the 5αH steroid.

Comparisons of relative potencies for effects on guinea pig atrial contractility and binding to beef heart Na^+,K^+-ATPase for a series of cardiotonic steroids gave different regression lines for the 5αH and 5βH series,[102] which further supports the idea that the two series interact with the receptor in different ways.

Rings A, B, and C of the steroid system all appear to adopt the chair conformation, but there is some question about ring D which has a greater degree of flexibility than the other rings. X-ray crystallographic studies[110,117] indicate that ring D can adopt envelope and half-chair conformations. However, it is possible that this reflects different crystal packing constraints and that one form predominates in solution.

b. Substituent Groups

Apart from the 17β side chain, the other principal substituents associated with high cardiotonic potency are the 14β-OH and the 3β-sugar moiety. The latter will be discussed in the next section.

Replacement of the 14β-OH with 14β-H reduces activity by about 90%.[87] The group is thus not "essential" but it makes a very important contribution to activity. In fact, it is more important for potency than the sugar moieties. To my knowledge, no evidence has been obtained identifying the role this group plays in potentiating the activity of the steroid. The group projects from the β face of the steroid (Figure 12) and may bind to the same receptor group that binds the 17β side chain, or it may exert a favorable influence on the conformation of the 17β side chain.

The 3β hydroxyl group does not seem to influence activity but serves as an attachment point for the sugar moieties. Further hydroxylation of the steroid reduces potency. For example, we found that the K_D values for the binding of digitoxin, digoxin, and gitoxin to beef heart Na^+,K^+-ATPase were 1.1, 3.6, and 15 $\times$ 10^{-9} M, respectively.[102]

De Pover and Godfraind[120] found that formylation of the 16β-OH of gitoxin increased activity by a factor of 41. This modification does not just offset the hydrophilic effect of the 16β-OH but increases activity to several times that of digitoxin. It was suggested that the formyl group contributes to binding, possibly by interacting with the same site as the lactone. Some form of dipole interaction (which could include hydrogen bonding) was suggested (Figure 19).

Fullerton et al.[110] found that the 16β-formyl and acetate altered the conformation of the lactone, bringing the carbonyl into a more favorable position for binding, but that this effect was not sufficient to explain the effect on potency.

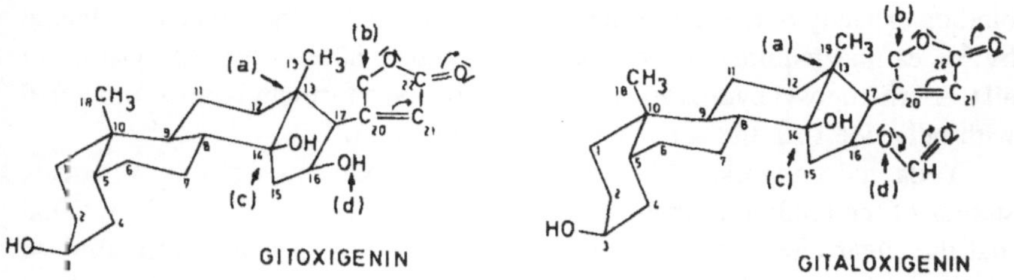

FIGURE 19. Molecular model and structure of gitaloxigenin. The arrows refer to (a) 13-methyl, (b) 17β-lactone ring, (c) 14-OH, and (d) 16β-OH or 16β-OCHO, respectively, for gitoxigenin and gitaloxigenin. (From De Pover, A. and Godfraind, T., *Naunyn-Schmiedeberg's Arch. Pharmacol.*, 321, 135, 1982. With permission.)

TABLE 4
Influence on Potency of some Sugar Residues Attached to the 3β Position of Digitoxigenin

Sugar substituent	Potency enhancement factor[a] (guinea pig atria)
Thevetose	27
Rhamnose	22
Monodigitoxose	15
Bisdigitoxose	10
Trisdigitoxose	9
Trisdigitoxose + acetyl + glucose[b]	3
Glucose	2
Mannose	1[c]
2′ Acetyl rhamnose	0.4
Triacetyl rhamnose	0.2

[a] Relative potencies from cumulative log concentration/response curves obtained using guinea pig atria.
[b] Trivial name: lanatoside C.
[c] Factor of 1 means no enhancement of genin potency.

Based on data from Brown, L. and Thomas, R. E., *Arzneim.-Forsch.*, 33, 814, 1983. With permission.

3. The Sugar Moieties

The sugar moieties, as such, have no digitalis-like activity but, when attached to the 3-OH, can modify greatly the activity of the parent genin. This is illustrated by the data in Table 4. As shown, an appropriate sugar can increase potency by a factor of 20 to 30, although some sugars, e.g., mannose, had no effect on potency, and certain substituted sugars decreased activity.

Acetylation of the 2′OH of digitoxigenin rhamnoside reduced activity by 98% giving a compound with less than half the activity of the parent genin. That such a relatively minor modification was able to offset much of the binding capacity of the rest of the molecule implies that the molecule is bound by forces that require a very close drug receptor interaction. Acetylation of all three rhamnosyl hydroxyls reduced activity even further giving a compound with only one fifth the activity of the parent genin.

Yoda and co-workers[121-124] were the first group to carry out systematic studies of the binding of cardiac glycosides to Na^+,K^+-ATPase. They found that the sugar closest to the genin made the most important contribution to binding. They also found that glycosides with 6-deoxy sugars (i.e., those with 5′ methyl groups) were the most potent of all cardiotonic steroids and concluded that the methyl group played a key role in binding the sugar to the receptor. They found that genins bind and dissociate rapidly from the receptor and concluded that the binding of glycosides takes place in two steps: a rapid step involving binding of the genin portion followed by a slow binding of the sugar residue. The presence of an appropriate sugar was also thought to account for the low dissociation rate constants of potent glycosides.

Brown and Thomas[85,100] confirmed the importance of the 5′-methyl group and identified the 4′-hydroxyl and the α-L-glycosidic link as being associated with high potency in monosides. The 2′-OH did not seem to be involved in essential binding, although if acetylated activity was lost, probably due to a steric factor. We concluded that the 3′-OH may or may not be important for binding depending on the structure of the sugar.[86,100] As will be discussed later, we believe that either the 3′ or 4′-OH can contribute to binding, although neither is essential.

As a general rule, the addition of an extra sugar unit to a monoside leads to a fall in activity. This is shown for the digitoxosides of digitoxigenin in Table 4. We have also observed a similar fall in activity for the glucosides of uzarigenin, and Dzimiri and Fricke[112] found somewhat similar results for the digitoxosides of digoxigenin: the mono and bis were roughly equipotent and the tris was less potent.

Fallon and Thomas[125] attached a series of nonsugar esters to the 3β position of digitoxigenin to produce side chains of the type:

$$ROC(CH_2)_nCOO\text{-}STEROID$$

of which, the most active compounds belonged to the β-methylglutarate group which had the following general structure:

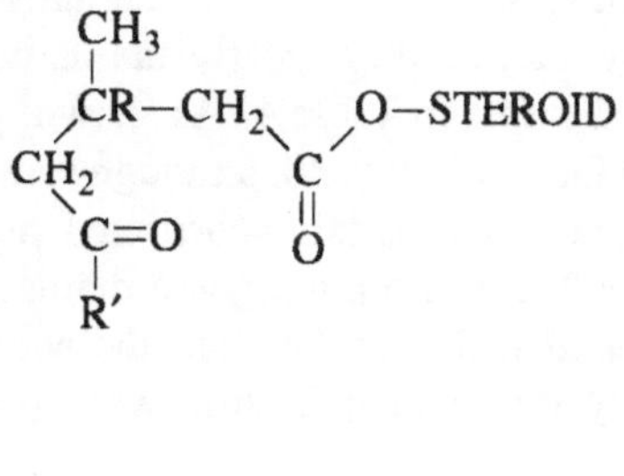

The most active compound was the methylglutarate methyl ester (R = H; R′ = CH_3) which had 3.5 times the potency of the parent digitoxigenin. It was thus more effective than glucose and mannose but less effective than digitoxose in potentiating the action of digitoxigenin. Since the molecule has no proton-donating groups, these are clearly not essential for potentiation of activity. That the side chain does not potentiate activity to the same extent as the 5′-methyl sugars could be due to its nonrigid structure which would decrease the statistical probability of the groups being in the correct orientation for binding or it could be that it lacks the equivalent of the sugar 3′- or 4′-OH groups.

The amides of the series were also quite active. For example, the amide of the dimethylglutarate (R = CH_3; R′ = NH_2) had three times the potency of the parent genin. Although the amide N-H can hydrogen bond, the corresponding methyl ester (R = CH_3; R′ = CH_3) was equipotent. Hydrogen binding through the amide does not, therefore, seem to be involved.

One of the most interesting findings made by Fallon and Thomas was that the free acids of the series (where R′ = 0^-) were all inactive. This was taken as evidence that an anionic group was present in the receptor near where the sugar binds. In 1974 we had drawn a similar conclusion to explain the activity of the 3,20-bisguanylhydrazones,[28] basing our argument on a report by Kyte[126] that an anionic site appeared to exist in the region of the receptor that binds the sugar moiety.

If such an anionic group is located in the sugar-binding site, it may bind to the hydrogens of the 3′ or 4′-OH of appropriate sugars. This would enhance potency but is not essential, as the work of Fallon and Thomas has shown.

It would seem, from what has been said, that the region of the digitalis receptor that binds the first sugar residue is sterically very limited and largely hydrophobic, except for the possible presence of a carboxylic acid group. Van der Waals forces could be the basis of the binding interaction but are enhanced by a strategically placed group capable of hydrogen bonding with either the 3′ or 4′-OH groups of sugar moieties.

An interesting sugar derivative is the amino glycoside, ASI-222 [(3β-*O*-(4-amino-4,6-dideoxy-β-D-galactopyranosyl) digitoxigenin]. This compound is more potent than ouabain and, on isolated rabbit atria, was found to have an improved therapeutic ratio as well as being capable of a greater increase in force of contraction before showing signs of toxicity.[127] Similar findings were obtained in whole anaesthetized dogs.[128] Recently it was reported that changing K^+ concentrations influenced the interaction of ASI-222 in a way that was different from digoxin. Also, infusion of K^+ into animals given arrhythmogenic doses reversed the arrhythmia in the case of digoxin but did not do so with ASI-222.[129] The increased potency shown by ASI-222 could be due to its ability to bind to the carboxylic group that may exist in the sugar-binding site. It is not easy to explain the other properties of the drug, especially in terms of the widely held view that the toxic and therapeutic

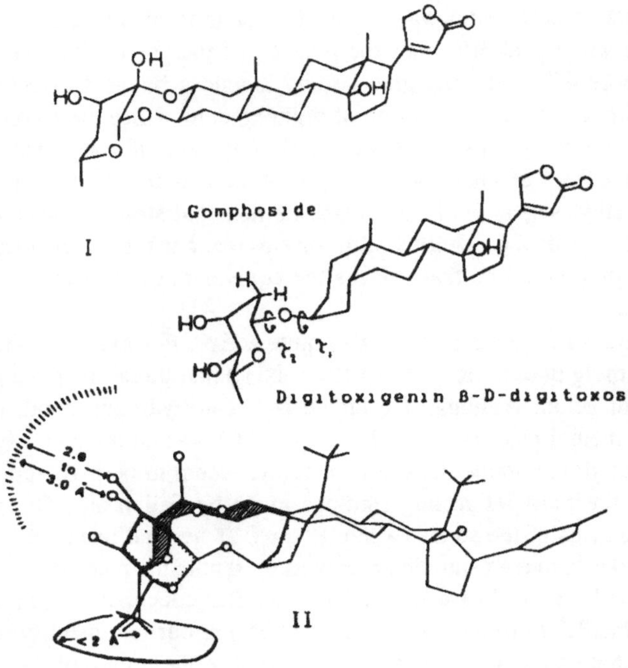

FIGURE 20. Structures of (I) gomphoside and a flexible cardiac glycoside digitoxigenin β-D-digitoxoside and (II) gomphoside and the flexible sugar digitoxigenin-α-L-rhamnoside shown superimposed. (Modified from Chiu, C. K. and Watson, T. R., *J. Med. Chem.*, 28, 509, 1985. With permission.)

effects of cardiac glycosides are both due to inhibition of myocardial Na^+,K^+-ATPase.

For a further discussion of reports of cardiotonic steroids with improved therapeutic ratios, see Reference 4.

One of the problems in studying the SAR of the sugar side chain is its flexibility around the glycosidic link. Although constraints to free rotation apply, these vary with different sugars and, in most cases, a reasonable degree of nonrestricted movement is possible. In 1986 we reported that certain 5αH glycosides from Asclepiadaceae species had sugars that were bi-linked to the steroid through positions 2 and 3, and thus could serve as useful reference compounds, especially since the most active glycoside, gomphoside (Figure 20) was 23 times more potent than digitoxigenin.[86]

Computer graphics based on potential energy calculations derived from X-ray crystallography were used to compare the structures of the 5αH bi-linked sugars with a selection of monolinked glycosides.[130,131] An interesting comparison is that of gomphoside with digitoxigenin rhamnoside and digitoxigenin monodigitoxoside (Figure 20). We found the potencies of these compounds, relative to digitoxigenin, to be 23, 22, and 15, respectively. Chiu and Watson[130] showed that both the flexible glycosides had preferred conformations which placed the 3′-OH and 5′-methyl groups in locations identical to those of the rigid gomphoside, but that the rotational freedom of the rhamnoside was more restricted which meant there was a greater statistical probability that it would have the preferred conformation of gomphoside. This was thought to account for the fact that the potency of the rhamnoside was higher than that of the digitoxoside. Figure 20 shows the preferred conformations of both molecules, and, although one is 5αH and the other 5βH, the 3′-OH and 5′-methyl groups show close coincidence. This not only confirms the importance of these groups but it also supports our previous conclusion that the configuration about the A/B ring junction was important mainly with respect to its influence on the orientation of the sugar side chain.

VI. CONCLUSIONS

There is a growing body of work which indicates that the therapeutic and toxic effects of cardiotonic steroids on the heart may be mediated by multiple mechanisms of which inhibition of Na^+,K^+-ATPase may be just one facet. It is clear that the concentrations of drug used in most experimental studies are far too high and this is probably the basis for the contentions that surround the mode of action of these drugs.

Regardless of the contribution that inhibition of Na^+,K^+-ATPase makes to the clinical effects of cardiac glycosides, there is no doubt that a receptor for digitalis exists on the external face of this enzyme. Because of the key role that this enzyme plays in the control of cell function (it accounts for about 25% of the ATP metabolized in an individual at rest), the interaction of cardiotonic steroids with Na^+,K^+-ATPase is a topic of major importance, especially since an endogenous inhibitor of Na^+,K^+-ATPase may be the basis of essential hypertension.

The sequencing of Na^+,K^+-ATPase, together with classical biochemical studies, have provided strong evidence that the digitalis receptor is associated with the extracellular H_1-H_2 and H_3-H_4 domains of the α subunit of Na^+,K^+-ATPase.

The receptor appears to lie in a hydrophobic cleft with binding interactions taking place between the 17β side chain, the steroid ring system and, if present, the sugar moiety. As yet, the nature of the forces that bind these regions remains speculative.

REFERENCES

1. **Haber, E. and Haupert, G. T., Jr.**, The search for a hypothalamic Na^+,K^+-ATPase inhibitor, *Hypertension*, 9, 315, 1987.
2. **Hamlyn, J. M., Harris, D. W., Clark, M. A., Rogowski, A. C., White, R. J., and Ludens, J. H.**, Isolation and characterization of a sodium pump inhibitor from human plasma, *Hypertension*, 13, 681, 1989.
3. **Kelly, R. A. and Smith, T. W.**, The search for the endogenous digitalis: an alternative hypothesis, *Am. J. Physiol.*, 256, C937, 1989.
4. **Thomas, R., Gray, P., and Andrews, J.**, Digitalis: its mode of action, receptor, and structure-activity relationships, in *Advances in Drug Research*, Vol. 19, Testa, B., Ed., Academic Press, New York, 1990, 311.
5. **Flegg, J. L. and Lakatta, E. G.**, How useful is digitalis in patients with congestive heart failure and sinus rhythm?, *Int. J. Cardiol.*, 6, 295, 1984.
6. **DiBianco, R., Shabetai, R., Kostuk, W., Moran, J., Schlant, R. C., and Wright, R. A.**, A comparison of oral milrinone, digoxin, and their combination in the treatment of patients with chronic heart failure, *N. Engl. J. Med.*, 320, 677, 1989.
7. **Lewis, R. P.**, Digitalis: a drug that refuses to die, *Crit. Care Med.*, 18, 5, 1990.
8. **Erdmann, E.**, The value of positive inotropy in acute and chronic heart failure, *J. Cardiovasc. Pharmacol.*, 14 (Suppl.), 36, 1989.
9. **Erdmann, E. and Böhm, M.**, Positive inotropic stimulation in the normal and insufficient human myocardium, in *Inotropic Stimulation and Myocardial Energetics*, Just, H. J., Holubarasch, Ch., and Scholz, H., Eds., Steinkopff Verlag, Darmstadt, 1989, 125.
10. **Repke, K.**, Metabolism of cardiac glycosides, in *Proceedings of the First International Pharmacological Meeting*, Vol. 3, Wilbrandt, W., Ed., Pergamon Press, Oxford, 1963, 47.
11. **Repke, K.**, Effect of digitalis on membrane adenosine triphosphatase of cardiac muscle, in *Drugs and Enzymes — Proceedings of the Second International Pharmacological Meeting*, Vol. 4, Pergamon Press, Oxford, 1965, 65.
12. **Repke, K. R. H. and Dittrich, F.**, Thermodynamics of information transfer from cardiotonic steroids to receptor transport ATPase, *Trends Pharmacol. Sci.* 1, 398, 1980.
13. **Gillis, R. A. and Quest, J. A.**, The role of the nervous system in the cardiovascular effects of digitalis, *Pharmacol Rev.*, 31, 19, 1980.
14. **Gillis, R. A. and Quest, J. A.**, Extracardiac effects of cardiac glycosides, in *Cardiac Glycosides 1785—1985*, Erdmann, E., Greef, K., and Skou, J. C., Eds., Steinkopff Verlag, Darmstadt, 1986, 347.
15. **Belz, G. G., Erbel, R., Schurmann, K., and Gilfrich, H. J.**, Dose-response relationship and plasma concentrations of digitalis glycosides in man, *Eur. J. Clin. Pharmacol.*, 13, 103, 1978.
16. **Grupp, G., Grupp, I. L., Ghysel-Burton, J., Godfraind, T., and Schwartz, A.**, Effects of a very low concentration of ouabain on contractile force of isolated guinea-pig, rabbit and cat atria and right ventricular papillary muscles: an interinstitutional study, *J. Pharmacol. Exp. Ther.*, 220, 145, 1982.
17. **Pace, D. G. and Gillis, R. A.**, Neuroexcitatory effects of digoxin in the cat, *J. Pharmacol Exp. Ther.*, 199, 583, 1976.
18. **Lown, B., Grayboys, T. B., Podrid, P. J., Cohen, B. H., Stockman, M. B., and Graughan, C. E.**, Effect of a digitalis drug on ventricular premature beat, *N. Engl. J. Med.*, 296, 301, 1977.
19. **Mason, D. T., Zelis, R., and Amsterdam, E. A.**, Unified concept of the mechanism of action of digitalis: influence of ventricular function and cardiac disease on hemodynamic response to fundamental contractile effect, in *Basic and Clinical Pharmacology of Digitalis*, Marks, B. H. and Weissler, A. M., Eds., Charles C Thomas, Springfield, IL, 1972, 1206.

20 **Arnold, S. B., Byrd, R. C., Meister, W., Melmon, K., Cheitlin, M. D., Bristoro, J. D., Parmley, W. W., and Chatterjee, K.,** Long term digitalis therapy improves left ventricular function in heart failure, *N. Engl. J. Med.*, 303, 1443, 1980.
21. **Thames, M. D., Vaickman, L. A., and Abboud, F. M.,** Sensitizaton of cardiac receptors (vagal efferents) by intracoronary acetylstrophanthidin, *Am. J. Physiol.*, 8, H628, 1980.
22. **Evans, D. E. and Gillis, R. A.,** Effect of ouabain and its interaction with diphenylhydantoin on cardiac arrhythmias induced by hypothalmic stimulation, *J. Pharmacol. Exp. Ther.*, 195, 577, 1975.
23. **Somberg, J. C.,** Localization of neurally mediated arrhythmogenic and coronary vasoconstrictor properties of digitalis, *Fed. Proc.*, 43, 2963, 1984.
24. **Somberg, J. C.,** Digitalis: neurally mediated arrhythmogenic and coronary vasoconstrictor properties, *J. Clin. Pharmacol.*, 25, 529, 1985.
25. **Heller, M.,** Interactions of cardiac glycosides with cells and membranes. Therapeutic and toxic doses of ouabain acting on sodium and calcium pumps in plasma membranes, *Biochem. Pharmacol.*, 37, 2293, 1988.
26. **Heller, M., Hallaq, H., and Panet, R.,** Interactions of cardiac glycosides with cells and membranes. IV. Effects of ouabain and bumetaide on $^{86}Rb^+$ influx in cultured cardiac monocytes from neonatal rats, *Biochim. Biophys. Acta*, 939, 595, 1988.
27. **Gervais, A., Lane, L. K., Anner, B. M., Lindenmayer, G. E., and Schwartz, A.,** A possible molecular mechanism of the action of digitalis. Ouabain action on calcium binding to sites associated with a purified sodium-potassium-activated adenosine triphosphatase from kidney, *Circ. Res.*, 40, 8, 1977.
28. **Thomas, R. E., Boutagy, J., and Gelbart, A.,** Synthesis and biological activity of semisynthetic digitalis analogs, *J. Pharm. Sci.*, 63, 1649, 1974.
29. **Schwartz, A., Lindenmayer, G. E., and Allen, J. C.,** The sodium-potassium adenosine triphosphatase, *Pharmacol. Rev.*, 27, 1, 1975.
30. **Thomas, R. E.,** Cardiac drugs, in *Berger's Medicinal Chemistry,* 4th ed., Vol. 3, Wolff, M. E., Ed., John Wiley & Sons, New York, 1980, chap. 38.
31. **Thomas, R. E., Brown, L., Boutagy, J., and Gelbart, A.,** The digitalis receptor: inferences from structure-activity relationship studies, *Circ. Res.*, 46, I-167, 1980.
32. **Greef, K., Ed.,** *Cardiac Glycosides, Part I: Experimental Pharmacology,* Springer-Verlag, Berlin, 1981.
33. **Greef, K., Ed.,** *Cardiac Glycosides, Part II: Pharmacokinetics and Clinical Pharmacology,* Springer-Verlag, Berlin, 1981.
34. **Hansen, O.,** Interaction of cardiac glycosides with (Na^+,K^+)-activated ATPase. A biochemical link to digitalis-induced inotropy, *Pharmacol. Rev.*, 36, 143, 1984.
35. **Repke, K. R. H. and Schönfeld, W.,** Na^+/K^+-ATPase as the digitalis receptor, *Trends Pharmacol. Sci.*, 5, 393, 1984.
36. **Smith, T. W., Antman, E. M., Friedman, P. L., Blatt, C. M., and Marsh, J. D.,** Digitalis glycosides: mechanisms and manifestations of toxicity. I, *Prog. Cardiovasc. Dis.*, 26, 413, 1984.
37. **Smith, T. W., Antman, E. M., Friedman, P. L., Blatt, C. M., and Marsh, J. D.,** Digitalis glycosides: mechanisms and manifestations of toxicity. II, *Prog. Cardiovasc. Dis.*, 26, 495, 1984.
38. **Smith, T. W., Antman, E. M., Friedman, P. L., Blatt, C. M., and Marsh, J. D.,** Digitalis glycosides: mechanisms and manifestations of toxicity. III, *Prog. Cardiovasc. Dis.*, 27, 21, 1984.
39. **Anner, B. M.,** The receptor function of the Na^+,K^+-activated adenosine triphosphatase system, *Biochem. J.*, 227, 1, 1985.
40. **Blaustein, M. P.,** The cellular basis of cardiotonic steroid action, *Trends Pharmacol. Sci.*, 6, 289, 1985.
41. **Lee, C. O.,** 200 Years of digitalis; the emerging central role of the sodium ion in the control of cardiac force, *Am. J. Physiol.*, 249 (Cell Physiol. 18), C367, 1985.

42. **Repke, K. R. H.,** New developments in cardiac glycoside structure-activity relationships, *Trends Pharmacol. Sci.*, 6, 275, 1985.
43 **Erdmann, E., Greef, K., and Skou, J. C., Eds.,** *Cardiac Glycosides 1785—1985*, Springer-Verlag, New York, 1986.
44. **Marban, E. and Smith, T. W.,** Digitalis, in *The Heart and the Cardiovascular System*, Fozzard, H. A. et al., Eds., Raven Press, New York, 1986, chap. 66.
45. **Grupp, G.,** Selective updates on mechanisms of action of positive inotropic agents, *Mol. Cell. Biochem.*, 76, 97, 1987.
46. **Schwartz, A., Grupp, G., Wallick, E., Grupp, I. L., and Ball, W. J.,** Role of Na^+,K^+-ATPase in the cardiotonic action of cardiac glycosides, *Prog. Clin. Biol. Res.*, 268B, 321, 1988.
47 **Godfraind, T.,** The therapeutic mode of action of cardiac glycosides, *Arch. Int. Pharmacodyn Ther.*, 206, 384, 1973.
48. **Godfraind, T. and Ghysel-Burton, J.,** Binding sites related to ouabain-induced stimulation or inhibition of the sodium pump, *Nature*, 265, 165, 1977.
49. **Ghysel-Burton, J. and Godfraind, T.,** Stimulation and inhibition of the sodium pump by cardioactive steroids in relation to their binding sites and their inotropic effect on guinea-pig isolated atria, *Br. J. Pharmacol.*, 66, 175, 1979.
50. **Lichtstein, D., Samuelov, S., Deutsch. J., Xu, H., Lutz, R. A., Chernick, S. S., and Chernick, S. S.,** The ouabain receptor in animal tissues and its endogenous ligand, *Klin. Wochenschr.*, 65(suppl. 8), 40, 1987.
51. **Bentfeld, M., Lüllman, H., Peters, T., and Proppe, D.,** Interdependence of ion transport and the action of ouabain in heart muscle, *Br. J. Pharmacol.*, 61, 19, 1977.
52. **Lüllman, H., Peters, T., Prillwitz, H.-H., and Ziegler, A.,** Cardiac glycosides with different effects in the heart, in *Cardiac Glycosides Receptors and Positive Inotropy*, Erdmann, E., Ed., Steinkopff, Darmstadt, 1984, 93.
53. **Sweadner, K. J.,** Isozymes of the Na^+/K^+-ATPase, *Biochim. Biophys. Acta*, 988, 185, 1988.
54. **Lingrel, J. B., Orlowski, J., Shull, M. M., and Price, E. M.,** Molecular genetics of Na^+,K^+-ATPase, *Prog. Nucleic Acid Res. Mol. Biol.*, 38, 37, 1990.
55. **Albers, R. W.,** Biochemical aspects of active transport, *Annu. Rev. Biochem.*, 36, 727, 1967.
56. **Post, R. L., Kume, S., Tobin, T., Orcutt, B., and Sen, A. K.,** Flexibility of an active centre in sodium plus potassium adenosine triphosphatase, *J. Gen. Physiol.*, 54(suppl.), 306, 1969.
57. **Yoda, S. and Yoda, A.,** ADP- and K^+-sensitive phosphorylated intermediate of Na^+,K-ATPase, *J. Biol. Chem.*, 261, 1147, 1986.
58. **Scheiner-Bobis, G., Fahlbusch, K., and Schoner, W.,** Demonstration of cooperating α subunits in working (Na^+ + K^+)-ATPase in the use of the MgATP complex analogue cobalt tetrammine ATP, *Eur. J. Biochem.*, 168, 123, 1987.
59. **Scheiner-Bobis, G., Zimmerman, M., Kirch, V., and Schoner, W.,** Ouabain binding site of (Na^+,K^+)-ATPase in right-side out vesicles has not an externally accessible SH group, *Eur. J. Biochem.*, 165, 653, 1987.
60. **Schoner, W., Pauls, H., and Patzelt-Wenczler, R.,** Biochemical characteristics of the sodium pump: indications for a half-of-sites reactivity of (Na^+ + K^+)-ATPase, in *Myocardial Failure*, Riecker, G., Weber, A., and Goodwin, J., Eds., Springer-Verlag, Berlin, 1977, 115.
61. **Scheiner-Bobis, G., Buxbaum, E., and Schoner, W.,** The catalytic subunits cooperate in (Na^+,K^+)-ATPase: demonstration with $MgPO_4$ and MgATP complex analogues, in *The Na^+,K^+-Pump, Part A: Molecular Aspects*, Skou, J. C., Nørby, J. G., Maunsbach, A. B., and Esmann, M., Eds., Alan R. Liss, New York, 1988, 219.
62. **Scheiner-Bobis, G., Esman, M., and Schoner, W.,** Shift to the Na^+ form of Na^+/K^+-transporting ATPase due to modification of the low-affinity ATP-binding site by $Co(NH_3)_4ATP$, *Eur. J. Biochem.*, 183, 173, 1989.

63. **Herbert, H., Skriver, E., and Maunsbach, A. B.,** Three-dimensional structure of renal Na^+,K^+-ATPase determined by electron microscopy of membrane crystals, *FEBS Lett.*, 187, 182, 1985.
64. **Jørgensen, P. L., Skriver, E., Hebert, H., and Maunsbach, A. B.,** Structure of the Na, K pump: crystallization of pure membrane-bound Na^+,K^+-ATPase and identification of functional domains of the α-subunit, *Ann. N.Y Acad Sci.*, 402, 297, 1982.
65. **Collins, J. H., Zot, A. S., Ball, W. J., Jr., Lane, L. K., and Schwartz, A.,** Tryptic digest of the α-subunit of lamb kidney ($Na^+ + K^+$)-ATPase, *Biochim. Biophys. Acta,* 742, 358, 1983.
66. **Shull, G. E., Schwartz, A., and Lingrel, J. B.,** Amino-acid sequence of the catalytic subunit of the ($Na^+ + K^+$) ATPase deduced from a complementary DNA, *Nature,* 316, 691, 1985.
67. **Shull, G. E., Greeb, J., and Lingrel, J. B.,** Molecular cloning of three distinct forms of the Na^+,K^+-ATPase alpha-subunit from rat brain, *Biochemistry,* 25, 8125, 1986.
68. **Shull, M. M. and Lingrel, J. B.,** Multiple genes encode the human Na^+,K^+-ATPase catalytic subunit, *Proc. Natl. Acad. Sci. U.S.A.*, 84, 4039, 1987
69. **Shull, G. E., Lane, L. K., and Lingrel, J. B.,** Amino-acid sequence of the β-subunit of the ($Na^+ + K^+$)ATPase deduced from a cDNA, *Nature,* 321, 429, 1986.
70. **Pedemonte, C. H. and Kaplan, J. H.,** Chemical modification as an approach to elucidation of sodium pump structure-function relations, *Am. J. Physiol., 258, (Cell Physiol.* 27), C1, 1990.
71. **Forbush, B., III,** Cardiotonic steroid binding to Na,K-ATPase, in *Current Topics in Membranes and Transport,* Vol. 19, Hoffman, J. F. and Forbush, B., III, Eds., Academic Press, New York, 1983, 167.
72. **Goeldner, M. P., Hirth, C. G., Rossi, B., Ponzio, G., and Lazdunski, M.,** Specific photoaffinity labeling of the digitalis binding site of the sodium and potassium ion activated adenosinetriphosphatase induced by energy transfer, *Biochemistry,* 22, 4685, 1983.
73. **Rossi, B., Goeldner, M., Ponzio, G., Hirth, C., and Lazdunski, M.,** New ouabain derivatives to covalently label the digitalis binding site, in *Current Topics in Membranes and Transport,* Vol. 19, Hoffman, J. F. and Forbush, B., III., Eds., Academic Press, New York, 1983, 271.
74. **Hall, C. C. and Ruoho, A. E.,** Ouabain-binding-site photoaffinity probes that label both subunits of Na^+,K^+ ATPase, *Proc. Natl. Acad. Sci. U.S.A.*, 77, 4529, 1980.
75. **Fortes, P. A. G.,** Anthroylouabain: a specific fluorescent probe for the cardiac glycoside receptor of Na^+,K^+-ATPase, *Biochemistry,* 16, 531, 1977.
76. **Sweadner, K. J.,** Two molecular forms of ($Na^+ + K^+$)-stimulated ATPase in the brain. Separation and difference in affinity for strophanthidin, *J. Biol. Chem.*, 254, 6060, 1979.
77. **Ng, Y. C. and Akera, T.,** Two classes of ouabain binding sites in ferret heart and two forms of Na^+-K^+-ATPase, *Am. J. Physiol.*, 252, H1016, 1987.
78. **Sweadner, K. J. and Farshi, S. K.,** Rat cardiac ventricle has two Na^+,K^+-ATPases with different affinities for ouabain: developmental changes in immunologically different catalytic subunits, *Proc. Natl. Acad. Sci. U.S.A.*, 84, 8404, 1987.
79. **Maixent, J.-M. and Lelièvre, L. G.,** Differential inactivation of inotropic and toxic digitalis receptors in ischemic dog heart, *J. Biol. Chem.*, 262, 12458, 1987.
80. **Maixent, J.-M., Charlemagne, D., de la Chapelle, B., and Lelièvre, L. G.,** Two Na^+,K^+-ATPase isoenzymes in canine cardiac myocytes, *J. Biol. Chem.*, 262, 6842, 1987.
81. **Price, E. M. and Lingrel, J. B.,** Structure-function relationships in the Na^+,K^+-ATPase α subunit: site-directed mutagenesis of glutamine-111 to arginine and asparagine-122 to aspartic acid generates a ouabain-resistant enzyme, *Biochemistry,* 27, 8400, 1988.
82. **Price, E. M., Rice, D. A., and Lingrel, J. B.,** Site-directed mutagenesis of a conserved, extracellular aspartic acid residue affects the ouabain sensitivity of sheep Na^+,K^+-ATPase, *J. Biol. Chem.*, 264, 21902, 1989.

83. **Price, E. M., Rice, D. A., and Lingrel, J. B.,** Structure-function studies of Na^+,K^+-ATPase, *J. Biol. Chem.*, 265, 6638, 1990.
84. **Schönfeld, W., Weiland, J., Lindig, C., Masnyk, M., Kabat, M. M., Kurek, A., Wicha, J., and Repke, K. R. H.,** The lead structure in cardiac glycosides is 5β,14β-androstane-3β,14-diol, *Naunyn-Schmiedeberg's Arch. Pharmacol.*, 329, 414, 1985.
85. **Brown, L. and Thomas, R. E.,** Comparison of the inotropic effects of some 5α-cardenolides on guinea pig left atria, *Arzneim.-Forsch.*, 34, 572, 1984.
86. **Brown, L., Thomas, R. E., and Watson, T.,** Cardiac glycosides with non-rotating steroid to sugar linkages: tools for the study of digitalis structure-activity relationships. *Naunyn-Schmiedeberg's Arch. Pharmacol.*, 332, 98, 1986.
87. **Shigei, T., Tsuru, H., Saito, Y., and Okada, M.,** Structure-activity relationship of the cardenolide, with special reference to the substituents and configuration at C-14 and C-15, *Experentia*, 29, 449, 1973.
88. **Jarreau, F. X., Koenig, J. J., and Fenard, S.,** A new inotropic aminosteroid: LND 623, *J. Mol. Cell. Cardiol.*, 15(suppl. 2), 44, 1983.
89. **Adamantidis, M. M., Honoré, E. R., and Dupuis, B. A.,** Electrical and mechanical effects of new aminosteroids on guinea-pig isolated ventricular muscles, *Br. J. Pharmacol.*, 95, 1063, 1988.
90. **Thomas, R. E., Boutagy, J., and Gelbart, A.,** Cardenolide analogs. V. Cardiotonic activity of semisynthetic analogs of digitoxigenin. *J. Pharmacol. Exp. Ther.*, 191, 219, 1974.
91. **Gelbart, A. and Thomas, R.,** Cardenolide analogues. VII. Synthesis and biological activity of some new steroidal guanylhydrazones, *J. Med. Chem.*, 21, 284, 1978.
92. **Gelbart, A., Boutagy, J., and Thomas, R.,** Cardenolide analogues. IX. Synthesis and biological activity of 17β-carboxyethylene- and 17β-cyanoethylene-14αH-steroids, *J. Med. Chem.*, 22, 287, 1979.
93. **Ng, Y.-C., Leung, W.-Y., and Akera, T.,** Cardiac Na^+,K^+-ATPase isoenzymes: sensitivity to prednisolone bisguanylhydrazone, *Eur. J. Pharmacol.*, 155, 93, 1988.
94. **Chow, E., Kim, R.-S. S., La Bella, F. S., and Queen, G.,** Ouabain receptor binding of hydroxyprogesterone derivatives, *Br. J. Pharmacol.*, 67, 345, 1979.
95. **Kim, R.-S. S., LaBella, F. S., Zunza, H., Zunza, F., and Templeton, J. F.,** Progesterone derivatives that bind to the digitalis receptor: structure-activity relationships, *Mol. Pharmacol.*, 18, 402, 1980.
96. **Labella, F. S., Bihler, I., Templeton, J., Kim, R.-S., Hnatowich, M., and Rohrer, D.,** Progesterone derivatives that bind to the digitalis receptor: effects on Na^+,K^+-ATPase and isolated tissues, *Fed. Proc.*, 44, 2806, 1985.
97. **Fullerton, D. S., Kitatsuji, E., Deffo, T., Rohrer, D. C., Ahmed, K., and From, A. H. L.,** Use of Prophet and MMs-X computer graphics in the study of the cardiac steroid receptor site of Na^+,K^+-ATPase, *Curr. Top. Membr. Transp.*, 19, 257, 1983.
98. **Templeton, J. F., Kumar, V. P. S., Bose, D., and Labella, F. S.,** Cardiac glycoside-like structure and function of 5β,14β-pregnanes, *J. Med. Chem.*, 32, 1977, 1989.
99. **Swynghedauw, B., Jarreau, F. X., Nittenberg, A., Mouas, C., Preteseille, M., and LeLievre, L.,** Effect of two new glycosides on contractility and on sarcolemmal Na^+,K^+-ATPase in rat heart, *J. Mol. Cell. Cardiol.*, 15(suppl. 2), 55, 1983.
100. **Brown, L. and Thomas, R. E.,** Comparison of the inotropic potencies of some synthetic and naturally occurring glycosides using isolated left atrium of guinea pig, *Arzneim.-Forsch.*, 33, 814, 1983.
101. **Beard, N. A., Rouse, W., and Somerville, A. R.,** Cardiotonic steroids: correlation of sodium-potassium adenosine triphosphate inhibition and ion transport *in vitro* with inotropic activity and toxicity in dogs, *Br. J. Pharmacol.*, 54, 65, 1975.
102. **Brown, L., Erdmann, E., and Thomas, R. E.,** Digitalis structure-activity relationship analyses, *Biochem. Pharmacol.*, 32, 2767, 1983.

103. **Ahmed, K., Rohrer, D. C., Fullerton, D. S., Kitatsuji, E., Deffo, T., and From, A. H. L.,** Interaction of Na^+,K^+-ATPase and digitalis genins: a general model for inhibitory activity, *J. Biol. Chem.*, 258, 8092, 1983.
104. **From, A. H. L., Fullerton, D. S., and Ahmed, K.,** Digitalis receptor sugar binding site characteristics: a model based upon studies of Na^+,K^+-ATPase preparations with differing digitalis sensitivities, *Mol. Cell Biochem.*, 94, 157, 1990.
105. **Smith, P., Brown, L., Boutagy, J., and Thomas, R.,** Cardenolide analogues. XIV Synthesis and biological activity of glucosides of C17β-modified derivatives of digitoxigenin, *J. Med. Chem.*, 25, 1222, 1982.
106. **Fallon, M. and Thomas, R.,** unpublished results.
107. **Franklin, T. J.,** Binding energy and the excitation of hormone receptors, in *Towards Understanding Receptors*, Lamble, J. W., Ed., Elsevier/North-Holland Biomedical Press, Amsterdam, 1981, 8.
108. **Beer, J., Kunze, R., Herrmann, I., Portius, H. J., Mirsalichova, N. M., Abubakirov, N. K., and Repke, K. R. H.,** The thermodynamic essence of the reversible inactivation of Na^+/K^+-transporting ATPase by various digitalis derivatives is relaxation of enzyme conformational energy, *Biochim. Biophys. Acta*, 937, 335, 1988.
109. **Repke, K.,** New developments in cardiac glycoside structure-activity relationships, *Trends Pharmacol. Sci.*, 6, 275, 1985.
110. **Fullerton, D. S., Ahmed, K., From, A. H. L., McParland, R. H., Rohrer, D. C., and Griffin, J. F.,** Modelling the cardiac steroid receptor, in *Molecular Graphics and Drug Design*, Burgen, A. S. V., Roberts, G. C. K., and Tute, M. S., Eds., Elsevier, Amsterdam, 1986, 257.
111. **Bohl, M. and Simon, Z.,** MTD approach to quantitative structure-activity relationships for cardiotonic steroids, *Z. Naturforsch.*, 40c, 858, 1985.
112. **Dzimiri, N. and Fricke, U.,** Lipophilicity and pharmacodynamics of cardiotonic steroids in guinea-pig isolated heart muscle preparations, *Br. J. Pharmacol.*, 93, 281, 1988.
113. **Kamernitzky, A. V., Reshetova, I. G., Ovchinnikov, A. A., Shamovsky, I. L., Massova, I. A., and Mirsalikhova, N. M.,** Structural basis of steroid compounds interaction with "digitalis"-receptor sites of Na,K-dependent ATPase, *J. Steroid Biochem.*, 32, 857, 1989.
114. **Vick, R. L.,** Effects of some steroid and non-steroid lactones on potassium exchange and physiological properties of the isolated perfused guinea pig ventricle, *J. Pharmacol. Exp. Ther.*, 125, 40, 1959.
115. **Thomas, R.,** Chemical and Metabolic Studies Related to the Study of the Mode of Action of the Cardiac Glycosides, Ph D. dissertation, The University of Sydney, Sydney, 1966.
116. **Repke, K. and Portius, H. J.,** Analysis of structure activity relationships in cardioactive compounds on the molecular level, in *Proceedings of the 25th International Congress of Pharmaceutical Sciences*, Vol. 1, Hanč, O. and Hubik, J., Eds., Butterworths, London, 1966, 39.
117. **Go, K. and Bhandary, K. K.,** Structural studies on the biosides of *Digitalis lanata:* bisdigitoxosides of digitoxigenin, gitoxigenin and digoxigenin, *Acta Cryst.*, B45, 306, 1989.
118. **Bohl, M. and Süssmilch, R.,** Calculations on molecular structure and electrostatic potentials of cardiotonic steroids, *Eur. J. Med. Chem.*, 21, 193, 1986.
119. **Megges, R., Portius, H. J., and Repke, K. R. H.,** Lactam analogs of cardioactive steroid lactones, 11th IUPAC Int. Symp. Chem. Natl. Prod., Symposium Abstracts, Vol. 2, 1978, 164.
120. **De Pover, A. and Godfraind, T.,** Influence of 16β formylation on Na^+,K^+-ATPase inhibition by cardiac glycosides, *Naunyn-Schmiedeberg's Arch. Pharmacol.*, 321, 135, 1982.

121. **Yoda, A., Yoda, S., and Sarrif, A. M.,** Structure-activity relationships of cardiotonic steroids for the inhibition of sodium- and potassium-dependent Na,K-ATPase. II. Association rate constants of various enzyme-cardiac glycoside complexes, *Mol. Pharmacol.*, 9, 766, 1973.
122. **Yoda, A.,** Association and dissociation rate constants of the complexes between various cardiac monoglycosides and Na,K-ATPase, *Ann. N.Y. Acad. Sci.*, 242, 598, 1974.
123. **Yoda, S., Sarrif, A. M., and Yoda, A.,** Structure-activity relationships of cardiotonic steroids for the inhibition of sodium- and potassium-dependent Na,K-ATPase. IV. Dissociation rate constants for complexes of the enzyme with cardiac oligodigitoxosides, *Mol. Pharmacol.*, 11, 647, 1975.
124. **Yoda, A. and Yoda, S.,** Structure-activity relationships of cardiotonic steroids for the inhibition of sodium- and potassium-dependent adenosine triphosphatase. V. Dissociation rate constants of digitoxin acetates, *Mol. Pharmacol.*, 11, 653, 1975.
125. **Fallon, M. and Thomas, R.,** unpublished work.
126. **Kyte, J.,** The titration of the cardiac glycoside binding site of the (Na^+ + K^+)-adenosine triphosphatase, *J. Biol. Chem.*, 247, 7634, 1972.
127. **Caldwell, R. W. and Nash, C. B.,** Pharmacological studies of a new 4-aminosugar cardiac glycoside, *J. Pharmacol. Exp. Ther.*, 197, 19, 1976.
128. **Cook, L. S., Caldwell, R. W., and Nash, C. B.,** Comparison of the cardiac effects of ASI-222 HCl, an aminosugar cardiac glycoside, and digoxin, *Arch. Int. Pharmacodyn. Ther.*, 227, 220, 1977.
129. **Songu-Mize, E., Gunter, J. L., and Caldwell, R. W.,** Comparative ability of digoxin and an aminosugar cardiac glycoside to bind to and inhibit Na^+,K^+-adenosine triphosphatase. Effect of potassium, *Biochem. Pharmacol.*, 38, 3689, 1989.
130. **Chiu, C. K. and Watson, T. R.,** Conformational factors in cardiac glycoside activity, *J. Med. Chem.*, 28, 509, 1985.
131. **Watson, T. R., Cheung, H. T. A., and Thomas, R. E.,** Structure activity studies of cardiac glycosides, in *Natural Products and Drug Development*, Krogsgaard-Larsen, P., Christensen, S. B., and Kofod, H., Eds., Munksgaard, Copenhagen, 1984, 337.

INDEX

A

B

C

D

E

F

O

P

Q

R